Soft Computing in Industrial Applications

Yukinori Suzuki, Seppo Ovaska,
Takeshi Furuhashi, Rajkumar Roy
and Yasuhiko Dote (Eds)

Soft Computing in Industrial Applications

With 294 Figures

Springer

Yukinroi Suzuki, D.Eng
Muroran Institute of Technology, Department of Computer Science & Systems
Engineering, 27-1, Mizumoto - cho, Muroran 050-8585, Japan

Seppo Ovaska, DSc
Helsinki University of Technology, Department of Electrical & Communications
Engineering, Otakaari 5A, FIN-02150 Espoo, Finland

Takeshi Furuhashi, D.Eng
Department of Information Electronics, Nagoya University, Furo-cho, Chikusaku,
Nagoya 464-8603, Japan

Raikumar Roy, PhD
Department of Enterprise Integration, School of Industrial & Manufacturing Science,
Building 53, Cranfield University, Bedford, MK43 0AL, UK

Yasuhiko Dote, PhD
Muroran Institute of Technology, Department of Computer Science & Systems
Engineering, 27-1, Mizumoto - cho, Muroran 050-8585, Japan

ISBN 1-85233-293-X Springer-Verlag London Berlin Heidelberg

British Library Cataloguing in Publication Data
Soft Computing in industrial applications
    1.Soft computing 2.Soft computing - Industrial applications
    3.Computational intelligence 4.Intelligent control systems
    I.Suzuki, Yukinori
    0026.3
    ISBN 185233293X

Library of Congress Cataloging-in-Publication Data
A catalog record for this book is available from the Library of Congress

Typesetting: Camera ready by contributors
Printed and bound at the Athenæum Press Ltd., Gateshead, Tyne & Wear
34/3830-543210 Printed on acid-free paper SPIN 10741543

Preface

This book contains recent theoretical innovations and a comprehensive collection of industrial applications in the emerging field of Soft Computing. Soft computing is a new form of artificial intelligence and it consists of four core methodologies: Fuzzy Computing, Neuro Computing, Evolutionary Computation, and Probabilistic Computing. These individual techniques are clearly complementary or synergistic rather than competitive. Therefore, it is a common practice to combine two or three methodologies when solving complex problems. Also the systematic fusion of soft computing and hard computing is a highly potential alternative to be considered. Soft computing methodologies are suitable for various real-world applications, because the available information and system knowledge are often imprecise, uncertain, or partially even incorrect. To handle such demanding conditions and obtain the required robustness with pure hard computing would typically be either very difficult or expensive. This book is a unique collection of technical articles providing a thorough overview of the state-of-the-art theory and industrial applications. The core articles on evolutionary computation, fuzzy computing, and neuro computing are of particular interest to researchers and practicing engineers.

Our book contains 58 selected articles that were presented during the 4th On-line World Conference on Soft Computing in Industrial Applications (WSC4) in September 1999. For WSC4, excellent papers were invited from two physical conferences on soft computing in industrial applications that were held in Japan and Finland in June, 1999 simultaneously. Therefore, WSC4 provided an opportunity to integrate physical and virtual conferences. Five universities from Japan, Finland, and the United Kingdom hosted this major event on the Internet (World-Wide-Web). The conference was organized by the World Federation on Soft Computing, and technically co-sponsored by the IEEE Industrial Electronics Society together with the IEEE Systems, Man, and Cybernetics Society. WSC4 was, indeed, a great success and it set a high reference level for future conferences in the Cyberspace. We had more than 450 registered participants and the authors represented 20 countries around the world. Thus this book provides a solid overview on the present status of soft computing research and development. In addition to fulfilling the expanding information requirements of researchers and engineers, Soft Computing in Industrial Applications forms an excellent basis for graduate seminars on computational intelligence.

The international team of co-editors would like to express their most sincere appreciation and gratitude to the authors and reviewers, as well as the publisher for making this book possible. Our book is a truly joint contribution of hard-working people around the world. The close cooperation over several time zones was feasible and efficient due to the existence of the Internet. We hope that the readers will benefit greatly from the numerous novel algorithms, methods, implementations, and applications described in this book.

Cyberspace
1st December, 1999

Prof. Yukinori Suzuki
Prof. Takeshi Furuhashi
Prof. Seppo J. Ovaska
Dr. Rajkumar Roy
Prof. Yasuhiko Dote

Contents

x

Foreword

Rapidly progressing microelectronics, software engineering, sensor, and communications technologies are offering us a flexible and continuously evolving basis to develop intelligent systems. Intelligent behavior can result from an ability to predict the environment coupled with the selection of an algorithm that permits the translation of each prediction into a suitable response. The new concept "Machine IQ" is often mentioned when intelligent systems are considered. In real world applications, relevant information that is typically available is imprecise, uncertain, and possibly incorrect. Soft computing deals advantageously with these difficulties, and makes it possible to achieve high efficiency, low total cost, robustness, and tractability. The core methodologies of soft computing are Fuzzy Logic, Artificial Neural Networks, Evolutionary Computation, and Probabilistic Computing. These methodologies provide a solid foundation for the conception, design, and applications of competitive systems with high Machine IQ.

This unique book, Soft Computing in Industrial Applications, contains the latest theoretical and application achievements in the emerging field of soft computing or computational intelligence. State-of-the-art algorithms and methods on fuzzy computing, neuro computing, evolutionary computation, probabilistic computing, chaos, and immune networks are presented with immediate connections to specific industrial applications. In addition, our book deals with successful applications of soft computing in signal processing, pattern recognition, system identification, optimization, image processing, fault diagnosis, prediction, and control engineering. Thus it provides a truly wide exposure of advanced technologies and developments. The dominating articles on fuzzy computing, neuro computing, and evolutionary computation, as well as their novel applications are of particular value and interest.

Our book consists of 9 chapters and 58 carefully reviewed and selected articles. Each of the chapters contains 3 – 10 articles. Chapter 1 includes 10 articles on Fuzzy Computing. Those articles describe nonlinear modeling, controller design, traffic signal control, and pattern recognition using fuzzy logic. Chapter 2 contains 10 articles on the theory and applications of Neuro Computing. The presented applications are dealing with nonlinear process modeling, pattern recognition, autonomous mobile robots, fault diagnosis, and nonlinear prediction. Chapter 3 is devoted to Evolutionary Computation. One of its 7 articles, "Evaluation of Virtual Cities Generated by Using a Genetic Algorithm," won the notable Best Paper and Presentation Award of WSC4. The next chapter, Chapter 4, is on Probabilistic Computing. In that chapter, novel computational methods based on probability theory are described. Hybrid Methods, Chaos, and Immune Networks are discussed in Chapter 5. There the main applications are related to prediction and optimization. Chapter 6 is a compact chapter with 3 papers on Rough Sets. These papers present interesting aspects of this new computational methodology. 5 articles on Image Processing are included in Chapter 7. Soft computing techniques are applied to image compression, watermarking, image segmentation, and texture analysis. Design of Human Interfaces is considered in Chapter 8 with a heterogeneous collection of applications from freehand curve identification to control of foundries. Finally Chapter 9, New Frontiers of Soft Computing, includes multi-agents, knowledge databases, and other complementary methodologies. Therefore, this book presents a truly comprehensive and wide mixture of advanced theories and successful industrial applications of soft computing.

Chapter 1: Fuzzy Computing

Papers:

Fuzzy Process Model Development with Missing Data
L. Collantes and R. Roy

Decomposed Fuzzy Models for Modelling and Identification of Dynamic Systems
M. Golob

Investigation of Least Square Fuzzy Identification via a Virtual Higher Resolution Fuzzy Model
K. M. Chow and A. B. Rad

Fuzzy Morphologies Revisited
M. Koeppen and K. Franke

A Design Method of Stable Non-separate Controller Using Symbolic Expressions
H. Yamamoto and T. Furuhashi

Adjustment of Identified Fuzzy Measures
O. Furuya and T. Onisawa

Design and Application of Block-Oriented Fuzzy Models – Fuzzy Hammerstein Model
J. Abonyi, R. Babuska, F. Szeifert, L. Nagy, and H. Verbruggen

Fuzzy Process Model Development with Missing Data

Luis Collantes, Rajkumar Roy[1]

[1] Department of Enterprise Integration, SIMS, Cranfield University
Cranfield, Beds MK43 0AL, United Kingdom, r.roy@Cranfield.ac.uk or r.roy@ieee.org

Abstract: The research has developed a fuzzy logic approach to handling missing data. A prototype fuzzy model was developed, using the FuzzyTech software, to assess the quality of the steel production in terms of composition, time, and temperature. As tools like FuzzyTech are not able to handle missing data, the research has introduced a fuzzy logic approach to decision making with less data. A number of workshops were carried out in the plant, and the aired experts's knowledge was the basis for the research's development. This paper will present the state of the art research on the application of artificial intelligence and statistical techniques for handling the missing data problem.

1 Introduction

The problem of missing data (incomplete feature vectors) is of great interest. In many situations it is important to know how to react if the available information is incomplete, due to sensors failing or sources of information becoming unavailable. As an example, when a sensor fails in a production process, it might no be necessary to stop everything if sufficient information is implicitly contained in the remaining sensor data.

In this particular application, a fuzzy model that assesses the quality of the steel production [1][2], data is extracted from the plant and automatically stored in text files. Each processing parameter has two different values, the aim and the real value.

After analysing the data it is observed that there are missing composition values in the files. When the aim value is missing then the code of practice must be applied, but when there is no real value it is impossible to give a default value.

Depending on the relative importance of the missed parameter, it is possible to assess the quality of the steel production by introducing a new input variable which assess the confidence of the fuzzy model score. This means that if a composition parameter like carbon is missing, then is not possible to qualify the steel production as it is one of the most important parameters. However, if it is the manganese parameter that is missing then it is still possible to realise the assessment giving a degree of confidence.

It was also observed that there are two types of missing data, full missing data and partial missing data. Full missing data exists when all the processing parameter values are missing in the database. The existence of full missing data is consistent with the parameter SAM_IND (quality of the sample) classification that identifies the sample as bad (Table 1).

Partial missing data on the other hand means that only some of the values in terms of composition are missing in the database, and in this case the data SAM_IND indicates a good sample. So even when the device indicates the sample as good there may be missing values.

These records, containing partial missing data, are the main object of the study and have been discovered in the tapping and flushing stages [3] of the secondary steel making process.

Table 1: Quality of the sample

HEAT_NO	Q_CODE	ACT_C	ACT_MN	ACT_P	ACT_S	SAM_IND	SLAG_FE	OXY LEVEL
36585	1439	0.05	0.11	0.01	0.0042	G	20.64	813.75
37394	1439	0.02	0.09	0.01	0.01	G	24.91	
38487	1439					B		

There are two possible solutions to overcoming the problem of missing data. Firstly, the missing parameters can be filled using default values, which are identified by studying patterns and trends within the database.

Secondly, the records can be assessed with the fuzzy model using a factor of confidence. This factor demonstrates how sure the fuzzy model is about the given score.

2 Sources of missing data

A workshop and meetings, with plant experts, were carried out to establish the reasons for the missing data in the database. A brainstorming session was carried out to try and describe the source of the missing data. The ideas were assembled in the following groups:

-Incorrect station: Samples have been assigned to the incorrect stations due to being measured too late.

-Human error

-Communication problems: This group includes ideas such as control instrumental limits, computer analysis failure, computer crashes and computer breakdown.

-Sampling Errors: The metal sample-failure during analysis or improper sampling and failed measurement.

-Others: Data out of range, no missing data but corrupt data and data treated before being stored in the database.

Although the experts believe that these are all feasible and logical reasons for missing data, there is no evidence available that could demonstrate any of them.

Because of the nature of the problem, it was assumed that there is no significant pattern causing the missing data.

3 Related Research

This section describes possible solution to the problem of missing data. Two main potential solutions were studied:

3.1 Filling missing data

The possibility of filling the incomplete records by looking for different trends and patterns in the database was studied. In this particular case it is very difficult to detect these kinds of patterns since within the steelmaking industry, and more in particular in the secondary steelmaking process, each evaluation is independent.

Problems in the underlying knowledge base can be investigated using neural network clustering techniques [4], for instance, finding similar patterns and trends in the database. An expert's knowledge can be refined using neural network style learning.

Some other approaches use the rule model, which evolves during the training process, to predict the most possible value for each one of the missing attributes [5]. Computing the so-called "best guess" value can complete the input vector. After training using the data set is finished, the constructed rule set is used to replace the missing value in the training data, repeating the process again with this new data set.

In the k-nearest neighbour method [6][7] a data set is kept (usually in memory) for comparison with new data items.

When a new record is presented for prediction, the "distance" between it and similar records in the data set is found, and the most similar (or nearest neighbours) are identified.

Another popular approach is called the "backpropagation" that uses neural networks. Structurally a neural network consists of a number of interconnected elements, called neurones, organised in layers that learn by modifying the connection strengths connecting the layers [8][9].

There are some iterative methods based on statistical analysis to discover information automatically [10] by forming hypotheses, making some queries and running some statistics, afterwards view results and perhaps modify the hypotheses, continuing this cycle until a pattern emerges.

Because of the nature of the problem, the application of these techniques was discarded.

3.2 Decision making with incomplete information

Research has carried on developing a fuzzy expert system that can evaluate and design solutions from a design space using a realistically small number of rules [11]. The system uses knowledge separation and integration techniques.

A new method of approximating decision concepts by condition concepts is to describe all the concepts by using fuzzy sets [12]. The central question in inductive learning is whether one can reproduce (exactly or approximately) the expert's classification of a given. set.

The main idea of rough approximation is to find a lower bound and an upper bound for a set. The quality of approximation is measured by the tolerance coefficient, since the fuzzy rough sets give information about only the strength of a relationship between the condition and the decision fuzzy sets.

Fusion of information from multiple sources for object recognition and classification is an important issue to develop. Fusion of information is often made more difficult by problems of uncertainty. As in most of the researches, the design uses the concept of confidence factor, which assesses the quality of the final decision [13].

There is also a proposal of a stochastic sampling [14], which converges to the optimal solution because of the number of samples and it can handle arbitrary pattern of noisy and missing data.

4 Missing data as a membership function

This section describes the methodology followed for making the software become able to identify the missing values that affect the confidence factors.

In order to consider missing data as a value, it is necessary to create a new membership function called "missing data" for each linguistic variable.

Keeping the definition points of the old membership functions, the range of the linguistic variable is extended to define the new membership function (figure 1) (figure 2).

By means of the new membership function it is possible to allocate a default value in place of the missing data, allowing the record to be assessed. A record is the set of parameters that have been measured in the steel.

The newly defined linguistic variable gives the opportunity to identify corrupt data, meaning data lying outside the old range

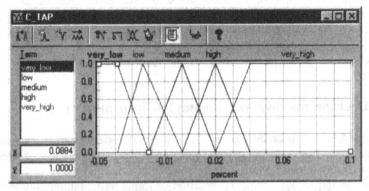

Figure 1: Linguistic variable "Carbon" in the old model

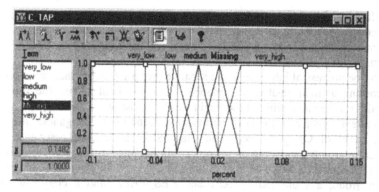

Figure 2: Linguistic variable "Carbon" in the new model. The range is extended

Using Excel as an interface, the default values are introduced into the database and the records are then evaluated.

5 Separation-Integration technique

Due to the generation of the new membership function (Missing), there is a huge increase in the number of rules, making the model difficult to maintain or expand. It was because of this fact that the separation-integration technique [11] was applied.

The basic idea is to decompose the problem into small problems and then solve each particular problem to get particular results, which can then be integrated in order to create the final score (figure 3).

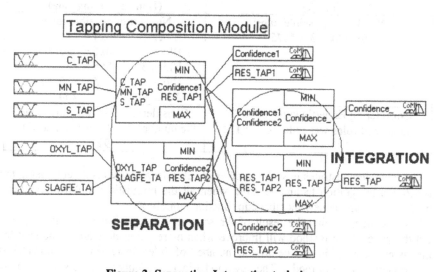

Figure 3: Separation-Integration technique

After several meetings with plant experts it was decided to group the different parameters according to their relative importance to the quality of the steel.

The separation process produces a score, and its associated confidence value, for each group of parameters.

For the process of integration, the separate group scores are then allocated weights according to their relative importance within the final score. The same method applies for the calculation of the final confidence factor.

Figure 4 shows how the input parameters, Carbon (C), Manganese (Mn), Sulphur (S), Level of Oxygen (Oxyl) and Slag (Slag) have been assembled into two different groups depending on the relative importance in terms of the quality of the steel. C, Mn and S are in the first group and Oxyl and Slag in the second group. For each group, a score (RES_TAP1 and RES_TAP2) and a confidence factor is obtained. The score is an estimation of the quality of the steel in terms of the composition parameters evaluated in each group, and the confidence factor is an estimation of the accuracy of the score.

After the separation process, both scores have to be integrated to get the final score (RES_TAP). At the same time the confidence factors have to be integrated to get the final confidence factor. Due to the different importance of the processing parameters in the assessment different weights were allocated to the individual solutions, but it was also taken into consideration that if one of the scores is very bad then the final score will be very bad.

For the purpose of an example the old tapping model is contrasted to the new one.

Table 2: Comparison between the two models

OLD MODEL (Everything is integrated in only one block)	NEW MODEL (Separation-Integration) (Handle missing data)
Number of possible rules: $5 \times 5 \times 5 \times 5 \times 3 = 1875$	Number of possible rules: Block 1 $6 \times 6 \times 4 = 144$ Block 2 $6 \times 6 = 36$ Block 3 $5 \times 5 = 25$ Block 4 $5 \times 5 = 25$ Total -------------- **230**
Considering Expert's knowledge the number of rules was reduced to : **174**	Considering Expert's knowledge the number of rules was reduced to: **92**

The table demonstrates how the number of rules decreases considerably using the separation-integration technique, even when all the linguistic variables have been modified adding the new membership function "missing".

It was concluded that the result of the separation-integration technique is good, but the yield of the method will improve when more parameters are introduced for the assessment. This is due to the number of rules increasing in geometrical progression in the old model. The new model increases only with the sum of a new block of rules. This ultimately leads to the reduction of the complexity of the model.

6 Creation of the new rules

New rules were necessary for each block. The rules created in blocks 1 and 2 are based on the old model. As an example, the creation of the rules for the first block, when Sulphur is missing, is described as follows:

- Let it be supposed that it is necessary to know which score and which confidence factor to assign to the following inputs. Carbon is medium (medium is the membership function) and Manganese is also medium.
- First of all every single rule, in which Carbon is medium and Manganese is medium, was selected from the old model (Table 3).

Table 3: Selection of rules from the old model

| IF | | | | | THEN | |
C_TAP	MN_TAP	OXYL_TAP	S_TAP	SLAGFE_TAP	DoS	RES_TAP
Medium	Medium	medium	Low	medium	0.80	good
Medium	Medium	medium	Medium	low	0.80	good
Medium	Medium	medium	Low	high	0.80	good
Medium	Medium	low	Low	high	1.0	average
Medium	Medium	high	Low	medium	0.80	good
Medium	Medium	high	Low	low	1.0	average
Medium	Medium	high	Low	high	1.0	average

- Once the rules have been selected the column with the missing parameter should be ignored.
- As it can be observed, when Carbon is medium and Manganese is medium the result is 4 times **GOOD** with a weighting of 0.80, and 3 times **AVERAGE** with a weighting of 1.0, so probabilistically the result will be between **GOOD** and **AVERAGE**.
- Due to the concentration of the results, for this specific case there is a rule with a good result. Hence, the final result has to be associated to quite a high confidence factor. It is also necessary to consider the possibility of extreme situations. This means that for very high or very poor Sulphur levels, the result should be very poor (the experts consider that when sulphur, or any other processing parameters, are extreme the final result will be poor or very poor independent of the other parameters). This is going to reduce the probability of getting a good result and is also going to increase the uncertainty, decreasing the value of the confidence factor.
- After considering all the factors, the final rule is as follows (Table 4):

Table 4: Final rule in the block 1 of the new model

| IF | | | THEN | | | |
C_TAP	MN_TAP	S_TAP	DoS	Confidence1	DoS	RES_TAP
Medium	Medium	Missing	1.0	**medium**	0.5	**good**

After several workshops with the experts it was also concluded that if Carbon (the most important processing parameter in the evaluation of the quality of the steel) is missing, then there is no possibility to assess the record. This also happens when 2 or more parameters are missing. Because of this, it was necessary to create a new membership function (for the output variable) called "No-Assessment", which identifies those records without enough data to be assessed.

7 Analysis of the results

This section explains a comparison experiment carried out between the old model (Table 5), the new model with all information (Table 6)and the new model with missing data (Table 7).

Table 5: Random selected records evaluated with the old model

HEAT_NO	Q_CODE	ACT_C	ACT_MN	SLAG_FE	OXY_LEV	ACT_S	SCORE
25163	1949	0.03	0.08	25.69	762.31	0.0100	20
25175	1949	0.02	0.11	18.37		0.02	
26012	1949	0.04	0.08	23.91	789.46	0.0100	37
26013	1949	0.06	0.17	14.38	544.45	0.0095	31
26922	1949			20.34	828.65		
27193	1655	0.03	0.07	25.79	860.09	0.0200	0
28956	1557	0.02	0.08	27.88		0.0061	
30050	1949	0.06	0.08	21.61	594.97	0.0200	53
30074	1949	0.04	0.11	20.35	862.39	0.0100	45

We can observe in Table 5 how there are records with missing values that the old model can not assess (Heat No: 25175, 26922 and 28956). As it can be seen in the following tables the new model can assess the records 25175 and 28956 because although the OXY_LEV is missing there is still information enough to evaluate these records. However, record 26922 will have a "No-Assessment" result due to insufficient information.

Table 6 also contains columns with partial results. The results of the first and second blocks give a quick overview of which group of parameters is the cause of the possible bad final result.

Table 6: Random selected records evaluated with the new model

HEAT_NO	CODE	ACT_C	ACT_MN	SLAG_FE	OXY_LEV	ACT_S	CONF_1	CONF_2	CONF	TAP1	TAP2	SCORE
25163	1949	0.03	0.08	25.69	762.31	0.01	92	67	92	50	35	26
25175	1949	0.02	0.11	18.37		0.02	92	33	67	11	50	11
26012	1949	0.04	0.08	23.91	789.46	0.01	92	67	92	63	48	56
26013	1949	0.06	0.17	14.38	544.45	0.0095	92	67	92	56	51	51
26922	1949			20.34	828.65		0	67	0	NO	58	NO
27193	1655	0.03	0.07	25.79	860.09	0.02	92	67	92	0	28	0
28956	1557	0.02	0.08	27.88		0.0061	92	40	85	6	1	1
30050	1949	0.06	0.08	21.61	594.97	0.02	92	67	92	62	82	62
30074	1949	0.04	0.11	20.35	862.39	0.01	92	67	92	77	46	46

The above records were assessed using the new model. The record **25175** has a result of **11** with a final confidence factor of **67%**. On the other hand the record **28956** has a result of **1** with a high confidence factor, **85%**. The reason for this high confidence factor is the accuracy of the bad result that the software has for the first block. For the record **26922** there is not enough information for an evaluation to be made, so therefore the score is **NO** ("No-Assessment").

Another important point to highlight is that although the scores are not the same as in the old model, the difference between them is not important. It is essential, however, to remember that the model gives us a qualitative evaluation.

Table 7: Random selected records evaluated with the new model (with different missing data)

HEAT_NO	CODE	ACT_C	ACT_MN	SLAG_FE	OXY_LEV	ACT_S	CONF_1	CONF_2	CONF	TAP1	TAP2	SCORE
25163	1949		0.08	25.69	762.31	0.0100	0	67	0	NO	35	NO
25175	1949	0.02	0.11	18.37		0.02	92	33	67	11	50	11
26012	1949	0.04	0.08	23.91	789.46	0.0100	92	67	92	63	48	56
26013	1949	0.06		14.38	544.45	0.0095	39	67	41	60	51	45
26922	1949			20.34	828.65		0	67	0	NO	58	NO
27193	1655	0.03	0.07		860.09	0.0200	92	46	76	0	23	0
28956	1557	0.02	0.08	27.88		0.0061	92	66	91	6	1	1
30050	1949	0.06	0.08	21.61	594.97	0.0200	92	67	92	62	82	62
30074	1949	0.04		20.35	862.39	0.0100	45	67	61	77	46	41

With these results it can be observed how some of the different confidence factors have changed by quite a large amount, but without essentially modifying the final score.

8 Validation of results

In order to validate the model several records, comparing results from the old model with those of the new model, were given to the expert. The following feedback was the outcome:

- The marks appear to be OK
- There was a disagreement with some of the confidence factors (they seem to be quite low). However, the rating of the confidence factor was taken from the results of the workshop, and does not agree with the expert's comment.
- The expert also commented that the final score seemed to be calculated from the previous marks, by simply taking the lowest value. Due to the different importance of the processing parameters, different weights were allocated to the individual solutions, but it was also taken into account that if one of the scores is very bad then the final score will be very bad.

9 Conclusions and further investigations

The research has developed a fuzzy logic based framework that enables the steel evaluation to take place with less information (missing data).

The research has demonstrated that the new model is able to handle missing data. It has also been observed that the model, using the separation-integration technique, becomes more flexible in terms of maintenance and possible extension.

The possibility of identifying bad group of parameters in a specific record also gives the advantage of being able to more tightly control them.

Both data analysis and expertise are used to design the final model. During the definition of the membership functions data analysis is mostly used, whereas during the generation of the rules and the validation of the model expertise is used.

For the future it would be interesting to analyse more deeply the nature of the missing data as well as identifying the different possible sources of missing data.

References

[1] Roy R., Collantes L., Berdou et al., 1999, *Fuzzy Process Modelling for the Secondary Steelmaking*. IEEE, Nafips'99, pp. 864-868.

[2] Collantes L., Roy R. and Madill J., 1999, *Fuzzy process evaluation Using a Fuzzy Expert System*. RASC'99, Springer-Verlag London.

[3] Berdou J., 1998, *The Development of a fuzzy process model for the steel industry*. Cranfield University.

[4] Lacher K., Adair and Hruska S., 1993, *Reconciling data-derived knowledge with expert rules using clustering*. Lectures notes in computer science, vol. 1280, pp.771-776

[5] Berthold M. and Huber K., 1996, *Missing values and Learning of fuzzy rules*. International Journal of Uncertainty, Fuzziness and Knowledge-Based systems.

[6] Keller J., Gray M. and Givens J., 1985, *A Fuzzy K-Nearest Neighbour Algorithm*. IEEE. vol SMC-15, no 4, pp.580-585

[7] Information Discovery inc., 1998, *A Characterization of Data Minig Technologies and Processes*. The Journal of Data Warehousing.

[8] Tukey J., 1973, *Exploratory Data Analysis*. McMillan, New York.

[9] Information Discovery inc., 1996, *Surveying Decision Support: New Realms of Analysis*. Database Programming and Design.

[10] Palm R. and Kruse R., 1996, *Methods for data analysis in classification and control*. Fuzzy sets and systems, vol 85, pp. 127-129.

[11] Roy R., Parmee I. And Purchase G., 1996, *Qualitative evaluation of engineering designs using fuzzy logic*. ASME Design Engineering Technical Conferences and Computer in Engineering Conference.

[12] Bodjanova S., 1997, *Approximation of fuzzy concepts in decision making*. Fuzzy sets and systems, vol. 85, pp. 23-29

[13] Loskiewicz-Buczak A. and Uhrig R., 1994, *Decision fusion by fuzzy set operations*. IEEE, pp. 1412-1417.

[14] Trespand V. and Hofmann R., 1998, *Non-Linear Time-series Prediction with Missing and Noisy Data*. Neural Computation, vol. 10 pp. 731-747.

Decomposed Fuzzy Models for Modelling and Identification of Dynamic Systems

Marjan Golob

University of Maribor, Faculty of Electrical Engineering and Computer Science, Institute of Automation, Laboratory for process Automation,
Smetanova ulica 17, 2000 Maribor, SLOVENIA, mgolob@uni-mb.si

Abstract: This paper presents an approach which is useful for the identification of discrete dynamic systems based on fuzzy relational models. If the number of input variables and fuzzy sets increases, a fuzzy system gets increasingly intractable. A concept based on the decomposition of multivariable rule-base is presented. Two decomposed fuzzy models based on the simplified inference break up method are proposed and applied to a dynamic systems modelling. Evolution of identification algorithms for the decomposed fuzzy model is suggested. A comparative study of the dynamic system identification with the conventional relational model and the decomposed relational model is presented for Box-Jenkins data.

1. Introduction

There are a wide variety of possible types of non-linear model including: model based on Volterra series; neural networks; and fuzzy models.

Recently many studies have been focused on fuzzy modelling and its application since it can describe non-linear system better then other modelling methods. Fuzzy systems are characterised by a rule-base specification. If the complexity of a rule-base increases, knowledge acquisition may become a tedious work because the number of rules increases with an increasing number of fuzzy variables. This problem is tackled by the decomposing a single large fuzzy system into subsystems.

This paper is concerned with one of possible modelling techniques: fuzzy relational modelling based on decomposition of relational matrix. An inference break-up method is used to decompose the common fuzzy model. The same method is used in [1] for design of the decomposed PID fuzzy controller.

The paper begins with a description of fuzzy model for discrete dynamic systems. The effects of increasing number of fuzzy input variables and fuzzy sets are discussed. The next section describes proposed decomposition scheme. Decomposed fuzzy models for dynamic systems based on the proposed inference break up method is presented in the section 4. Fuzzy identification algorithms for decomposed fuzzy models are discussed in the next section. An identification of the decomposed fuzzy models is evaluated using the Box-Jenkins furnace data. Conclusions are made in the final sections.

2. A Fuzzy Model for Dynamic Systems

A discrete-time model for a SISO dynamic system may be written as

$$y(k) = f\left(y(k-1),\dots,y(k-n_y),u(k-\tau),\dots,u(k-\tau-n_u)\right) \qquad (1)$$

where $f(.)$ is non-linear function, $y(k)$ is system output variable, $u(k)$ is system input variable, τ is input time delay and n_y, n_u presents the system orders. The main task in system modelling is to determine the best function approximation for unknown non-linear function $f(.)$. Like some types of neural networks also certain fuzzy systems are universal function approximators. Neural networks are usually used if the little is known about the system behaviour and a lot of training data is available. If the system behaviour can be described linguistically with if-then rules, fuzzy systems are used. A common fuzzy model for SISO dynamic system is shown in figure 1, although the generalisation to multiple input multiple output system is obvious.

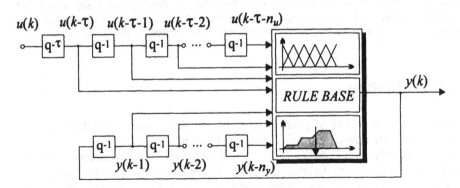

Figure 1: The common fuzzy model for the SISO dynamic system

The q^{-1} represents a shift-operator. Note that the fuzzy system used in model of SISO dynamic system presented in figure 1 is static approximation of non-linear function f. Dynamic elements of the model are placed out of fuzzy system.

Several modelling methods based on fuzzy reasoning have been proposed in recent years. They can be classified into the fuzzy relational equation based fuzzy systems [2], the fuzzy linear functional systems [3] and the fuzzy basis function systems[4].In order to give a introduction to decomposed fuzzy systems, it is necessary to describe the notation for a fuzzy system based on fuzzy relational equations.

Let a vector $x=(x_1, x_2, \dots, x_n)$ be an input vector over the input universe of discourse $x \in X \subseteq R^n$ and y is an output signal over the output universe of discourse $Y \subseteq R$. Fuzzy system are a non-linear mapping $f: x \to y$. Note that in case of the fuzzy model for SISO dynamic system presented in figure 1, elements of input vector x are { $u(k-\tau)$, $u(k-\tau-1)$ ... $u(k-\tau-n_u)$, $y(k-1)$, $y(k-2)$, ... $y(k-n_y)$ }, dimension of input vector x is $n=n_u+n_y+1$ and the output signal is $y=y(k)$. A fuzzy set A defined in a universe of discourse X is expressed by its membership function $\mu_A: X \rightarrow [0,1]$ where $\mu_A(x)$ expresses the extent to which x fulfils the category specified by A:

$$A = \{ x, \mu_A(x) \}, \quad \forall x \in X$$

For an intuitive description of the fuzzy mapping the input and output universes of discourse are divided by fuzzy sets that represents linguistic terms. Based on these linguistic terms, the mapping is described by a set of if-then rules like

$$\text{if } x_1 \text{ is } A_1^{(j)} \text{ and } x_2 \text{ is } A_2^{(j)} \text{ and } \dots \text{ and } x_n \text{ is } A_n^{(j)} \text{ then } y \text{ is } B^{(j)}. \qquad (2)$$

$A_i^{(j)}$ with $i \in \{1,2,...,n\}$ and $j \in \{1,2,...,m\}$ is the fuzzy set in the premise of rule j for input i. $B^{(j)}$ with $j \in \{1,2,...,m\}$ is the fuzzy set in the conclusion of rule j that determines the output signal. m is the total number of rules. Possible fuzzy sets per input i are $A_i^{(j)} \in \{A_{i,1}, A_{i,2},..., A_{i,n_{Ai}}\}$ and the $B^{(j)} \in \{B_1, B_2,..., B_{n_B}\}$ are possible fuzzy sets for the output. n_{Ai} and n_B are total numbers of different fuzzy sets for the inputs and the output variable.

The basic concept of evaluating a fuzzy system is compositional rule of inference as is presented by Zadeh [5]. The rule base of a fuzzy system is realised as a fuzzy relation between the fuzzy inputs and the fuzzy output. The fuzzy sets in the premises of a rule j are combined into a multidimensional input fuzzy set:

$$A^{(j)}(x) = A_1^{(j)} \times A_2^{(j)} \times...\times A_n^{(j)} \quad \text{with}$$
$$\mu_{A^{(j)}}(x) = \mu_{A_1^{(j)}}(x) \wedge \mu_{A_2^{(j)}}(x) \wedge...\wedge \mu_{A_n^{(j)}}(x) \qquad (3)$$

where $\times$ denotes a Cartesian product realised by a fuzzy *and* operator $\wedge$ which is a *t-norm*, e.g. *min* operator.

The rule j described with (2) is presented with fuzzy relation $R^{(j)}$

$$R^{(j)}(x,y) = A^{(j)}(x) \rightarrow B^{(j)}(y) \quad \text{with} \quad \mu_{R^{(j)}}(x,y) = \mu_{A^{(j)}}(x) \rightarrow \mu_{B^{(j)}}(y). \qquad (4)$$

The symbol $\rightarrow$ denotes arbitrary fuzzy implication. The fuzzy relation of all rules in rule-base is aggregated with a fuzzy *or* operator $\vee$ which is a fuzzy *t-conorm*:

$$R(x,y) = R^{(1)}(x,y) \vee R^{(2)}(x,y) \vee...\vee R^{(m)}(x,y). \qquad (5)$$

In the general case, the compositional rule of inference processes fuzzy sets as inputs and generates a fuzzy set as an output. When the fuzzy input is the n

dimensional fuzzy set $A'(x)$, then the output fuzzy set $B'(y)$ will be calculated by the fuzzy inference

$$B'(y) = A'(x) \circ R(x,y) \text{ with } \mu_{B'}(y) = \sup_{x \in X}(\mu_{A'}(x) \wedge \mu_R(x,y)), \qquad (6)$$

where $\circ$ denotes the composition operator. By combining equations (4) and (6) we get

$$B'(y) = \bigvee_{j=1}^{m} \bigwedge_{i=1}^{n}\left[A_i'(x_i) \circ \left(A_i^{(j)}(x_i) \wedge B^{(j)}(y) \right) \right] \qquad (7)$$

and if the *t-norm* and *t-conorm* are distributive, the next simplification is possible [6]:

$$B'(y) = \bigvee_{j=1}^{m}\left[\alpha^{(j)}(x) \wedge B^{(j)}(y) \right] \text{ with } \alpha^{(j)}(x) = \bigwedge_{i=1}^{n}\left[A_i'(x_i) \circ A_i^{(j)}(x_i) \right] \qquad (8)$$

where $\alpha^{(j)}$ denotes the degree of fulfilment of rule j. If the numerical values are used as inputs, the computation of fuzzy matching between the premises fuzzy sets $A_i^{(j)}$ and the fuzzy set of the actual input variable is simplified. With the singleton x' as an input vector, the calculation of the degree of fulfilment reduces to

$$\alpha^{(j)}(x') = \bigwedge_{i=1}^{n} A_i^{(j)}(x'_i). \qquad (9)$$

In most applications the fuzzy output $B'(y)$ has to be defuzzified with one of possible defuzzification methods, e.g. *centre of gravity* method (*COG*):

$$y' = defuzz(B'(y)) = \frac{\displaystyle\int_{y \in Y} \mu_{B'(y)} \cdot y\, dy}{\displaystyle\int_{y \in Y} \mu_{B'(y)}\, dy}. \qquad (10)$$

The realisation of *COG* is based on numerical approximation of the integration. If the fuzzy sets $B^{(j)}$ are restricted to singletons type fuzzy sets $b^{(j)}$, the *COG* defuzzification method is simplified to *centre of singleton* method (*COS*):

$$y' = \frac{\displaystyle\sum_{j=1}^{m} \alpha^{(j)}(x') \cdot b^{(j)}}{\displaystyle\sum_{j=1}^{m} \alpha^{(j)}(x')}. \qquad (11)$$

The representation of the rule base with complex rules, which are composed by n conditions in the premise of rules, leads to $n+1$ dimensional discrete fuzzy relation matrix. The fuzzy composition results in handling large multidimensional matrices and is difficult to perform. If a fuzzy system is used for function approximation, it is necessary to cover the whole input space with rules. In this case the total number of rules m is determined by the number of inputs n and the number of fuzzy sets per input n_{Ai} as the product $m = n_{A1} \cdot n_{A2} \cdot \ldots \cdot n_{An}$. This leads to the problem of an exponential growth of the number of rules m with the number of inputs n and consequently to a multidimensionality of fuzzy relation matrix. To overcome this problem, it is proposed to decompose the fuzzy system in sense to break up the multidimensional compositional rule of inference.

3. Decomposed Fuzzy Systems based on an Inference Break Up Method

Complex rules are rules where there are more then one antecedent variables and more then one facts in premise part of the rule. Consider the rule of the form:

$$\text{if } x_1 \text{ is } A_1 \text{ and } x_2 \text{ is } A_2 \text{ then } y \text{ is } B \tag{12}$$

then, with respect to (4), the fuzzy relation takes the following form:

$$R(x_1, x_2, y) = \left[A_1(x_1) \times A_2(x_2)\right] \to B(y). \tag{13}$$

Mizumoto [7] has been showed for eight implication functions that the break up of complex rules into simple rules is possible. The proofs, Demirli and Turksen [8] have been presented are more general. They are valid for general *S-type* and *R-type* implications.

As will be shown in the following, break up of complex rule base presented with fuzzy relation based on conjunction type of implication function (*T-norms* implication, e.g. *min* $\wedge$ implication, used by Mamdani [9]), is possible. The rules of the form (13), with *min* conjunction operator, *Mamdani* implication and *max* aggregation operator, can be broken up into conjunction of two rules:

$$\left[A_1(x_1) \wedge A_2(x_2)\right] \to B(y) = \left[A_1(x_1) \to B(y)\right] \wedge \left[A_2(x_2) \to B(y)\right]. \tag{14}$$

With respect to (4) the broken rule, presented with fuzzy relation R matrices is

$$R(x_1, x_2, y) = R_1(x_1, y) \wedge R_2(x_2, y). \tag{15}$$

It is important that the compositional rule of inference with the broken up rules yields the same conclusions as the compositional rule of inference with the complex rule, therefore

$$B'(y) = \left[A_1'(x_1) \wedge A_2'(x_2)\right] \circ R(x_1, x_2, y) = \left[A_1'(x_1) \circ R_1(x_1, y)\right] \wedge \left[A_2'(x_2) \circ R_2(x_2, y)\right] =$$
$$= B_1(y) \wedge B_2(y)$$

(16)

It is now shown that, the inference of a rule base with more then one rule can be broken up into the inference of a number of rule bases with simple rules. When considering a rule base with two rules and rules as above, the rule base break up results in:

$$B'(y) = \left[A_1'(x_1) \wedge A_2'(x_2)\right] \circ \left\{ \left(\left[A_1^{(1)}(x_1) \wedge A_2^{(1)}(x_2)\right] \to B^{(1)}(y)\right) \vee \right.$$
$$\left. \left(\left[A_1^{(2)}(x_1) \wedge A_2^{(2)}(x_2)\right] \to B^{(2)}(y)\right)\right\}$$

(17)

$$= \left[A_1'(x_1) \circ \left\{\left[A_1^{(1)}(x_1) \to B^{(1)}(y)\right] \vee \left[A_1^{(2)}(x_1) \to B^{(2)}(y)\right]\right\}\right] \wedge$$
$$\left[A_2'(x_2) \circ \left\{\left[A_2^{(1)}(x_2) \to B^{(1)}(y)\right] \vee \left[A_2^{(2)}(x_2) \to B^{(2)}(y)\right]\right\}\right] \wedge$$
$$\left[\left\{A_1'(x_1) \circ \left[A_1^{(1)}(x_1) \to B^{(1)}(y)\right]\right\} \vee \left\{A_2'(x_2) \circ \left[A_2^{(2)}(x_2) \to B^{(2)}(y)\right]\right\}\right] \wedge$$
$$\left[\left\{A_1'(x_1) \circ \left[A_1^{(2)}(x_1) \to B^{(2)}(y)\right]\right\} \vee \left\{A_2'(x_2) \circ \left[A_2^{(1)}(x_2) \to B^{(1)}(y)\right]\right\}\right]$$

and description with fuzzy relation R matrices is

$$B'(y) = \left[A_1'(x_1) \wedge A_2'(x_2)\right] \circ \left[R^{(1)}(x_1, x_2, y) \vee R^{(2)}(x_1, x_2, y)\right] =$$
$$= \left[A_1'(x_1) \circ \left\{R_1^{(1)}(x_1, y) \vee R_1^{(2)}(x_1, y)\right\}\right] \wedge \left[A_2'(x_2) \circ \left\{R_2^{(1)}(x_2, y) \vee R_2^{(2)}(x_2, y)\right\}\right] \wedge$$
$$\left[\left\{A_1'(x_1) \circ R_1^{(1)}(x_1, y)\right\} \vee \left\{A_2'(x_2) \circ R_2^{(2)}(x_2, y)\right\}\right] \wedge$$
$$\left[\left\{A_1'(x_1) \circ R_1^{(2)}(x_1, y)\right\} \vee \left\{A_2'(x_2) \circ R_2^{(1)}(x_2, y)\right\}\right]$$

(18)

The result of the inference break up can easily be extended to a large number of rules or more complex rules. The number of parts resulting from the inference break up can become quite large (in the case of n variables in the premise of the m rules, the number of cross-inferences is n^m), but many simplifications are possible. First simplification of the inference of a simple rule base can be achieved when the rule base can be divided into a set of a simple rule bases which do not interact (overlap of membership functions of fuzzy sets in rules premises). When the fuzzy premises of the rules are assumed to take fuzzy sets which form a fuzzy partition (fuzzy sets are convex and normalised with no more then two overlapping membership functions), then it can be derived that the number of necessary simple rule bases is $2 \cdot n_B - 1$ where n_B is the number of fuzzy sets defined on the domain of the output variable.

To understand better the linguistic description and mathematical presentation of the break up fuzzy system, a block diagram form is proposed in Figure 1.

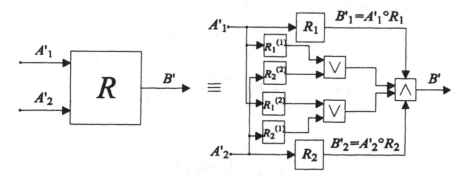

Figure 2: The block diagram of the compositional rule of inference with the decomposed rule base of two complex rules.

The R denotes $\cdot R^{(1)}(x_1, x_2, y) \wedge R^{(2)}(x_1, x_2, y)$, R_1 denotes $R_1^{(1)}(x_1, y) \wedge R_1^{(2)}(x_1, y)$ and R_2 denotes $R_2^{(1)}(x_2, y) \wedge R_2^{(2)}(x_2, y)$.

An interesting possibility concerning the inference break up is a processor which performs compositional rule of inference of input fuzzy set and two-dimensional fuzzy relation. A speed-up of the inference can be achieved by using optimised code to perform the composition of a fuzzy set and a fuzzy relation. The code optimisation can be applied to software or hardware implementation.

For reason to reduce a number of rules, a next simplification is made. From all cross-inferences in decomposed system presented in Figure 2 only two direct inferences (R_1 and R_2) are selected. The compositional rule of inference (18) is reduced to

$$B'_s(y) = \left[A_1'(x_1) \circ \left\{ R_1^{(1)}(x_1, y) \vee R_1^{(2)}(x_1, y) \right\} \right] \wedge \left[A_2'(x_2) \circ \left\{ R_2^{(1)}(x_2, y) \vee R_2^{(2)}(x_2, y) \right\} \right] \quad (19)$$

Using the inclusion operator $\subset$, the output fuzzy set $B'(y)$ from equation 18 can be written as $B'(y) \subset B'_s(y)$. The inclusion sign means that the output fuzzy set $B'(y)$ is subset of the set obtained by neglecting the some cross-inferences. In many practical cases these neglected terms may not be very significant. Usefulness of the simplified decomposed fuzzy structure for development and real-time realisation of the simple decomposed fuzzy PID controller has been presented in [1].

For n inputs and m rules, the general structure of the simplified decomposed compositional rule of inference would take the form

$$B'_s(y) = \bigwedge_{i=1}^{n}\left[A_i'(x_i) \circ R_i\right] \text{, with } R_i = \bigvee_{k=1}^{m}\left[R_i^{(k)}(x_i,y)\right]. \tag{20}$$

The block-diagram is presented in Figure 3.

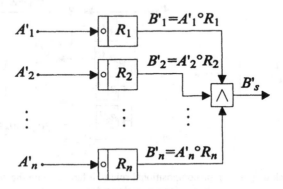

Figure 3: The block-diagram of the general structure of the simplified decomposed compositional rule of inference

The multivariable structure of the fuzzy system which has n inputs contains n two-dimensional compositional rule of inference blocks and an fuzzy intersection block. Advantages of the block diagram presentation of a multivariable fuzzy system are in the fact that this kind of presentation allows evaluation of each inference contribution with regard to overall performance of the system.

4. Decomposed Fuzzy Models for Dynamic Systems based on the Inference Break Up Method

To reduce the dimension of the fuzzy relation was one of the main reasons for the decomposed fuzzy approach. The compositional rule of inference of the fuzzy model for SISO dynamic system presented in Figure 1 is realised with the n_u+n_y+2 dimensional fuzzy relation. With respect to the simplified decomposed compositional rule of inference presented in Figure 3, a decomposed fuzzy model for SISO dynamic system with n_u+n_y+1 two-dimensional fuzzy relations is suggested. The *decomposed fuzzy model* (DFM) for SISO dynamic systems is shown in Figure 4.

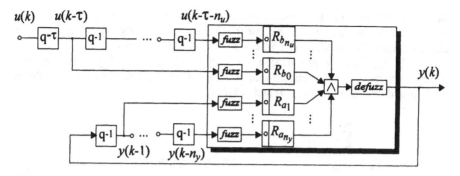

Figure 4: The decomposed fuzzy model for a SISO dynamic system

By combining Equations (6), (20) and (10) the output of the decomposed fuzzy model (*DFM*) results in

$$B'_s(y) = \bigwedge_i \left[A_i'(x_i) \circ R_i \right] \qquad i \in \left\{ b_0 \cdots b_{n_u}, a_1 \cdots a_{n_y} \right\}$$

$$R_i = \bigvee_{j=1}^{m_i} \left[R_i^{(j)}(x_i, y) \right] \qquad x_i \in \left\{ u(k-\tau) \cdots u(k-\tau-n_u), y(k-1) \cdots y(k-n_y) \right\}$$

$$y(k) = defuzz(B'_s(y)) \tag{21}$$

If the Mamdani implication is used, i.e. the implication operator is the *min*, then the rule aggregation operator is *max* and the intersection block is *min*. The fuzzy output $B'_s(y)$ of the decomposed system is an intersection of particular fuzzy outputs and should be understand as a fuzzy weighting average of particular fuzzy outputs. To simplify the realisation of the decomposed fuzzy controller (4) a decomposition of the entire fuzzy system inclusively the deffuzification procedure is proposed, where output is calculated as average of defuzzified outputs of the particular decomposed fuzzy systems. The structure of the *simplified decomposed fuzzy model* (*SDFM*) for SISO dynamic system is shown in Figure 5.

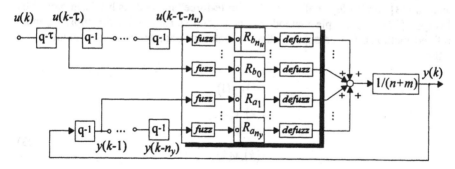

Figure 5: The simplified decomposed fuzzy model (SDFM) for the SISO dynamic system

Note, that the dynamic structure in Figure 5 is similar to the structure of a linear auto-regressive model, i.e. discrete linear ARX model [10]. Parameters a_i and b_i of discrete linear ARX model are substituted with simple fuzzy systems.

If the *Larsen implication*, i.e. *algebraic product* [6], and the *sum* aggregation operator are used, the output of simplified decomposed fuzzy model is (with regards to equations (8), (9) and (11)) as follows:

$$y'_i = \frac{\sum_{j=1}^{m_i} \alpha^{(j)}(x'_i) \cdot b^{(j)}}{\sum_{j=1}^{m_i} \alpha^{(j)}(x'_i)} \qquad i \in \left\{ b_0 \cdots b_{n_u}, a_1 \cdots a_{n_y} \right\} \tag{22}$$

$$x_i \in \left\{ u(k-\tau) \cdots u(k-\tau-n_u), y(k-1) \cdots y(k-n_y) \right\}$$

$$y(k) = \frac{1}{n_u + n_y + 1} \sum_i y'_i$$

5. Fuzzy Identification Algorithms for Decomposed Fuzzy Models

There has been a plethora of methods for identification of rules and tuning rule bases of fuzzy systems. A most of these are useful for modelling of dynamic systems with decomposed fuzzy models. The fuzzy modelling usually is accomplished by a two-step procedure. The first step is identification of rules, which are usually described with fuzzy relation. The second step consists of a tuning of the membership functions of input and output fuzzy sets.

There have been many approaches for relational fuzzy model identification reported in literature as described by Czogala and Pedrycz [11], Pedrycz [12], Xu and Lu [13] and Ridley at al. [14]. The "brute-force" method, as has been named by Pedrycz [12], is a simple method and is a good starting-point for applying more refined methods. Let us suppose that the relevant data set consist of input-output pairs

$$\left\{ \begin{array}{c} x(1), y(1) \\ x(2), y(2) \\ \vdots \\ x(k), y(k) \\ \vdots \\ x(m), y(m) \end{array} \right. \tag{23}$$

defined over the input universe of discourse X and output universe of discourse Y. Let input space be partitioned into n_A reference sets A and output space be partitioned into n_B reference sets B, with membership functions which satisfy requirements of being normal, convex and completely covering the spaces. Then for input value x_k the n_A reference sets generate a input fuzzy variable $X_k = \{\mu_1(x_k), \mu_2(x_k) \cdots \mu_{n_A}(x_k)\}$ and for output value y_k the n_A reference sets generate a output fuzzy variable $Y_k = \{\mu_1(y_k), \mu_2(y_k) \cdots \mu_{n_B}(y_k)\}$. In "brute-force" approach, we calculate the fuzzy relation of the model as a fuzzy union of the partial results of the fuzzy relational equation:

$$R = \bigcup_{k=1}^{m} X_k \times Y_k . \qquad (24)$$

More recent works has shown that a method developed by Ridley at al. [14] is efficient, particularly when significant noise is present in the identification data. The identification is based on the probabilistic fuzzy relational matrix R with elements:

$$R[i_A, i_B] = \frac{\sum_{k=1}^{m} X_k[i_A] \cdot Y_k[i_B]}{\sum_{k=1}^{m} X_k[i_A]} \qquad i_A = 1 \cdots n_A, i_B = 1 \cdots n_B . \qquad (25)$$

The predicted value from relational model is computing as follows:

$$\hat{Y}_k[i_B] = \sum_{i}^{n} X_k[i_A] \cdot R[i_A, i_B] \qquad i_B = 1 \cdots n_B . \qquad (26)$$

The second step of a fuzzy modelling is a tuning of the membership functions. Thanks to supervised learning methods, it is possible to optimise both the antecedent and consequent parts of a linguistic rule-based fuzzy systems. A many of learning methods are derived from back propagation algorithm.

In next section the presented identification methods applied to the decomposed fuzzy model for SISO dynamic system will be tested by well-known identification problem, namely the Box-Jenkins gas furnace [15].

6. Numerical example

The Box and Jenkins's gas furnace is a popular example of the system identification. The Box-Jenkins data set consists of 296 input-output observations, where the input $u(t)$ is the rate of the gas flow into the furnace and the output $y(t)$ is the CO_2 concentration in the outlet gases.

The equation (1) is one of a general forms of the SISO discrete-time fuzzy models. As is well-known, the identification problem usually involves both the structure identification and the parameter estimation. Obviously, the structure identification for the system under study is to determine the delay time τ and the orders n_y, n_u. In this work, the $n_y=1$, $n_u=1$ and $\tau=4$ are chosen. Then, with respect to Figure 1 and the equations (1), (4), (5), (6) and (10), the output of the common fuzzy model (CFM) is calculated as follows:

$$\hat{y}(k) = defuzz\left\{\left[fuzz\{u(k-4)\} \wedge fuzz\{y(k-1)\}\right] \circ R\left(u_{k-4}, y_{k-1}, y_k\right)\right\}. \qquad (27)$$

With respect to Figure 4, the output of the decomposed fuzzy model (DFM) is

$$\hat{y}_d(k) = defuzz\left\{\left[fuzz\{u(k-4)\} \circ R_U\left(u_{k-4}, y_k\right)\right] \wedge \left[fuzz\{y(k-1)\} \circ R_Y\left(y_{k-1}, y_k\right)\right]\right\} \qquad (28)$$

and finally, the output of the simplified decomposed fuzzy model (SDFM), is calculated as

$$\hat{y}_s(k) = \frac{1}{2}\left\{defuzz\left[fuzz\{u(k-4)\} \circ R_U\left(u_{k-4}, y_k\right)\right] + defuzz\left[fuzz\{y(k-1)\} \circ R_Y\left(y_{k-1}, y_k\right)\right]\right\} \qquad (29)$$

All three fuzzy systems was realised with *max-min* compositional rule of inference and with *sum-prod* compositional rule of inference. The first one was realised with the *min* conjunction operator, the Mamdani implication function, and *max* disjunction as rules aggregation operator. The second inference is based on *prod* conjunction, Larsen implication function and *sum* disjunction as rules aggregation operator. The membership functions, defined on the input variable space and the output variable space, are shown in Figure 6. It is fairly intuitive that the accuracy of the approximation produced by a relational model will increase with increasing the number of reference sets.

In order to compare the performance quality of fuzzy models, an mean square error criterion is used:

$$J_{MSE} = \frac{1}{296}\sum_{k\,1}^{96}\left[y(k) - \hat{y}(k)\right] \qquad (30)$$

In Table 1, the results obtained with the *CFM* (27), the *DFM* (28), and *SDFM* (29) fuzzy models are presented.

We can see that the results obtained with *decomposed fuzzy models (DFM)* and *(SDFM)* are worse then results obtained with the *common fuzzy model (CFM)*. This is due to the simplification of the fuzzy equation and the reduction of the number of rules in rule bases. The results indicate that it could be interesting to study the identification methods for tuning the parameters of the membership functions. It could be expected that fine tuning of membership functions will results in better identification results.

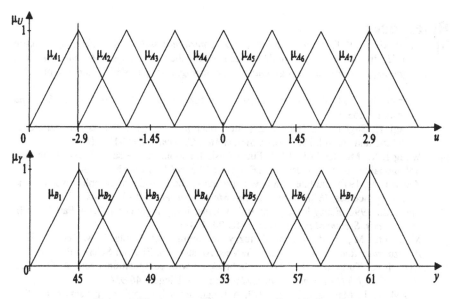

Figure 6: The input and the output membership functions.

Table 1: Identifications results for Box-Jenkins gas furnace

Fuzzy model	Inference method	No. of rules	MSE
CFM	*Max-min*	49	0.42
	Sum-prod	49	0.21
DFM	*Max-min*	14	0.73
SDFM	*Sum-prod*	14	0.57

7. Conclusions

In this paper we have presented two fuzzy models for SISO dynamic systems based on the decomposed compositional rule of inference. Both models based on the simplification of the decomposed model structure. Advantages of the proposed models are; the reduction of number of rules in rule base, simple composition of a fuzzy sets and a two-dimensional fuzzy relation, code optimisation and hardware inference realisation is possible, and close relationship between structure of the proposed model and the discrete linear ARX model. Further researches will be directed to improve identification methods for decomposed fuzzy models.

References

[1]   Golob M., 1999, Decomposition of a fuzzy controller based on the inference break-up method, In: R. Roy, T. Furuhashi and P.K. Chawdhry (Eds.) *Advances in soft computing : engineering design and manufacturing*, Springer-Verlag London, pp. 215-227.

[2]   Pedrycz W., 1984, Identification algorithm in fuzzy relational systems, *Fuzzy Sets and Systems* 13, 153-167.

[3]   Takagi T., Sugeno M., 1985, Fuzzy identification of systems and its applications to modelling and control, *IEEE trans. System Man Cybernet.* SMC-15, 116-132.

[4]   Wang L.X., Mendel J.M., 1992, Fuzzy basis functions, universal approximation, and orthogonal least-squares learning, *IEEE trans. Neural Networks* NN-3, 807-814.

[5]   Zadeh L.A., 1973, Outline of a new approach to the analysis of complex systems and decision processes, *IEEE Trans. Systems Man Cybernet.* SMC-3 , 28-44.

[6]   Lee C.C., 1990, Fuzzy logic in control systems: fuzzy logic controller – Part I and II, *IEEE Trans. System Man Cybernet.* SMC-20, 404-435.

[7]   Mizumoto M., 1985, Extended fuzzy reasoning, in: Gupta M.M. et al (Eds) *Approximate Reasoning in Experts Systems*, Amsterdam, Elsevier Science Publishers.

[8]   Demirli K., Turksen I. B., 1992, Rule break up with compositional rule of inference, *Proc. of the 1th IEEE Conf. on Fuzzy Systems*, San Diego, 946-956.

[9]   Procyk T. J., Mamdani E. H., 1979, A Linguistic self-organising process controller, *Automatica* 15, 15-30.

[10] Ljung L., Soderstrom T., 1983, *Theory and Practice of Recursive Identification*, The MIT Press, London.

[11] Czogala E., Pedrycz W., 1981, On identification in fuzzy systems and its applications in control problems, *Fuzzy Sets and Systems* 6, 73-83.

[12] Pedryzc W., 1989, *Fuzzy Control and Fuzzy Systems*, Research Studies Press LTD. Taunton, England.

[13] Xu C.W,, Lu Y.Z., 1987, Fuzzy model identification and self-learning for dynamic systems. *IEEE Trans. Systems Man Cybern.* 17(4) 68-689.

[14] Ridley J.N., Shaw I.S., Kruger J.J., 1988, Probabilistic fuzzy model for dynamic systems, *Electronics Letters* 24 (14), 890-892.

[15] Box G.E.P., Jenkins G.M., 1970, *Time Series Analysis, Forecasting and Control*, Holden Day, San Francisco.

Investigation of Least Square Fuzzy Identification via a Virtual Higher Resolution Fuzzy Model

K.M. Chow and A.B. Rad
The Hong Kong Polytechnic University, Department of Electrical Engineering
Email: eekmchow@ee.polyu.edu.hk; eeabrad@polyu.edu.hk

Abstract: In this paper, the memory requirement and computational burden of using recursive least square method for parameters updating of fuzzy rule table will be investigated. Conventionally, the entire fuzzy rule table is needed to update in each sampling interval which will lead the size of the covariance matrix will be very large. In order to circumvent this problem, a virtual higher resolution fuzzy model is adopted to minimize the size of the covariance matrix and hence to speed up the computation. A simulation study of a non-linear system identification is implemented to demonstrate the performance of the proposed algorithm.

1. Introduction

Recently, fuzzy modeling has gained a lot of attention for offering promising solutions to the problem of system identification. There have been various methods for tuning the fuzzy systems. For example, the popular techniques are: tuning the scaling factor by fuzzy tuner [6], tuning the consequence fuzzy set by Recursive Least-Square (RLS) method [1,9,13], genetic algorithm [12,15], tuning the TSK fuzzy model by gradient descent method [9,17]. Other solutions such as tuning various parameters (e.g. centre of fuzzy set, width of fuzzy set and output consequence's fuzzy set of the rule base) of the fuzzy system [2,7-9,15] are also reported in literature.

It is well known that the size of covariance matrix that is used for Recursive Least Square (RLS) method is based on the number of parameters being updated. In the last decade, many researchers applied the least square method for updating the parameters of the fuzzy rule table [1,9,13]. However, if the size of the fuzzy rule table becomes large, the size of the covariance matrix will be increased exponentially and the computational burden will be very high.

In this paper, for the sake of reducing the size of covariance matrix, which will be used for RLS method, a Virtual Higher-Resolution Fuzzy Model (VHR-FM) [4,5] will be used for circumvent this situation. Each cell of the rule-base will be expanded into 2^n cells (where n is number of input variables). Therefore, the Ordinary Fuzzy System (O-FS) is now implicitly divided into $\prod_{i=1}^{n} N_i$ Separated Local Fuzzy Systems (SLFS) (where N_i is the number of fuzzy subset of input i). Then, the size of the covariance matrix can be reduced significantly.

The rest of this paper is organized as follows: In Section 2, fuzzy indexing method and VHR fuzzy model will be briefly discussed. Section 3 contains the formulation of fuzzy identification algorithm with RLS update. The memory requirement and computational burden by using RLS method for VHR-fuzzy model will be discussed in section 4. In Section 5, computer simulation studies demonstrate the performance of the proposed algorithm. Finally, some conclusions and further works are presented in Section 6.

2. Granulation of the VHR-Fuzzy Rule Table

The granulation of the VHR-Fuzzy Rule Table (VHR-FRT) consists of 2 parts:
1) Fuzzy indexing method [5,10]
2) VHR-fuzzy set [4,5]

2.1 Fuzzy Indexing Method

The fuzzy indexing method is used to determine which 2^n (the most and where n is the number of input variable) rules in the fuzzy system will fired in each sampling interval. In general, the fuzzification, fuzzy inferencing and defuzzification of Ordinary Fuzzy Model (O-FM) can be written as follows:

Let,

x_i be the input i and N_i is the number of fuzzy subset of input i.

A_i^k be a consistent, complete, triangular fuzzy set and denote k^{th} fuzzy set of input i.

$$D_{A_i^k} = sup\,(A_i^k)$$

$I_i = k \mid x_i \in [D_{A_i^k}, D_{A_i^{k+1}}), k = 1...N_i$ are the fuzzy index of input i.

$$\tilde{X}_i = \{\mu_{A_i^k}(x_i) \mid k = 1...N_i\}$$ are the fuzzy vector of input i.

$$\tilde{X}_i^{I_i} = \mu_{A_i^{I_i}}(x_i)$$

$\bar{y}^{jk}$ be the centre of fuzzy set B^{jk} and it can be indexed by fuzzy index I_i.

Assume to use singleton fuzzifier, product inferencing engine and centre average defuzzifier, the output y of the fuzzy model can be written as:

$$y = \frac{\sum\limits_{\phi_1=0}^{1} \cdots \sum\limits_{\phi_n=0}^{1} \bar{y}^{(I_1+\phi_1)\ldots\ldots(I_n+\phi_n)} \prod\limits_{i=1}^{n} \tilde{X}_i^{(I_i+\phi_i)}}{\sum\limits_{\phi_1=0}^{1} \cdots \sum\limits_{\phi_n=0}^{1} \prod\limits_{i=1}^{n} \tilde{X}_i^{(I_i+\phi_i)}} \quad \ldots\ldots\ldots\ldots\ldots\ldots\ldots (1)$$

2.2 VHR-Fuzzy Set

Now, each fuzzy subset is further divided into two fuzzy sub-subsets (Fig.1) and each fuzzy subset is defined as follows:

For each $a,b,c \in \Re$ and $a<b<c$, Let $\mu_{a,b}(x): \Re \to \Re \neq 0 \mid x \in [a,b)$ & $\mu_{b,c}(x): \Re \to \Re \neq 0 \mid x \in [b,c]$ be the membership functions. For example, if we consider a fuzzy system with *2* inputs and *1* output and each input variable consists of *3* fuzzy subsets, the fuzzy rule table, which is defined by traditional method, will be *3x3*. It is noted that the new higher-resolution rule-base is formed as a *6x6* fuzzy rule table

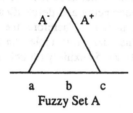

Fuzzy Set A

Figure 1: VHR-fuzzy set

Table 1: Virtual higher resolution fuzzy rule table

			X_I					
			A_1		A_2		A_3	
			A_1^-	A_1^+	A_2^-	A_2^+	A_3^-	A_3^+
X_2	A_1	A_1^-	Z	Z	PS	PS	PS	PS
		A_1^+	Z	Z	PS	PS	PS	PS
	A_2	A_2^-	NS	NS	Z	Z	PS	PS
		A_2^+	NS	NS	Z	Z	PS	PS
	A_3	A_3^-	NS	NS	NS	NS	Z	Z
		A_3^+	NS	NS	NS	NS	Z	Z

This virtual fuzzy rule-base can be regarded as decomposition of one fuzzy consequence (cell) from the Ordinary-Fuzzy Rule Table (O-FRT) into 2^n fuzzy consequences of that cell. Where n is number of input variables. For example, each rule in ordinary fuzzy model of a two inputs and single output fuzzy system will be expanded into the following 4 rules:

R_i: If x_1 is A_j^- and x_2 is A_k^-, then y is B^{j-k-}

        If x_1 is A_j^- and x_2 is A_k^+, then y is B^{j-k+}

        If x_1 is A_j^+ and x_2 is A_k^-, then y is B^{j+k-}

        If x_1 is A_j^+ and x_2 is A_k^+, then y is B^{j+k+}

$i = 1,2,3,\ldots\ldots 9, j,k = 1,\ldots,3$

Consequently, the VHR-FRT, shown in Table 1, can easily be constructed. This implies that in the initial design phase, O-FRT produced by any method can be used to fill the table and then expand it to form VHR-FRT. Note also that while the number of fuzzy subsets of each input variable is increased twice as much, the number of times of matching, fuzzification, fuzzy inferencing and defuzzification remain the same [5,10] when we use the fuzzy indexing method. Therefore, the computational burden on the fuzzy model is not affected. The fuzzy rule table can now be regarded as mainly containing 4 main (triple layered) and 5 auxiliary (double layered) individual Separated Local Fuzzy Model (SLFM).

3. Update Law

Since the number of fuzzy subsets of each input variable remain unchanged by using VHR-FM. Therefore, the equation (1) can be used for formulating the equation of VHR-FM with a minor change. From equation (2), we can see that we only need to multiply the fuzzy index of the output consequence by 2.

$$y = \frac{\sum_{\phi_1=0}^{1} \ldots \sum_{\phi_n=0}^{1} \bar{y}^{(I_1 \times 2 + \phi_1)\ldots\ldots(I_n \times 2 + \phi_n)} \prod_{i=1}^{n} \tilde{X}_i^{(I_i+\phi_i)}}{\sum_{\phi_1=0}^{1} \ldots \sum_{\phi_n=0}^{1} \prod_{i=1}^{n} \tilde{X}_i^{(I_i+\phi_i)}} \quad \ldots\ldots\ldots\ldots\ldots\ldots\ldots(2)$$

Moreover, the update of the VHR-FRT can easily be carried out by standard least-squares technique (see the Appendix).

The step-by-step procedures for implementation of the proposed fuzzy identifier are as follows:

1. Select the universe of discourse, the shape and number of the membership functions of the fuzzy linguistic variables.
2. Design the center of the input variables' membership functions.
3. If no *a priori* information is available, go to 5
4. Fill the cell of O-FRT from *a priori* expert knowledge.
5. Expand the O-FRT to form VHR-FRT.
6. Design the scaling factor for the input variables.
7. In each sampling interval, calculate the corresponding fuzzy index for each input variable.
8. Fuzzify the input variables with the fuzzy sets, which are indexed by fuzzy indices.
9. Locate the corresponding SLFS by the fuzzy indices.
10. Update the corresponding SLFS by RLS method by using its corresponding covariance matrix $P^{I_1 \cdots I_n}$ and parameter vector $\theta^{I_1 \cdots I_n}$. Notice that the VHR-FI has its own covariance matrix P within each SLFS.
11. Go back to step 7.

4. Memory Requirement & Computational Burden by RLS

Conventionally, the size of the covariance matrix P is depended on the size of the O-FRT. Therefore, it needs to manipulate a high dimension matrix computation at each sampling interval if the table size is large (For example, consider a 2 inputs and single output ordinary fuzzy system and each input variable contains 5 fuzzy subsets, the covariance matrix P will be 25x25). However, if we use the VHR-FM, we only need to handle low dimension matrix computation since each covariance matrix $P^{I_1 I_2}$ in each SLFS is only 4x4. In that example, the VHR-FM contains 25 SLFS that mean 25 covariance matrixes need to be stored. As long as the memory requirement needed to store the covariance matrix P for both VHR-FM and O-FM are concerned, the VHR-FM requires smaller memory space than the O-FM. Since the total memory space need to store the O-FM's covariance matrix is 625 (25x25) and the VHR-FM only need 4x4x25=400. In general, the memory requirement of VHR-FM and the O-FM are govern by the following equation:

$$\prod_{i=1}^{n} N_i + \left(\prod_{i=1}^{n} N_i\right)^2 = \prod_{i=1}^{n} N_i + \prod_{i=1}^{n} N_i^2 \quad \dots\dots\dots\dots\dots\dots\dots\dots\dots\dots\dots\dots(3)$$

$$\prod_{i=1}^{n} (N_i \times 2) + (2^n)^2 \prod_{i=1}^{n} N_i = \prod_{i=1}^{n} (N_i \times 2) + 2^{2n} \prod_{i=1}^{n} N_i \quad \dots\dots\dots\dots\dots\dots\dots(4)$$

The first term in equation (3) is the total memory space need to store the fuzzy consequence for the O-FM. The second term in (3) is the total memory space need to store the covariance matrix P for the O-FM.

The first term in equation (4) is the total memory space need to store the fuzzy consequence for the VHR-FM. The second term in (3) is the total memory space need to store all the covariance matrixes $P^{I_1 \cdots I_n}$ for the VHR-FM.

Therefore, the increase or decrease in memory space usage by VHR-FM when we compare with O-FM can be written as:

$$V(n, N_i) = \prod_{i=1}^{n} (N_i \times 2) + 2^{2n} \prod_{i=1}^{n} N_i - \prod_{i=1}^{n} N_i - \prod_{i=1}^{n} N_i^2 \quad \text{...............................}(5)$$

Note: negative means decrease memory space by using VHR-FM and vice-versa.

From equation (5), we can observed that if N_i is set to more than 5 and n is more than 2, equation (5) will be always negative and implied that decrease in memory space usage.

5. Simulation Studies

The Virtual High-Resolution Fuzzy Identifier (VHR-FI) is implemented to identify the following non-linear system:

$$y(k+1) = \frac{2y(k)y(k-1) + y(k-2)}{1.1 + y(k-1)y(k-2) + y(k)^3} + u(k)$$

The entire elements of fuzzy rule table are initially filled with zero and the covariance matrixes P are set to 1000 at start. The excitation $u(k)$ is $sin(0.1k)$ and the VHR-FRT is updated in every $0.1sec$. Moreover, the performance index is used as integral of squared error (ISE) for measuring the performance of the two different fuzzy identifiers. Both of the Fuzzy Identifiers (Ordinary Fuzzy Model Fuzzy Identifier (OFM-FI) and VHR-FI) use 5 fuzzy subsets in each input variable. In Fig. 2a, the output response of the VHR-FI is shown. We can see that the response of VHR-FI and the $ISE_{VHR-FI}=0.914$ is better than OFM-FI ($ISE_{OFM-FI}=1.924$) (Fig. 2b). Figure 2c shows the residual of the VHR-FI and Figure 2d shows the residual of the OFM-FI. We can see that the error between the actual output and the VHR-FI output is smaller than the OFM-FI.

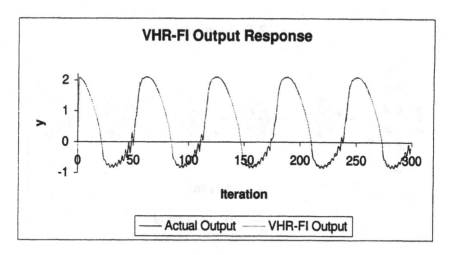

Figure 2a: VHR-FI output response

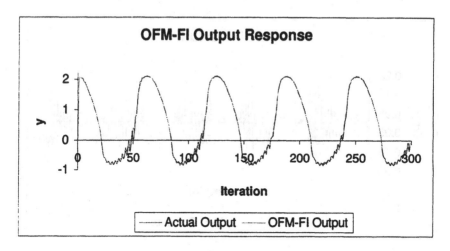

Figure 2b: OFM-FI output response

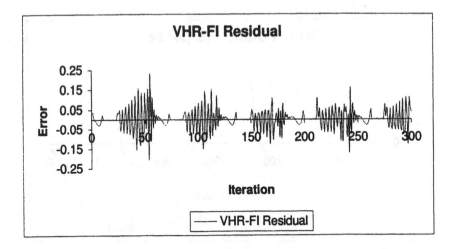

Figure 2c: VHR-FI residual

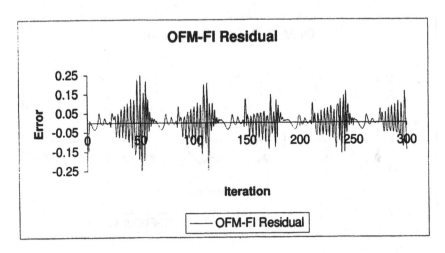

Figure 2d: OFM-FI residual

6. Conclusions

A new approach for obtaining a virtual higher resolution fuzzy model is reported and the results are compared with the conventional approaches. Simulation study and experimental results demonstrate overall improvement in the response. Moreover, the covariance matrix P and the memory required to store the covariance matrix are largely reduced.

References:

1. A. Garcia-Cerezo, A. Mandow and M. J. Lopez-Baldan, "Fuzzy Modelling Operation Navigation Behaviors", Proceeding of the Sixth IEEE International Conference on Fuzzy Systems, Barcelona, Spain, July 1-5, 1997, pp1339-1345.
2. Francois Guély and Patrick Siarry, "Gradient Descent Method for Optimizing Various Fuzzy Rule Bases", Proceeding 1993 Second IEEE International Conference on Fuzzy Systems, p1241-1246.
3. Gee Yong Park, Poong Hyun Seong, "Towards Increasing the learning Speed of Gradient Descent Method in Fuzzy System", Fuzzy Sets and Systems 77(1996) 229-313.
4. K.M. Chow and A.B. Rad, "System Identification via a Virtual Higher-resolution Fuzzy Model with Recursive Least Square Update", International Workshop on Soft Computing in Industry '99, June 16-18, 1999, Muroran, Hokkaido, Japan, pp246-251.
5. K.M. Chow and A.B. Rad, "System Identification via a Virtual Higher-Resolution Fuzzy Model", Intelligent Automation and Soft-Computing (Impress).
6. Hung-Yuan Chung, Bor-Chin Chen, Jin-Jye Lin, "A PI-type fuzzy controller with self-tuning scaling factor", Fuzzy Sets and Systems 93 (1998) 23-28.
7. Li-Xin Wang and Jerry M. Mendel, "Back-Propagation Fuzzy System as Non-linear Dynamic System Identifiers", IEEE International Conference on Fuzzy Systems, 1992 pp1409-1412.
8. Li-Xin Wang, "Design and Analysis of Fuzzy Identifiers of Nonlinear Dynamic System", IEEE Transactions on Automatic Control. Vol.40, no.1, January 1995, pp11-23.
9. Li-Xin Wang, A Course in Fuzzy Systems and Control, Prentice-Hall International, Inc.-16
10. K.M. Passino and S. Yurkovich, Fuzzy Control, Addison Wesley, 1998.
11. Minglu Zhang, Shangxian Peng and Qinghao Meng, "Neural network and fuzzy logic techniques based collision avoidance for a mobile robot", Robotica (1997) volume 15. pp 627-632.
12. P.T. Chan, W.F. Xie and A.B.Rad, "Tuning of Fuzzy Controller for an Open-Loop Unstable System: A Genetic Approach", Fuzzy Set and Systems, 1998 (In press).
13. Pekka Isomursu and Seppo Kemppainen, "Practical Adaptation of the Least-Mean-Square Method for Tuning of Fuzzy Logic Controllers", Intelligent Information Systems, 1994 Proceeding of the 1994 Second Australian and New Zealand Conference, pp209-213.
14. Sabi J. Asseo, "Terrain Following/ Terrain Avoidance Path Optimization using the method of Steepest Descent", Proceeding of the IEEE 1988 National Aerospace and Electronics Conference – NAECON 1988, p1128-1136.
15. T.C. Chin, X.M. Qi, "Genetic Algorithm for Learning the Rule Base of Fuzzy Logic Controller", Fuzzy Sets and Systems 97(1998) 1-7.

Fuzzy Morphologies Revisited

Mario Köppen, Katrin Franke

Department Pattern Recognition
Fraunhofer IPK Berlin, Pascalstr. 8-9, 10587 Berlin
email: mario.koeppen@ipk.fhg.de, franke@ipk.fhg.de

Abstract. In this paper, several approaches to fuzzy morphology are proposed and discussed. These approaches are based on the fundamental attitude, that fuzzy morphology should be an image processing discipline, and not a means for just defining fuzzy hedges on fuzzy sets (like "very", "few", "most"). Also, there is no deduction process leading to exactly one and only one fuzzy morphology, but an inductive process, leading to the co–existence of a big familiy of fuzzy morphologies. Among the fuzzy concepts used for defining a fuzzy dilation considered are fuzzy membership values, t- and s-norms, the extension principle, the fuzzy integral, fuzzy subsethood and ordered weighted averaging. All of these fuzzy morphologies fulfills the key concepts of Serra for generalized morphologies, but give a unique class of image operations as well.

1 Introduction

Mathematical morphology can be considered as a theoretical and practical means for analyzing spatial structures. It comprises a versatile toolset of techniques for image processing, whose usefulness has been proven for the processing of binary images and grayscale images as well. Operations of mathematical morphology are image-to-image transformations based on a structuring element, which acts like a probe sensitive for structural information. As a result of the operation, some image features might be enhanced, suppressed or preserved [14].

Since its beginnings, the generalization of mathematical morphology to new fields has puzzled many researchers. The most prominent attempts were devoted to the concepts of higher dimensions, multivariate data and, what is the major concern of the presented work, fuzzy logic. In this scope, the term "fuzzy morphology" was quickly patterned, and a multitude of possible definitions followed. The most important survey on all fuzzy morphologies presented so far was given in [2], and from the carefull discussion given there, the term seemed to have acquired its final meaning. But, the proposals for fuzzy morphologies did not stop [7] [8] [11] [5]. And, what is the more important, the new proposals did not fit into the framework given in [2] due to a simple reason, which will be given next.

In general, there are two fundamental paradigms underlying a fuzzy extension of mathematical morphology. The first one, and this is the one discussed in [2], is the definition of morphological operations, which act on fuzzy sets.

This work has the value of providing morphological aspects of fuzzy hedges. The second one considers fuzzy morphology as an image processing approach, following the traditional perspective of morphological image processing. As new approaches are mostly due to the second paradigm, they will not fit into the fuzzy set based framework. From the image processing community, the fuzzy set approach has been criticized due to its inability for defining even a single new image processing operation (see [5]). This is also the viewpoint shared in this study, since the apparent successes of mathematical morphology stems from image processing, and the question of detecting structure in fuzzy membership functions (only few of them might give structural miracles, since most of them are manually designed) is not considered of comparable importance.

This paper is organized as follows. Section 2 shortly recalls mathematical morphology and summarizes the requirements for its generalisation. Section 3 gives some choices for fuzzyfying traditional approaches, which might apply to mathematical morphology as well. From this, a variety of fuzzy morphologies can be derived. Some of them are discussed and their effect on images is exemplified. The paper ends with a conclusion and the reference.

2 Mathematical Morphology

Mathematical morphology could be considered as a family of image processing operations, which are related to the detection, enhancement or filtering of structures in images. This operations employ a so-called structuring element B, which acts like a probe within images. Mathematical morphology can be further subdivided into fields like granulometry, immersion operations, filtering, fractal image processing, sceletonization approaches or convex hull approximation. For a comprehensive introduction consider [14] [12] [13].

All of this fields are in some manner linked to the basic definition of the operations dilation and erosion with a structuring element B. Definitions exist for the case of binary and grayscale images. Be f a (two-dimensional) image function, defined on a finite subdomain of $Z \times Z$, which assigns a grayvalue (in most cases an integer from the set $\{0, 1, 2, \ldots, g_{max}\}$) to each point (x, y) of the image domain (grayscale image), or be A just a subset of the image domain (binary image), then, binary dilation of A by the structuring element B is defined by

$$A \oplus B = \bigcup_{a \in A} a + B \tag{1}$$

and grayscale dilation is given with

$$(A \oplus B)(z) = \max_{z=x+y} \{f(z) \mid x \in A, y \in B\} \tag{2}$$

Since a binary image could be considered as a grayscale image with the image function

$$f(x, y) = \begin{cases} g_{max} & : & (x, y) \in A \\ 0 & : & (x, y) \notin A \end{cases},$$

thus transforming equation 1 into 2, it is sufficient to consider the grayscale morphology in the following. For the purpose of generalization, we follow the approach given in [17] and introduce the δ set operator:

$$(A \oplus B)(z) = \operatorname*{argmax}_{z=x+y} \delta\{f(z) \mid x \in A, y \in B\} \tag{3}$$

If δ is choosen as the identity operator, the operation simply selects the maximum grayvalue out of the set of all neighbours of the point z, which are linked to z via offsets of the (inverted) structuring element B. In this case, B could be considered as a mask with a central point positioned over z. From all images positions covered by the mask, the one with the maximum grayvalue replaces the actual point.

Complimentary to the dilation $\oplus$ is the erosion operation $\ominus$, which selects the minimum out of the neighbours covered by the mask.

The issues related to generalising this appraoch to other data classes (color images, image sequences, multiscalar images) are generally mistaken as a deduction process. The procedure given in [2] and referred to as logical construction, is as follows: defining requirements, stating principles governing the construction, *deriving* basic definitions (dilation and erosion) based on these principles and satisfying as many requirements as possible.

Each generalized morphology is classified according to the chosen set of required properties. However, these sets differ. Serra [13] considered the following key ideas important for a generalized morphology:

- an order relationship,
- a supremum or infimum pertaining to that order and
- the possibility of admitting an infinity of operands.

These are the most "relaxed" requirements for a generalized morphology.

In the scope of [2], much more (all together 14) were given, and they were even subdivided into required and optional properties. Among them are: translation invariance, compatibility with homothetics, local knowledge, semi-continuity, idempotence, fitting characterization, relationship with cuts a.s.o.

Also, for multivariate data, it is considered to be important for the generalized (i.e. multivariate) morphologies to be vector–preserving (see [17]), what means, that the operation of dilation must not produce image values, which were not present in the original image. Also, the dilation being exchangeable with the maximum operation is generally considered as an important requirement. A propose of a class of generalized dilations, which give the same results, if restricted to the local Pareto sets, will be presented in [9].

However, generalization is never performed by deduction, but by induction. And from induction one is always faced with the inductive ambiguity, what was pointed out by Watanabe [19]. From this, it will never be possible to justify a generalized morphology to be the one and only correct one. The set based foundation of mathematical morphology allows for the co-existence of a large number of classes of generalized morphologies.

The essential aspect of the given requirements is to hint on possible application fields of the operations. For example, a non vector–preserving morphology is of restricted usability for color image processing, because it would disturb the structural color appearance of the image by possibly introducing color artifacts. A non–contiguous dilation might be hardly targeted for fast algorithms, a.s.o.

The discussion just given also applies to fuzzy morphologies. As will be demonstrated in the next section, there are several possible definitions of fuzzy dilation as an *image processing operation*.

In order to justify the generalized dilations, the following simplified subset of requirements is used in the following:

1. Whether the generalized dilation is vector–preserving, i.e. if the dilation selects one image function value out of the set of data values within the neighborhood $z + B$ of a pixel at position z.
2. A generalized dilation should be an increasing operation, i.e.

$$p \oplus a \geq p,$$

where the meaning of $\geq$ is according to the inherent key idea of sorting. It could also be said, that a dilation should be exchangeable with the maximum operation.
3. When $\oplus_B$ assigns standard binary morphology, the generalized dilation should be compatible according to this operation, i.e.

$$(p \oplus a) \oplus b = p \oplus (a \oplus_B b).$$

4. A fuzzy morphology should become a standard morphology, if its fuzzyness is stripped of.

3 Fuzzy Concepts

The possible concepts of fuzzy set theory leading to new, generalized morphologies are given in the following subsections.

3.1 Fuzzyfications of crisp data

A crisp data value is considered as having membership degree of a fuzzy set. Hence, a is replaced by $\mu(a)$. This is the most simple way for fuzzyfying a concept. It was used in [11] to extend binary morphology to fuzzy morphology by considering the subset A, which gives the binary image, as a fuzzy set. However, this is commonly considered as a grayvalue image, and this fuzzy approach leads directly to grayscale morphology, therefore all given requirements are known to be fulfilled.

3.2 T–Norms and S–Norms

In fuzzy set theory, t-norms and s-norms are used to model the logical connectives *and* and *or*. A mapping $T : [0,1] \times [0,1] \to [0,1]$ is considered to be a t-norm, if it is symmetric, associative, non–decreasing in each argument and if $T(a,1) = a$ for all $a \in [0,1]$. The s-norm is opposed to the t-norm by fulfilling $S(a,0) = 0$ for all $a \in [0,1]$.

From definition, the conventional min and max operations are t- and s-norms, respectively. Other definitions were collected in [18]. Some of these definitions for t-norms and s-norms are given in tables 1 and 2.

T-Norm	S-Norm
Standard	
$\min(a,b)$	$\max(a,b)$
Algebraic Product / Algebraic Sum	
ab	$a + b - ab$
Bounded Sum / Difference	
$\max[0, a+b-1]$	$\min[1, a+b]$
Drastic Section / Drastic Union	
$T(a,b) = \begin{cases} a \text{ f"ur } b = 1 \\ b \text{ f"ur } a = 1 \\ 0 \text{ sonst} \end{cases}$	$S(a,b) = \begin{cases} a \text{ f"ur } b = 0 \\ b \text{ f"ur } a = 0 \\ 1 \text{ sonst} \end{cases}$
Hamacher, 1978	
$\dfrac{ab}{\gamma+(1-\gamma)(a+b-ab)}$	$\dfrac{a+b+(\gamma-2)ab}{\gamma+(\gamma-1)ab}$
Rem.: $\gamma > 0$	
Frank, 1979	
$\log_s\left[1 + \dfrac{(s^a-1)(s^b-1)}{s-1}\right]$	$1 - \log_s\left[1 + \dfrac{(s^{1-a}-1)(s^{1-b}-1)}{s-1}\right]$
Rem.: $s > 0$, $s \neq 1$	
Dubois und Prade, 1980	
$\dfrac{ab}{\max[a,b,\alpha]}$	$1 - \dfrac{(1-a)(1-b)}{\max[1-a,1-b,\alpha]}$
Rem.: $\alpha \in (0,1)$	
Yager, 1980	
$1 - \min\left[1, [(1-a)^w + (1-b)^w]^{1/w}\right]$	$\min\left[1, [a^w + b^w]^{1/w}\right]$
Rem.: $w > 0$	
Dombi, 1982	
$\dfrac{1}{1+\left[\left(\frac{1}{a}-1\right)^{\lambda}+\left(\frac{1}{b}-1\right)^{\lambda}\right]^{1/\lambda}}$	$\dfrac{1}{1+\left[\left(\frac{1}{a}-1\right)^{-\lambda}+\left(\frac{1}{b}-1\right)^{-\lambda}\right]^{-1/\lambda}}$
Rem.: $\lambda > 0$. For $ab = 0$ the T-Norm is 0, for $(1-a)(1-b) = 0$ the S-Norm is 1.	
Weber, 1983	
$\max\left[0, \dfrac{a+b+\lambda ab-1}{1+\lambda}\right]$	$\min\left[1, a+b - \dfrac{\lambda ab}{1+\lambda}\right]$
Rem.: $\lambda > -1$	
Yu, 1985	
$\max[0, (1+\lambda)(a+b-1) - \lambda ab]$	$\min[1, a+b+\lambda ab]$
Rem.: $\lambda \leq 1$	

Table 1. Overview of different definitions of T- and S-Norms for values $a, b \in [0,1]$.

Each of these definitions gives the basis for a fuzzy dilation (s-norm) and its dual erosion, where the minimum operation is replaced by a t-norm. Each of these fuzzy erosions is unique, and highly flexible due to the additional degree of freedom given by the parameter values. This approach leads to a fuzzy dilation, which is not vector–preserving, increasing and which is not compatible to binary morphology in general. By stripping of the fuzzyness this approach to fuzzy morphology becomes a standard morphology as well.

T- und S-Norms by Schweizer and Sklar, 1963
$T(a,b) = \max[0, a^p + b^p - 1]^{1/p}$ $S(a,b) = 1 - \max[0, (1-a)^p + (1-b)^p - 1]^{1/p}$ Rem.: $p \neq 0$
$T(a,b) = 1 - [(1-a)^p + (1-b)^p - [(1-a)(1-b)]^p]^{1/p}$ $S(a,b) = [a^p + b^p - (ab)^p]^{1/p}$ Rem.: $p > 0$
$T(a,b) = \exp\left[-(\mid \ln a \mid^p + \mid \ln b \mid^p)^{1/p}\right]$ $S(a,b) = 1 - \exp\left[-(\mid \ln(1-a) \mid^p + \mid \ln(1-b) \mid^p)^{1/p}\right]$ Rem.: $p > 0$
$T(a,b) = \dfrac{ab}{[a^p + b^p - (ab)^p]^{1/p}}$ $S(a,b) = 1 - \dfrac{(1-a)(1-b)}{[(1-a)^p + (1-b)^p - [(1-a)(1-b)]^p]^{1/p}}$ Rem.: $p > 0$

Table 2. Four definitions for T- and S-norms by Schweizer and Sklar with values $a, b \in [0, 1]$ (1963).

3.3 The extension principle

The extension principle, as given e.g. in [23], is a means for generalizing a mapping f from crisp sets U to V to the fuzzy calculus. If $v = f(u)$ is this mapping and $F = \mu_1 u_1 + \mu_2 u_2 + \ldots + \mu_n u_n$ is a fuzzy set, the image of F under f is given by

$$f(F) = \mu_1 f(u_1) + \mu_2 f(u_2) + \ldots + \mu_n f(u_n).$$

Similarily, if f maps $U \times V$ to W, and F and G are the corresponding fuzzy sets, the extension principle gives:

$$f(F, G) = \sum_{i=1}^{n} (\mu_F(v_i) \wedge \mu_G(u_i)) f(u, v).$$

By applying the extension principle to the definition of binary morphology, given with equation 1, one gets a new definition for a fuzzy morphology

$$(A \oplus B)(z) = \max_{z=x+y} \min(A(x), B(y)).$$

Here it is assumed for the structuring element to be a "non–flat" one, i.e. the structuring element has an image function $B(x)$, too. This approach to fuzzy morphology was considered in [5]. But it is equal to the definition, which consideres Bloch [2] as the most natural extention to fuzzy morphology based on α-cuts. It is very similar to the conventional threedimensional mathematical morphology, as can be found in [14]. The dilation with a structuring element $B(x)$ is normally given as

$$(A \oplus B)(z) = \max_{z=x+y} \{f(z) + B(y) \mid x \in A, y \in B\},$$

what changes to the extension principle based fuzzy morphology, if the addition is replaced by the minimum operation. This fuzzy morphology is not vector–preserving, since it could select a function value of $B(x)$. It is not increasing and not compatible, too. If the function values of $B(x)$ are restricted to $\{0, 1\}$, i.e. the structuring element becomes flat, the dilation mutates to the standard grayscale dilation.

The approach can be further generalized by replacing the maximum and minimum operations by s- and t-norms [2].

3.4 Fuzzy Integral

The calculus of fuzzy integral and fuzzy measure, especially the fuzzy λ-measure, was introduced in [15] and later revised (and simplified) by [16]. It will not be given in detail here. In [7], Grabisch made an important remark about the similarity of mathematical morphology and fuzzy integral by generally considering the dilation and erosion operation as fusion operation with nonlinear weighting of coalitions. The approach was further studied in [8]. The fuzzy dilation is defined by a non–flat structuring element $B(x)$, employing its function values as fuzzy densities, on which the fuzzy λ-measure can be based on. The corresponding fuzzy dilation is not vector–preserving, not increasing and not compatible. In [8], the proof is given, that it transforms to standard grayscale dilation, if the function values of the structuring element approaches either 0 or 1.

3.5 Fuzzy Subsethood

According to the fuzzy point-as-set approach of Kosko [10], fuzzy sets could be considered as points in the n-dimensional unit square (or unit cube) by using the membership degrees as coordinates. If the parallels to the coordinate axis are drawn, a hyperrectangle is constructed this way. Hence, each fuzzy set corresponds to a hyperrectangle in the unit square. This approach gives the way for a redefinition of the term subsethood. Initially, a fuzzy set A was considered as a subset of fuzzy set B, when for all membership values $a_i \leq b_i$ was fulfilled [22]. Kosko extended the concept to degrees of subsethood. This degree is derived geometrically from the assigned hyperrectangles. It equals the ratio of the volume of the intersection of both hyperrectangles to the volume of one hyperrectangle. Thus, even the whole set is a subset of each of its subsets to a certain degree[1].

More formally: if $F = \sum_i \mu_F(u_i)u_i$ and $G = \sum_i \mu_G(v_i)v_i$ are two fuzzy sets, the degree of membership of G in F is given by

$$\mu(G \subseteq F) = \frac{\prod_i \mu_F(u_i) \wedge \mu_G(v_i)}{\prod_i \mu_G(v_i)}.$$

From fuzzy subsethood, a fuzzy dilation and its dual erosion can be defined, which acts on multivariate data (e.g. color images). The data points within the local scope of a flat structuring element (i.e. for the position z the set of all points $z + b$ with $b \in B$) are considered as fuzzy sets. In equation 3, the δ operator is specified as the minimum value of mutual degrees of subsethood. The following argmax operation selects the image function value of exactly that position, which is a subset of any other position to the lowest degree. Replacing argmax with argmin and minimum with maximum gives the corresponding fuzzy dilation. This dilation is vector–preserving, and it is increasing in the Pareto sense (the result value of the dilation will never be dominated by the initial value). If data become onedimensional, it goes over to conventional grayscale morphology. However, it is not compatible.

3.6 Ordered Weighted Averaging

The ordered weighted averaging operators (OWA), initially proposed in [20], generalize morphological operations as well as ranking and averaging operations. Given a set of n weights w_i and a set of n values x_i to order, the OWA is defined by sorting the values in decreasing order and calculating

$$w \oplus_{owa} x = \sum_i w_i x_{j_i},$$

where j_i stands for the index of value x_i after the sorting. If w is choosen as $(1, 0, \ldots, 0)$, the OWA operator specialises to the standard grayscale dilation. By choosing w as $(w_1, w_2, \ldots, w_k, 0, \ldots, 0)$, OWA becomes the so–called restricted morphology, which could also be considered as fuzzy ranking. The operation is a fuzzy dilation, which is not vector–preserving, not increasing, but compatible. If k goes to 1 and w_1 to 1.0, this fuzzy dilation becomes the standard grayscale dilation.

Based on the idea of the OWA, several other operations have been defined, as the generalized OWA, the Weighted Median Aggregation or the Ordered Weighted Minimum operation [21]. Most of them can be used to replace the maximum or minimum operation in the definition of dilation or erosion, giving an alternative approach to fuzzy morphology. In general, they are not vector-preserving, not increasing and not compatible, but they give a grayscale morphology, if the weights are restricted appropriately.

---

[1] This was considered as fuzzy foundation of probability by Kosko.

4 Conclusion

From the examples given in the last section, there are many fuzzy concepts, which may be used in order to define generalized morphologies. Each of these morphologies could be referred to as fuzzy morphology, but this does not help for being specific about what is meant by this term. Also, none of the fuzzy morphologies can outperform the other due to a single criteria, which would allow for stating, which definitions best suits the requirements of a generalized morphology. In this point, we strictly disagree with the position presented in [2], which includes considering fuzzy morphology as a non image processing discipline, which encourages the use of exactly one and only one "true" fuzzy morphology, and which attempts to define generalized morphologies by deduction, instead of induction, which is the usual logical mode of generalization.

By demonstrating, how fuzzy morphologies can be based on fuzzy concepts like membership functions, t- and s-norms, the extension principle, fuzzy integral, fuzzy subsethood or OWA, it becomes obvious, that there is no "royal road" to one and only one fuzzy morphology. All of them give a unique set of usefull and applicable image processing operations. This is a common side-effect of inductive reasoning, the so-called inductive ambiguity.

For the fuzzy morphologies it came out, that most of them are not vector-preserving (the only non-trivial example for a vector-preserving morphology is the one based on fuzzy subsethood). Also, they do rarely give an increasing and compatible [15] operation. But, since they are derived from standard binary or grayscale morphology formulas, in all cases they reduce to a non-fuzzy morphology, if their fuzzyness is stripped of. But all of them refers to the key concepts of Serra by implementing a sort of ranking, a supremum due to this ranking, and by admitting an infinity of operands. By this, they really comprise generalized morphologies.

We hope to revive the discussion on fuzzy morphologies in future works by this little study.

References

1. V. Barnett. *The ordering of multivariate data.* J.R.Statist.Soc.A, Vol. 139, Part 3, p. 318-355, 1976.
2. I.Bloch, H. Maître. *Fuzzy mathematical morphologies: A comparative study.* Pattern Recognition 28 (9), pp. 1341-1387, 1995.
3. Ch. Busch, M. Eberle. *Morphological operations for color-coded Images.* Proc. EUROGRAPHICS'95, Computer Graphics Forum, Vol. 14, No. 3, pp. C193-204, 1995.
4. M.L. Cormer, E.J. Delp. *An empirical study of morphological operators in color image enhancement.* Proc. SPIE Conf. Image Processing Algorithms and Techniques III, Vol. 1657, pp. 314-325, 1992.
5. P.D. Gader. *Fuzzy spatial relations based on fuzzy morphology.* Proc. FUZZ-IEEE'97, Barcelona, Spain, pp. 1179-1183, 1997.

6. D.E. Goldberg. *Genetic algorithms in search, optimization, and machine learning.* Addison-Wesley, Reading, MA, 1989.

7. M. Grabisch. *Mathematical morphology and fuzzy logic.* Proc. IIZUKA'94, Iizuka, Japan, pp. 349-350, 1994.

8. S. Grossert, M. Köppen, B. Nickolay. *A new approach to fuzzy morphology based on fuzzy integral and its application in image processing.* Proc. ICPR'96, Vienna, pp. 625-639, 1996.

9. M. Köppen, Ch. Nowack, G. Rösel. *Fuzzy-subsethood based color image processing.* Proc. FNS'99, Leipzig, Germany, 1999.

10. B. Kosko. *Neural Networks and Fuzzy Systems.* Prentice Hall, 1991.

11. M.C. Maccarone, V. DiGesu, M. Tripiciano. *An algorithm to compute medial axis of fuzzy images.* Proc. SCIA'95, Uppsala, Sweden, pp. 525-532, 1995.

12. J. Serra. *Image Analysis and Mathematical Morphology.* Academic Press, New York, 1982.

13. J. Serra. *Image Analysis and Mathematical Morphology: Theoretical Advances.* Academic Press, New York, 1988.

14. P. Soille. *Morphological Image Analysis - Principles and Applications.* Springer Verlag, Berlin a.o., 1999.

15. M. Sugeno. *Theory of fuzzy integral and its applications.* Tokyo Institute of Technology, Ph.D. thesis, 1974.

16. H. Tahani, J.M. Keller. *Information fusion in computer vision using the fuzzy integral.* IEEE Trans. SMC, 20(3), pp. 733-741, 1990.

17. H. Talbot, C. Evans, R. Jones. *Complete Ordering and Multivariate Mathematical Morphology.* In: Henk J.A.M. Heijmans, Jos B.T.M. Roerdink (Eds.): Mathematical Morphology and its Applications to Image and Signal Processing. Kluwer Academic Publishers, pp. 27-34, 1998.

18. H.R. Tizhoosh. *Fuzzy-Bildverarbeitung.* Springer Verlag, Berlin a.o., 1998 (in German).

19. S. Watanabe. *Pattern recognition: human and mechanical.* John Wiley and Sons, 1985.

20. R.R. Yager. *On ordered weighted averaging aggregation operators in multi-criteria decision making.* IEEE Trans. SMC 18, pp. 183-190, 1988.

21. R.R. Yager, J. Kacprzyk. *The ordered weighted averaging operators - Theory and applications.* Kluwer Academics Publishers, 1997.

22. L.A. Zadeh. *Fuzzy Sets.* Information and Control, 8(3), pp. 338-353, 1965.

23. L.A. Zadeh. *Calculus of fuzzy restrictions.* In: L.A. Zadeh (ed.). Fuzzy sets and their applications to cognitive and decision process. Academic Press, pp. 1-39, 1975.

A Design Method of Stable Non-separate Controller Using Symbolic Expressions

Hidehiro YAMAMOTO and Takeshi FURUHASHI

Depertment of Information Electronics of Nagoya University
Furo-cho, Chikusa-ku, Nagoya, 464-8603, Japan
E-mail: hidehiro@bioele.nuee.nagoya-u.ac.jp
E-mail: furuhashi@nuee.nagoya-u.ac.jp

Abstract This paper presents a design method of stable fuzzy controller on symbolic level of fuzzy control system. The validity of granularization of control system is guaranteed by satisfying the 'non-separate' condition. Reachability is specially defined for driving a necessary condition for the design of stable 'non-separate' controller. Simulations are done to show the feasibility of the controller.

Keywords: Fuzzy control, Stability analysis, Petri net, Construction of fuzzy controller

1 INTRODUCTION

Fuzzy inference can describe control rules using if-then rules and it can incorporate expert's know-how. Continuous values of the input and output of the control system are granularized, and each granule is labeled with a symbol. These granules and corresponding symbols are good tools for experts to describe their know-how. It was difficult to guarantee the stability of fuzzy control system. Stability analysis on the fuzzy control system has been remarkably advanced in recent years.[1-7] Hasegawa, Furuhashi, and Uchikawa[5-7] have proposed a stability analysis method of fuzzy control system using a state transition matrix based on a bipartite directed multigraph of Petri net. The stability analysis of fuzzy control system is made possible by calculating the matrices. A theorem of asymptotic stability of fuzzy control system was derived in [7]. Furuhahi, Yamamoto, Peters, Pedrycz[8] presented a condition of tokens, for the validity of the stability analysis of the fuzzy control system. They introduced neural network representation of the generalized fuzzy Petri net model[9]. This condition bridges between the stability analysis on the symbolic level and the actual behavior of the control system on the continuous values. This paper presents a new design method of stable fuzzy controller. A reachable matrix is introduced to derive a necessary condition fro the design of stable fuzzy controller satisfying the non-separate condition.

2 NEURAL NET REPRESENTATION OF GENERALIZED FUZZY PETRI NET

A fuzzy control system is represented by the fuzzy Petri net model[7]. A fuzzy rule R^j is expressed as

$$R^j : \text{if } x_i \text{ is } A_{j1,1} \text{ and } x_2 \text{ is } A_{j2,2} \text{ and } \ldots$$
$$\ldots \text{and } x_n \text{ is } A_{jn,n} \text{ then } y \text{ is } B_j \tag{1}$$

where $A_{j1,1}$, $A_{j2,2}$, $\ldots$, $A_{jn,n}$ and B_j are fuzzy variables. $x_i (i = 1, \ldots, n)$ are input variables and y is the output. This fuzzy rule is represented by a neural network shown in Fig.2[8]. This figure shows a network for fuzzy rules

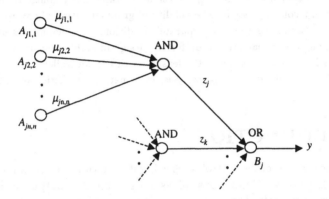

Figure 1: Fuzzy Petri net and its neural network reparesentation

having the same output variable. The input output relationships of AND-, OR-neurons in this network are given by

$$z_j = \mathop{T}_{i=0}^{n} \mu_{ji,i} \tag{2}$$

$$z_j = \mathop{S}_{i=0}^{m} z_i \tag{3}$$

where, $\mu_{ji,i}$ is the grade of membership of $A_{ji,i}$, T, S denote t-norm and s-norm, respectively. A fuzzy control system is modeled with the above representation. Table 1(a) and 1(b) are examples of fuzzy rules of a model of controlled object and those of its fuzzy controller, respectively. The controlled object is a nonlinear one whose output y_{k+1} at time $k+1$ is determined with its input u_k and the output y_k at time k. the controller gives a manipulated variable u_k from the command y_k^* and the output of the controlled object y_k. Fig.2 shows the membership functions used for the controller and the model of the controlled object. The labels of membership functions 'N', 'Z', 'P' are "*Negative*", "*Zero*", and "*Positive*", respectively.

Table 1: An example of fuzzy control system

(a)Fuzzy model of the controlled object

y_{k+1}		y_k	
		N	Z	P

u_k		N	Z	P
	N	N	N	Z
	Z	Z	Z	Z
	P	Z	P	P

(a)Fuzzy controller

u_k		y_k	
		N	Z	P

y^*_k		N	Z	P
	N	N	N	N
	Z	Z	Z	Z
	P	P	P	P

The neural network representation of this fuzzy control system is shown in Fig.3. In this network, the basic configuration of the neural network shown in Fig.2 is employed, and the neural networks of the fuzzy controller and the controlled object are combined.

The behavior of the control system is described by the firing of neurons and change of marking in the network. Those units in the layers of control command, manipulated variable u, and output of the controlled object y are labeled from left to right as N, Z, P. The unit in z_c, z_p-layers have the function in eq.(2). The units in u, y-layer have those in eq.(3). Fig.4 shows an example of transition of tokens in the case where the command $y^* = 0.2$ and the initial output of the controlled object $y_0 = 0.5$. The grade of membership of "y^* is Z" is 0.8, and that of "y^* is P" is 0.2. Both of the grades of "y is Z and P" are 0.5. Each unit has a token with the amount corresponding to the grade, e.g. unit 'Z' in y^*-layer has a token of 0.8. This amount is expressed with the area of black circle as shown in Fig.4(a). The first transition is the firing of fuzzy rules of the fuzzy controller. The input neurons are 'Z' and 'P' in both of y^* and y-layer. In this case the following four rules of the controller are fired:

If $y^*$is Zand y_k is Z then u_k is Z

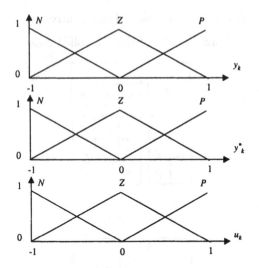

Figure 2: Membership function

If y^* is P and y_k is Z then u_k is P

If y^* is Z and y_k is P then u_k is Z

If y^* is P and y_k is P then u_k is P

Four units in the z_c-layer are fired as shown in Fig.4(a). Assume that the AND-wise operation in eq.(2) is multiplication. The amounts of tokens in the units in the z_c-layer are given from left to right as 0, 0, 0, 0, 0.4, 0.1, 0, 0.4 and 0.1 The output units are OR-wise operation in eq.(3) is summation, they receive the tokens of 0.8 and 0.2 after this 1st transition as shown in Fig.4(b).

3 STABILITY ANALYSIS

3.1 Granularized Control System

The granularized behavior of the fuzzy control system can be represented with bipartite directed multi-graphs[7] of Petri net. These behavior are given by utilizing the symbolic representation of the granules of the inputs and outputs of the control system. The firing of transition corresponds to the firing of fuzzy rule. The place represents one of the states of the control system before/after a transition. The place is made so that each transition. The place is made so that each transition has one input and one output. The process of converting production rules into a bipartite directed multi-graph is well described in [7]. It is known by applying the analysis in [7] that the control system in Table 1(a) and 1(b) is asymptotically stable.

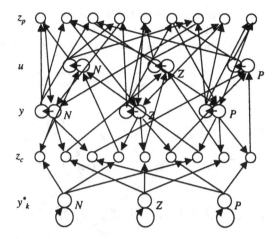

Figure 3: Neural network representation

3.2 Validity of Stability Analysis

The stability analysis in [7], which is based on the symbolic representation of the granularized system, sacrifices the rigorous analysis. The symbolic system does not reflect the types of the AND-wise operation in eq.(2) and the OR-wise operation in eq.3. The actual numerical value of the control system is determined by these operations and the actual behavior might differ from the results of the analysis on the discretized system. There should be a guarantee for the validity of the discretization.

A state of 'non-separate' of tokens in the NN model was introduced in [8]. This condition is only for the units in y^*, y and u layers, and does not apply to those in z_c, z_p layers.

Definition 1 (Non-separate-tokens)
Tokens are non-separated when they are distributed in only one unit or in two neighboring units in a layer.

These non-separate tokens are given by the fuzzy rules fired at the same time which have only one label or at most two neighboring labels in the consequence.

Theorem 1 (Non-separate condition)
Assume that the initial allocation of tokens in the neural network representation of fuzzy control system is non-separated. Tokens will not be separated if output units of the fired transition units at any time are limited to only one or two neighboring units in an output layer.

(Proof of Theorem 1)
Initial tokens are nonseparated. Only one or two units in a layer have tokens.

54

Possibly fired transition units are $2n$ where n input variables (n-input layers) for the transition layer. Output units receiving tokens from $2n$ fired transition units are one or two neighboring units. These tokens are nonseparated. □

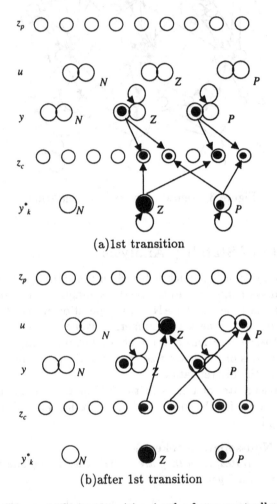

(a)1st transition

(b)after 1st transition

Figure 4: State transition in the fuzzy controller
$(y_k^* = 0.2;\ y_0 = 0.5)$

Theorem 2 (Validity of stability analysis)

If the symbolic system is asymptotically stable for any command values, and if the tokens in the neural network are non-separate at any transitions, two neighboring discretized values settle into one or two neighboring values, and the intermediate value between the two neighboring discretized values.

(Proof of Theorem 2)
If the discretized system is asymptotically stable, any discretized initial value

settles into a discretied value. If tokens are nonseparated, any neighboring tokens will not be separated. Any neighboring discretized tokens will also not to be separated, and will settle into one or two neighboring discretized values. The intermediate value is expressed with the two neighboring tokens with amounts in the range of [0,1], or one token with the always moves into the intermediate point between the two neighboring discretized values. If the two neighboring discretized values converge to a discretized value, the intermediate point coincides with the discretized value. □

4 DESIGN OF FUZZY CONTROLLER

In the previous section, we introduced the non-separate condition which guarantees the validity of the stability analysis of fuzzy control system on the symbolic level. This section presents a new design method for the stable fuzzy controller on the symbolic level. This method is to search for a controller satisfying the non-separate condition.

4.1 Reachability

For derivation of this design method, a new matrix called "reachable matrix" is introduced in this paper. "Reachability" is also defined. This definition of reachability can be interpreted as a subclass of reachability of Petri net. this paper discusses the reachability to be specifically defined in this paper for deriving the design method. The discussions in this section are on the symbolic level of the single input-single output and the first order system. the continuous values of the controlled object are granularized and labeled. An example is shown in Table 2. This table shows a case where input u_k and

Table 2: An example of fuzzy model of controlled object

y_{k+1}		y_k				
		NB	N	Z	P	PB
u_k	NB	NB	N	N	N	Z
	N	N	N	N	Z	Z
	Z	N	Z	Z	Z	P
	P	N	Z	P	P	PB
	PB	Z	Z	P	PB	PB

output y_k at time k are divided into five subspaces, and the granules have symbols 'NB', 'N', 'Z', 'P', 'PB' for "*Negative Big*", "*Negative*", "*Zero*", "*Positive*", "*Positive Big*".

Definition 2 (Reachability)
The controlled object is reachable from symbol B to another symbol A when the output of the system can reach A from B within finite iterations.

In the case in Table 2, the controlled object is reachable from 'NB' to 'PB' at the third iteration. The initial value $y_0 = $'$NB$'. By giving a constant input $u_k = $'$PB$', the output changes as $y_1 = $'$Z$', $y_2 = $'$P$', and $y_3 = $'$PB$'. But this system is not reachable to 'NB' from any state. There is no way from any of 'N' - 'PB' to 'NB' with any input.

Definition 3 (Globally Reachable)
The controlled object is globally reachable when the system is reachable to symbol A from any state within finite iterations.

The controlled object in Table 2 is globally reachable to 'N', 'Z', 'P', and 'PB', but not to 'NB'.

Let us introduce a reachable matrix between the symbols. The system in Table 2 is reachable within one iteration to 'NB', 'N' and 'Z' from 'NB'. This reachability can be expressed with a vector $r = (1\ 1\ 1\ 0\ 0)^T$. T is the transpose of vector. The reachability of this system from all the symbols of output within one sampling time is given by the following reachable matrix R:

$$R = \begin{pmatrix} 1 & 0 & 0 & 0 & 0 \\ 1 & 1 & 1 & 1 & 0 \\ 1 & 1 & 1 & 1 & 1 \\ 0 & 0 & 1 & 1 & 1 \\ 0 & 0 & 0 & 1 & 1 \end{pmatrix}. \tag{4}$$

From a symbol e.g. 'NB', the reachable symbols at one iteration are derived as

$$R \begin{pmatrix} 1 \\ 0 \\ 0 \\ 0 \\ 0 \end{pmatrix} = \begin{pmatrix} 1 & 0 & 0 & 0 & 0 \\ 1 & 1 & 1 & 1 & 0 \\ 1 & 1 & 1 & 1 & 1 \\ 0 & 0 & 1 & 1 & 1 \\ 0 & 0 & 0 & 1 & 1 \end{pmatrix} \begin{pmatrix} 1 \\ 0 \\ 0 \\ 0 \\ 0 \end{pmatrix} = \begin{pmatrix} 1 \\ 1 \\ 1 \\ 0 \\ 0 \end{pmatrix} \tag{5}$$

It is found that the power of R gives the reachable symbols from anyone of the symbols.

Theorem 3 (Reachability of system)
The system is globally reachable to any target symbols iff all the elements of R^{n-1} of n-dimensional square matrix R are unity.

In the case of the controlled object in Table 2,

$$R = \begin{pmatrix} 1 & 0 & 0 & 0 & 0 \\ 1 & 1 & 1 & 1 & 1 \\ 1 & 1 & 1 & 1 & 1 \\ 1 & 1 & 1 & 1 & 1 \\ 1 & 1 & 1 & 1 & 1 \end{pmatrix}. \tag{6}$$

This system is not reachable to 'NB'.

4.2 Construction of fuzzy controller

It is possible to control the output of the system, which is globally reachable from any initial symbol, to any desirable target on the symbolic level. This means that we can design a stable fuzzy controller in the sense in [7]. However, if we could not construct a controller, this control system would not have a guarantee for the validity of stability of the level of actual continuous values. We can derive a necessary condition for the controlled object to obtain a stable 'non-separate' controller. The 'non-separate' controller means the controller which satisfies the non-separate condition.

Theorem 4 (Necessary condition for controlled object)
*If a stable 'non-separate' controller can be designed, the fuzzy model of the controlled object has the rules in Table 3, where * denotes anyone of labels.*

Table 3: The rule set which
the controlled object must have

u_k \ y_{k+1}		y_k				
		NB	N	Z	P	PB
	NB	NB	NB	*	*	*
	N	N	N	N	*	*
	Z	*	Z	Z	Z	*
	P	*	*	P	P	P
	PB	*	*	*	PB	PB

(proof)
(1)Reachability

If the controlled object has the above rules, the reachable matrix is expressed with

$$R = \begin{pmatrix} 1 & 1 & * & * & * \\ 1 & 1 & 1 & * & * \\ * & 1 & 1 & 1 & * \\ * & * & 1 & 1 & 1 \\ * & * & * & 1 & 1 \end{pmatrix}. \tag{7}$$

This equation is the case where the labels of y_k and u_k are five each. The matrix powered by four is calculated as

$$R^4 = \begin{pmatrix} 1 & 1 & 1 & 1 & 1 \\ 1 & 1 & 1 & 1 & 1 \\ 1 & 1 & 1 & 1 & 1 \\ 1 & 1 & 1 & 1 & 1 \\ 1 & 1 & 1 & 1 & 1 \end{pmatrix}. \tag{8}$$

It is known that the controlled object is globally reachable to any symbol. This means that we can design an asymptotically stable controller on the symbolic level.

(2)Non-separate controller

From the rules of the controlled object in Table 3 a portion of controller shown in Table 4 can be designed. These rules satisfy the non-separate condition. If

Table 4: A portion of a stable 'non-separate' controller

u_k		y_k				
		NB	N	Z	P	PB
y^*_k	NB	NB	NB	*	*	*
	N	N	N	N	*	*
	Z	*	Z	Z	Z	*
	P	*	*	P	P	P
	PB	*	*	*	PB	PB

anyone of the designated labels of the controlled object differ from the rules in Table 3, some of the rules in Table 4 should be modified and any modification will violate the non-separate condition.

From the above (1) and (2), the set of rules in Table 3 is a necessary condition for a stable 'non-separate' controller.

4.3 Simulation

Table 5 shows an example of fuzzy model of a controlled object which satisfies the necessary condition in Theorem 4. The membership functions are

Table 5: An example of fuzzy model of the controlled object

y_{k+1}		y_k				
		NB	N	Z	P	PB
u_k	NB	NB	NB	N	N	Z
	N	N	N	N	N	Z
	Z	N	Z	Z	Z	Z
	P	N	Z	P	P	P
	PB	Z	Z	P	PB	PB

Table 6: A stable 'non-separate' controller
for the controlled object in Table 5

u_k		y_k				
		NB	N	Z	P	PB
y^*_k	NB	NB	NB	NB	N	N
	N	N	N	N	N	Z
	Z	N	Z	Z	Z	Z
	P	Z	Z	P	P	P
	PB	P	P	P	PB	PB

assumed to be given as shown in Fig.5. A stable 'non-separate' controller was easily designed as shown in Table 6. Asterisks * in Table 4 were decided so that the non-separate condition is satisfied. Some degrees of freedom existed in this decision. This freedom could be limited by introducing a criterion, e.g. quickest, and an automatic controller design. Figure 6 shows a simulation result of the control system in Table 5 and Table 6. Product-Sum-Center of Gravity method was used for the fuzzy inference. The output y settled at the target value $y^*_k = 1.6$.

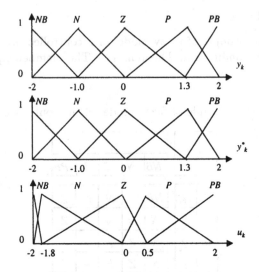

Figure 5: Membership function used for the simulation

5 CONCLUSIONS

This paper presented a new design method of stable 'non-separate' fuzzy controller. A necessary condition for controlled object were derived. Simulation results showed that the design on the symbolic level was also valid on the level of continuous values.

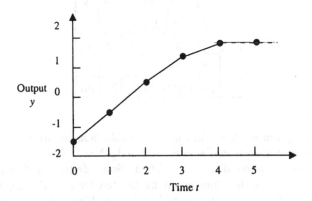

Figure 6: Simulation result
of the control system in Table 5 and 6
$(y_k^* = 1.8)$

References

[1] S. KITAMURA and T. KUROZUMI, "Extended Circle Criterion and Stability Analysis of Fuzzy Control Systems", Proceedings for the International Fuzzy Eng. Symp. '91, 2, pp634-643(1991).

[2] T. HOJO , T. TERANO, S. MASUI, "Stability Analysis of Fuzzy Control Systems Based on Phase Plane Analysis", Journal of Japan Society for Fuzzy Theory and Systems, 4, No.6, pp.1133-1146(1992), (in Japanese).

[3] K. TANAKA and M. SUGENO "Stability Analysis of Fuzzy Control Systems Using Lyapunov's Direct Method", Proceedings of NAFIP'90, pp133-136(1990).

[4] K. TANAKA and M. SUGENO "Stability Analysis and Design of Fuzzy Control Systems", Fuzzy Sets and Systems, Vol45, No.2, pp.135-156(1992).

[5] T. HASEGAWA and T. FURUHASHI and Y. UCHIKAWA, "Stability analysis of fuzzy control systems using petri nets", Proceedings of 1996 Biennial Conference of the North American Fuzzy Information Processing Society NAFIPS, pp.97-191(1996)

[6] T. HASEGAWA and T. FURUHASI and Y.UCHIKAWA, "Approximated Discrete System of Fuzzy Control System and Stability Analysis", Proceedings of Fifth IEEE International Conference on Fuzzy System, pp.2155-2161(1996).

[7] T. HASEGAWA and T. FURUHASHGI " Stability Analysis of Fuzzy Control System Simplified as a Discrete System", Control and Cybernetics, vol.27, No.4, pp.565-577(1998)

[8] T. FURUHASHI and H. YAMAMOTO and J. PETERS and W. PEDRYCZ "Fuzzy Control Stability Analysis Using Generalized Fuzzy Petri Net Model", Journal of Advanced Computational Intelligence, vol.3, No.2, pp.99-105(1999)

[9] W. PEDRYCZ, F. GOMIDE, "A Generalized Fuzzy Petri Net model", IEEE Tranz. on Fuzzy Systems, Vol.2, no.4, pp295-301(1994)

[10] W. BRAUER, G. ROZENBERG and A. SALOMA, "Petri Nets", Springer-Verlag

Adjustment of Identified Fuzzy Measures

Osamu Furuya[1], Takehisa Onisawa[2]

[1] Master's Program in Science and Engineering, Onisawa Lab.,
University of Tsukuba 1-1-1,
Tennodai, Tsukuba, 305-8573 Japan
E-mail: osamu@onisawa-gw.esys.tsukuba.ac.jp
[2] Institute of Engineering Mechanics and Systems, University of Tsukuba
1-1-1, Tennodai, Tsukuba, 305-8573 Japan
E-mail: onisawa@esys.tsukuba.ac.jp

ABSTRACT : A fuzzy measures and fuzzy integrals model is often applied to a human evaluation model. For the construction of the human evaluation model it is necessary to identify fuzzy measures by the use of input-output data. In this paper an algorithm for the adjustment of the identified fuzzy measures is discussed. The algorithm considers super-additivity or sub-additivity of fuzzy measures. A user interface is designed to help adjust fuzzy measures so that the errors between the data and the outputs of the evaluation model are made uniform. The contradictory data can be found out by the present algorithm and the user interface. In order to construct a better human evaluation model the contradictory data are fed back to human subjects for a re-evaluation. The fuzzy measures are identified again using the re-evaluation data. The effectiveness of the adjustment of fuzzy measures is shown by evaluation examples about apartments. In order to confirm whether the evaluation model fits human evaluation or not, errors between the model outputs and questionnaire, and the analysis results of obtained fuzzy measures are considered.

1. INTRODUCTION

A human evaluation model of a multi-attribute object has often the form of the overall evaluation of the object based on each attribute evaluation. Fuzzy measures and fuzzy integrals have been applied to the human evaluation model of a multi-attribute object[1-5], since fuzzy measures do not necessarily assume independence among attributes and additivity[6], and these properties of fuzzy measures are appropriate for the construction of a human evaluation model.

Applying fuzzy measures and integrals to the human evaluation model of a multi-attribute object, there is a problem how fuzzy measures are obtained. Usually fuzzy measures are identified using input-output data obtained by, e.g., a questionnaire, and many identification algorithms have been proposed[1][3].

In these identification algorithms the least mean square error between the data and the outputs of the fuzzy measures and integrals model is usually used as the identification index. Validity of the algorithm is confirmed from the viewpoint that identified fuzzy measures are approximately equal to the fuzzy measures which are used in order to generate test data, i.e., the input-output data. From the application point of view, however, it is necessary to verify whether the evaluation model with identified fuzzy measures fits human evaluation or not, since we have no fuzzy measures to be compared with identified fuzzy measures in a practical application. And the least mean square error does not necessarily lead to the construction of a good human evaluation model. In this respect it is necessary to adjust identified fuzzy measures. Moreover, there may be contradictory data in the questionnaire data. That is, fuzzy measures are identified using data including contradictory data.

In this paper the algorithm for the adjustment of identified fuzzy measures is proposed for the construction of a better human evaluation model. The algorithm considers super-additivity/sub-additivity of fuzzy measures. A user interface is designed to help adjust fuzzy measures so that the errors between the data and the outputs of the evaluation model are made uniform. Furthermore, contradictory data are found out using the user interface and are fed back to human subjects for re-evaluation in a questionnaire. The fuzzy measures are identified again using the re-evaluation data. The effectiveness of the adjustment of fuzzy measures is shown by the evaluation examples about apartments.

2. FUZZY MEASURES AND INTEGRALS

2.1 Fuzzy Measures

Let X be a finite set and $P(X)$ be a power set of X. A set function $g : P(X) \rightarrow [0,1]$ with the following properties is called a fuzzy measure.

$$g(\phi) = 0, \tag{1}$$

$$g(X) = 1, \tag{2}$$

$$A \subset B \subset X \Rightarrow g(A) \leq g(B), \tag{3}$$

where ϕ is an empty set. Fuzzy measures have only monotonicity (3) but have not necessarily additivity. Therefore, fuzzy measures are appropriate for the measure of human subjective evaluation. And one of the following three equations holds good for $\forall A, \forall B \subset X, A \cap B = \phi$.

$$g(A \cup B) = g(A) + g(B) \quad : \quad \text{additivity} \tag{4}$$

$$g(A \cup B) > g(A) + g(B) \quad : \quad \text{super-additivity} \tag{5}$$

$$g(A \cup B) < g(A) + g(B) \quad : \quad \text{sub-additivity} \tag{6}$$

2.2 Fuzzy Integrals

Let $f : X \to [0,1]$ be a measurable function and let us assume $H_\alpha = \{x \mid f(x) > \alpha, \alpha \in [0,1]\}$. The Choquet integral of the function f based on fuzzy measures is defined by

$$(c)\int f(x)dg = \int_0^1 g(H_\alpha)d\alpha. \qquad (7)$$

When the function is a simple function

$$f(x) = \sum_{i=1}^{n}(\alpha_i - \alpha_{i-1})\chi_{A_i}(x), \qquad (8)$$

the Choquet integral is defined by

$$(c)\int f(x)dg = \sum_{i=1}^{n}(\alpha_i - \alpha_{i-1})g(A_i), \qquad (9)$$

where $0 = \alpha_0 \leq \alpha_1 \leq ..... \leq \alpha_n \leq 1, A_1 \supset A_2 \supset ..... \supset A_n$ and χ_{A_i} is a characteristic function of a set A_i.

Equation (9) implies the shaded area shown in **Fig.1**.

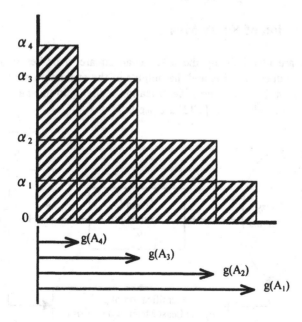

Fig. 1   Choquet Integral of Simple Function

3. EVALUATION MODEL AND IDENTIFICATION OF FUZZY MEASURES

3.1 Evaluation Model

In the subjective evaluation model of a multi-attribute object the overall evaluation is obtained by the subjective importance weighted evaluation of each attribute. In this paper the human evaluation model is considered as shown in **Fig. 2**, where $f(x_i)$ is the evaluation of the i-th attribute x_i, $g's$ are fuzzy measures and Z is the overall evaluation, obtained by the Choquet integral of f based on fuzzy measures.

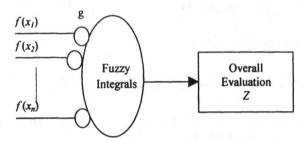

Fig. 2 Human Evaluation Model

3.2 Identification of Fuzzy Measures

Fuzzy measures are identified by the least mean square error between the data obtained by, e.g., a questionnaire, and the outputs of the evaluation model. **Fig. 3** shows the concept of fuzzy measures identification model. In this paper the fuzzy measures identification algorithms[1][3] are used.

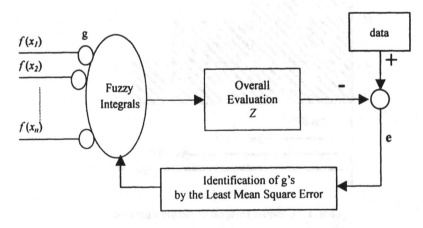

Fig. 3 Fuzzy Measures Identification Model

4. ADJUSTMENT OF IDENTIFIED FUZZY MEASURES

It is not certain whether the evaluation model with identified fuzzy measures only by the least mean square error fits human evaluation or not from the practical point of view, since only one index such as the least mean square error is used, and a man does not evaluate all attributes consistently, that is, some contradictory data may be included in questionnaire data. Therefore, it is necessary to adjust identified fuzzy measures. Furthermore, if some contradictory data are observed, the re-evaluation of the data is necessary.

4.1 Adjustment Guide

In this paper the following guides are considered for the adjustment.
(1) Intuitive adjustment by linguistic terms.
(2) Adjustment considering the characteristics of fuzzy measures, i.e., their super-additivity or sub-additivity.
(3) Assumption that the evaluation model with identified fuzzy measures reflects human evaluation to some extent, though not completely.
The guide (1) is based on the consideration that linguistic terms are appropriate for the expressions of the difference between the output of the model and human evaluation. The guide (2) is based on the consideration that emphasis of the characteristics of identified fuzzy measures can have much influence on the adjustment of the fuzzy measures. The guide (3) means that if the evaluation model with fuzzy measures differs from human evaluation completely, only adjustment of fuzzy measures cannot reduce the difference. Neglect of the guide (3) leads to a problem out of the scope of this paper.

4.2 Adjustment Algorithms

The following equations are used as the foundations of the adjustment.

○adjustment of fuzzy measure for increasing

$$g(X_i) = g'(X_i) + (1 - Z) \times word, \qquad (10)$$

○ adjustment of fuzzy measure for decreasing

$$g(X_i) = g'(X_i) + Z \times word, \qquad (11)$$

where i is an attribute number, $g'(X_i)$ is a fuzzy measure before the adjustment, $Z \in [0,1]$ is the model output, and *word* means the numerical value corresponding to the linguistic term which is used to adjust the fuzzy measure. The numerical value is positive when the fuzzy measure is increased, and the value is negative when the fuzzy measure is decreased. In this paper the numerical values as shown in Table 1 are used. Eqs. (10) and (11) imply that the fuzzy measure is increased a little or is decreased much when the model output is large, and that the fuzzy measure is increased much or is decreased a little when the output is small. These equations consider the guide (1).

Table 1　Word and Corresponding Values

Words		Values	Words		Values
Much Larger	(ML)	0.5	Much Smaller	(MS)	-0.5
Larger	(L)	0.3	Smaller	(S)	-0.3
a Little Larger	(LL)	0.1	a Little Smaller	(LS)	-0.1

Considering the guide (2), the following equations are defined.

○ adjustment of fuzzy measure for increasing
$$g(X_i) = g'(X_i) + (1-Z) \times word \times (1 + a(X_i)), \qquad (12)$$

○ adjustment of fuzzy measure for decreasing
$$g(X_i) = g'(X_i) + Z \times word \times (1 + a(X_i)). \qquad (13)$$

Because fuzzy measures must satisfy monotonicity, if $g(X_i) > g(Y_i)$ for $X_i \subset Y_i$ by Eq. (12) or (13), then the fuzzy measure is adjusted such as $g(Y_i) = g(X_i)$. In Eqs. (12) and (13), $a(X_i)$ is defined by

$$a(X_i) = \frac{f_i - f_{i+1}}{\max_i f_i} \times \frac{g_{ave,i} - \min_i g_{ave,i}}{\sum_i (g_{ave,i} - \min_i g_{ave,i})}, \qquad (14)$$

where $g_{ave,i}$ considers super-additivity/sub-additivity of fuzzy measures defined by

$$g_{ave,i} = \begin{cases} g(X_i) - \dfrac{\sum\limits_{j=1}^{k} \sum\limits_i g(X_i^{jj})}{k}, & |X_i| \neq 1, \\[4mm] 0, & |X_i| = 1, \end{cases} \qquad (15)$$

where $g(X_i^{jl})$ is the fuzzy measure of X_i^{jl} satisfying $X_i^{jl} \subset X_i$, which are disjoint sets of X_i, $|X_i|$ is the cardinal number of a set X_i, k is the number of the combination of sets X_i^j satisfying $X_i^j \subseteq P(X_i)$, and l is the number of sets X_i^{jl} composing a set X_i^j. The concept of the partition of a set X_i is shown in **Fig. 4**.

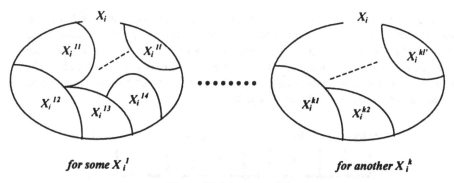

for some X_i^1   *for another X_i^k*

Fig.4 Disjoint sets of X_i

The $g_{ave,i}$ means the degree of super-additivity of the fuzzy measure $g(X_i)$, since the difference between $g(X_i)$ and the arithmetic means of $g(X_i^{jl})$ with respect to j is considered. Therefore, $g_{ave,i} = 0$ when $|X_i| = 1$. And with respect to $a(X_i)$, the larger the f_i and the larger the degree of super-additivity, the larger the $a(X_i)$. That is, $a(X_i)$ considers the characteristics of fuzzy measures.

4.3 Restriction of the Number of Adjustment Times

The restriction of the number of adjustment times is considered in order to avoid the problem that the adjustment is not converged when the same fuzzy measures are adjusted many times. The restriction is imposed on the fuzzy measure adjusted once by the following equations.

○ adjustment of fuzzy measure for increasing

$$g(X_i) = g'(X_i) + \alpha^m \times (1 - Z) \times word \times (1 + a(X_i)), \qquad (16)$$

○ adjustment of fuzzy measure for decreasing

$$g(X_i) = g'(X_i) + \alpha^m \times Z \times word \times (1 + a(X_i)), \qquad (17)$$

where $\alpha \in [0,1]$, and m is the number of adjustment times.

5. EVALUATION EXPERIMENTS

5.1 Evaluation of Apartments

Ten apartments are evaluated by 11 students. The rent, the plan of a house, an apartment building quality, and the environment around the apartment building are considered as the attributes of the evaluation of apartments. In the questionnaire, the apartments are evaluated with one of 11 scale evaluations as shown in **Fig. 5.** The questionnaire results are expressed numerically with 0.1 notch by the normalization the worst evaluation as 0 and the best one as 1.

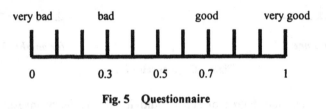

Fig. 5   Questionnaire

5.2 Identification of Fuzzy Measures

Fuzzy measures of each subject are identified by the use of each subject's questionnaire data and the identification algorithms[1][3]. **Table 2** shows the output example of the evaluation model of subject 1, his questionnaire data of evaluation of ten apartments, and the errors between the model outputs and his questionnaire data.

Table 2   Example of the Evaluation Model of Subject 1

Number	Model Outputs	Questionnaire Data	Errors
1	0.256	0.3	-0.044
2	0.068	0	0.068
3	0.249	0.2	0.049
4	0.242	0.2	0.042
5	0.649	0.7	-0.051
6	0.528	0.5	0.028
7	0.67	0.8	-0.130
8	0.301	0.3	0.001
9	0.309	0.3	0.009
10	0.623	0.7	-0.077

5.3 Adjustment of Identified Fuzzy Measures

Identified fuzzy measures are adjusted by the use of the user interface as shown in **Fig. 6**. In the right upper side the *number* means the object number, and the *adjustment* means the word chosen for the adjustment. In the left side of the user interface, the output values of the model with identified fuzzy measures are presented. The errors between the outputs of the model and the questionnaire data are presented in the right lower side of the human interface. Errors are shown visually and the changes of errors by the adjustment are found out easily through the interface. Fuzzy measures are adjusted so as to make errors almost uniform. The adjustment guides are as follows:

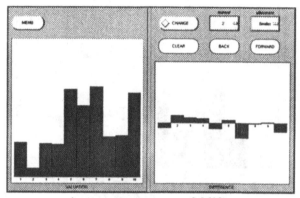

(root mean square error 0.0608)
Fig. 6   Example of User Interface

(1) The data with the largest error is adjusted among data with the error larger than 0.1.
(2) When all data errors are smaller than 0.1, the adjustment is finished.
(3) If, after the adjustment of the data of object i, the error of the data of another object j becomes the largest, then the data of object j is adjusted.
(4) If the step (3) is repeated between objects i and j, the data of object i and object j are regarded as contradictory data. If the convergence of the errors, however, are found, these data are not regarded as contradictory.
(5) If contradictory data are observed, adjustment is continued ignoring the errors of these contradictory data.

The adjustment example about the data of subject 1 shown in Fig. 6 is presented in **Table 3**. Identified fuzzy measures are adjusted by the following; (Object 7, a little larger), (Object 7, a little larger), (Object 2 smaller). It is found that absolute values of all errors are smaller than the threshold value 0.1. **Fig. 7** shows the adjustment results. The root mean square error of the model with adjusted measures is a little larger than that of the model with identified ones since fuzzy measures are identified by the least mean square error but are not adjusted by them.

Table 3 Changes of Errors about Data of Subject 1

	Object 1	2	3	4	5	6	7	8	9	10
Words — Initial Errors	-0.044	0.068	0.049	0.042	-0.051	0.028	-0.130	0.001	0.009	-0.077
Object 7 LL	-0.023	0.089	0.072	0.042	-0.051	0.028	-0.101	0.001	0.016	-0.074
Object 7 LL	-0.013	0.102	0.093	0.042	-0.051	0.028	-0.084	0.001	0.022	-0.071
Object 2 S	-0.019	0.097	0.080	0.042	-0.051	0.028	-0.097	0.001	0.020	-0.078

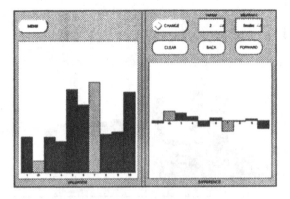

(root mean square error 0.0610)
Fig. 7   Adjustment Result

5.4 Contradictory Data

A man often evaluates some attributes inconsistently.   Therefore, questionnaire data do not necessarily include contradictory data.   If questionnaire data include contradictory data at the identification of fuzzy measures, the errors between questionnaire data and the outputs of the model become large. This paper tries to find out contradictory data using the user interface.   The contradictory data are re-evaluated by the subject.   And this paper tries to reconstruct the human evaluation model with fuzzy measures identified using re-evaluation data.   **Fig. 8** shows the outputs of the model of subject 2 and the errors between the model outputs and his questionnaire data.   Identified fuzzy measures are adjusted as shown in **Table 4**.   It is found from Table 4(1) that the absolute value of the error of object 9 is the largest among the absolute values of the errors of objects 4, 5, and 9 larger than 0.1.   As shown in Table 4(1), in the third and the fourth adjustments, it is found that the data of object 4 and that of object 5 are contradictory to each other.   The data of object 4 is ignored, since the absolute value of the error of object 4 is smaller than that of object 5 and it is easier to adjust the data with large error than to adjust the data with small error. Ignoring object 4, in the third and the fourth adjustments as shown in Table 4(2), it is found that the data of object 5 and that of object 9 are also contradictory to each other. The data of object 5 is ignored because of its initial small error. Finally the errors of all objects except for objects 4 and 5 become smaller than 0.1 as shown in Table 4(3).

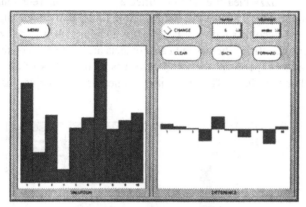

(root mean square error 0.0694)
Fig. 8　Output Model of Subject2

Table 4　Adjustment of Contradictory Data
(1) First Step

	object 1	2	3	4	5	6	7	8	9	10
Initial Words Errors	0.046	0.025	0.007	-0.103	0.109	-0.011	-0.070	0.002	-0.131	0.024
Object 9 L	0.137	0.051	0.023	-0.103	0.126	-0.011	-0.049	0.060	-0.098	0.024
Object 1 LS	0.083	0.038	-0.025	-0.103	0.126	-0.011	-0.068	0.060	-0.098	0.024
Object 5 S	0.083	0.028	-0.025	-0.158	0.069	-0.039	-0.068	0.003	-0.121	0.024
Object 4 L	0.083	0.028	-0.025	-0.085	0.110	-0.025	-0.068	0.012	-0.121	0.024

(2) Second Step : ignoring object 4

	object 1	2	3	4	5	6	7	8	9	10
	0.046	0.025	0.007		0.109	-0.011	-0.070	0.002	-0.131	0.024
Object 9 L	0.137	0.051	0.023		0.126	-0.011	-0.049	0.060	-0.098	0.024
Object 1 LS	0.083	0.038	-0.025		0.126	-0.011	-0.068	0.060	-0.098	0.024
Object 5 S	0.083	0.028	-0.025		0.069	-0.039	-0.068	0.003	-0.121	0.024
Object 9 L	0.149	0.054	-0.015		0.117	-0.039	-0.049	0.062	-0.101	0.024

(3) Third Step : ignoring objects 4 and 5

	Object 1	2	3	4	5	6	7	8	9	10
	0.046	0.025	0.007			-0.011	-0.070	0.002	-0.131	0.024
Object 9 L	0.137	0.051	0.023			-0.011	-0.049	0.060	-0.098	0.024
Object 1 LS	0.083	0.038	-0.025			-0.011	-0.068	0.060	-0.098	0.024

5.5 Re-evaluation of Contradictory Data

If only apartments having contradictory evaluation data are evaluated again, there is a possibility that the subject has a bias toward the objects. Therefore, some apartments including objects 4 and 5 are evaluated again by subject 2 in the questionnaire. Fuzzy measures are identified again using the re-evaluation data in the questionnaire. **Fig. 9** shows the results. It is found that comparing Fig. 9 with Fig. 8, the errors become small. It is necessary to adjust identified fuzzy measures again in order to confirm whether indicated objects are evaluated consistently or not. In this evaluation experiment the absolute values of the errors of all re-evaluation data are smaller than 0.1. Therefore, it is not necessary to adjust fuzzy measures again.

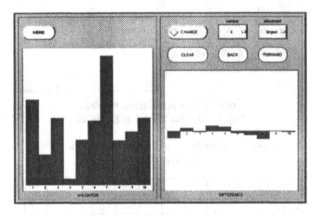

(root mean square error 0.038)
Fig. 9   Results of Re-evaluation

5.6 Remarks

Table 5 shows the comparison of least mean square errors before the adjustment, those after the adjustment, and those after the re-evaluation. It is found that root mean square errors after the adjustment are not smaller than those before the adjustment. Nevertheless, almost all subjects answer that the human evaluation model with adjusted fuzzy measures fits their evaluation well. Furthermore, it is found that least mean square errors after the re-evaluation are smaller than those before the re-evaluation. Subjects answer that human evaluation model with fuzzy measures identified using re-evaluation data fits their evaluation better than the

Table 5   Comparison of Errors

	subject 1	2	3	4	5	6	7	8	9	10	11
Before	0.061	0.069	0.073	0.062	0.067	0.094	0.048	0.062	0.069	0.059	0.144
After	0.061	0.073	0.084	0.063	0.069	0.098	0.048	0.068	0.069	0.062	0.157
Re-evaluation	—	0.038	0.071	0.06	—	0.068	—	0.056	0.069	0.031	0.145

model with fuzzy measures identified using the data before re-evaluation. If contradictory data are found out and the re-evaluation is performed, the human evaluation model with re-identified fuzzy measures fits human subjects well. Besides errors between model outputs and questionnaire data it is also necessary to analyze obtained fuzzy measures in order to confirm whether the model fits human evaluation or not. In this paper the change of $g_{ave,i}$'s at the adjustment of fuzzy measures is analyzed. **Table 6** shows changes of $g_{ave,i}$'s of subject 1 and those of subject 3. These subjects have no contradictory data. It is found that signs of $g_{ave,i}$ are not changed at the adjustment, and that the degree of super-additivity and that of sub-additivity are not changed. This means that super-additivity and sub-additivity are preserved during the adjustment. It is also found that when the signs of $g_{ave,i}$'s are changed, $g_{ave,i}$'s are nearly equal to zero, that is, fuzzy measures are additivity. This is why almost all subjects answer that the human evaluation model with adjusted fuzzy measures fits their evaluation well.

Table 6   Change of g$_{ave,i}$'s of Subject 1 and Subject 3

Subject 1

attributes	12	13	14	23	24	34	123	124	134	234
initial value	0.408	-0.045	0.264	0.037	0.002	0.124	0.316	0.387	0.037	-0.004
Object 7 LL	0.408	-0.045	0.231	0.037	-0.031	0.131	0.316	0.371	0.011	0.042
Object 7 LL	0.408	-0.045	0.201	0.037	-0.061	0.135	0.316	0.356	-0.013	0.052
Object 2 S	0.408	-0.045	0.219	0.037	-0.043	0.135	0.316	0.365	0.001	0.045

Subject 3

attributes	12	13	14	23	24	34	123	124	134	234
initial value	-0.096	-0.255	-0.044	-0.171	0.310	0.144	-0.295	0.472	0.395	0.150
Object 8 LL	-0.090	-0.330	-0.044	-0.171	0.310	0.144	-0.272	0.387	0.330	0.150

On the other hand **Table 7** shows changes of $g_{ave,i}$'s of subject 2 and those of subject 9, who have contradictory data. It is found that signs of $g_{ave,i}$'s are changed after the re-evaluation or that the degree of super-additivity and that of sub-additivity are changed more largely in the re-evaluation than Table 6. From the analysis results it is found that it is necessary to re-evaluate contradictory data for the construction of a better human evaluation model.

Table 7 Change of $g_{ave,i}$'s of Subject 2 and Subject 9

Subject 2

attributes	12	13	14	23	24	34	123	124	134	234
initial value	-0.046	0.374	0.161	-0.070	-0.025	0.345	-0.019	0.261	0.087	0.115
Object 9 L	-0.046	0.374	0.161	-0.118	-0.033	0.345	-0.175	0.096	0.087	0.031
Object 1 LS	-0.046	0.374	0.248	-0.118	0.003	0.431	-0.175	0.152	0.131	0.023
Object 5 S	0.019	0.307	0.294	-0.105	0.119	0.431	-0.143	0.257	0.182	0.107
Object 9 L	0.047	0.307	0.294	-0.071	0.119	0.431	-0.142	0.149	0.182	0.061
4,5 re-evaluation	0.000	0.414	0.000	0.430	0.000	0.596	0.254	0.000	0.414	0.340

Subject 9

attributes	12	13	14	23	24	34	123	124	134	234
initial value	0.223	0.000	-0.072	0.319	0.060	0.158	0.281	0.307	-0.039	0.022
Object 6 LS	0.267	0.000	-0.029	0.319	0.060	0.158	0.313	0.329	-0.037	0.022
Object 1 LS	0.267	0.000	0.012	0.275	0.045	0.155	0.324	0.363	-0.006	0.034
4 re-evaluation	0.220	0.000	-0.031	0.500	0.000	0.249	0.540	0.583	-0.085	0.508

6. CONCLUSIONS

This paper mentions the adjustment of identified fuzzy measures in order to make a human evaluation model with fuzzy measures and integrals fit to human evaluation. The adjustment is based on the consideration of super-additivity/sub-additivity of fuzzy measures. Fuzzy measures are adjusted so that the errors between the data and the outputs of the evaluation model are made uniform. In the adjustment process contradictory data can be also found. With respect to the contradictory data, objects are evaluated and fuzzy measures are identified again. The effectiveness of the adjustment of identified fuzzy measures is shown by evaluation experiments about apartments. In this paper the effectiveness of the adjustment is considered by the errors between the model outputs and the questionnaire data and by the analysis of obtained fuzzy measures, that is, the analysis of the changes of $g_{ave,i}$'s. This study is applicable to the design of, e.g., a human interface.

REFERENCES

[1] K.Ishii and M.Sugeno, "A Model of Human Evaluation Process Using Fuzzy Measure", *Int. J. of Man-Machine Studies*, Vol.22. pp.19-38, 1985

[2] T.Onisawa, M.Sugeno, Y.Nishiwaki, et al, "Fuzzy Measure Analysis of Public Attitude towards the Use of Nuclear Energy", *Fuzzy Sets and Systems*, Vol.20, pp.259-289, 1986

[3] T.Mori and T.Morofushi, "An Analysis of Evolution Model Using Fuzzy Measure and the Choquet Integral", Proc. of the 5th Fuzzy Systems Symposium, pp.207-212, 19899(in Japanese).

[4] Y.S.Sohn and T.Onisawa, "Design of Human Interface with Fuzzy Measures and Integrals Model", *Transactions of the Society of Instrument and Control Engineers*, Vol.35, 1999(To Appear, in Japanese).

[5] M.Grabisch, "k-order Additive Discrete Fuzzy Measures and their Representation", *Fuzzy Sets and Systems*, Vol.92, pp.167-189, 1997.

[6] Japan Society for Fuzzy Theory and Systems ed., *Lecture Fuzzy 3, Fuzzy Measures*, Nikkan Kogyo Shinbunsha, Tokyo, 1993(in Japanese).

Design and Application of Block-Oriented Fuzzy Models – Fuzzy Hammerstein Model

Janos Abonyi[1,2], Robert Babuska[1]
Ferenc Szeifert[2], Lajos Nagy[2], and Henk Verbruggen[1]

[1] Delft University of Technology, Department of Information Technology and Systems,
Control Engineering Laboratory, P.O.Box 5031 2600 GA Delft, The Netherlands
[2] University of Veszprém, Department of Chemical Engineering Cybernetics, P.O.Box 158
H-8201 Veszprém, Hungary

Abstract: Recently, data driven fuzzy modelling has drawn a great deal of attention not only from academia but also from industry. However, a dynamic fuzzy model might be difficult to develop when training data do not contain sufficient information about the process nonlinearity. In order to avoid this problem and to reduce the model complexity, this paper presents a fuzzy Hammerstein (FH) modelling approach, where a static fuzzy model is integrated with a linear dynamic model in series. A constrained recursive least-squares algorithm is used for the simultaneous identification of the steady-state and the dynamic part of the FH model. To demonstrate the proposed modelling approach, an electrical water-heater process is used as an example.

1. Introduction

A critical step in the application of model based control algorithms is the development of a suitable model of the process dynamics. Recently fuzzy modelling of dynamic systems has drawn a great deal of attention.

The Nonlinear AutoRegressive with eXogenous input (NARX) model is frequently used with many nonlinear identification methods [1]. As all system identification strategies, the NARX modelling has also several weaknesses. Problems associated with system identification in high dimensions are caused by the *course of dimensionality* [2]. Therefore, due to the exponentially increasing memory and information requirements, the usage of these types NARX fuzzy models for complex, high-order dynamic processes is impractical. Moreover, this approach does not take into account the fact that most of industrial processes can be mildly perturbed around the nominal operation point. This results the transient data do not contain sufficient non-linear information about the process nonlinearity.

Therefore, it is highly desirable to incorporate the whole body of *a priori* knowledge into the construction of dynamic fuzzy models in the reduction of the dimension of its input space. Simplified non-linear block-oriented fuzzy models can be used alternatives.

It has been shown that nonlinear effect encountered in most industrial processes, distillation columns, pH neutralisation processes, heat-exchangers, electro-mechanical systems, etc. can be effectively modelled as a combination of a nonlinear static element and a gain-independent dynamic part [3, 4]. Probably the best-known members of this class of models are the Hammerstein and Wiener model [5, 6].

These "next-step-beyond-linear-models" have proven suitable for grey-box modelling where it is assumed that the steady-state behaviour of the process is known *a priori* [7]. The disadvantage of this approach is that – as a process gets more complex in its physical description – the first-principles steady-state model tends to become increasingly complicated and computationally intensive. It requires nonlinear equation-solving techniques and iterative numerical searches to handle it. Moreover, first-principles steady-state models can be rarely obtained.

Therefore, in this study, the steady-state fuzzy model is identified with the help of linguistic rules and data gathered from the process.

The identification of a block-oriented model is a challenging task [5]. Due to the nature of block oriented models the identification of the model can be simplified if the non-linear steady-state data is available, because these model structures permit the steady-state behaviour to be specified independently of the dynamics [8].

In the presented approach only transient data are used in model identification. Moreover, the steady-state nonlinearity and the linear dynamic are identified simultaneously. This leads to nonlinear identification problem [3]. Because, the proposed algorithm is intended to be easily realisable, the identification problem is transformed to constrained linear least-squares estimation [9] of the FH model parameters.

An example is presented where the FH model is applied to a simulation model of a laboratory water-heater process. The model is obtained from input-output measurements by using the proposed constrained identification algorithm. Simulation results show that when the FH model identification technique is used, not only good dynamic modelling performance is achieved, but also the steady-state behaviour of the system is well modelled.

2. The FH Model Structure

The FH model is used to approximate a Hammerstein system, which consists of a series connection of a memoryless nonlinearity and linear dynamics, see Figure 1.

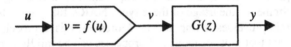

Figure 1: A series combination of a static nonlinearity and a linear dynamic system

The NARX representation of this simplest block-oriented model is:

$$y(k+1) = \sum_{i=1}^{n_v} a_i y(k-i+1) + \sum_{i=1}^{n_u} b_i \{f(u(k-i-n_d+1))\} \tag{1}$$

where $y(k),...,y(k-n_y+1)$ and $u(k-n_d),...,u(k-n_u-n_d+1)$ are the lagged outputs and inputs of the dynamical system, and $f(.)$ is the memoryless nonlinearity. If the gain of the linear system is equal to one,

$$\frac{\sum_{i=1}^{n_u} b_i}{1-\sum_{i=1}^{n_y} a_i} = 1 \tag{2}$$

the $f(.)$ function describes the steady-state behaviour of the system $y_s = f(u_s, \mathbf{x})$, where u_s denotes a steady-state input and $\mathbf{x}$ denotes the vector of operating parameters having effects to the steady-state output, y_s.

The steady-state behaviour of the plant is represented by a Singleton or zero-order TS fuzzy model formulated by a set of rules as follows:

$$r_{i_1,...,i_n}: \ \textbf{IF} \ z_1 \ \text{is} \ A_{1,i_1} \ \text{and} \ ... \ \text{and} \ z_n \ \text{is} \ A_{n,i_n} \ \textbf{THEN} \ y_s = d_{i_1,...,i_n} \tag{3}$$

where $r_{i_1,...,i_n}$ denotes the fuzzy rule, $A_{j,i_j}(z_j)$ is the $i_j = 1,2,...,M_j$-th antecedent fuzzy set referring to the $j \in [1-n]$-th input universe, whose membership functions are denoted by the same symbols as the fuzzy values, where M_j is the number of the fuzzy sets on the j-th input domain.

For a given input, $\mathbf{z} = [u_s, \mathbf{x}]$, the output of the fuzzy model, y_s, is inferred by computing the weighted average of the rule consequents:

$$y_s = \frac{\sum_{i_1=1,i_n=1}^{M_1,M_n} w_{i_1,...,i_n} d_{i_1,...,i_n}}{\sum_{i_1=1,i_n=1}^{M_1,M_n} w_{i_1,...,i_n}} \tag{4}$$

where $\sum_{i_1=1,i_n=1}^{M_1,M_n} \equiv \sum_{i_1=1}^{M_1}...\sum_{i_n=1}^{M_n}$ and the weight, $w_{i_1,...,i_n} > 0$, represents the overall truth value (degree of fulfilment) of the $i_1,...,i_n$-th rule calculated based on the degrees of membership values:

$$w_{i_1,...,i_n} = \prod_{j=1}^{n} A_{j,i_j}(z_j) \qquad j = 1,2,...,n, \ i_j = 1,2,...,M_j \tag{5}$$

In this paper, triangular membership functions are applied (Figure 2), however the results are independent of the type of the membership functions.

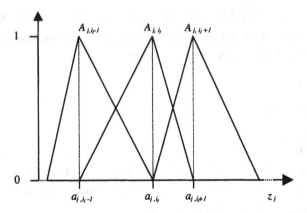

Figure 2: The membership functions used

The triangular membership functions are defined as follows:

$$A_{j,i_j}(z_j) = \frac{z_j - a_{j,i_j-1}}{a_{j,i_j} - a_{j,i_j-1}}, \quad a_{j,i_j-1} \leq z_j < a_{j,i_j}$$

$$A_{j,i_j}(z_j) = \frac{a_{j,i_j+1} - z_j}{a_{j,i_j+1} - a_{j,i_j}}, \quad a_{j,i_j} \leq z_j < a_{j,i_j+1} \tag{6}$$

$$A_{j,i_j}(z_j) = 0, \quad \text{othewise}$$

Because the product operator is used for and connective and the sum operator is used for aggregation, the complete rule base results:

$$\sum_{i_1=1,i_n=1}^{M_1,M_n} w_{i_1,\dots,i_n} = 1 \tag{7}$$

Therefore, at given observations (4) can be simplified to:

$$y_s = \sum_{i_1=1,i_n=1}^{M_1,M_n} \left[\left(\prod_{j=1}^{n} A_{j,i_j}(z_j) \right) \cdot d_{i_1,\dots,i_n} \right] \tag{8}$$

3. The Fuzzy Hammerstein Model Identification

The identification of a block-oriented model is a challenging task. Due to the nature of block oriented models the identification of the model can be simplified if the non-linear steady-state data is available, because these model structures permit the steady-state behaviour to be specified independently of the dynamics.

This section describes a new method, when only transient data are used in model identification. Moreover, the steady-state nonlinearity and the linear dynamic are identified simultaneously.

For sake of simplicity in notation let denote the rule indexes in the fuzzy model as follows:

$$j = i_n + \sum_{l=1}^{n-1} (i_l - 1) \prod_{k=l+1}^{n} M_k \tag{9}$$

where $j = 1 \dots N_r$, where N_r denotes the number of the rules.

By using this notation the FH model can be formulated in the form:

$$y(k+1) = \sum_{i=1}^{n_y} a_i y(k-i+1) + \sum_{i=1}^{n_u} b_i \sum_{j=1}^{N_r} w_j (u(k-n_d-j+1), \mathbf{x}) \cdot d_j \tag{10}$$

The aim of the identification algorithm is the determination of the parameters a_i, b_i, d_j, while the antecedent part of the fuzzy model (the number and the cores of the fuzzy sets) is assumed to be designed *a priori*.

For the determination of the antecedent part of the fuzzy model, fuzzy clustering can be used [10], because more fuzzy sets are needed where the slope of the nonlinearity changes significantly than in places where the process behaviour is close to linear.

The identification of b_i and d_j parameters leads to a nonlinear optimisation problem, because of their product in (10).

In this paper, this problem is transformed to constrained linear least-squares estimation [9], by introducing a new parameterisation: $p_{i,j} = b_i d_j$ and nonlinear constraints based on the fact that $\dfrac{p_{i,k}}{p_{j,k}} = \dfrac{b_i}{b_j}$:

$$\frac{p_{i,k}}{p_{j,k}} \frac{p_{j,l}}{p_{i,l}} = 1 \tag{11}$$

This parameterisation is similar that Eskinat has been suggested in the identification of Hammerstein systems described by polynomial nonlinearity [3].

Hence, the reparameterised model becomes linear in its variables, a_i, $p_{i,j}$:

$$y(k+1) = \sum_{i=1}^{n_y} a_i y(k-i+1) + \sum_{i=1}^{n_u} \sum_{j=1}^{N_r} p_{i,j} w_j (u(k-n_d-j+1), \mathbf{x}), \tag{12}$$

this pseudo-linear model can be formulated in a predictor form:

$$y(k) = \varphi^T (k-1) \theta(k-1) \tag{13}$$

$$\theta(k-1) = \{a_1, \dots, a_{n_y}, p_{1,1}, \dots, p_{n_u, N_r}\} \tag{14}$$

$$\varphi^T (k-1) = \{y(k-1), \dots, y(k-n_y), w_1 (u(k-n_d-1), \mathbf{x}) \dots, w_{N_r} (u(k-n_d-1), \mathbf{x}) \dots$$
$$w_1 (u(k-n_d-n_u), \mathbf{x}) \dots, w_{N_r} (u(k-n_d-n_u), \mathbf{x})\} \tag{15}$$

During the identification/adaptation the following cost function is minimised:

$$E = [\hat{y}(k) - \varphi^T (k-1) \theta(k-1)]^2 \tag{16}$$

where $\hat{y}(k)$ denotes the measured process output.

The unconstrained weighted recursive least squares (RLS) estimate of θ is:

$$\theta(k) = \theta(k-1) + \frac{P(k-2)\varphi^T(k-1)\left[\hat{y}(k) - \varphi^T(k-1)\theta(k-1)\right]}{\alpha + \varphi^T(k-1)P(k-2)\varphi(k-1)} \quad (17)$$

$$P(k-1) = \frac{1}{\alpha}\left[P(k-2) + \frac{P(k-2)\varphi(k-1)\varphi^T(k-1)P(k-2)}{\alpha + \varphi^T(k-1)P(k-2)\varphi(k-1)}\right] \quad (18)$$

where $\alpha = \begin{bmatrix}0.9 & 1\end{bmatrix}$ and P denotes the forgetting factor and the covariance matrix, respectively.

In order to get a Hammerstein model, the identified parameters have to satisfy the constraints defined by (11). The simplest solution of this problem is to apply linear equality constraints in the form of:

$$M \cdot \theta^c = k \quad (19)$$

The optimal constrained solution, θ^c, can be obtained by optimal projection of the parameter vector resulted form the unconstrained optimisation (17) [9]:

$$\theta^c = \theta - P \cdot M^T\left[\left(M \cdot \theta \cdot M^T\right)^{-1}(M \cdot \theta - k)\right]. \quad (20)$$

Unfortunately, this method can not be directly applied, because the constraints represented by (11) are nonlinear. Therefore, an iterative solution is applied, where in iteration the following linear constraints are derived from the linearization of (11):

$$\frac{p'_{i,k}}{p'_{j,k}\,p'_{i,l}}\,p_{j,l} + \frac{p'_{j,l}}{p'_{j,k}\,p'_{i,l}}\,p_{i,k} - \frac{p'_{i,k}}{\left(p'_{j,k}\right)^2}\frac{p'_{j,l}}{p'_{i,l}}\,p_{j,k} - \frac{p'_{i,k}}{p'_{j,k}}\frac{p'_{j,l}}{\left(p'_{i,l}\right)^2}\,p_{i,l} = 1 - \frac{p'_{i,k}}{p'_{j,k}}\frac{p'_{j,l}}{p'_{i,l}} \quad (21)$$

where the ' superscript denotes the solution resulted in the previous iteration step.

The identification algorithm is implemented through the following steps:
1. Generation of the unconstrained solution by using (17) and (18)
2. Generation of the linear constraints by using (21) and (19)
3. Obtain the constrained solution by (20)
4. If the constrained solution is not equal to the previously generated constrained solution, go to step 2., otherwise the optimal projection has been obtained.

There is no proof for the convergence of the presented algorithm. However, we experienced that the algorithm converges after few (approximately four) iterations.

4. Example: Identification of an Electrical Water-heater

A simulation example is used to illustrate the advantages of the proposed identification method. A simulation model of a laboratory water-heater is considered.

4.1 Process Description

The schematic diagram of the water-heater is shown in Figure 3. The water comes from the water pipeline into the heater through a control valve (CV). After measuring the flow rate, F, the water passes through a pair of metal pipes containing a cartridge heater. The outlet temperature, T_{out}, of the water can be varied by adjusting the heating signal, u, of the cartridge heater. The detailed description of the process and its first principle model can be found in [11].

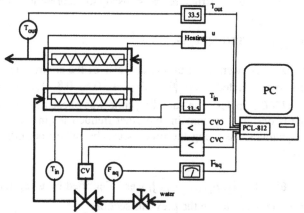

Figure 3: The scheme of the physical system

The physical system can be decomposed into three main elements, which are in strong interaction with each other. These elements are the cartridge-heater (CH), the streaming water (w), and the wall of the pipe (W). Another element is the environment (env). Considering these elements, three heat balances can be obtained:

$$V_{CH} \cdot \rho_{CH} \cdot C_{p_{CH}} \cdot \frac{\partial T_{CH}}{\partial t}(t, z) = Q(u) - \alpha_1 \cdot A_1 \cdot (T_{CH} - T_w)$$

$$V_w \cdot \rho_w \cdot C_{p_w} \cdot \frac{\partial T_w}{\partial t}(t, z) + (F \cdot \rho \cdot C_p)_w \cdot \frac{\partial T_w}{\partial z}(t, z) = \alpha_1 \cdot A_1 \cdot (T_{CH} - T_w) - \alpha_2 \cdot A_2 \cdot (T_w - T_W)$$

$$V_W \cdot \rho_W \cdot C_{p_W} \cdot \frac{\partial T_W}{\partial t}(t, z) = \alpha_2 \cdot A_2 \cdot (T_w - T_W) - \alpha_{env} \cdot A_{env} \cdot (T_W - T_{env})$$

(22)

where $z \in [0, L]$, where L is the length of the pipe and $Q(u)$ is the performance of cartridge heater:

$$Q = Q_M \left[u - \frac{\sin(2\pi \cdot u)}{2\pi} \right],$$

(23)

where Q_M is the maximal performance, and u is the heating signal in electrical voltage.

The aim of the identification is the prediction of the dynamic behaviour of the output temperature of the water, $T_{out} = T_w(t, z = L)$ varied by adjusting the heating signal, u. The detailed description and values of parameters used in this study are given in [12].

4.2 The Fuzzy Hammerstein Model Identification

The FH model is constructed form process measurements. The system can be modelled approximately as a second-order system with a time-delay. This results in the following fuzzy Hammerstein model structure:

$$y(k+1) = a_1 y(k) + a_2 y(k-1) + b_1 f\big(u(k-3)\big) + b_2 f\big(u(k-4)\big) \tag{24}$$

Because in this case-study the water-heater operates at constant flow rate, $F=70l/h$, the following fuzzy model structure was used to represent the steady-state nonlinearity, $y_s = f(u_s)$:

$$r_j: \textbf{ IF } u_s \textbf{ is } A_j \textbf{ THEN } y_s = d_j \tag{24}$$

Six antecedent fuzzy sets were used on the universe of the control signal. The cores of the fuzzy sets were selected to be $\{0 \quad 0.2 \quad 0.4 \quad 0.6 \quad 0.8 \quad 1\}$. The identification data contains $N=1500$ samples. The sampling time was chosen to two seconds. The excitation signal is designed to contain important frequencies in the expected range of the process dynamics, assuming the process can be mildly perturbed around the operation points.

The parameters, $\theta^c = \{a_1, a_2, p_{1,1}, \ldots, p_{2,6}\}$, were identified by using the constrained identification method presented in the previous section.
Because the linear gain independent part of the Hammerstein model has a unity gain,

$$\frac{b_1 + b_2}{1 - (a_1 + a_2)} = 1 \quad \text{and} \quad \frac{p_{1,k}}{p_{2,k}} = \frac{b_2}{b_1}, \tag{25}$$

form θ^c, the parameters of the fuzzy Hammerstein model can easily obtained.

Hence, the transfer function of the identified linear dynamic model and the the fuzzy model represents steady-state nonlinearity are:

$$G(z) = \frac{b_1 z^{-1} + b_2 z^{-2}}{1 - a_1 z^{-1} - a_2 z^{-2}} z^{-2} = \frac{0.0093 z^{-3} + 0.0019 z^{-4}}{1 - 1.8 z^{-1} + 0.8112 z^{-2}}, \tag{26}$$

$$d = \{14.49 \quad 15.16 \quad 20.74 \quad 29.06 \quad 34.63 \quad 35.18\}. \tag{27}$$

The developed fuzzy model was validated in a separate validation data set. The modelled and the simulated outlet temperature are shown in Figure 4. As Figure 4 shows the identified FH model gives good modelling performance in the whole operating range, while the linear model identified for the operating region $y_s=22.7506$, $u_s=0.4477$

$$G_{lin}(z) = \frac{0.3219 z^{-3} + 0.0862 z^{-4}}{1.0000 - 1.8170 z^{-1} + 0.8302 z^{-2}} \tag{28}$$

gives good modelling performance only at the narrow region where it was identified.

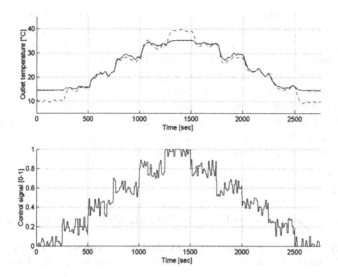

Figure 4: The measured (—) and the simulated process output.
(--) Hammerstein model, (··) linear model

Figure 5 compares the real steady-state behaviour of the system and the nonlinear part of the identified fuzzy Hammerstein model. It can be seen that the resulted fuzzy steady-state model gives a good description of the steady-state behaviour of the process, even through it was identified without any steady-state input-output data.

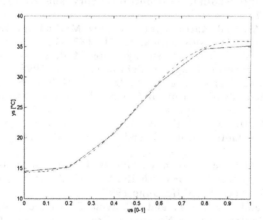

Figure 5: The identified (—) and the real (- -) steady-state behaviour of the process

5. Conclusions

A new approach to data driven identification of fuzzy models has been presented. The new structure identifies a Hammerstein type block-oriented model instead of a general NARX model. The advantage of the proposed method that the new fuzzy

Hammerstein (FH) model handles the steady-state part of the nonlinear system separately from the dynamical. Therefore, the resulted fuzzy model does not suffer from the course of dimensionality. In the presented approach, only transient data can be used for model identification. Moreover, the steady-state nonlinearity and the linear dynamic are identified simultaneously by using a constrained linear least-squares algorithm. Simulation results have been obtained for a laboratory water-heater process. It has been shown that when the FH model identification technique is used, not only good dynamic modelling performance is achieved, but also the steady-state behaviour of the system is well modelled. This indicates the usability of this new model in building model-based controllers.

References

[1] Babuška R., Fuzzy Modelling for Control, Kluwer Academic Publishers, Boston, 1998

[2] Bossely K.M., Neurofuzzy Modelling Approaches in System Identification, Ph.D. thesis, University of Southampton, Southampton, 1997

[3] Eskinat E. and Johnson S. H. and Luyben W., Use of Hammerstein Models in Identification of Nonlinear Systems, AIChE Journal, 37, 2, 255-268, 1991

[4] Pottman M. and Pearson R.K., Block-Oriented NARMAX Models with Output Multiplicities, AIChE Journal, 44, 1, 131-140, 1998

[5] Duwaish H.Al. and Karim N.M. and Chandrasekar V., Use of Multilayer Feedforward Neural Networks in Identification and Control of Wiener Model, IEE Proceedings of Control Theory and Applications, 143, 3, 255-258, 1996

[6] Duwaish H.Al. and Karim N.M., A New Method for Identification of Hammerstein Model, Automatica, 33, 10, 1871-1875, 1997

[7] Ramchandran S., Consider Steady-State Models for Process Control, Chemical Engineering Progress, February, 75-81, 1998

[8] Abonyi J. and Nagy L. and Szeifert F., Hybrid Fuzzy Convolution Modelling and Identification of Chemical Process Systems, Int. Journal of Systems Science (to appear)

[9] Timmons W.D. and Chizeck H.J. and Katona P.G., Parameter-Constrained Adaptive Control, Ind. Eng. Chem. Res., 36, 4894-4905, 1997

[10] Babuška R. and Verbruggen H., Fuzzy Identification of Hammerstein Systems, In Proceedings Seventh IFSA World Congress, volume II, 348-353, Prague, Czech Republic, June 1997.

[11] Bódizs, Á. and Szeifert F. and Chován T., Convolution Model Based Predictive Controller for a Nonlinear Process, Ind. Eng. Chem. Res., 38, 154-161, 1999

[12] Abonyi J. and Bódizs Á. and Nagy L. and Szeifert F., Hybrid Fuzzy Convolution Model Based Predictor Corrector Controller, Computational Intelligence for Modelling Control and Automation, 265-270, IOS Press Holland, ISBN 9-051-99474-5, 1999

Evolving Fuzzy Detectives:
An Investigation into the Evolution of Fuzzy Rules

Peter J. Bentley

Department of Computer Science, University College London,
Gower Street, London WC1E 6BT, UK
P.Bentley@cs.ucl.ac.uk

Abstract: This paper explores the use of genetic programming to evolve fuzzy rules for the purpose of fraud detection. The fuzzy rule evolver designed during this research is described in detail. Four key system evaluation criteria are identified: intelligibility, speed, handling noisy data, and accuracy. Three sets of experiments are then performed in order to assess the performance of different components of the system, in terms of these criteria. The paper concludes: 1. that many factors affect accuracy of classification, 2. intelligibility and processing speed mainly seem to be affected by the fuzzy membership functions and 3. noise can cause loss of accuracy proportionate to the square of noise.

1   Introduction

It is easy to spot a thief if he wears a black mask and carries a bag over his shoulder with the word 'swag' written boldly across it. Most crimes are perhaps not as obvious as this amusing movie cliché, but there will be, more often than not, physical evidence left behind which will incriminate the wrongdoer. However, there is a type of crime far subtler in its implementation. This clandestine activity relies upon deception, concealment and mendacity. It is known as *fraud*, and it impacts on every aspect of our financial world - from insurance to social security benefits to pensions. Its exposure requires a different type of detection. There are no fingerprints left to be found by forensics - but in the computer databases there are other types of fingerprints stored unknowingly by the fraudsters. These fingerprints are small patterns of data, hidden amongst vast archives of information.

Until recently, often the only way to identify such fraud was for experts to study each data item and mentally apply a set of learned rules. For example, if the home insurance claim is for a large amount and if there have been many such claims from that address in a short space of time, then perhaps the claim is fraudulent. Identifying such 'fingerprints' in data is a laborious and slow task, and is dependent on the fingerprint characterising a true fraudulent activity. However, with the advent of data mining and machine learning techniques, such detection can now be performed by computer.

This paper explores the use of evolutionary computation and fuzzy logic in combination to perform machine learning or *pattern classification*, with an emphasis on the capabilities required for the detection of fraud. The next section briefly describes the background to this area. Section 3 gives details of the fuzzy rule evolver and its different components. Section 4 outlines the evaluation requirements identified after consultation with our collaborating company. The fifth section describes and analyses three sets of experiments that were performed on the system and section 6 concludes.

2 Background

Machine Learning, pattern classification and data mining are huge fields in Computer Science, with countless different techniques in use or under investigation. This paper concentrates on a single approach: the use of *fuzzy logic* with *genetic programming* to classify data.

Fuzzy sets were introduced by Lofti Zadeh in 1965 [15]. Designed to allow the representation of 'vagueness' and uncertainty that conventional set theory disallowed, the sets and their manipulation by logical operators led to the development of the field known as Fuzzy Logic [3]. Despite the name, fuzzy techniques are actually capable of greater precision compared to classical approaches [6]. Fuzzy controllers have been used with considerable success: examples include controllers for elevators, subway trains, and even fuzzy autofocus systems for cameras [11].

Another appeal of fuzzy logic is its *intelligibility*. Fuzzy rules use linguistic identifiers such as 'high', 'short' and 'inexpensive'. Because all humans tend to think in such vague terms, the specification and understandability of such rules becomes simple, even to someone unaware of the mechanisms behind this technique [6]. The combination of representation of uncertainty, precision, and intelligibility has motivated the use of fuzzy logic in pattern classification [3], and indeed, forms the motivation for its use in this research.

Fuzzy logic can be combined or hybridized with many other techniques, including evolutionary algorithms. Some have developed fuzzy-evolutionary systems [12] where fuzzy logic is used to tune parameters of an evolutionary algorithm. Others use evolutionary-fuzzy approaches, where evolution is employed to generate or affect fuzzy rules [9,10]. This paper describes the latter approach, and makes use of Genetic Programming (GP).

John Koza developed GP for the purposes of automatic programming [7] (making computers program themselves). GP differs from other EAs in three main respects: solutions are represented by tree-structures, crossover normally generates offspring by concatenating random subtrees from the parents, and solutions are evaluated by *executing* them and assessing their function.

Like all evolutionary algorithms (EAs), GP maintains *populations* of solutions. These are evaluated, the best are selected and 'offspring' that inherit features from their 'parents' are created using crossover and mutation operators. The new solutions are then evaluated, the best are selected, and so on, until a good solution has evolved, or a specific number of generations have passed.

EAs are often used for pattern classification problems [8], but although the accuracy can be impressive, it is often difficult to understand the evolved method of classification. By evolving fuzzy rules it is possible to get the best of both worlds - accurate and intelligible classification [9].

3 The Fuzzy Rule Evolver

The system developed during this research comprises two main elements: a Genetic Programming (GP) search algorithm and a fuzzy expert system. Figure 1 provides an overview of the system.

Data is provided to the system in the form of two comma-separated-variable (CSV) files: training data and test data. In each file the first row contains the labels for each column (e.g. "Length, Width, Height, Cost"). These labels are used by the system in the rules that it evolves. The next Sn rows (or *data items*) are known to be of the 'suspicious' class, the remaining data comprises one or more other classes. There may be up to 256 values in each data item, as long as the training and test data sets are consistent (normally training and test data files are constructed by splitting a single data file into two). The system assumes that the data is numerical and that there are no missing values (a data pre-processing program has been developed to assign numbers for alphanumeric entries and to fill missing values with random values). All training takes place on the training data set; testing of evolved rules takes place on both training and test data sets.

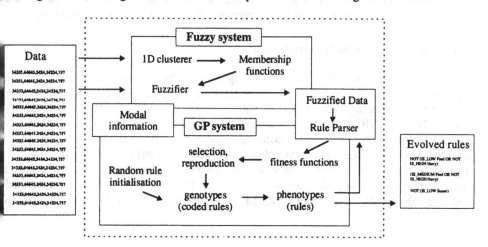

Figure 1 Block diagram of the Evolutionary-fuzzy system.

3.1 Clustering

When started, the system first clusters each column of the training data into three groups using a one-dimensional clustering algorithm. A number of clusterers are implemented in the system, including C-Link, S-Link, K-means [5] and a simple numerical method (in which the data is sorted, then simply divided into three groups with the same number of items in each group). This paper investigates the last two of these methods in the system. Once selected by the user, the same clusterer is used for all learning and testing of the data.

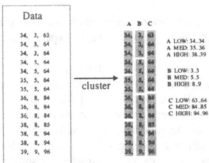

Figure 2: Data is clustered column by column to find the fuzzy membership function ranges.

After every column of the data has been successfully clustered into three, the minimum and maximum values in each cluster are found, see fig. 2. These values are then used to define the domains of the membership functions of the fuzzy expert system.

3.2 Fuzzy Membership Functions

Three membership functions, corresponding to the three groups generated by the clusterer, are used for each column of data. Each membership function defines the 'degree of membership' of every data value in each of the three fuzzy sets: 'LOW', 'MEDIUM' and 'HIGH' for its corresponding column of data. Since every column is clustered separately, with the clustering determining the domains of the three membership functions, every column of data has its own, unique set of three functions.

The system can use one of three types of membership function: 'non-overlapping', 'overlapping', and 'smooth', see figure 3. The first two are standard trapezoidal functions, the third is a set of functions based on the arctangent of the input in order to

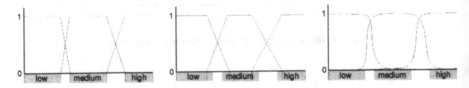

Figure 3: The three types of membership functions used by the system: non-overlapping (left), overlapping (middle), smooth (right).

provide a smoother, more gradual set of 'degree of memberships'.

Each type of membership function uses the results from the clusterer in a different way to determine the domains of the functions. The 'non-overlapping' functions give a membership of 1.0 for the fuzzy set corresponding to the cluster that the value falls in, and 0.0 for all other fuzzy sets. For example, a value that falls in the lowest cluster would be fuzzified into (1.0, 0.0, 0.0) for the low, medium and high fuzzy sets, respectively. Since, for this application, it is normal for all values to fall within one of the three clusters, it is extremely rare for values to be between two clusters. Hence, the 'non-overlapping' functions almost never allow a value to be a member of more than one fuzzy set at a time (i.e. the fuzzy sets are, to all intents and purposes, non-overlapping). In contrast, the 'overlapping' functions place the 'knees' and 'feet' of each function at three quarters of the values provided by the clusterer, see figure 3 (middle). This has the effect of ensuring that the three fuzzy sets overlap - a value towards the outer extent of the low fuzzy set might thus be fuzzified into (0.8, 0.2, 0.0) for low, medium and high fuzzy sets respectively. Finally, the 'smooth' functions increase the level of overlap still further. For example, a value in the centre of the low fuzzy set might be fuzzified into (0.98 0.02, 0.0), while a value towards the outer extent of the low fuzzy set might be fuzzified as (0.96, 0.4, 0.0).

Whichever set of membership functions are selected, they are then shaped according to the clusterer and used to fuzzify all input values, resulting in a new database of fuzzy values. The GP engine is then seeded with random genotypes (coded rules) and evolution is initiated.

3.3 Evolving Rules

The implementation of the GP algorithm is perhaps best described as a genetic algorithmist's interpretation of GP, since it employs many of the techniques used in GAs to overcome some of the problems associated with simple GP systems. For example, this evolutionary algorithm uses a crossover operator designed to minimise the disruption caused by standard GP crossover, it uses a multiobjective fitness ranking method to allow solutions which satisfy multiple criteria to be evolved, and it also uses binary genotypes which are mapped to phenotypes.

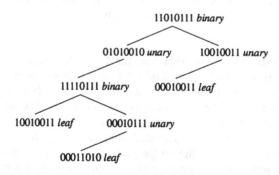

Figure 4: An example genotype used by the system.

Genotypes and Phenotypes

Genotypes consist of variable sized trees, where each node consists of a binary number and a flag defining whether the node is binary, unary or a leaf, see figure 4. At the start of evolution, random genotypes are created (usually containing no more than 3 binary and 4 unary nodes). Genotypes are mapped onto phenotypes to obtain fuzzy rules, e.g. the genotype shown in fig. 4 maps onto the phenotype:

"(IS_MEDIUM (Height OR IS_LOW Age) AND IS_MEDIUM Age)".

Currently the system uses two binary functions: 'OR' and 'AND', four unary functions: 'NOT', 'IS_LOW', 'IS_MEDIUM', 'IS_HIGH', and up to 256 leaves (column labels such as "Length", "Width", "Height", "Cost"). Depending on the type of each node, the corresponding binary value is mapped to one of these identifiers and added to the phenotype.

Rule Evaluation

Every evolved phenotype (or fuzzy rule) is evaluated by using the fuzzy expert system to apply it to the fuzzified training data, resulting in a defuzzified score between 0 and 1 for every fuzzified data item. (Section 3.4 describes this in full.) This list of scores is then assessed by four fitness functions which provide separate fitness values for the phenotype, designed to:

i.   minimise the number of misclassified items (where a misclassified item is a 'normal' data item with a score > 0.5).

ii. maximise the difference between the average scores for correctly classified 'suspicious' items and the average scores for 'normal' items (where a correctly classified suspicious item is a data item in the first Sn of the training set with score > 0.5).

iii. maximise the sum of scores for 'suspicious' data items.

iv. penalise the length of any rules that contain more than four identifiers (binary, unary, or leaf nodes).

The first function ensures as few misclassifications as possible. The second forces evolution to distinguish between 'suspicious' and 'normal' classes of data, while the third demands that 'suspicious' items are given higher scores than 'normal' ones. The final function ensures that all evolved rules are short - serving the dual purpose of preventing bloat and increasing the readability of the final output.

Rule Generation

Using these four fitness values for each rule, the GP system then employs the SWGR multiobjective optimisation ranking method [2] to determine how many offspring each pair of rules should have. (Fitnesses are scaled using the effective ranges of each function, multiplied by importance values and aggregated. Rules with higher overall fitnesses are given higher ranking values, and hence have an increased probability of producing offspring.) Child rules are generated using one of two forms of crossover. The first type of crossover emulates the single-point crossover of genetic algorithms by finding two random points in the parent genotypes that resemble each other, and splicing the genotypes at that point. By ensuring that the same type of nodes, in approximately the same places, are crossed over, and that the binary numbers within the nodes are also crossed, an effective exploration of the search space is provided without excessive disruption [1]. The second type of crossover generates child rules by combining two parent rules together using a binary operator (an 'AND' or 'OR'). This more unusual method of generating offspring (applied approximately one time out of every ten instead of the other crossover operator) permits two parents that detect different types of 'suspicious' data to be combined into a single, fitter individual. Mutation is also occasionally applied, to modify randomly the binary numbers in each node by a single bit.

The GP system employs population overlapping, where the worst $Pn\%$ of the population are replaced by the new offspring generated from the best $Pm\%$. Typically values of $Pn = 80$ and $Pm = 40$ seem to provide good results. The population size was normally 100 individuals.

Modal Evolution

Each evolutionary run of the GP system (usually only 15 generations) results in a short, readable rule which detects some, but not all, of the 'suspicious' data items in the training data set. Such a rule can be considered to define one mode of a multimodal problem. All items that are correctly classified by this rule (recorded in the modal database, see figure 1) are removed and the system automatically restarts, evolving a new rule to classify the remaining items. This process of modal evolution continues until every 'suspicious' data item has been described by a rule. However, any rules that misclassify more items than they correctly classify are removed from the final rule set by the system.

3.4 Assessment of Final Rule Set

Once modal evolution has finished generating a rule set, the complete set of rules (joined into one by disjunction, i.e., 'OR'ed together) is automatically applied to the training data and test data, in turn. Information about the system settings, number of claims correctly and incorrectly classified for each data set, total processing time in seconds, and the rule set are stored to disk.

3.5 Applying Rules to Fuzzy Data

The path of evolution through the multimodal and multicriteria search space is guided by fitness functions. These functions use the results obtained by the Rule Parser - a fuzzy expert system that takes one or more rules and interprets their meaning when they are applied to each of the previously fuzzified data items in turn.

This system is capable of two different types of fuzzy logic rule interpretation: traditional fuzzy logic, and *membership-preserving* fuzzy logic, an approach designed during this research. Depending on which method of interpretation has been selected by the user, the meaning of the operators within rules and the method of defuzzification is different.

Traditional Fuzzy Logic Rule Parser

Traditional fuzzy logic involves finding 'degrees of membership' in the fuzzy sets for each value in the current data item, then using operators to select which membership value should be selected and used in combination. So, given a data item comprising two fuzzified values:

A(0.0, 0.2, 0.8)

B(0.1, 0.9, 0.0)

and a fuzzy rule:

(IS_LOW A AND IS_MEDIUM B)

the traditional fuzzy rule parser takes the degree of membership of A for fuzzy set LOW and the degree of membership of B for the fuzzy set MEDIUM, and calculates which of the two is smaller. So in this case, the result of applying the rule is 0.0. Table 1 describes the behaviour and syntax of each of the fuzzy operators.

Table 1: Traditional fuzzy operators.

Operator	Result
IS_LOW <a, b, c>	A
IS_MEDIUM <a, b, c>	B
IS_HIGH <a, b, c>	C
NOT a	1-a
(a AND b)	min(a,b)
(a OR b)	max(a,b)

This fuzzy grammar imposes certain constraints upon allowable solutions. For example, the argument to 'IS_LOW', 'IS_MEDIUM' or 'IS_HIGH' must always consist of a fuzzy vector: $<Low_{membership}, Medium_{membership}, High_{membership}>$. The arguments to 'AND', 'OR' and 'NOT' functions must always be single-valued results obtained from the application of one or more of the functions.

As is clear from the example phenotype given in section 3.3.1, evolved rules do not always satisfy the constraints imposed by fuzzy grammars. However, rather than impose these damaging constraints on evolution, such grammatically incorrect rules are corrected by the rule parser. (Work performed during this research showed that using mapping to satisfy constraints in a GP system is one of the more effective approaches [14].)

Table 2: Mapping performed by the Rule Parser.

Operator	Result
<a, b, c>	IS_HIGH <a, b, c>
IS_LOW a	a
IS_MEDIUM a	a
IS_HIGH a	A

Functions requiring a fuzzy vector, but receiving a single value do nothing. Functions requiring a single value, but receiving a fuzzy vector, apply 'IS_HIGH' by default in order to generate the single value. Table 2 describes this behaviour in full. Consequently, when interpreted by the fuzzy rule parser, the rule in section 3.3.1 equates to:

"((IS_HIGH Height OR IS_LOW Age) AND IS_MEDIUM Age)".

Defuzzification of the final output value is unnecessary (although it is possible to impose a scaling, or non-linear function to transform the output in some way). It was decided simply to use a one-to-one function for defuzzification (i.e., return the output of the fuzzy rule as the defuzzified value).

Membership-Preserving Fuzzy Logic Rule Parser

The alternative behaviour of the rule parser preserves the three membership values within data items, even after the application of operators such as 'IS_LOW'. This is done in an attempt to permit rules to use all the information found by the clusterer, and thus hopefully to reduce the number of rules needed to classify data. In addition, the operators are designed to be more conducive to evolution by allowing multiple operators to have combined effects without constraints on syntax. For example:

IS_HIGH IS_HIGH y

is now equivalent to

IS_VERY_HIGH y

It should be noted, however, that the English descriptors for these operators does not always fully encompass their behaviour in a rule. Table 3 shows the new behaviours of the operators.

Table 3: Membership-preserving fuzzy operators.

Operator	Result
<a, b, c>	<a, b, c>
IS_LOW <a, b, c>	Conc. <c, b, a>
IS_MEDIUM <a, b, c>	Conc. <0, max(a,c), b>
IS_HIGH <a, b, c>	Conc. <a, b, c>
NOT <a, b, c>	<c, b, a>
(<a, b, c> AND <d, e, f>)	min(<a, b, c>,<d, e, f>)
(<a, b, c> OR <d, e, f>)	max(<a, b, c>,<d, e, f>)

Where Conc. *concentrates* the vector (making the
largest value larger and the other two values smaller).

Because this novel approach preserves all three membership values during the application of all operators (although the values may be intensified or reduced), the final result is also a vector comprising three values. To obtain a single, defuzzified value, three defuzzification functions are applied, using the vector to define three trapezoidal shapes, see figure 5. The shapes are then 'piled up on top of each other' and the centre of mass calculated (using overlapping shapes results in a loss of information). A centre of mass falling in the centre results in an output of 0.5, falling to the right gives a score between

0.5 and 1.0, and if the centre of mass falls to the left, the final defuzzified value is between 0.5 and 0, see figure 6.

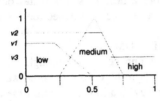

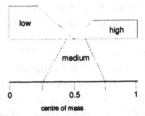

Figure 5: Defuzzifying the three membership values <v1, v2, v3>

Figure 6: Finding the centre of mass during defuzzification.

The membership-preserving (M-P) fuzzy logic is designed to make use of overlapping membership functions. Indeed, for non-overlapping functions, the behaviour of the M-P fuzzy operators becomes largely identical to the traditional operators.

4 Criteria for Fraud Detection

The research described here is being carried out with the eventual aim of the detection of 'suspicious' home insurance claims. This difficult real-world classification task does not simply involve finding the most accurate method for distinguishing between ordinary and dubious data items. There are, in fact, more important criteria for evaluating the performance of a technique. Table 4 shows the four capabilities considered to be most important by our collaborating company Lloyds/TSB, with importance rankings.

Table 4: Important features of a good fraud-detection system.

Feature	Importance
Intelligibility of classification rules	1
Speed of classification	2
Handling noisy data	2
Accuracy of classification	3

It may be surprising to note that accuracy is considered less important than intelligibility. However, for this type of application, an expert must review all suggestions made by the classifier (wrongly accusing anyone of fraud is a serious and potentially libellous activity, so the computer should be used only to identify the possibility of suspicion to experts). If the person cannot find an easily understandable explanation of why a particular data item has been labelled as 'suspicious', then the result is of little use, regardless of the reported accuracy of classification.

Speed of classification is also essential, for most real-world financial problems of this type involve an enormous quantity of data. Increasingly it is becoming necessary for learning techniques to be performed in real time (as new data arrives), but at the very east, the detection method must be fast enough to keep up to date, and also fast enough to justify its use at all.

The ability to handle noisy data was ranked equal in importance with speed. Input errors, omitted data, or conversion problems may cause noise in the data. Although such noise is unlikely to affect more than a small percentage of values in the data, it is clearly important that the classifier is not misled by any occurrence of noise. Other important considerations include minimising the misclassifications by the system - it is considered better to miss a few dubious data items than to misclassify normal data. It is clearly not good for customer relations if too many people are wrongly investigated for potential wrong-doing. This is the reason for the inclusion of the first fitness function, described in section 3.3.2.

5 Experiments

5.1 Objectives

With the requirements for a good fraud-detection system in mind, this section describes a series of experiments designed to evaluate these key capabilities of the system. The experiments investigate three aspects of the system: the effect of using different membership functions and fuzzy operators, the effect of using different clusterers, and the ability of the system to cope with noisy data. For all three sets of experiments, the intelligibility of results, processing time, and accuracy of detection are assessed.

5.2 Experimental Setup

To allow comparison of this system with other techniques reported in the literature, the fuzzy rule evolver was applied to two standard data sets for all experiments: the Iris and Wisconsin Breast Cancer data sets.

The Iris data is "perhaps the best known database to be found in the pattern recognition literature"[1], and comprises a simple domain of 150 instances in three classes, each of 50 items. Data items have four attributes; there are no missing values. Because the 'Setosa' class is linearly separable from the other two classes, for all experiments the system was set the harder task of detecting the 'Virginica' class from the 'Versicolour' and 'Setosa' classes combined. Training and test data files were prepared by splitting the data set into two (taking alternate data items for each file). Misclassification rates for this data set are normally reported as 0% for the 'Setosa' class and "very low" for the other classes in the literature e.g. [4].

The Wisconsin Breast Cancer data is a more complex data set, comprising 699 instances in two classes: 'Malignant' (241 data items) and 'Benign' (458 items). There are 16 missing values in the data, which were filled by random numbers. The training and test data sets were constructed by splitting the file into two, taking alternate values. (For the sake of symmetry, one 'Malignant' item was discarded and two 'Benign' items

---

According to the information provided by UCI with the data.

duplicated, resulting in two sets of 350 data items, each with 120 'Malignant'.) Results reported in the literature include accuracies of 93.5%, 95.9% [13], and 92.2% [16].

50 trials were run for each experiment, with the average and best accuracies reported here. Percentage accuracy of detection was found by calculating:

$$100 - \frac{100(MisclassifiedItems + Unclassifieditems)}{TotalItems}$$

Intelligibility was measured in terms of the average number of rules evolved - the fewer the rules, the more intelligible the result. Average processing speed was measured in seconds (and includes the negligible time taken to apply the completed rule set to both data sets).

The fitness functions reported in section 3.3.2 were used without change for all experiments. Importance rankings [2] were set as 0.5, 2.0, 1.0 and 0.5 for fitness functions one to four, respectively. Mutation of a single bit occurred with a probability of 0.001 in each genotype. Population sizes of 100 were used, and each modal evolutionary run was for exactly 15 generations. The K-Means clusterer was used in the system (unless otherwise stated). Experiments were run on a PC with a 233Mhz AMD K6 processor.

5.3 Experiment 1: Investigating the Effect of Membership Functions

The objective of the first set of experiments was to examine the effects of different membership functions (and different ways of using the information contained in the membership functions) on the ability of the system to detect data items with good intelligibility, speed, and accuracy. Four different system set-ups were investigated: traditional fuzzy logic with non-overlapping and overlapping membership functions, and M-P fuzzy logic with overlapping and smooth membership functions. (Traditional fuzzy logic does not work well with the level of overlap provided by the smooth functions, and the M-P fuzzy logic with non-overlapping functions behaves in the same way as traditional fuzzy logic, so these set-ups are not investigated here.) Table 5 shows the results obtained from 50 runs of each system set-up for each data set.

As shown in Table 5, for both data sets the average accuracy appears to fall as the level of overlap of membership functions is enlarged. However, there is clearly a quite dramatic increase in intelligibility (a reduced number of rules) as the overlap increases. This is illustrated by two example solutions evolved by the system for the Wisconsin Breast Cancer data set. Figure 7 shows a typical 12-rule set evolved when using traditional fuzzy logic and non-overlapping membership functions. Figure 8 shows a typical single rule evolved when using M-P fuzzy logic with smooth membership functions. It should be apparent that the latter is substantially more intelligible than the former. Not only that, but by reducing the number of rules, far more effective feature-selection takes place (i.e., instead of using all ten fields in the data, the single rule shows that only two are required).

Table 5: Mean and best accuracy rates, processing times and intelligibility of solutions when using different membership functions and fuzzy operators. (Accuracy values in normal intensity indicate results for the training set, bold values show results for the test data set.)

System:	Iris data				Cancer data			
	Av. Accuracy	Best accuracy	Av. time	Av. # of rules	Av. accuracy	Best accuracy	Av. time	Av. # of rules
FL, non-overlapping MFs	96.67% **95.79%**	**97.3%**	25.6 secs	2.94	98.6% **94.07%**	**96.0%**	317 secs	9.88
FL, overlapping MFs	91.3% **94.69%**	**96%**	13.6 secs	1.50	93.01% **90.44%**	**96.29%**	335 secs	7.16
M-P FL, overlapping MFs	96.05% **90.67%**	**90.67%**	15.1 secs	1.04	91.19% **86.93%**	**92.57%**	175 secs	4.58
M-P FL, overlapping smooth MFs	82.69% **82.59%**	**88%**	16.2 secs	1	95.14% **95.71%**	**95.71%**	162 secs	1

It is clear that accuracy is reduced as the number of rules that classify the data is reduced. The exception to this is the MP-FL system with smooth MFs applied to the cancer data, which generated both accurate results with a very intelligible single (and simple) rule, e.g. fig 8. However, this result seems likely to be more the exception than the rule – the accurate result may well be due to a fortunate combination of placement of membership functions, and the combination of the three fuzzy membership values for this particular problem. Nevertheless, the result certainly indicates that it is possible to classify real-world problems with both accuracy and intelligibility.

```
Adhesion
(ClumpThickness AND CellShape)
(CellSize AND Chromatin)
(ClumpThickness AND EpithCellSize)
(CellSize AND ClumpThickness)
(IS_LOW Samplecode AND BareNuclei)
(IS_MEDIUM NormalNucleoli AND ClumpThickness)
(BareNuclei AND EpithCellSize)
(Mitoses AND ClumpThickness)
(ClumpThickness AND BareNuclei)
(IS_MEDIUM NormalNucleoli AND BareNuclei)
(ClumpThickness AND IS_LOW EpithCellSize)
```

Figure 7: A 12-rule set evolved using traditional fuzzy interpretation by the rule parser and non-overlapping functions.

```
IS_HIGH (CellSize OR BareNuclei)
```

Figure 8: A single rule evolved using M-P fuzzy interpretation by the rule parser and smooth functions.

As Table 5 shows, processing times fell as the level of overlap of membership functions was increased. This speedup is readily explainable: as the overlap of MFs was increased, the number of rules evolved by the system fell, and since each rule is the result of one modal evolutionary run of 15 generations, system speed is proportionate to the number of rules evolved during classification. The longest learning time for the 7000-value Cancer set took around five and half minutes in these experiments. However, once learned, the time taken to apply the rules to the data is less than one second.

5.4   Experiment 2: Investigating the Effects of Clusterers

The objective of this second set of experiments was to determine the impact of using different clusterers in terms of the three performance measures of intelligibility, speed and accuracy. Two extremes of clusterer were employed: the basic method (described in section 3.1) and the substantially more advanced K-Means approach. For these tests, the system used traditional fuzzy logic and non-overlapping membership functions.

Tables 6 shows the results obtained from 50 runs of each system set-up for each data set. As can be seen from the average and best accuracy percentages, the basic clustering does result in slightly reduced performance of classification. The performance loss is perhaps surprisingly low, though, when it is recalled how simple the basic clustering method is, compared to the K-Means approach. Different rules and different numbers of rules were evolved when using each type of clusterer, as shown by the other results in Table 6. There does not seem to be any clear correlation between intelligibility or processing speed and the type of clusterer used.

5.5   Experiment 3: Investigating the Effects of Noise on the System

The objective of this final set of experiments was to evaluate the change of performance of the system as levels of noise in the data sets was increased. Noise was cumulatively added to both data sets (and both the training and test files of each) in steps of 2%. This was achieved by scaling one randomly chosen value in every fifty by a random value. The experiments investigate levels of noise up to 10% (i.e. one in ten values is wrong). Observed levels of noise in the data sets for which this system is designed are around 1%. These experiments were performed with the system using traditional fuzzy logic and non-overlapping membership functions.

Table 7 shows the results obtained from 50 runs of the system for both data sets, at six different levels of noise. Generally, the results show a gradual decrease in accuracy for both data sets. At ten percent noise the accuracy for the Iris data does increase, but this is likely to be chance and the fact that the number of values in the set is insufficient for a 2% noise differential to affect a significant number of data items. However, the accuracy falloff for the larger, Wisconsin Breast Cancer data set is particularly revealing.

Figure 9 shows the rate at which accuracy falls for classification of items in the training and test data sets as noise levels increase. Note the way accuracy falls linearly for the training data, but appears to fall proportionate to the square of the percentage of noise in the test data. This large decrease in performance is likely to be caused by the noise reducing the homogeneity of the training and test sets, so rules evolved for the training set work less and less well for the test set.

Table 6: Mean and best accuracy rates, processing times and intelligibility of solutions when using different clusterers. (Accuracy values in normal intensity indicate results for the training set, bold values show results for the test data set.)

System:	Iris data				Cancer data			
	Av. Accuracy	Best accuracy	Av. time	Av. # of rules	Av. accuracy	Best accuracy	Av. time	Av. # of rules
Using *basic cluster*	97.84%		14.9 secs		97.3%		419 secs	
	92.4%	**93.3%**		**1.38**	**93.62%**	**95.43%**		**11.3**
Using *k-means*	96.67%		25.6 secs		98.6%		317 secs	
	95.79%	**97.3%**		**2.94**	**94.07%**	**96.0%**		**9.88**

Table 7: Mean and best accuracy rates , processing times and intelligibility of solutions for different noise levels. (Accuracy values in normal intensity indicate results for the training set, bold values show results for the test data set.)

Noise level:	Iris data				Cancer data			
	Av. accuracy	Best accuracy	Av. time	Av. # of rules	Av. accuracy	Best accuracy	Av. time	Av. # of rules
0%	96.67%		25.6 secs		98.6%		317 secs	
	95.79%	**97.3%**		**2.94**	**94.07%**	**96.0%**		**9.88**
2%	94.67%		23.0 secs		98.51%		347 secs	
	97.33%	**97.33%**		**1.0**	**93.33%**	**95.71%**		**9.66**
4%	92.29%		18.5 secs		97.56%		380 secs	
	87.47%	**94.67%**		**2.10**	**93.87%**	**95.43%**		**8.46**
6%	74.19%		19.3 secs		96.98%		454 secs	
	84.11%	**94.67%**		**1.90**	**89.54%**	**92.29%**		**9.82**
8%	73.95%		19.3 secs		95.67%		423 secs	
	84.83%	**94.67%**		**1.82**	**86.16%**	**90.57%**		**8.86**
10%	74.93%		19.5 secs		95.41%		523 secs	
	87.55%	**96.0%**		**1.96**	**78.02%**	**84.86%**		**13.6**

Upon consideration, such reduced effectiveness of rules may be manifested in two ways. Firstly rules become 'misled' by the noise (perhaps because of overfitting by too many excessively specific rules) and thus do not generalise well to the test data set. Secondly, the noise disrupts the clusterers, so that the clustering for training and test sets becomes increasingly different. The resulting LOW, MEDIUM and HIGH fuzzy sets for the test and training data become increasing disparate, reducing the effectiveness of the rules further.

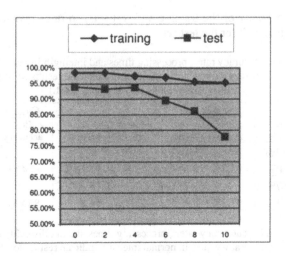

Figure 9: Average accuracy for increasing levels of noise in training and test cancer data. These effects are not so obvious for the Iris data where only a linear decrease in accuracy is evident, perhaps because of the small number of data points in this set or because few rules were used, helping rules to generalise without being misled by noise.

Table 7 also shows the number of rules and processing times for different levels of noise. From these results, there does not appear to be any clear correlation between levels of noise in the data and intelligibility or speed.

6 Conclusions

This paper has investigated the use of a genetic programming system to evolve fuzzy rules for the purpose of detecting 'suspicious' data amongst 'normal' data. The system contains many novel elements, including a crossover operator designed to minimise disruption, binary genotypes, and a new method for interpreting fuzzy rules designed to preserve all fuzzy set membership values.

Consultation with our collaborating company, Lloyds/TSB resulted in a set of evaluation criteria for the system: intelligibility, speed, handling noise and accuracy. With these aspects in mind, three sets of experiments were performed on the system, using two standard data sets to permit comparison with the literature. The first test investigated the effect of membership functions on the system. By increasing the overlap between fuzzy membership functions and by preserving the information held in the membership values, the results showed that the number of rules needed to classify data could be reduced. This reduction often led to a decrease in accuracy of classification, but this was offset by the dramatically increased intelligibility of output, faster processing time, and better feature selection. The second test investigated the effects of using different clusterers in the system. It was found that a basic clusterer slightly reduced the accuracy of the system, compared to the more complex K-Means approach. The choice of clusterer did not seem to have any consistent effect on intelligibility of output or processing speed. The final test investigated the ability of the system to cope with increasing levels of noise in the data. As one would expect, accuracy of classification was detrimentally affected as noise increased. Interestingly, the intelligibility and processing speed showed no clear trend for increasing levels of noise.

Together, these experiments show:

- many factors affect accuracy of classification
- intelligibility and processing speed only seem to be affected by the type and use of membership functions - noise and the choice of clusterer seems unimportant.
- noisy data causes at best a linear drop in accuracy, and at worst, a fall proportionate to the square of input noise.

As with most real-world problems, there is no clearly defined 'best solution' to the problem of detecting fraud by computers. This paper has examined one approach, and has shown that, with the appropriate system components enabled, the use of GP to evolve fuzzy rules can provide intelligible, accurate classification quickly, even for noisy data.

Future Work

The system will be applied to a set of home insurance data provided by Lloyds/TSB in order to assess its abilities to detect 'suspicious' claims. Since the experiments reported here indicate that different settings of the system provide useful classifications for different data sets, an obvious solution is the use of decision aggregation using a committee approach, allowing the best solution generated by the different models employed by the system to be provided automatically.

Acknowledgments

Thanks to Hugh Mallinson for his implementation of the clustering algorithms and some of the fitness functions used in the system. Data was provided by the Data Set Repository at the Information and Computer Science Dept., University of California, Irvine. The breast cancer database originated from University of Wisconsin Hospitals, Madison, assembled by Dr. William H. Wolberg:
tp://ftp.ics.uci.edu/pub/machine-learning-databases/breast-cancer-wisconsin/breast-cancer-wisconsin.names
Insurance data has been kindly provided by Lloyds/TSB. This project is being performed in collaboration with Searchspace Ltd and Lloyds/TSB, and is funded by the EPSRC, ref: GR/L3/708.

References

[1] Bentley, P. J. & Wakefield, J. P., 1996, Hierarchical Crossover in Genetic Algorithms. In Proceedings of the 1st On-line Workshop on Soft Computing (WSC1), pp. 37-42), Nagoya University, Japan.

[2] Bentley, P. J. & Wakefield, J. P., 1997, Finding Acceptable Solutions in the Pareto-Optimal Range using Multiobjective Genetic Algorithms. Chawdhry, P.K.,Roy, R., &

Pant, R.K. (eds) Soft Computing in Engineering Design and Manufacturing. Springer Verlag London Limited, Part 5, 231-240.

[3] Bezdek, J. C. and Pal, S. K. (Ed.s), 1992, *Fuzzy Models for Pattern Recognition.* IEEE Press, New York.

[4] Dasarathy, B.V., 1980, Nosing Around the Neighborhood: A New System Structure and Classification Rule for Recognition in Partially Exposed Environments. *IEEE Transactions on Pattern Analysis and Machine Intelligence*, Vol. PAMI-2, No. 1, 67-71.

[5] Hartigan, J. A, 1975, *Clustering algorithms.* Wiley, NY.

[6] Kosco, B., 1994, *Fuzzy Thinking, the new science of fuzzy logic.* Flamingo. Harper Collins Pub., London.

[7] Koza, J., 1992, *Genetic Programming: On the Programming of Computers by Means of Natural Selection.* MIT Press.

[8] Koza, J. et al., 1998, *Genetic Programming '98: Proceedings of the Third Annual Genetic Programming Conference.* Morgan Kaufman Pub., CA.

[9] Mallinson, H. and Bentley, P.J., 1999, Evolving Fuzzy Rules for Pattern Classification. In Proc. of the Int. Conf. on Computational Intelligence for Modelling, Control and Automation - CIMCA'99.

[10] Marmelstein, R. E. and Lamont, G. B., 1998, Evolving Compact Decision Rule Sets. In Koza, John R. (ed.). *Late Breaking Papers at the Genetic Programming 1997 Conference,* Omni Press, pp. 144-150.

[11] Mc. Neill, D. and Freiberger, P., 1993, *Fuzzy Logic.* Touch Stone Pub.

[12] Pedrycz, W. (Ed.), 1997, *Fuzzy Evolutionary Computation.* Kluwer Academic Publishers, MA.

[13] Wolberg, W. H., and Mangasarian, O. L., 1990, Multisurface method of pattern separation for medical diagnosis applied to breast cytology. In *Proceedings of the National Academy of Sciences*, 87,9193--9196.

[14] Yu, T. and Bentley, P., 1998, Methods to Evolve Legal Phenotypes. In Proceedings of the Fifth Int. Conf. on Parallel Problem Solving From Nature. Amsterdam, Sept 27-30, 1998, pp. 280-282.

[15] Zadeh, L. A., 1965, Fuzzy Sets. *Journal of Information and Control*, v8, 338-353.

[16] Zhang, J. (1992). Selecting typical instances in instance-based learning. In *Proceedings of the Ninth International Machine Learning Conference*, pp. 470--479. Aberdeen, Scotland: Morgan Kaufmann.

Fuzzy Logic Two-phase Traffic Signal Control for Coordinated One-way Streets

Jarkko Niittymäki and Riku Nevala

1. Helsinki University of Technology, Transportation Engineering P.O.Box 2100, FIN-02015, HUT, jarkko.niittymaki@hut.fi, rnevala@cc.hut.fi

Abstract: The opportunities offered by fuzzy logic for controlling traffic are the subject of our FUSICO-research project. The specific goal of this study was to compare the differences in various traffic signal control algorithms when simulating two consequent one-way intersections with no turning traffic. The compared algorithms used in the simulations were coordinated fixed-time signal group control, traffic-actuated gap seeking non-coordinated signal group control, traffic actuated fuzzy logic signal group control (FUSICO) and a combination of standard traffic actuated control and FUSICO-controller. This study was our first attempt to use fuzzy methods in coordinated traffic signals. The results were promising. All previous results of FUSICO-project have indicated better overall efficiency than traditional vehicle-actuated control.

1. Introduction

The opportunities offered by fuzzy logic for controlling traffic are the subject of our FUSICO-research project. Particular areas of interest include the theoretical analysis of fuzzy traffic signal control, developing generalized fuzzy rules for traffic signal control using linguistic variables, validating fuzzy control principles, and developing a fuzzy adaptive signal controller [1]. Since the computer controlled traffic systems were adopted in large cities in 1960's, many theoretical papers mainly about the optimal control of signals have been published. In this paper, we clarify that the fuzzy logic controller is valid in the signal control of two successive traffic intersections of the urban street.

The specific goal of this study was to compare the differences in various traffic signal control algorithms when simulating two consequent one-way intersections with no turning traffic. Simulations were conducted using HUTSIM - traffic simulation software developed at HUT (http://www.hut.fi/Units/Transportation/HUTSIM/). The compared algorithms used in the simulations were coordinated fixed-time signal group control, traffic-actuated gap seeking non-coordinated signal group control, traffic actuated fuzzy logic signal group control (FUSICO) and a combination of standard traffic actuated control and FUSICO-controller.

2. Fuzzy Control

A brief introduction to fuzzy rule-based control is presented in this section. At the basis of fuzzy logic is the representation of linguistic descriptions as membership functions, which indicate the degree to which a value belongs to the class labeled by the linguistic descriptions.

Fuzzy control rules are typically expressed in the following form:

$$\text{If } X_1 \text{ is } A_{i,1} \text{ and } X_2 \text{ is } A_{i,2} \text{ then } U \text{ is } B_i$$

where X_1 and X_2 are the inputs to the controller, U is the output, A's and B's are membership functions, and the subscript i denotes the rule number. For example, a rule for traffic signal control may state "If A is a few and Q is a few then EXT is short". Given input values X_1 and X_2, the degree of fulfillment of rule is given by the minimum of the degrees of satisfaction of the individual antecedent clauses, i.e.,

$$DOFi = Min \; \{A_{i,1}(x_1), A_{i,2}(x_2)\}$$

Because the decision rules are fuzzy, we need some kind of defuzzification method to achieve a crisp output for the final control action. In general, there is no systematic procedure to choose a defuzzification strategy. The defuzzification process involves a mapping from a space of fuzzy control actions into a space of non-fuzzy control actions. This procedure is inverse to that of the fuzzifier. The initial data value, y, consists of the membership values of the current output with respect to all the output fuzzy subsets of the output space,

$$y_0 = \text{defuzzifier}(y),$$

where y_0 is the non-fuzzy control output and defuzzifier is the defuzzification operator. The commonly used defuzzification strategies are max criterion method (MC), the mean of maximum method (MOM) and the center of gravity method (COG) [2]. A new defuzzification method has been introduced in our project, maximal fuzzy similarity, which has a stronger mathematical background than the other defuzzification methods have.

3. Control Algorithm

A fuzzy logic controller in a single intersection of one-way streets is composed of fuzzy control statements as follows:

After minimum green (5 s)

	if A is zero then terminate immediately
or	*if A is a few and Q is a few then EXT is short (3 s)*
or	*if A is mt (a few) and Q is any then EXT is medium (6 s)*
or	*if A is many and Q is any then EXT is long (9 s).*

After the first extension (ext₁ + min gr)

           if A is zero then terminate immediately
or   *if A is a few and Q is a few then EXT is short (3 s)*
or   *if A is medium and Q is any then EXT is medium (6 s)*
or   *if A is many and Q is any then EXT is long (9 s).*

After the second extension (ext₁ + ext₂ + min gr)

           if A is zero then terminate immediately
or   *if A is a few and Q is a few then EXT is short (3 s)*
or   *if A is medium and Q is lt(medium) then EXT is medium (6 s)*
or   *if A is many and if Q is lt(medium) then EXT is long (9 s)*

After the third extension. (ext₁ + ext₂ + ext₃+ min gr)

....

After the nth extension (ext₁ + ext₂ + ext₃ +...+extₙ + min gr)

           if A is zero then terminate immediately
or   *if A is mt(a few) and Q is a few then EXT is short (3 s)*
or   *if A is medium and Q is lt (a few) then EXT is medium (6 s)*
or   *if A is many and Q is lt(a few) then EXT is long (9 s).*

An example of fuzzy inference with COG-defuzzification method is presented in Figure 1.

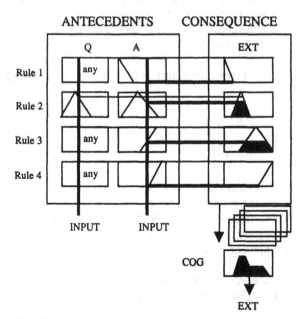

Figure 1: Example of fuzzy inference of two-phase control

4. Simulation Study

In the area of traffic signal control, microscopic computer simulation can be used both in practice, for the comparison of actual planning and design alternatives, and in theoretical work, for the research and development of new control methods and strategies. In both cases, the main advantages of simulation are the possibility to test different alternatives exactly in the same traffic situation in the office. An underlying assumption is that the simulation system in use is able to describe correctly the basic functions and interactions between the vehicles (actually between driver-vehicle-elements), the traffic environment and the signal control.

Simulations provide great amount of detailed data about vehicle movements, like the average delay of vehicles, length of the queues, amount of emissions, fuel consumption and the percentage of stopped vehicles. Simulation also gives the possibility to analyze the signal controller functions, like the lengths of the green phases, the phase sequence and so on. The results of this study include average delays and queue lengths, 95% and 5% percentiles for these two, average green lengths and the performance indexes for signal groups, intersections, whole area and for both main and secondary roads. The effect of the intersection spacing has also been studied.

Totally 72 cases were simulated with 4 different control modes, 6 different traffic volumes (major flows 400, 800 and 1200 with minor flows 200 and 400 veh/h) and 3 different intersection distances (200m, 500m, 1000m). The control methods were:

1) Coordinated fixed timing
2) Isolated vehicle-actuated timing in both intersections
3) Isolated fuzzy timing in both intersections
4) Vehicle-actuated in the first intersection and fuzzy timing in the second intersection.

The fixed timing was calculated for each case individually using the Webster-algorithm [3]

$$C_0 = \frac{1.5L + 5}{1 - \sum_{i=1}^{n} Y_i} \tag{1}$$

where C_0 = optimum cycle time, s
   L = total lost time / cycle, s
   Y_i = maximum value of the ratios of approach volumes to saturation flows for phase i
   n = number of phases.

In our test case, we had three different distances between intersections. The offset time (time gap between beginning green signals in major direction) was adjusted using the simulation. The offset times used for different distances were 15 s for 200 m, 36 s for 500 m and 66 s for 1000 m.

The vehicle actuated control algorithm was planned using the three different detector locations (d5, d60, d120). The extension gaps were 0,5 s in d5 and 3,6 s in d60 and d120. The maximum and minimum green times were as practical as possible.

In our fuzzy application only one detector was located per each approach lane. The location of that one was 120 m (D120). This means that we knew how many vehicles were approaching to the stop line within next 8–10 s. The membership functions used were the same that were successfully used in the previous simulation study of isolated two-phase intersection.

5. Results of Simulation Study

One of the most important jobs of the transportation engineer is to determine how well transportation system is functioning. The usual measure of effectiveness (MoE) used to evaluate the performance of signalized intersections is the average delay. However, the optimal performance of the signalized intersections is the combination of time value, environmental effects and traffic safety.

The SOAP-model [4] uses the several measures of effectiveness to determine optimal performance, including vehicle delay and the number of stops made by the vehicles traveling through the traffic system. These two measures are combined to form a performance index (P.I.)

$$P.I. = d + \alpha S \qquad (2)$$

Where P.I. = the performance index
d = delay, s
α = fixed parameter
S = number of stops of vehicles traveling through the system

Basically, the number of stops gives a rough approximation for environmental measures. The idea (shown in P.I.-index) is to determine the sanction for each stop. The sanction includes the costs of the extra pollution and the extra fuel consumption, which have occurred because of the extra acceleration and deceleration. The performance indexes of simulated cases are shown in Figure 2.

The results indicate that at present, the FUSICO-controller is able to handle intersection that has a load factor smaller than 0.9. During congested traffic, the algorithm used in FUSICO still needs development. In our case, the maximum amount of extensions was 5, which means 50 s maximum green for one signal group. Otherwise, the optimum timing using the Webster-formula was 83 s. The problem can be handled by adding some control intervals more or by increasing the duration of a single extension. The second solution is simpler, but the first one provides a control method, that is more adjustable to the traffic situation.

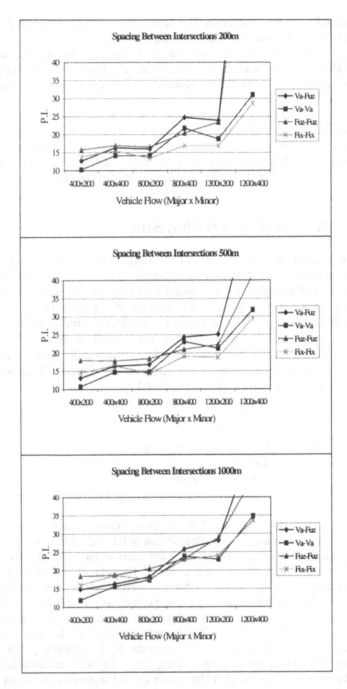

Figure 2: Summary of P.I. results in simulated cases.

The two non-coordinated traffic actuated controllers are able to handle worse conditions than the FUSICO-controllers. Still, the operation of a VA-controller relies totally on the pre-set maximum green lengths for different signal groups, where FUSICO-controller is more independent.

For pre-known traffic situations with no strong random differences in incoming traffic volumes a pre-set fixed interval signal control with visually verified co-ordination is able to operate at an optimum level. The optimally set fixed-time controllers can be used in this study as a benchmark for the rest of the control logic.

A model with both a VA controller and FUSICO controller was also built. The major differences in the operation of these two systems (VA controller operating at much higher green times as an FUSICO-controller) leads into an unwanted situation where the queues lengthen especially during high traffic volumes.

On the other hand, the results are also affected by the fact that the FUSICO-controller tends not to give as much priority to the main road as e.g. the hand-made fixed time controllers. This can be seen in Figure 3.

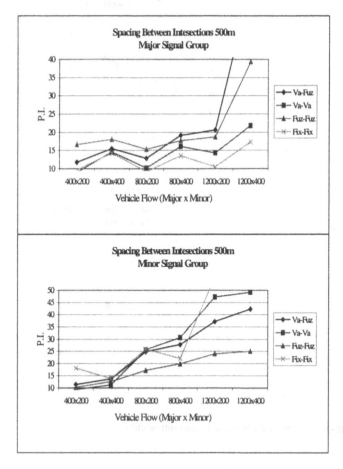

Figure 3: P.I.-results of major and minor flows with 500 m intersection distance.

The result indicates that our fuzzy algorithm works quite equitable, which is good if you are just minimizing delays in one intersection. The average delays in the first intersection are quite similar to the performance index: the minor flow delay of the fuzzy controller is clearly the lowest, while the delays of the major flow are a little bit higher. The results of minor flow in the second intersection are also good, but the results of major flow are not so good. The delays of the second intersection are presented in Figure 4.

Despite of the good performance in minor streets, the delays of the major flow in the second intersection means that the main goal of this study, coordination between controlled intersections, has not fully been reached. The fuzzy control method made for isolated intersections still needs some adjustments to provide better coordination.

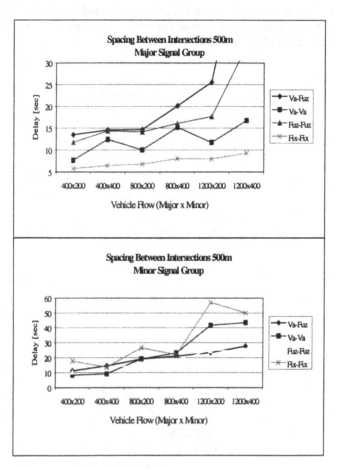

Figure 4: Average delays in the second intersection.

6. Discussion

Coordinated signal control gives preference to progressive traffic flow along the street. In contrast with isolated intersections, the signals must operate as a system. The aim is that coordinated control recognizes that a signal releases platoons that travel to the next signal. This kind of cooperation was not working well enough in our first fuzzy application, but despite that the results were promising.

We have two ideas to develop fuzzy coordinated traffic signal system: to give more attention for major flow signal groups, called split-development, or to formulate new fuzzy rules for system operating. The first one can be solved on two ways: by detector location or by membership function adjusting. According to our experience, the membership function adjusting is not effective way, but the learning algorithm of neural networks and genetic algorithms are available.

The second idea, new fuzzy rules, is probably more effective way to ensure sufficient coordination. Usually, the coordinated control systems are based on a time relationship between the beginning green at one intersection and the beginning green at the next intersection (OFFSET). The fuzzy rules are formed to imitate the decisions made by an expert, for example a policeman standing in the second intersection. This gives us a new perspective: you are not dependent on estimated offset times anymore. Instead, the system operates like the policeman. If the policeman sees a platoon arriving from the first intersection, it is most likely that he will let it pass the second intersection without stopping it. Only when there is a queue long enough waiting at the minor street the policeman might stop the platoon. After the green phase of the first intersection has finished and the platoon has passed the second intersection, the policeman will let the vehicles from the minor street to go on. In a situation, when there is no platoons on main street and the queues on minor street are not long, the policeman makes his decisions based on the arriving vehicles and vehicles waiting for green phase.

However, the second intersection still works ostensibly independent and the coordinated green phase of the second intersection will be started based on platoon recognition. Despite of the ostensible independence, the platoon recognition is a function that replaces the fixed offset-based green phase starting mode, creates the missing connection between the intersections and finally provides the coordination. The vehicles arriving to the detector are considered as a platoon, if the number of vehicles is more than two and the gap between the first and last (third) car is small enough. The membership functions will be formed for the time gap and for the queue (Q) of the minor street. The defuzzification method used will be the maximal fuzzy similarity. The fuzzy decision-making table is presented in Table 1.

Table 1: Fuzzy decision-making table of coordinated one-way streets

	Q Small	Q Medium	Q Long
PLATOON	MainGreen	MainGreen	MinorGreen
No-PLATOON	Basic Fuzzy	Basic Fuzzy	MinorGreen

Maingreen means the green phase of the coordinated direction and Minorgreen the green phase of the conflicting direction. If there is no platoon arriving and the coordination is unnecessary, the same rules as in the simulation study are used.

7. Conclusions

In this study, we showed examples of two intersections without interchanging. The conclusions of this study can be compressed into three statements. The importance of coordination is highest on small intersection distances, though a good coordination can be used on as high as 1000 meters intersection distances. On its current parameterization, the FUSICO-controller is not able to handle the congested traffic, but this might be solved by extending the queue parameter resolution on the fuzzy set. During small traffic volume the FUSICO operates best when comparing the queue management on intersections.

This study was our first attempt to use fuzzy methods in coordinated traffic signals. The results were promising. All previous results of FUSICO-project have indicated better overall efficiency than traditional vehicle-actuated control. Our hypothesis is that fuzzy control has the most to offer when applied to more complicated environments, such as a coordinated traffic signals. The coordinated fuzzy control based on platoon recognition is our next attempt to solve the problems of coordination. The simulations with the HUTSIM-simulators will start on July 1999, and first results will be available on August.

The algorithm and principles shown in this study are very simple. The traffic signal control systems can be constructed by cheap microcomputers as we did in Oulunkylä, Helsinki, 1998. Next fuzzy traffic signals will be installed to Vantaa, Lahti and Jyväskylä in Finland.

References

[1] T. Eloranta, Fuzzy Coordinated Traffic Signal Control, Helsinki University of Technology, Transportation Engineering, Special Assignment, Nov. 1998.

[2] J. Niittymäki, Isolated traffic Signals – Vehicle Dynamics and Fuzzy Control, Helsinki University of Technology, Transportation Engineering, Publication 94, Jan. 1998.

[3] F. Webster, Traffic Signal Settings, Road Research Laboratory, Road Research Technical Paper No. 56, London, IV+44p, 1958.

[4] FHWA–SA–95–032, Traffic Control Systems Handbook, Final Report, Feb. 1996.

Clustering Models Extracting Dynamic and Non-Dynamic Changes for 3-Way Data

Mika Sato-Ilic

University of Tsukuba, Inst. of Policy and Planning Sciences, Tenodai 1-1-1, Tsukuba 305-8573, Japan, E-mail: mika@sk.tsukuba.ac.jp

.**Abstract:** This paper focuses on data consisting of the similarities of objects over times, called 3-way data. For example, the similarities of objects observed at several times is such a data.

This paper proposes a clustering model for 3-way data which can obtain two results at the same time. The first result shows a clustering result through times - this is the non-dynamic result on the times. The second result shows the dynamic changes of the clustering situations over the times.

1. Introduction

With conventional methods for 3-way data the results show only the changes of weights for each cluster over different times (INDCLUS[2], INDASCAL[1]) or shows each of the results which are dynamic and non-dynamic results as I mentioned above. (shown in [5],[6])

However, in real world situations, obtaining both results at the same time could be more convenient for some data analysts. For example, several medical data consisting of patients, variables (they are often medical measurements), and times may need techniques which would get non-dynamic result through times for 3-way data, because such a result may contribute to some estimations of good treatments over times. On the other hand, if researchers need to obtain the exact results for each time and how it changes exactly, then they should use dynamic techniques. In the above example, the analyses of more detailed changing situations of patients is adaptable for this method. However, this result would not contribute to determining a good treatment, because this technique can not extract patterns of changing situations of patients over times as clusters on this result. Such a case is not only for medical data, most data which take the time aspect needs both results to investigate the latent structures of the data. So, I propose a clustering technique which will give these two results

2. Additive Fuzzy Clustering Model

In this section, I will explain an additive model which will be extended for the proposed model.

Clustering models are used to obtain suitable clusters by fitting a structural model into a given similarity data. The additive clustering model (ADCLUS) [7] is the representative model which was proposed by Shepard and Arabie. This model is intended to find the structure of a similarity relationship between a pair of objects by clusters.

In order to develop the ADCLUS, the following two points of view were introduced. The first is to reduce the number of clusters that are required in order to get a sufficiently good fit. The second point is to avoid a discrete optimization problem. In this paper, I introduce the notion of fuzzy clustering for relaxation of the restriction of the model.

In fuzzy clustering, a fuzzy cluster is defined as a fuzzy subset of a set of objects and the fuzzy grade of each object represents it's degree of belongingness. The degree of belongingness of object i to cluster k is denoted by u_{ik}. To avoid situations in which objects do not belong to any clusters, I assume that

$$u_{ik} \geq 0, \ \sum_{k=1}^{K} u_{ik} = 1. \tag{2.1}$$

The simplest extension of the ADCLUS using the fuzzy grade, will be given,

$$s_{ij} = \sum_{k=1}^{K} u_{ik} u_{jk} + \varepsilon_{ij}, \tag{2.2}$$

where s_{ij} $(0 \leq s_{ij} \leq 1; \ i,j = 1, 2, \cdots, n)$ is the observed similarity between objects i and j, K is the number of clusters and ε_{ij} is an error. This model means that a degree of contribution to the similarity s_{ij} is given by the product $u_{ik} u_{jk}$ when the objects i and j belong to the cluster k with the grade u_{ik} and u_{jk}, respectively. However, the degree of contribution to the similarity is not always the product u_{ik} and u_{jk}. Generally, it would be given as a function of u_{ik} and u_{jk}, say $\rho(u_{ik}, u_{jk})$, which would be called the aggregation function. I proposed the model using the aggregation function as follows:

$$s_{ij} = \sum_{k=1}^{K} \rho(u_{ik}, u_{jk}) + \varepsilon_{ij}, \tag{2.3}$$

where $\varepsilon_{ij} (= \varepsilon_{ji})$ is an error term. The aggregation function is assumed to satisfy the following conditions:

$$0 \leq \rho(u_{ik}, u_{jl}) \leq 1, \ \rho(u_{ik}, 0) = 0, \ \rho(u_{ik}, 1) = u_{ik}, \tag{2.4}$$

$$\rho(u_{ik}, u_{jl}) \leq \rho(u_{sk}, u_{tl}) \text{ whenever } u_{ik} \leq u_{sk}, u_{jl} \leq u_{tl}, \tag{2.5}$$

$$\rho(u_{ik}, u_{jl}) = \rho(u_{jl}, u_{ik}), \tag{2.6}$$

where i, j, s, t are suffixes for objects, k, l are suffixes for clusters, and they satisfy $1 \leq i, j, s, t \leq n, \ 1 \leq k, l \leq K$.

3. Dynamic Additive Fuzzy Clustering Model

The data is observed by the values of similarity with respect to n objects for T times, and the similarity matrix of t-th time is shown by $S^{(t)} = \{s_{ij}^{(t)}\}$. Then a $Tn \times Tn$ matrix $\tilde{S}$ is denoted as follows:

$$
\tilde{S} =
\begin{bmatrix}
S^{(1)} & S^{(12)} & S^{(13)} & \cdots & S^{(1T)} \\
S^{(21)} & S^{(2)} & S^{(23)} & \cdots & S^{(2T)} \\
\vdots & \vdots & \vdots & \vdots & \vdots \\
S^{(T1)} & S^{(T2)} & S^{(T3)} & \cdots & S^{(T)}
\end{bmatrix},
\tag{3.1}
$$

where the diagonal matrix is the $n \times n$ matrix $S^{(t)}$. $S^{(rt)}$ is an $n \times n$ matrix and the element is defined as

$$
s_{ij}^{(rt)} \equiv m(s_{ij}^{(r)}, s_{ij}^{(t)}),
$$

where $s_{ij}^{(t)}$ is the (i, j)th element of the matrix $S^{(t)}$. $m(x, y)$ is an average function from the product space $[0, 1] \times [0, 1]$ to $[0, 1]$ and satisfies the following conditions:

(1) $\min\{x, y\} \leq m(x, y) \leq \max\{x, y\}$
(2) $m(x, y) = m(y, x)$:(symmetry)
(3) m(x,y) is increasing and continuous

Moreover, from (1), $m(x, y)$ satisfies the following condition:

(4) $m(x, x) = x$:(idempotency)

The examples of the average function are shown in Table 3.1.

Table 3.1. Examples of Average function

harmonic mean	$\dfrac{2xy}{x + y}$
geometric mean	$\sqrt{xy}$
arithmetic mean	$\dfrac{x + y}{2}$
dual of geometric mean	$1 - \sqrt{(1 - x)(1 - y)}$
dual of harmonic mean	$\dfrac{x + y - 2xy}{2 - x - y}$

For the element $\tilde{s}_{ij}$ of $\tilde{S}$, a new fuzzy clustering model which is named the dynamic additive fuzzy clustering model is as follows:

$$
\tilde{s}_{ij} = \sum_{k=1}^{K} \rho(u_{i^{(t)}k}, u_{j^{(t)}k}) + \varepsilon_{ij}, \ 1 \leq i, j \leq Tn, \ i = nt + i^{(t)},
\tag{3.2}
$$

where K is a number of clusters and $u_{i^{(t)}k}$ shows a degree of belongingness of an object i to a cluster k for time t, and $1 \leq i^{(t)} \leq n$, $1 \leq t \leq T$.

4. Dynamic Additive Fuzzy Clustering Model for Asymmetric Similarity Data

If the observed similarity is asymmetric, then the model (3.2) cannot be used. In order to apply the asymmetric data to the model (3.2), we have to reconstruct the super-matrix (3.1). Suppose the upper triangular matrix of $S^{(t)}$ is denoted as $S^{(tU)}$ and the lower triangular matrix is $S^{(tL)}$. Then we assume the super-matrix as:

$$\tilde{\tilde{S}} = \begin{bmatrix} S^{(1U)} & S^{(1U)(1L)} & \cdots & S^{(1U)(TL)} \\ S^{(1L)(1U)} & S^{(1L)} & \cdots & S^{(1L)(TL)} \\ S^{(2U)(1U)} & S^{(2U)(1L)} & \cdots & S^{(2U)(TL)} \\ \vdots & \vdots & \vdots & \vdots \\ S^{(TL)(1U)} & S^{(TL)(1L)} & \cdots & S^{(TL)} \end{bmatrix},$$

where $s_{ij}^{(rU)(tL)} \equiv m(s_{ij}^{(rU)}, s_{ij}^{(tL)})$, and $s_{ij}^{(rU)}$ is (i,j)th element of the matrix $S^{(rU)}$. Then model (3.2) is rewritten as

$$\tilde{\tilde{s}}_{ij} = \sum_{k=1}^{K} \rho(u_{i(c)d_k}, u_{j(c)d_k}), \tag{4.1}$$

$$1 \leq i,j \leq 2Tn, \quad i = n(c-1) + i^{(c)}, \quad 1 \leq i^{(c)} \leq n, \quad 1 \leq c \leq 2T.$$

$\tilde{\tilde{s}}_{ij}$ is (i,j)th element of $\tilde{\tilde{S}}$ and $u_{i(c)d_k}$ is a degree of belongingness of an object i to a cluster k, and c and d are suffixes for representing time and the direction of the asymmetric proximity data respectively. The relationship is shown as follows:

$$d = \begin{cases} c \equiv 0 \pmod 2 & , \quad d = L \text{ and } t = \frac{c}{2} \\ c \equiv 1 \pmod 2 & , \quad d = U \text{ and } t = \frac{c+1}{2} \end{cases},$$

where $1 \leq t \leq T$. L and U show the lower and upper triangular matrixes, respectively, that is, asymmetric relationship in the same time. t shows tth time, and $c \equiv 0 \pmod 2$ means that c is congruent to 0 modulo 2.

5. Dynamic Additive Fuzzy Clustering Model for Asymmetric Similarity Data

I propose a model using the asymmetric aggregation operators. The supermatrix is defined as follows:

$$\tilde{S} = \begin{bmatrix} S^{(1)} & S^{(12)} & S^{(13)} & \cdots & S^{(1T)} \\ S^{(21)} & S^{(2)} & S^{(23)} & \cdots & S^{(2T)} \\ \vdots & \vdots & \vdots & \vdots & \vdots \\ S^{(T1)} & S^{(T2)} & S^{(T3)} & \cdots & S^{(T)} \end{bmatrix}, \tag{5.1}$$

where, $S^{(t)}$ is an asymmetric similarity matrix for t-th time and $S^{(t)} = \{s_{ij}^{(t)}\}$, $s_{ij}^{(t)} \neq s_{ji}^{(t)}$. $S^{(rt)}$ is defined by $S^{(r)}$ and $S^{(t)}$ as follows:

$$s_{ij}^{(rt)} \equiv \gamma(s_{ij}^{(r)}, s_{ij}^{(t)}), \quad \gamma(s_{ij}^{(r)}, s_{ij}^{(t)}) \neq \gamma(s_{ij}^{(t)}, s_{ij}^{(r)}),$$

where $\gamma(\cdot, \cdot)$ is an asymmetric aggregation operator.[6] Denoting the asymmetric aggregation function by $\gamma(x, y)$, this is defined as follows: Suppose that $f(x)$ is a generating function of t-norm and $\phi(x)$ is a continuous monotone decreasing function satisfying

$$\phi : [0, 1] \to [1, \infty], \quad \phi(1) = 1.$$

Then we define the asymmetric aggregation operator $\gamma(x, y)$ as:

$$\gamma(x, y) = f^{[-1]}(f(x) + \phi(x)f(y)). \tag{5.2}$$

Then the model is shown as follows:

$$\tilde{\tilde{s}}_{ij} = \sum_{k=1}^{K} \gamma(u_{i^{(t)}k}, u_{j^{(t)}k}), \quad 1 \leq i, j \leq Tn, \quad i = (n-1)t + i^{(t)}, \tag{5.3}$$

where, $1 \leq i^{(t)} \leq n$, $1 \leq t \leq T$. $u_{ik}^{(t)}$ is a degree of belongingness in t-th time of an object i to a cluster k and $\tilde{\tilde{s}}_{ij}$ is (i, j)-th element of the matrix $\tilde{\tilde{S}}$ in (5.1).

6. Dynamic and Non-Dynamic Clustering Model

The data is observed as similarities of objects over times and the similarity data matrix at t-th time is shown as follows:

$$S^{(t)} = (s_{ij}^{(t)}), \quad t = 1, \cdots, T,$$

where $s_{ij}^{(t)}$ is a similarity between two objects i and j at t-th time. The super matrix $\tilde{S}$ is defined as

$$\tilde{S} \equiv (S^{(1)}, S^{(2)}, \cdots, S^{(T)}).$$

Then the model is denoted as

$$\tilde{s}_{ij} = \sum_{k=1}^{K} \sum_{l=1}^{K} w_{kl} \rho(u_{ik}, v_{jl}) + \varepsilon_{ij}, \tag{6.1}$$

where $i = 1, \cdots, n$, $j = 1, \cdots, nT$, $k, l = 1, \cdots, K$. K is the number of clusters and u_{ik} shows a clustering result which is the degree of belongingness of an object i to a cluster k over all times. v_{jl} shows another result of the clustering, that is clustering results on different times fixed by the same clusters. u_{ik} and v_{jl} are assumed to satisfy these conditions:

$$\sum_{k=1}^{K} u_{ik} = 1, \ \sum_{l=1}^{K} v_{jl} = 1, \ \sum_{i=1}^{n} u_{ik} > 0, \ \sum_{j=1}^{nT} v_{jl} > 0.$$

w_{kl} shows similarities of clusters k and l and $w_{kk} = 1$, $w_{kl} = w_{lk}$. The degree ρ satisfies the conditions (2.4), (2.5) and (2.6). In this case, I can obtain two results represented by u_{ik} and v_{jl} for the same clusters. So, I can compare these two results and this is a merit to get these results at the same time.

7. Numerical Example

The observed data is the number of students who moved from one prefecture to another prefecture in order to enter universities over three years. By standardizing for each prefecture, we can treat this data as an asymmetric similarity data among prefectures. Using a result which I have obtained by a former clustering technique [5], I selected ten prefectures which were included in a same cluster and this cluster showed remakable movements among all prefectures in Japan.

Figure 5.1 shows the clustering results using model (5.3). In this case, I used the asymmetric aggregation operator $\gamma(x, y) = x^m y / (1 - y + x^{m-1}y)$ which is created by the generator function of the Hamacher prod., $f(x) = (1 - x)/x$ and $\phi(x) = 1/x^m$, $(m = 2)$. Figure 5.2 shows the movements of these results on different times with respect to three axes which show the three clusters fixed on the different times. From this, we can see remakable movements of Osaka, Okayama and Wakayama. Os(t) shows Osaka's result at the t-th time point and Ok(t) and W(t) are Okayama and Wakayama, respectively. Osaka was clearly belongs to Cluster 3 at the first time, however, moved to a vagueness situation of clustering, that is the classification is not clear at the second and the third time points. Wakayama also has the same tendency of the movement. On the other hand, Okayama has opposite change, that is vagueness classification to clear.

Figure 5.3 shows the results for dynamic clustering shown by v_{jl} in (6.1). From this result, I can find that Kyoto and Osaka prefectures created cluster 1, but in second time, Osaka prefecture moved to cluster 3. During the period from time 2 to time 3, I can not find any changes. This result may be caused by the drastic change of policy for the locations of private universities during the period from time 1 to 2. In particular, the movements of Osaka and Kyoto prefectures which have large capacities of students for entering universities seem to have big influences for determining the same clusters during this period.

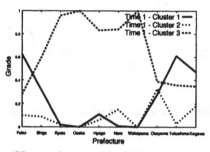

Clustering at the First Time

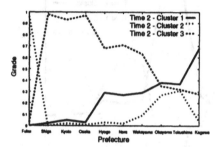

Clustering at the Second Time

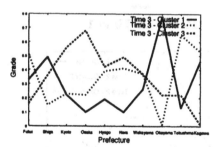

Clustering at the Third Time

Generator Function of Hamacher Prod. : $\gamma(x,y) = x^m y / (1 - y + x^{m-1} y)$
$f(x) = (1 - x)/x$, $\phi(x) = 1/x^m$, $(m = 2)$

Figure 5.1 Clustering of Dynamic Asymmetric Model

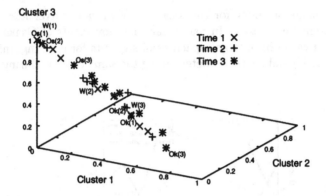

Figure 5.2 Dynamic Clustering of Asymmetric Model

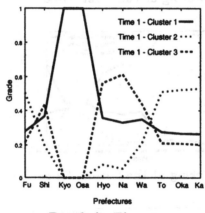

Result in Time 1

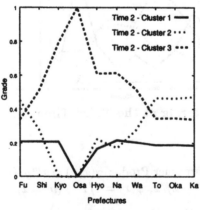

Result in Time 2

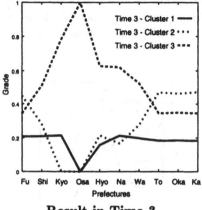

Result in Time 3

Fu = Fukushima, Shi = Shiga, Kyo = Kyoto, Osa = Osaka, Hyo = Hyogo,
Na = Nara, Wa = Wakayama, Oka = Okayama, To = Tokushima, Ka = Kagawa

Figure 5.3 Dynamic Clustering

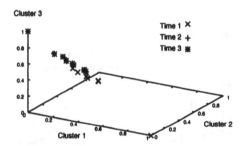

Figure 5.4 Dynamic Changes of Grade

Figure 5.4 shows the clustering situations in the solution's space. The three axes show the three clusters and each dot shows the position vector of the prefectures with respect to each grade of clusters. In this clustering model, all of the clusters are the same, so the result of the grade on different times can be comparable. Figure 5.5 shows the result of non-dynamic clustering shown by u_{ik} in (6.1). From this result, we can see the sharing situation of the grade and can get the abstract of this clustering result. Figure 5.6 also show this. In this figure, the dots named average show the average points with respect to all results shown in Figure 5.3 and the dots named non-dynamic clustering show the result shown in Figure 5.5. Both results show that the results of grade does not have large dispersion without Osaka and Kyoto in this case of average.

Table 5.1 shows the similarities of clusters. In this case, I get almost unit matrix as the result. From this, I may believe that the salience of each cluster is almost the same. However, there is a way not to assume w_{ii} is 1, this is now under consideration.

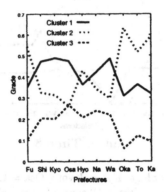

Fu = Fukushima, Shi = Shiga, Kyo = Kyoto, Osa = Osaka, Hyo = Hyogo, Na = Nara, Wa = Wakayama, Oka = Okayama, To = Tokushima, Ka = Kagawa

Figure 5.5 Non-Dynamic Clustering

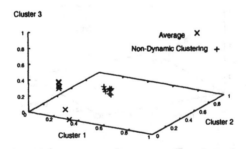

Figure 5.6 Comparison of Non-Dynamic Clustering and Average of Dynamic Clustering

Table 5.1 Similarities of Clusters

Cluster	C_1	C_2	C_3
C_1	1.0000	0.0000	0.0003
C_2	0.0000	1.0000	0.0000
C_3	0.0001	0.0002	1.0000

8. Conclusion

In this paper, I proposed a clustering method which will get two results under the same clusters. One of the results shows the result over times and the other one shows the result of each time. So, results depended on time and do not depended on time, in other words, dynamic and non-dynamic results can be obtained at the same time and these results are comparable due to the same clusters. This means that this technique can get solutions in different parameter spaces, but in the same coordinate. Also, this technique can treat asymmetric data. If I assume that W is diagonal, U and V are orthogonal in (6.1), then this method can be contracted to the usual singular value decomposition. [4] This technique is attempting to relax these conditions.

Moreover, the form of model (6.1) is similar to a model named GENNCLUS [3] which is an extended model of ADCLUS [7]. However, in the case of GENNCLUS, we can not use 3-way data, so the result is clearly different from GENNCLUS, and the degree of belongingness can be obtained as real feasible solutions in the proposed model.

References

[1] J.D. Carroll and J.J. Chang (1970), "Analysis of individual differences in multidimensional scaling via an N-way generalization of "Eckart-Young" decomposition", *Psychometrika, 35,* 283-319.

[2] J.D. Carroll and P. Arabie (1983), "INDCLUS: An individual differences generalization of the ADCLUS model and MAPCLUS algorithm", *Psychometrika, 48,* 157-169.

[3] W.S. Desarbo (1982), "GENNCLUS: New Models for General Nonhierarchical Clustering Analysis", *Psychometrika, 47,* 449-475.

[4] J.C. Gower (1977), "The Analysis of Asymmetry and Orthogonality", Recent Developments in Statistics, J.R. Barra et al. ed., *North-Holland,* 109-123.

[5] M. Sato and Y. Sato (1997), "Time Dependency of Fuzzy Clustering Model", *IEEE International Conference on Intelligent Processing Systems,* 963-968.

[6] M. Sato-Ilic (1998), "A Clustering Model on Similarities for Dynamic Changes,*IEEE International Conference on Systems, Man and Cybernetics,* 2756-2761.

[7] R.N. Shepard and P. Arabie (1979), "Additive Clustering: Representation of Similarities as Combinations of Discrete Overlapping Properties", *Psychological Review, 86,* 2, 87-123.

References

Chapter 2: Neuro Computing

Papers:

Identification of Nonlinear Multivariable Process by Neural Networks: Open-loop and Closed-Loop Case Studies
L. S. Coelho and A. A. R. Coelho

Modelling Batch Learning of Restricted Sets of Examples
K. Y. M. Wong, Y. W. Tong, and S. Li

Neural Networks with Hierarchically Structured Information and its Unlearning Effects
T. Shirakura, H. Fukada, and K. Shindo

Discussion of Reliability Criterion for US Dollar Classification by LVQ
T. Kosaka, N. Taketani, S. Omatu, and K. Ryo

Facility Location Using Neural Networks
F. Guerrero, S. Lozano, K. A. Smith, and I. Eguia

Building Maps of Workspace for Autonomous Mobile Robots Using Self-Organizing Neural Network
K. Hori, Y. Hashimoto, S. Wang, and T. Tsuchiya

Neural Network Parameter Estimation and Dimensionality Reduction in Power System Voltage Stability Assessment
S. Repo

Identification of Nonlinear Multivariable Processes by Neural Networks: Open-Loop and Closed-Loop Case Studies

Leandro dos Santos Coelho and Antonio Augusto Rodrigues Coelho

Federal University of Santa Catarina, Department of Automation and Systems
P.O. Box 476, ZIP CODE 88040.900, Florianópolis, SC, BRAZIL
E-mail: {lscoelho, aarc}@lcmi.ufsc.br

Abstract: Nonlinear complexities and unknown uncertainty of models for multivariable process are difficult problems in identification tasks. Methods of open-loop and closed-loop multivariable identification based on neural networks are evaluated in this paper. Neural networks of multilayer perceptron and radial basis function types are employed. Radial basis function neural network is configured with training by k-means clustering (centers), nearest neighbor algorithm (widths) and pseudo-inverse (weights adjustment) techniques. Multilayer perceptron neural network employs the back-propagation algorithm through of a supervised learning. Two multivariable processes are assessed in this paper. The *MISO* (multi-input, single-output) identification in closed-loop of a horizontal balance process, consisting of two propellers driven by two *DC* motors is realized. The *MISO* identification configuration is realized with two inputs and a single output. Another process is a *MIMO* (multi-input, multi-output) nonlinear system with two inputs and two outputs and governed by difference equations. Experimental results and the performance of neural networks approaches for identification task of nonlinear multivariable processes, in estimation and validation phases, are assessed and discussed.

1. Introduction

The knowledge of the behavior of real systems by using dynamic models is important in many fields of science and engineering. System identification can be described, as the art and the science of building mathematical models of dynamic system and signals based on observed inputs and outputs. The practical importance of plant model identification in open-loop and closed-loop have been recognized for many years [1],[2]. A number of methods have been developed and analyzed for dealing with multivariable processes, such as recursive least squares, gradient methods, extended least squares and approximate maximum likelihood [1]. Dynamics of multivariable processes are typically unknown and complex (nonlinear and time-varying), where the generation of accurate models by conventional methods, such as linear and nonlinear regressions are impractical. Due to the inherent complexity of many real systems the application of traditional techniques is limited. In such instances more sophisticated (so called intelligent) modeling approaches are required. The field of computational intelligence has evolved from attempts to enhance the performance of identification and control methods by incorporating features from human intelligence, such as adaptation, learning, and flexibility capabilities [3],[4].

Design methodologies based on artificial neural networks (*ANNs*) have emerged in recent years as a promising way for multivariable nonlinear identification and control problems [4],[5],[6]. *ANNs* have the ability to map nonlinear relationships without a priori information about the process or system model. Any types of *ANNs* can approximate any continuous nonlinear multivariate function with any degree of accuracy [7]. Others relevant capabilities of *ANNs* in the control application context are: massively parallel distributed processing architecture allowing fast processing for large scale dynamic systems, adaptation, learning, generalization, and ability of fault tolerance and associative memory.

This paper presents the features, design and results of applying multivariable identification procedures based on neural networks. For the case studies two *ANNs* approaches with different types of mapping structures are available: global mappings and local mappings. The multilayer perceptron *ANN* (*MLP*) employs global mappings and radial basis function *ANN* (*RBF*) realizes group of local mappings. Two multivariable processes are assessed. An experimental nonlinear multi-input and single-output system — horizontal balance process — consisting of two propellers driven by two DC motors is presented and analysed. Another process consists of a *MIMO* nonlinear system with two inputs and two outputs governed by difference equations. The remainder of the paper is organized as follows: First, a brief review of the aspects of *RBF* and *MLP* designs are shown in section 2 and 3, respectively. In section 4, the performance of open-loop and closed-loop multivariable identification techniques are provided. Finally, in section 5, some concluding remarks are made.

2. Multilayer Perceptron Neural Network

Conventional perceptrons are feed-forward neural networks with a supervised learning algorithm. The limitation of the simple perceptron is that it can only classify linearly separable patterns. The patterns can only be classified by drawing a hyperplane. The development of multilayer perceptrons and the back error-propagation algorithm (*BP*) overcome the limitations of a simple perceptron. *MLPs* provide an arrangement for *ANN* implementations, by mean of nonlinear relationships between, firstly, the network inputs to outputs and, secondly, the network parameters to output. This network consists of a number of neuron layers, n, linking its input vector, by means of the equation:

$$y = \phi_n(W_n \phi_{n-1}(W_{n-1} ... \phi_1(W_1 x + b_1) + ... + b_{n-1}) + b_n) \tag{1}$$

where W_i is the weight matrix associated with the ith layer, ϕ_i is the nonlinear operator associated with the ith layer and b_i indicates threshold or bias values associated with each node in the *i*th layer [9]. In this study ϕ_i is a hyperbolic tangent for all and is given by:

$$\phi(x) = tanh(x) = \frac{e^x - e^{-x}}{e^x + e^{-x}} \qquad (2)$$

MLP is a non-dynamic *ANN* more commonly utilized in processes identification and control tasks. A simple way for introducing dynamics in the *MLP* consists of using an input vector, *X*, composed of delayed inputs and outputs of the process — tapped-delay-lines method. The *MLP* training for identification activities is realized in the off-line case, with supervised learning through the *BP* via generalized delta rule. In order to minimize this error function, the *BP* uses a gradient search technique called generalized delta rule. The *BP* performs the steepest descent on a surface in a weight space whose height at any point in weight space is equal to the measured error. This algorithm provides an efficient way for calculating the partial derivatives of the squared output error with respect to the weights by propagating the backward of the error to the previous layer [9]. The *BP* is basically constituted of two phases through the different layers of the *ANN* (forward and backward phases), synthesized by the following steps:

i) initialize randomly the weights of all connections;
ii) apply the training set (data of the process);
iii) propagate the neuron output of every layer of the *MLP* for the next layer neurons passing through the neuron activation function (forward step);
iv) specify the desired outputs and calculate all errors in all layers;
v) correct the weights by local gradient (backward step);
vi) repeat the step (iii) and (v) while the quadratic error function of the output does not meet the tolerance value of an acceptable error select by the user;
vii) utilize a test set to analyze the performance of the training.

3. Radial Basis Function Neural Network

The radial basis function model was traditionally used for strict interpolation in multidimensional space. The locally-tuned and overlapping receptive field is a well-known structure that has been studied in regions of cerebral cortex, the visual cortex, and others [10]. Based on the biological receptive fields, Moody & Darken [11] proposed a network structure called *RBF*. More recently, *RBFs* have been employed in system identification and control. *RBF* corresponds to a particular class of function approximators which can be trained, using a set of samples. A *RBF* with *n* inputs and *m* outputs implements a mapping $f_r : \Re^n \to \Re^m$ represented as:

$$f_r(x) = \Lambda_0 + \sum_{j=1}^{Me} \Lambda_j \Phi(\|x - c_j\|) \qquad (3)$$

where $x \in \Re^n$; $\Phi(\bullet)$ is a basis function; $\|\bullet\|$ denotes a Euclidean norm; $\Lambda_j \in \Re^m$; $j=1,...,Me$, are weight vectors; $\Lambda_0 \in \Re^m$ is the constant vector and $c_j \in \Re^m$; $j=1,...,Me$, are centers of the radial basis function [12] also

$$\Lambda_j = [\lambda_{mi}]^T, \quad j=1,...,Me \tag{4}$$

So, the equation (3) can be written in the decomposed form, where:

$$f_{ri}(x) = \Lambda_{i0} + \sum_{j=1}^{Me} \Lambda_{ij} \Phi(\|x - c_j\|), \quad i=1,...,m \tag{5}$$

The functional form, $\Phi(\bullet)$, is fixed, while the centers, c_j, are adjusted in order to k-means clustering algorithm. If the set inputs, $x(k)$, and the corresponding desired outputs, $y_d(k)$, for $k=1,...,m$ and $j=1,...,Me$, can be determined by using the recursive least-squares algorithm, gradient descent techniques, genetic algorithms or other optimization algorithms. In this case, the pseudo-inverse technique is used. The functions chosen for hidden layer neurons, $\Phi(\bullet)$, are Gaussian functions given by:

$$\Phi(v) = exp\left(-\frac{v^2}{2\sigma^2}\right) \tag{6}$$

The literature has been presenting other functions, such as: Cauchy function, thin-plate-spline, multiquadratic, and inverse multiquadratic. The determination of the parameter σ^2 is selected by the *p-nearest neighbor* algorithm [11]. Some papers have demonstrated the functional equivalence between *RBFs* and a special class of Takagi- Sugeno type fuzzy inference systems [3],[10]. These two models are motivated from different origins (*RBFs* from physiology and fuzzy inference systems from cognitive science). However, they share common features not only in their operations on data, but also in their learning process to achieve desired mappings. Both methods share a lot of advantages, such as: quick learning, local functioning and the use of optimization procedure with two steps. They also show common drawbacks, such as: curse of dimensionality and problems of overfitting.

4. Identification of Nonlinear Multivariable Processes

The adopted performance indices for evaluating of mathematical models in case studies are given by following equations:

- sum of squared error: $SSE_i = \sum_{k=1}^{N} [y_i(k) - \hat{y}_i(k)]^2$ (7)

- multiple correlation coefficient: $R_i^2 = 1 - \dfrac{\sum_{k=1}^{N} [y_i(k) - \hat{y}_i(k)]^2}{\sum_{k=1}^{N} [y_i(k) - \bar{y}_i]^2}$ (8)

where i is the index corresponding the process output, $y(k)$ is the process real output, $\bar{y}_i$ is the average of N measured samples from the process output, i. A value of R^2 closed to 1.0 indicates an exact adequacy of the model from measured data. An R^2 between 0.9 and 1.0 may be considered sufficient for many practical applications in control systems. Next, the results for two case studies of multivariable identification by *MLP* and *RBF* are presented.

4.1 Case Study 1 — *MIMO* Nonlinear System

The unknown process to be identified is a *MIMO* nonlinear system governed by the difference equation [13]:

$$y_1(k+1) = \frac{y_1(k)}{1 + y_1^2(k) + y_2^2(k-1)} + u_1(k-1)u_2(k-1) + u_1(k)u_2(k) \tag{9}$$

$$y_2(k+1) = \frac{y_1(k-1)y_2(k)}{1 + y_2^2(k-1)} + u_1(k-1)u_2(k) + u_2(k-1)u_2(k) \tag{10}$$

The data set for estimation (samples from 1 to 1000) and validation (samples from 1001 to 2000) phases are generated through of composition of white noise and step response signals and are presented in figure 1.

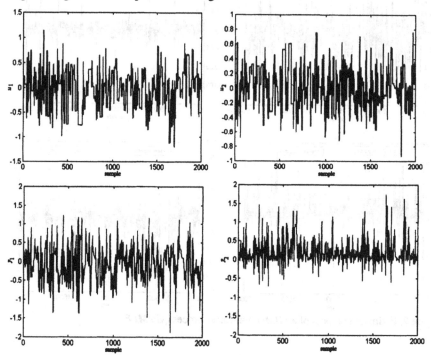

Figure 1. Input and output signals of a *MIMO* nonlinear process.

The adopted *RBF* have four inputs $\{y_1(k-1), y_2(k-1), u_1(k), u_2(k)\}$ and two outputs $\{y_1(k), y_2(k)\}$. *MLP* also adopted the same configuration of number of inputs and outputs. Table 1 presents the open-loop identification of the *MIMO* simulated process, respectively.

Table 1. Results of a *MIMO* process identification by *RBF* and *MLP*.

MLP				RBF			
30 neurons in hidden layer				10 Gaussian functions in hidden layer			
estimation phase		validation phase		estimation phase		validation phase	
SSE_1	SSE_2	SSE_1	SSE_2	SSE_1	SSE_2	SSE_1	SSE_2
2.4×10^{-5}	3.0×10^{-4}	2.6×10^{-5}	2.2×10^{-4}	4.2×10^{-6}	2.4×10^{-4}	4.2×10^{-6}	2.1×10^{-4}
R_1^2	R_2^2	R_1^2	R_2^2	R_1^2	R_2^2	R_1^2	R_2^2
0.9999	0.9986	0.9999	0.9987	0.9999	0.9988	0.9999	0.9988

Figures 2 and 3 show the best results obtained by *MLP* and *RBF* in open-loop identification of the *MIMO* simulated process, respectively.

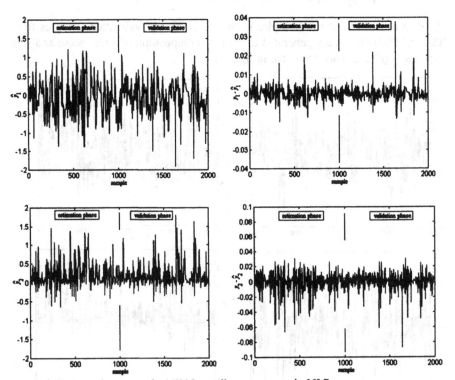

Figure 2. Estimated outputs of a *MIMO* non-linear process via *MLP*.

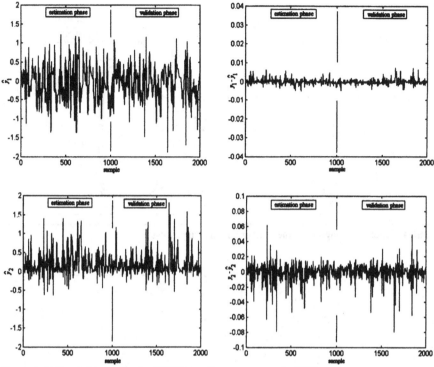

Figure 3. Estimated outputs of a *MIMO* non-linear process via *RBF*.

The *MLP* and *RBF* present similar results. However, the *RBF* present better results than *MLP* in terms of accuracy. The *RBF* is linearly related to its weights and learning in a *RBF* is faster than *MLP* with *BP*.

4.2 Case Study 2 — Horizontal Balance Experimental Process

The identification of dynamic systems by using data measurements of closed-loop experimental conditions is a problem which is highly relevant in many applications. In practice there are many situations in which identification in open-loop is difficult or simply not feasible. This includes for example the case of plants having integrator on its behavior or being unstable in open-loop, as well as the case of plants subject to significant drift in open-loop operation. Identification in closed-loop is also used when a controller is designed on the system and a validation of the current tuning or a redesign of the controller is considered [1],[2]. The horizontal balance process, employed in this case study, consists of a beam that horizontally oscillates and it is positioned by two propellers driven by *DC* motors (figure 4). The control variables are the tension, $u_1(k)$ and $u_2(k)$, supplied to the power amplifier of each *DC* motor coupled with each one of the two propellers, being maintained constant the voltage associated to the other motor [14]. The data for closed-loop identification are obtained via multivariable control design in multiloop *PID* configuration. Figure 5 presents the closed-loop identification data and identification configuration by *ANNs*.

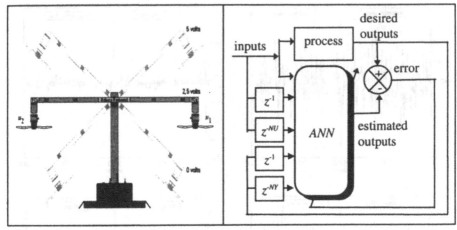

Figure 4. Schematic diagram of the horizontal balance process.

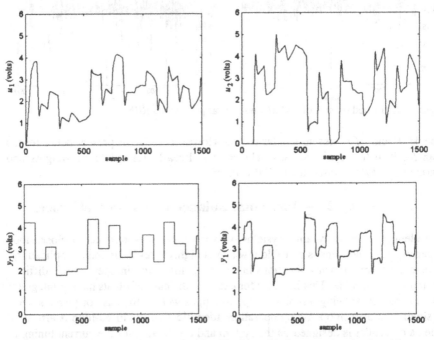

Figure 5. Input, reference and output signals of horizontal balance process in closed-loop.

RBF and *MLP* are configured, in this experiment, for series-parallel multivariable identification [4] and have three inputs $\{y_1(k-1), u_1(k), u_2(k)\}$ e one output $\{y_1(k)\}$. The learning rate and momentum term of *MLP* are 0.02 and 0.001, respectively. The results of identification of the horizontal balance process are presented in Table 2.

Table 2. Results of identification of the horizontal balance process via *RBF* and *MLP*.

ANNs	estimation phase			validation phase		
MLP	neurons in hidden layer					
indices	16	17	18	16	17	18
SSE_1	0.3308	0.2645	0.2600	0.2038	0.1733	0.1750
R_1^2	0.9836	0.9869	0.9871	0.9668	0.9718	0.9715
RBF	Gaussian functions in hidden layer					
indices	6	7	8	6	7	8
SSE_1	0.1270	0.2143	0.0858	0.0449	0.0587	0.0477
R_1^2	0.9938	0.9895	0.9958	0.9927	0.9904	0.9922

Results in Table 2 are similar for the *MLP* and *RBF*. The *RBF* presents better results by analysing the R_1^2 and SSE_1 indices in estimation and validation phases. However, all the configurations show results with $R_1^2 > 0.97$ and therefore are excellent results for identification problems. The best results for *MLP* (figure 6) and *RBF* (figure 7) are with 18 neurons in hidden layer and 8 Gaussian functions, respectively.

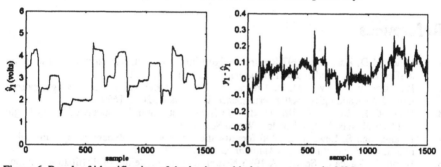

Figure 6. Result of identification of the horizontal balance process via *MLP*.

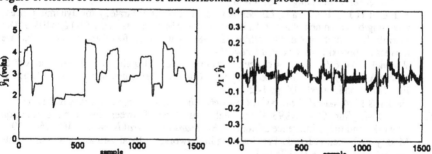

Figure 7. Result of identification of the horizontal balance process via *RBF*.

5. Conclusion

In this paper, theoretical aspects and analysis of practical results of *ANNs* were presented. The *ANNs* of *MLP* and *RBF* types were employed. The methods of open-loop and closed-loop multivariable identification based on *ANNs* were adequate and presented excellent results. *ANNs* are particularly well suited to multivariable applications due to their ability to map interactions and cross-couplings readily whilst incorporating many inputs and outputs. *RBF* has usually better capability for functional representation and perform once than *MLP* in terms of accuracy for system identification problems. Another characteristic is that the response of the *RBF* is linearly related to its weights, learning in a *RBF* is expected to be faster that *MLP*. A negative aspect of an *ANN* is when extreme problems as stability and approximation analysis are concerned, and, further, it is difficult to decide about the network structure [8]. However, *RBFs* offer considerable promise in terms of proving network stability and robustness, mainly because of the way in which the network can be readily defined by the set of equations. Based on two fundamental principles — *principle of data reduction* and *principle of parsimony* — should be realized an extension of this work dealing with the treatment of any relevant aspects in nonlinear identification. Another approach will be the cross correlation analysis of neural mathematical models and application of regularization theory and orthogonalization procedure for configuration of *RBFs*.

References

1. Landau I D, Karimi A 1997 Recursive algorithms for identification in closed-loop: an unified approach and evaluation *Automatica* 33(8):1499-1523
2. Hjalmarsson H, Gevers M, De Bruyne F 1996 For model-based control design, closed-loop identification gives better performance *Automatica* 32(12):1659-1673
3. Harris C J, Morre C G, Brown M 1993 Intelligent Control: Aspects of Fuzzy Logic and Neural Nets, Vol. 6, London, World Scientific
4. Jagannathan S, Lewis F L 1996 Identification of nonlinear dynamical systems using multilayered neural networks *Automatica* 32(12):1707-1712
5. Baratti R, Vacca G, Servida A 1995 Neural network modeling of distillation columns *Hydrocarbon Processing* 74(6):35-38
6. Van Can H J L, Hellinga C, Luyben K C A M et al 1996 Strategy for dynamic process modeling based on neural networks in macroscopic balances *AIChE J* 42(12):3403-3418
7. Hornik K, Stinchcombe M, White H 1989 Multilayer feedforward networks are universal approximators *Neural Networks* 2:359-366
8. Warwick K 1996 An overview of neural networks in control applications, In: Zalzala A M S, Morris A S (eds) *Neural Networks for Robotic Control: Theory and Applications*, Chapter 1. Ellis Horwood Limited, UK, pp 1-25
9. Haykin S 1994 *Neural Networks: A Comprehensive Foundation*, Piscataway, IEEE Press
10. Jang J -S, Sun C -T 1993 Functional equivalence between radial basis function networks and fuzzy inference systems *IEEE Trans on Neural Networks* 4(1):156-159
11. Moody J, Darken C J 1989 Fast learning in networks of locally-tuned processing units *Neural Computation* 1(2):281-294
12. Xiaohong C, Feng G, Jixin Q et al 1996 A nonlinear adaptive controller based on RBF networks *IEEE Int Conf on Systems, Man and Cybernetics*, Beijing, China, pp 661-666
13. Wang D, Chai T 1994 Multivariable adaptive control of unknown nonlinear dynamic systems using neural networks *33rd IEEE CDC*, Lake Buena Vista, FL, pp 2500-2505
14. Coelho L S, Coelho A A R 1999 Fuzzy PID controllers: structures, design principles and application for nonlinear practical process, In: Roy R, Furuhashi T, Chawdhry P K (eds), *Advances in Soft Computing: Eng Design and Manufacturing*, Springer, London, pp 147-159

Modelling Batch Learning of Restricted Sets of Examples

K. Y. Michael Wong, Y. W. Tong and S. Li

Department of Physics, Hong Kong University of Science and Technology,
Clear Water Bay, Kowloon, Hong Kong.
Email: {phkywong, phtong, phlisong}@ust.hk

Abstract: An important issue in neural computing concerns the description of learning dynamics with macroscopic dynamical variables. Recent progress on *on-line* learning only addresses the often unrealistic case of an infinite training set. For restricted training sets, previous studies have so far been limited to asymptotic dynamics or simple learning rules. We introduce a new framework to model batch learning of restricted sets of examples, widely applicable to *any* learning cost function, and fully taking into account the temporal correlations introduced by the recycling of the examples. Here we illustrate the technique using the Adaline rule learning random or teacher-generated examples.

1. Introduction

The dynamics of learning in neural computing is a complex problem on both the macroscopic and microscopic levels. Microscopically, it involves the iteration of a large number of weights connecting the nodes between successive layers, as well as the evolution of the activations for the training examples on each node. The interest on the macroscopic level is thus to describe this multi-variate process with macroscopic dynamical variables, so that insights on the design of efficient learning algorithms can be obtained.

Recently, much progress has been made on modelling the dynamics of *on-line* learning, in which an independent example is generated for each learning step [1, 2]. Since statistical correlations among the examples can be ignored, the dynamics can be described by instantaneous dynamical variables, facilitating a simple description. Phase transitions in the network can be found, and the optimization of the learning rate can be studied.

However, on-line learning represents an ideal case in which the network has access to an almost infinite training set, whereas in many applications, the collection of training examples may be costly. A restricted set of examples introduces extra temporal correlations during learning, and the dynamics is much more complicated. Early studies briefly considered the dynamics of Adaline learning [3, 4, 5], and has recently been extended to *linear* perceptrons learning nonlinear rules [6, 7]. Recent attempts, using the *dynamical replica theory*, have been made to study the learning of restricted sets of examples

[8]. Yet, its mathematical treatment is complicated and depends on *ad hoc* assumptions, and so far the progress is limited to simple learning rules such as Hebbian learning.

In this paper, we introduce a new framework to model batch learning of restricted sets of examples. It fully takes into account the temporal correlations during learning, and is therefore exact for large networks. The framework is widely applicable to any learning rule which minimizes an *arbitrary* cost function by gradient descent. We illustrate the technique using the Adaline learning rule applied to a set of random examples and a set of examples generated by a teacher network. The theory is confirmed by an excellent fit with simulation results. Convergence times for learning are discussed, the effects of weight decay and early stopping on generalization errors and test errors are considered.

The paper is organized as follows. In Section 2 we present the formulation of the learning problem. Section 3 outlines the approach by the cavity method for the general case, which is applied to the solvable case of the Adaline rule in Section 4. Results are presented in Section 5. The presentation is concluded in the final section.

2. Formulation

Consider the single-layer perceptron with $N \gg 1$ input nodes $\{\xi_j\}$ connecting to a single output node by the weights $\{J_j\}$. The proposed framework can be applied to networks with binary or continuous inputs and outputs. Here for convenience we assume that the inputs ξ_j are Gaussian variables with mean 0 and variance 1, and the output state S is a function $f(x)$ of the *activation* x at the output node, i.e.

$$S = f(x); \quad x = \vec{J} \cdot \vec{\xi}. \tag{1}$$

The network is assigned to "learn" $p \equiv \alpha N$ examples which map inputs $\{\xi_j^\mu\}$ to the outputs $\{S_\mu\}$ ($\mu = 1, \ldots, p$). In the case of random examples, S_μ are random binary variables, and the perceptron is used as a storage device. In the case of teacher-generated examples, S_μ are the outputs generated by a teacher perceptron $\{B_j\}$, namely

$$S_\mu = f(y_\mu); \quad y_\mu = \vec{B} \cdot \vec{\xi}^\mu. \tag{2}$$

Batch learning by gradient descent is achieved by adjusting the weights $\{J_j\}$ iteratively so that a certain cost function in terms of the activations $\{x_\mu\}$ and $\{y_\mu\}$ is minimized. Hence we consider a general cost function

$$E = -\sum_\mu g(x_\mu, y_\mu). \tag{3}$$

The precise functional form of $g(x, y)$ depends on the adopted learning algorithm. For the case of binary outputs, $f(x) = \text{sgn} x$. Early studies on the learning dynamics considered Adaline learning [3, 4, 5], where $g(x, y) = -(S-x)^2/2$ with $S = \text{sgn} y$. For recent studies on Hebbian learning [8], $g(x, S) = xS$.

To ensure that the perceptron is regularized after learning, it is customary to introduce a weight decay term. Furthermore, to avoid the system from getting trapped in local minima, it is customary to add noise in the dynamics. Hence the gradient descent dynamics for batch learning is given by

$$\frac{dJ_j(t)}{dt} = \frac{1}{N} \sum_\mu g_x(x_\mu(t), y_\mu)\xi_j^\mu - \lambda J_j(t) + \eta_j(t), \tag{4}$$

where $g_x(x, y)$ represents the partial derivative of $g(x, y)$ with respect to x, λ is the weight decay strength, and $\eta_j(t)$ is noise at temperature T with

$$\langle \eta_j(t) \rangle = 0 \quad \text{and} \quad \langle \eta_j(t)\eta_k(s) \rangle = \frac{2T}{\sqrt{N}}\delta_{jk}\delta(t - s). \tag{5}$$

3. The Cavity Method

The cavity method is the starting point of our work [9]. It has been used in studying the physical properties of magnetic and disordered systems. For neural networks, it has been used to study the *steady-state* properties of learning [10, 11]. The *dynamics* of learning has been studied for perceptrons with *discrete* weights, using a generating function approach, which is mathematically equivalent to the cavity method [12]. However, the cavity method has not been applied to the *dynamics* of learning with *analog* weights and general learning rules, especially in the transient regime.

The method uses a self-consistency argument to consider what happens when a new example is added to a training set. The central quantity in this method is the *cavity activation*, which is the activation of the new example on a node for a perceptron trained without that example. Since the original network has no information about the new example, the cavity activation is a Gaussian variable. Specifically, denoting the new example by the label 0, its cavity activation at time t is

$$h_0(t) = \vec{J}(t) \cdot \vec{\xi}^0. \tag{6}$$

For large N, $h_0(t)$ is a Gaussian variable. For random examples, its covariance is given by the correlation function $C(t, s)$ of the weights at times t and s,

$$\langle h_0(t)h_0(s) \rangle = \vec{J}(t) \cdot \vec{J}(s) \equiv C(t, s), \tag{7}$$

where we have made use of the independence of the random variables ξ_j^0 and ξ_k^0 for $j \neq k$. For teacher-generated examples, the distribution is further specified by the teacher-student correlation $R(t)$, given by

$$\langle h_0(t)y_0 \rangle = \vec{J}(t) \cdot \vec{B} \equiv R(t). \tag{8}$$

Now suppose the perceptron incorporates the new example at the batch-mode learning step at time s. Then the activation this new example at a subsequent time $t > s$ will no longer be a random variable. Furthermore, the activations of the original p examples at time t will also be adjusted from $\{x_\mu(t)\}$ to $\{x_\mu^0(t)\}$ because of the newcomer, which will in turn affect the evolution of the activation of example 0, giving rise to the so-called Onsager reaction effects. This makes the dynamics complex, but fortunately for large $p \sim N$, we can assume that the adjustment from $x_\mu(t)$ to $x_\mu^0(t)$ is small, and perturbative analysis can be applied.

Suppose the weights of the original and new perceptron at time t are $\{J_j(t)\}$ and $\{J_j^0(t)\}$ respectively. Then a perturbation of (4) yields

$$
\left(\frac{d}{dt} + \lambda\right)(J_j^0(t) - J_j(t)) = \frac{1}{N}g_x(x_0(t), y_0)\xi_j^0
$$
$$
+ \frac{1}{N}\sum_{\mu k}\xi_j^\mu g_{xx}(x_\mu(t), y_\mu)\xi_k^\mu(J_k^0(t) - J_k(t)). \quad (9)
$$

The first term on the right hand side describes the primary effects of adding example 0 to the training set, and is the driving term for the difference between the two perceptrons. The second term describes the secondary effects due to the changes to the original examples caused by the added example, and is referred to as the Onsager reaction term.

The equation can be solved by the Green's function technique, yielding

$$
J_j^0(t) - J_j(t) = \sum_k \int ds\, G_{jk}(t, s)\left(\frac{1}{N}g_0'(s)\xi_k^0\right), \quad (10)
$$

where $g_0'(s) = g_x(x_0(s), y_0)$ and $G_{jk}(t, s)$ is the *weight Green's function* satisfying

$$
G_{jk}(t, s) = G^{(0)}(t - s)\delta_{jk} + \frac{1}{N}\sum_{\mu i}\int dt'\, G^{(0)}(t - t')\xi_j^\mu g_\mu''(t')\xi_i^\mu G_{ik}(t' - s), \quad (11)
$$

$G^{(0)}(t - s) \equiv \Theta(t - s)\exp(-\lambda(t - s))$ is the bare Green's function, and Θ is the step function. The weight Green's function describes how the effects of example 0 propagates from weight J_k at learning time s to weight J_j at a subsequent time t, including both primary and secondary effects. Hence all the temporal correlations have been taken into account.

For large N, the equation can be solved by the diagrammatic approach. The weight Green's function is self-averaging over the distribution of examples and is diagonal, i.e. $\lim_{N\to\infty} G_{jk}(t, s) = G(t, s)\delta_{jk}$, where

$$
G(t, s) = G^{(0)}(t - s) + \alpha\int dt_1 \int dt_2\, G^{(0)}(t - t_1)\langle g_\mu'(t_1)D_\mu(t_1, t_2)\rangle G(t_2, s). \quad (12)
$$

$D_\mu(t, s)$ is the *example Green's function* given by

$$
D_\mu(t, s) = \delta(t - s) + \int dt'\, G(t, t')g_\mu''(t')D_\mu(t', s). \quad (13)
$$

This allows us to express the generic activations of the examples in terms of their cavity counterparts. Multiplying both sides of (10) and summing over j,

$$x_0(t) - h_0(t) = \int ds G(t,s) g_0'(s). \tag{14}$$

This equation is interpreted as follows. At time t, the generic activation $x_0(t)$ deviates from its cavity counterpart because its gradient term $g_0'(s)$ was present in the batch learning step at previous times s. This gradient term propagates its influence from time s to t via the Green's function $G(t,s)$. Statistically, this equation enables us to express the activation distribution in terms of the cavity activation distribution, thereby getting a macroscopic description of the dynamics.

To solve for the Green's functions and the activation distributions, we further need the fluctuation-response relation from statistical dynamics, given by

$$C(t,s) = \alpha \int dt' G^{(0)}(t-t') \langle g_\mu'(t') x_\mu(s) \rangle + 2T \int dt' G^{(0)}(t-t') G(s,t'). \tag{15}$$

Finally, for teacher-generated examples, the teacher-student correlation is related to the Green's function via

$$R(t) = \alpha \int dt' G^{(0)}(t-t') \langle g_\mu'(t') y_\mu \rangle. \tag{16}$$

The evolution of the training and generalization errors of the perceptron can be computed from these dynamical parameters.

4. A Solvable Case

The cavity method can be applied to the dynamics of learning with an arbitrary cost function. When it is applied to the Hebb rule, it yields results identical to [8]. Here for illustration, we present the results for the Adaline rule. This is a common learning rule and bears resemblance with the more common back-propagation rule. Theoretically, its dynamics is particularly convenient for analysis since $g''(x) = -1$, rendering the weight Green's function time translation invariant, i.e. $G(t,s) = G(t-s)$. In this case, the dynamics can be solved by Laplace transform, yielding

$$G(t) = \Theta(1-\alpha)(1-\alpha)e^{-\lambda t} + \int_{k_{min}}^{k_{max}} \frac{dk}{2\pi(k-\lambda)} \sqrt{(k_{max}-k)(k-k_{min})} e^{-kt}, \tag{17}$$

$$C(t,s) = \int_{k_{min}}^{k_{max}} \frac{dk}{2\pi k^2} \sqrt{(k_{max}-k)(k-k_{min})} \gamma(k)(1-e^{-kt})(1-e^{-ks}), \tag{18}$$

with k_{max} and k_{min} being $\lambda + (\sqrt{\alpha} \pm 1)^2$ respectively, and $\gamma(k) = 1$ and $1 + 2(k - \lambda - 1)/\pi$ for random and teacher-generated examples respectively.

For teacher-generated examples, the teacher-student correlation is given by

$$R(t) = \sqrt{\frac{2}{\pi}} \int_{k_{\min}}^{k_{\max}} \frac{dk}{2\pi k} \sqrt{(k_{\max} - k)(k - k_{\min})}(1 - \exp(-kt)). \qquad (19)$$

These expressions apply to the case of zero temperature T. Expressions for nonzero temperatures are omitted here because of space considerations.

We can now find the joint distribution $p(x,y)$ for teacher-generated and random examples. For teacher-generated examples, the teacher activation distribution $p(y)$ is a Gaussian with mean 0 and variance 1. The conditional distribution $p(x|y)$ is also Gaussian, with the mean $\langle x(t) \rangle$ and variance $\sigma(t)^2$ given by

$$\langle x(t) \rangle = \int_{k_{\min}}^{k_{\max}} \frac{dk}{2\pi \alpha k} \sqrt{(k_{\max} - k)(k - k_{\min})}(1 - e^{-kt})\mathrm{sgn}y$$

$$+ \int dt' K(t - t')R(t')y, \qquad (20)$$

$$\sigma(t)^2 = \int dt_1 K(t-t_1) \int dt_2 K(t-t_2)C(t_1, t_2) - \left[\int dt' K(t-t')R(t')\right]^2, \quad (21)$$

where

$$K(t-t') = \delta(t-t') - \int_{k_{\min}}^{k_{\max}} \frac{dk}{2\pi\alpha} \sqrt{(k_{\max} - k)(k - k_{\min})}e^{-k(t-t')}\Theta(t-t'). \quad (22)$$

For random examples the distribution $p(x|y)$ is also Gaussian, with the mean and variance given by the first terms of (20) and (21) respectively, and the interpretation that sgn$y$ is the desired output.

To monitor the progress of learning, we are interested in three performance measures: (a) *Training error* ϵ_t, which is the probability of error for the training examples. It is given by $\epsilon_t = \langle \Theta(-x\mathrm{sgn}y) \rangle_{xy}$, where the average is taken over the joint distribution $p(x,y)$. (b) *Test error* ϵ_{test}, which is the probability of error when the inputs ξ_j^μ of the training examples are corrupted by an additive Gaussian noise of variance Δ^2. This is a relevant performance measure when the perceptron is applied to process data which are the corrupted versions of the training data. It is given by $\epsilon_{test} = \langle H(x\mathrm{sgn}y/\Delta\sqrt{C(t,t)}) \rangle_{xy}$, where $H(x)$ is the probability that a Gaussian variable, with mean 0 and variance 1, is larger than x. When $\Delta^2 = 0$, the test error reduces to the training error. (c) *Generalization error* ϵ_g, which is the probability of error for an arbitrary input ξ_j when the teacher and student outputs are compared. It is given by $\epsilon_g = \arccos[R(t)/\sqrt{C(t,t)}]/\pi$.

5. Results

To verify the theoretical predictions, simulations were done with $N = 500$ and using 50 samples for averaging. Figure 1(a) shows the evolution of the average activation for random examples. The agreement is excellent.

To study the convergence rate of learning, we compute the *convergence time*, which is defined as the time for the average activation to reach half its asymptotic value. As shown in Fig. 1(b), the simulation and theoretical results agree remarkably.

In the limit of few and numerous examples, the convergence times τ are respectively given by

$$\lim_{\alpha \to 0} \tau = \frac{\ln 2}{1 + \lambda}, \quad \text{and} \quad \lim_{\alpha \to \infty} \tau = \frac{\ln 2}{\alpha}. \tag{23}$$

Thus for few examples, the convergence rate is determined by the weight decay strength, whereas for numerous examples, the convergence rate is determined by the size of the training set.

In early studies, the time scale for learning dynamics is described by the *relaxation time*, which however is more appropriate for asymptotic dynamics rather than the transient behaviour. Comparing the convergence time with the relaxation time in [3], we see that they are qualitatively the same for most situations. The exception is the case of few examples and weak weight decay, in which the convergence time approaches a constant, whereas the relaxation time diverges as λ^{-1}.

This difference arises because learning is much easier during the transient stage, when it is dominated by a significant growth of the projection onto a highly degenerate subspace in the limit of few examples. On the other hand, when learning has reached steady-state, the perceptron resides in the highly degenerate subspace [3]. Relaxation of learned information in this degenerate subspace can only be achieved by weight decay. Hence the relaxation time diverges when the weight decay vanishes.

Figure 2(a) shows the evolution of the generalization error at $T = 0$. When the weight decay strength varies, the steady-state generalization error is minimized at the optimum

$$\lambda_{opt} = \frac{\pi}{2} - 1, \tag{24}$$

which is independent of α. It is interesting to note that in the cases of the linear perceptron, the optimal weight decay strength is also independent of α and only determined by the output noise and unlearnability of the examples [5, 7]. Here the student is only provided the coarse-grained version of the teacher's activation in the form of binary bits, causing an effect similar to that of output noises.

For $\lambda < \lambda_{opt}$, the generalization error is a non-monotonic function in learning time. Hence the dynamics is plagued by *overtraining*, and it is desirable to introduce *early stopping* to improve the perceptron performance. Similar behaviour is observed in linear perceptrons [5, 6, 7].

Figure 2(b) compares the generalization errors at the steady-state and the early stopping point. It shows that early stopping improves the performance for $\lambda < \lambda_{opt}$, which becomes near-optimal when compared with the best result at $\lambda = \lambda_{opt}$. Hence early stopping can speed up the learning process without significant sacrifice in the generalization ability. However, it cannot outperform the optimal result at steady-state. This agrees with a recent empirical observation that a careful control of the weight decay may be better than early stopping in optimizing generalization [13].

In the search for optimal learning algorithms, an important consideration is the environment in which the performance is tested. Besides the generalization performance, there are applications in which the test examples have inputs correlated with the training examples. Hence we are interested in the evolution of the test error for a given additive Gaussian noise Δ in the inputs. Again, there is an optimal weight decay parameter λ_{opt} which minimizes the test error. Furthermore, when the weight decay is weak, early stopping is desirable.

The optimal weight decay λ_{opt} for the test error depends on input noise. For random examples, $\lambda_{opt} = \alpha\Delta^2$. This is different from the training error, which is minimized when weight decay vanishes. Hence weight decay is not needed when the training error is optimized, but when the perceptron is applied to process increasingly noisy data, weight decay becomes more and more important in performance enhancement. For teacher-generated examples, to the lowest order approximation, $\lambda_{opt} \propto \Delta^2$ remains valid for sufficiently large Δ^2. In the limit of small Δ^2, λ_{opt} vanishes as Δ^2 for $\alpha < 1$, whereas λ_{opt} approaches a nonzero constant for $\alpha > 1$.

It is interesting to consider the weight decay λ_{ot} below which overtraining occurs for the test error. For random examples, λ_{ot} coincides with λ_{opt}. For teacher-generated examples, again, to the lowest order approximation, $\lambda_{ot} \propto \Delta^2$ for sufficiently large Δ^2. However, unlike the case of generalization error, the line for the onset of overtraining does not coincide exactly with the line of optimal weight decay. In particular, for an intermediate range of input noise, the optimal line lies in the region of overtraining, so that the optimal performance can only be attained by tuning *both* the weight decay strength and learning time. However, at least in the present case, computational results show that the improvement is marginal.

6. Conclusion

Based on the cavity method, we have introduced a new framework for modelling the dynamics of learning, which is applicable to *any* learning cost function, making it much more versatile than exisiting theories. It takes into full account the temporal correlations generated by the use of a restricted set of examples, which is more realistic in many situations than theories of on-line learning. Compared with early work on Adaline learning [3, 4], which focuses more on the asymptotic dynamics, we have a better understanding on the transient behaviour and the convergence time. Compared with recent work on Hebbian learning [8], which is based on certain self-averaging assumptions, our theory develops naturally from the stochastic nature of the cavity activations. Hence our theory has the best potential to extend to more sophisticated multilayer networks of practical importance.

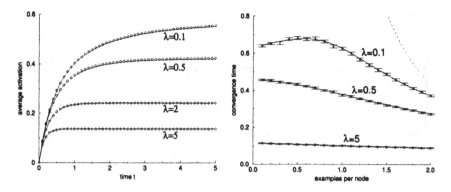

Figure 1: (a) The evolution of the average activation at $\alpha = 1.5$ for different weight decay strengths λ. Theory: solid line, simulation: symbols. (b) Dependence of the convergence time on the examples per node α for different weight decay strengths λ. The dashed line is 0.3 times the relaxation time in [3] for $\lambda = 0.1$.

The method enables us to study the effects of weight decay and early stopping. It shows that the optimal strength of weight decay is determined by the imprecision in the examples, or the level of input noise in anticipated applications. For weaker weight decay, the generalization performance can be made near-optimal by early stopping. Furthermore, depending on the performance measure, optimality may only be attained by a combination of weight decay and early stopping. Though the performance improvement is marginal in the present case, the question remains open in the more general context.

We consider the present work as only the beginning of a new area of study. Many interesting and challenging issues remain to be explored. For example, while the dynamics in the present work corresponds to the limit of very low learning rate, it is interesting to generalize the method to dynamics with discrete learning steps of finite learning rates.

Acknowledgments

This work was supported by the Research Grant Council of Hong Kong.

References

[1] D. Saad and S. Solla, *Phys. Rev. Lett.* **74**, 4337 (1995).

[2] D. Saad and M. Rattray, *Phys. Rev. Lett.* **79**, 2578 (1997).

[3] J. Hertz, A. Krogh and G. I. Thorbergssen, *J. Phys. A* **22**, 2133 (1989).

[4] M. Opper, *Europhys. Lett.* **8**, 389 (1989).

[5] A. Krogh and J. A. Hertz, *J. Phys. A* **25**, 1135 (1992).

[6] S. Bös and M. Opper, *J. Phys. A* **31**, 4835 (1998).

[7] S. Bös, *Phys. Rev. E* **58**, 833 (1998).

[8] A. C. C. Coolen and D. Saad, Preprint KCL-MTH-98-08 (1998).

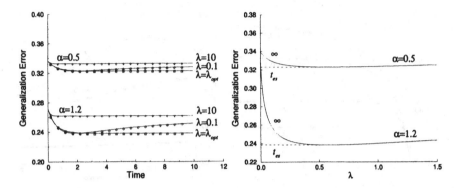

Figure 2: (a) The evolution of the generalization error at $T = 0$ for $\alpha = 0.5, 1.2$ and different weight decay strengths λ. (b) Comparing the generalization error at the steady state (∞) and at the early stopping point (t_{es}) for $\alpha = 0.5, 1.2$ and $T = 0$.

[9] M. Mézard, G. Parisi and M. Virasoro, *Spin Glass Theory and Beyond* (World Scientific, Singapore) (1987).

[10] K. Y. M. Wong, *Europhys. Lett.* **30**, 245 (1995).

[11] K. Y. M. Wong, *Advances in Neural Information Processing Systems* **9**, 302 (1997).

[12] H. Horner, *Z. Phys. B* **86**, 291 (1992); *Z. Phys. B* **87**, 371 (1992).

[13] L. K. Hansen, J. Larsen and T. Fog, *IEEE Int. Conf. on Acoustics, Speech, and Signal Processing* **4**, 3205 (1997).

Neural Networks with Hierarchically Structured Information and its Unlearning Effects

Takayuki Shirakura, Hidemi Fukada and Koichi Shindo

Department of Humanities and Social Sciences, Iwate University
Morioka 020-8550, Japan
shira@iwate-u.ac.jp

Abstract

We investigate Hopfield networks with hierarchically structured information. Two different learning rules are applied. One is the one presented by Parga and Virasoro (we call it the PV model). The other is the Hebbian learning rule adapted for only the memory patterns in the lowest level. Although these two models has both a memory retrieval phase and a concept retrieval phase, it is shown that the properties in these phases are quite different between two models. Especially, it is numerically seen that the unlearning is efficient for concept patterns as well as memory patterns in the PV model, but not for concept patterns in the other model. In order to investigate these difference, we also perform a signal-to-noise ratio analysis.

1 Introduction

The study of the associative memory with hierarchically structured information was begun by Parga and Virasoro (PV) [1] who extended the Hopfield network [2] with their learning rule. Patterns with hierarchical correlations were considered as follows: for simplicity we limit our discussion to the case of a two-level hierarchy. We call patterns in the higher level the concept patterns and those in the lower level the memory patterns. The concept patterns $\{\xi_i^\mu\}, \mu = 1, \ldots, p$, are randomly fixed to the values ± 1, where i denotes the neuron's number and μ the concept pattern's number. We consider $p = \alpha^{(F)} N$ concept patterns, where N is the number of neurons. For each concept pattern, a finite number of memory patterns $\{\xi_i^{\mu\nu}\}, \nu = 1, \ldots, s$

, is generated. Their components are independent random variables determined by the distribution

$$P(\xi_i^{\mu\nu}) = [(1 + \xi_i^\mu b)\delta(\xi_i^{\mu\nu} - 1) + (1 - \xi_i^\mu b)\delta(\xi_i^{\mu\nu} + 1)]/2 \qquad (1)$$

where $0 < b < 1$. It was shown [1, 3] that if the coupling J_{ij} between neurons i and j for all pairs of neurons is set to

$$N J_{ij}^{(PV)} = \sum_{\mu\nu}(\xi_i^{\mu\nu} - b\xi_i^\mu)(\xi_j^{\mu\nu} - b\xi_j^\mu)/(1 - b^2) + \sum_\mu \xi_i^\mu \xi_j^\mu, \qquad (2)$$

the network can store both the concept and memory patterns equivalently and a storage capacity is the same as the one of the original Hopfield model, that is, $\alpha_c^{(PV)} \simeq 0.14$ [4], where $\alpha^{(PV)} = p(1 + s)/N$.

On the other hand, Fontanari[5] applied the Hebbian learning rule for the memory patterns only in the same problem, that is,

$$N J_{ij}^{(F)} = \sum_{\mu\nu} \xi_i^{\mu\nu} \xi_j^{\mu\nu}. \qquad (3)$$

Although the learning is done for the memory patterns only, Fontanari showed that the retrieval of the concept patterns occurs for large s. [5, 6]

Here we investigate the difference of properties of the retrieval phases between the above two models. Especially, we study the unlearning effects for both the concept and memory patterns in both models.

2 Hebbian learning for memory patterns only

In this section we discuss the case of the learning rule eq. (3). In this case, Fontanari [5] calculated a particular class of solutions which has macroscopic correlations with one concept pattern only, by means of the replica trick and under the replica symmetric assumption.[4] We define the concept and memory overlaps as follows:

$$m^\mu \equiv < \sum_i \xi_i^\mu < S_i >_T /N > \qquad (4)$$

$$m^{\mu\nu} \equiv < \sum_i \xi_i^{\mu\nu} < S_i >_T /N > \qquad (5)$$

where $S_i(= \pm 1)$ denotes the state of the i-th neuron and $< \dots >_T$ and $< \dots >$ stand for the thermal average at a temperature T and the average over

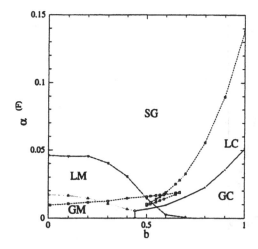

Figure 1: A phase diagram in the learning rule (3) when $s = 3$. There are five phases which correspond to regions of locally or globally stable memory states (LM, GM), locally or globally stable concept states (LC, GC) and the spin glass state (SG).

the pattern distributions, respectively. We study the same class of solutions as Fontanari, that is, $m^1 \neq 0, m^{1\nu} \neq 0, \nu = 1, \ldots, s, m^\mu = 0, m^{\mu\nu} = 0$ for $\mu > 1$. In this case, we have two types of solutions at $T = 0$ and $\alpha^{(F)} = 0$. One is $m^{11} > m^1 > m^{1\nu} > 0, \nu = 2, \ldots, s$, which we call the memory solution. The other one is $m^1 > m^{11} \geq m^{1\nu} > 0, \nu = 2, \ldots, s$, which we call the concept solution. Fontanari derived the general equations of these solutions for any s under the replica symmetric assumption, and then applied them for large s, using a Gaussian approximation.

Here we apply them for small s and calculate a phase diagram in the $\alpha^{(F)} - b$ plane at zero temperature $T = 0$. At $\alpha^{(F)} = 0$, we can easily show from the eqs. (2.9a,b) in the ref. [5] that in the case of $s = 2$ the memory solution is globally stable for all values of b, and for $s \geq 3$ there is a critical value of b above which only the concept solution can exist and which is given by $\sqrt{1/(s-1)}$. In Figure 1, we show a $(\alpha^{(F)}, b)$ phase diagram in the case of $s = 3$ [7], which is calculated from the eqs. (3.7)-(3.13) in the ref. [5]. We can see five phases which correspond to regions of locally or globally stable memory states (LM, GM), locally or globally stable concept states (LC, GC) and the spin glass state (SG).

We numerically examine the unlearning effect [8, 9, 10] at several points

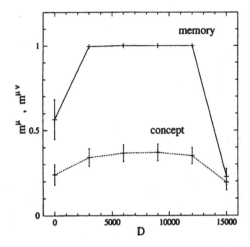

Figure 2: Average values of the self overlaps m^μ and $m^{\mu\nu}$ at $T = 0$ as a function of the number of unlearning steps D [10], when $N = 400, s = 3, b = 0.3, \alpha^{(F)} = 0.08$ and the learning rule (3). The number of samples is $N_s = 200$. Error bars denote the standard deviations of the distributions, not the standard errors.

in the phase diagram. In Figure 2, we show typical behaviors of the unlearning effect for both the concept and memory solutions. We can see that the unlearning is efficient for the memory solutions similarly to those in the original Hopfield model [10], but not for the concept solutions.

3   PV's learning rule

In this section we discuss the case of a modified PV's learning rule

$$N J_{ij} = \sum_{\mu\nu} (\xi_i^{\mu\nu} - b\xi_i^\mu)(\xi_j^{\mu\nu} - b\xi_j^\mu)/(1 - b^2) + a \sum_\mu \xi_i^\mu \xi_j^\mu, \qquad (6)$$

instead of the eq.(2). We introduce an asymmetric parameter a between the concept and memory learnings. The case of $a = 1$ reduces to the original PV's learning. In the original case ($a = 1$), it was shown [3] that a storage capacity is the same as the one in the original Hopfield model, that is, $\alpha_c^{(PV)} \simeq 0.14$. For $a \neq 1$, it is natural that a storage capacity is defined by $\alpha^{(F)} \equiv p/N$, not $\alpha^{(PV)}$, because we consider the case of $p \to \infty, s$: finite with $N \to \infty$. So we use $\alpha^{(F)}$.

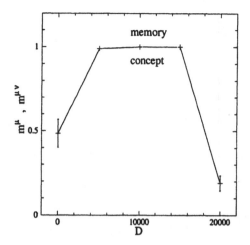

Figure 3: Average values of the self overlaps m^μ and $m^{\mu\nu}$ at $T = 0$ as a function of the number of unlearning steps D [10], when $N = 400, s = 3, b = 0.3, \alpha^{(F)} = 0.08$ and the learning rule (2).

We calculate the free energy under the replica symmetric assumption by means of the standard method [4], and get the following results at $T = 0$:

1. As shown previously [1, 3], when $a = 1$, both the concept and memory solutions satisfy the same equations as the single memory retrieval solution in the original Hopfield model with $\alpha = \alpha^{(PV)}$, and so their ground state energies are degenerate. We can see numerically that the unlearning is efficient for both solutions equivalently. In Figure 3, we show a typical example of the unlearning effect in the case of $a = 1$.

2. When $a > 1$, the memory states are only locally stable, not globally stable, for all values of b and small $\alpha^{(F)}$. When $a > 2$, the memory states can exist only for $b < 1/(a - 1)$ even at $\alpha^{(F)} = 0$. On the other hand, the concept states are globally stable for all values of b and small $\alpha^{(F)}$. In Figure 4, we show a $(\alpha^{(F)}, b)$ phase diagram when $s = 2$ and $a = 2$. It is numerically seen that the unlearning first corrects the unbalance of the memory and concept learnings, and then becomes to behave similarly to the case of $a = 1$. We show a typical example in Figure 5.

3. When $a < 1$, the concept states are only locally stable, not globally

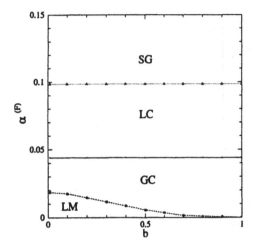

Figure 4: A phase diagram in the learning rule (6) when $s = 2$ and $a = 2$. The meanings of signs are the same as in Fig. 1.

stable. In Figure 6, a $(\alpha^{(F)}, b)$ phase diagram is shown when $s = 2$ and $a = 0.5$. The unlearning effect is similar to the case of $a > 1$ with an exchange of the roles of the memory and concept states.

4 Signal-to-Noise Ratio Analysis

In order to understand the difference of the unlearning effects between the two models, we perform the Signal-to-Noise (SN) ratio analysis. From the analytical reason, we use the paramagnetic (PM) unlearning presented by Nokura [?], instead of the Random Shooting (RS) unlearning in the previous sections. Numerically we can easily confirm the similar behaviors in the PM unlearning to those in the RS one. In the PM unlearning, the coupling $\tilde{J}_{ij}$ is given by

$$\begin{aligned}
\tilde{J}_{ij} &= J_{ij} - \epsilon < S_i S_j >_T /N \\
&\simeq J_{ij} - \epsilon(\beta J_{ij} + \beta^2 \sum_k J_{ik} J_{kj})/N,
\end{aligned} \tag{7}$$

where β is $1/T$, and the high-temperature expansion was used to the second-order of β. Here we want to compare the SN ratios of the local fields between for J_{ij} and $\tilde{J}_{ij}$, that is, $h_i^{\mu(\nu)} = \sum_j J_{ij} \xi_j^{\mu(\nu)}$ and $\tilde{h}_i^{\mu(\nu)} = \sum_j \tilde{J}_{ij} \xi_j^{\mu(\nu)}$.

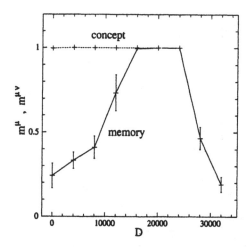

Figure 5: Average values of the self overlaps m^μ and $m^{\mu\nu}$ at $T = 0$ as a function of the number of unlearning steps D [10], when $N = 400, s = 3, b = 0.3, \alpha^{(F)} = 0.08$ and the learning rule (6) with $a = 3$.

First we consider the case of $\alpha^{(F)} \to 0(N \to \infty)$. For the original PV learning rule, there are no noise terms for both the patterns in the case of $\alpha^{(F)} = 0$, similarly to the original Hopfield model with random patterns. Introducing the asymmetry $a \neq 1$, noise terms appear for the memory retrieval only, and the memory retrieval becomes impossible for $b > b_c = 1/(a - 1)$ when $a > 2$. After the PM unlearning, the critical value b_c is shown to become large, which means that the PM unlearning stabilizes the memory retrieval.

On the other hand, in the Hebbian learning rule for the memory patterns only, there are noise terms for both the patterns even in the case of $\alpha^{(F)} = 0$. In the memory retrieval, we get $h_i^{11} = \xi_i^{11} + b^2 \sum_{\nu(\neq 1)} \xi_i^{1\nu}$ before the PM unlearning, and so a critical value b_c is given by $b_c = \sqrt{1/(s - 1)}$ which agree with the Fontanari's result. [5] After the PM unlearning, it is easily shown that the critical value b_c becomes large, that is, the PM unlearning stabilizes the memory retrieval.

In the concept retrieval, it is shown that the SN ratio does not change between before and after the PM unlearning. This fact may expain little effects of the unlearning for the concept retrieval in Figure 2. If we introduce

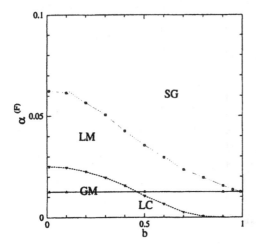

Figure 6: A phase diagram in the learning rule (6) when $s = 2$ and $a = 0.5$. The meanings of signs are the same as in Fig. 1.

the Hebbian learning for the concept patterns to the eq. (3),

$$NJ_{ij} = \sum_{\mu\nu} \xi_i^{\mu\nu} \xi_j^{\mu\nu} + a \sum_{\mu} \xi_i^{\mu} \xi_j^{\mu}, \qquad (8)$$

the SN ratio is shown to be improved proportionally to $a(> 0)$. We can confirm the improvement numerically. In Figure 7, we show a typical result in the RS unlearning when $a = 1$. Of course we should investigate the case of $\alpha^{(F)} \neq 0$ to compare the numerical results with the SN ratio analysis. When $\alpha^{(F)} \neq 0$, calculations show a little improvement of the SN ratio for the concept retrieval even at $a = 0$. The detailed results including the case of $\alpha^{(F)} \neq 0$ will be published elsewhere.

5 Summary and discussion

We have investigated two models with different learning rules for the same hierarchically structured information.

In the model which applies the Hebbian learning rule for the memory patterns only, the concept retrieval occurs when s is large or b is close to 1, although no learning for the concept patterns has been performed explicitly. However it has been seen numerically that the unlearning has almost no effect for the concept retrieval.

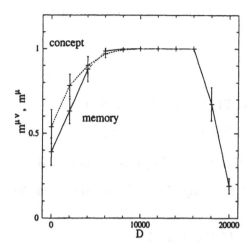

Figure 7: Average values of the self overlaps m^μ and $m^{\mu\nu}$ at $T = 0$ as a function of the number of unlearning steps D [10], when $N = 400, s = 3, b = 0.3, \alpha^{(F)} = 0.08$ and the learning rule (8) with $a = 1$.

In the PV model, the Hebbian learning for the concept patterns and the differences between the memory and concept patterns has been performed. In the original case given by PV ($a = 1$), both the patterns can be retrieved equivalently below a critical storage capacity and the unlearning is efficient in the same way for both the patterns. If we introduce an asymmetry ($a \neq 1$) between the concept learning and the difference learning, the equivalence for both the patterns breaks, that is, one becomes locally stable and the other becomes globally stable for all values of b and small $\alpha^{(F)}$. However the unlearning recovers the symmetry, that is, after some unlearning time, both the patterns become to behave equivalently, similarly to the case of $a = 1$.

Even in the former case, if we introduce the learning for the concept patterns (the eq.(8)), it is seen (see Figure 7) that the behavior after some unlearning steps becomes similar to those in the original PV model. The PV learning rule is not local, because it includes the parameter b. On the other hand, the eq. (8) is a local learning rule. Therefore the eq. (8) is more reasonable for biological networks.

Finally, in concerns with real learning, although the results of the phase diagrams suggest that concepts can be formed without any explicit consciousness, the ones of the unlearning suggest that learning with some consciousness for concepts is more efficient from the point of view of the un-learning effect (the REM sleep [8]).

References

[1] N. Parga and M. A. Virasoro, J. Phys. (Paris) 47 (1986) 1857.

[2] J. J. Hopfield, Proc. Natl. Acad. Sci. USA 79 (1982) 2554.

[3] S. Bos, R. Kuhn and J. L. van Hemmen, Z. Phys. B 71 (1988) 261.

[4] D. J. Amit, H. Gutfreund and H. Sompolinsky, Phys. Rev. Lett. 55 (1985) 1530.

[5] J. F. Fontanari, J. Phys. (Paris) 51 (1990) 2421.

[6] P. R. Krebs and W. K. Theumann, J. Phys. A 26 (1993) 3983.

[7] K. Toya, M. Okada and K. Fukushima (private communication) presented the almost same phase diagram for the same model by means of a different method, and found a bi-stable phase for the first time.

[8] F. Crick and G. Mitchinson, Nature 304 (1983) 111.

[9] J. J. Hopfield, D. I. Feinstein and R. G. Palmer, Nature 304 (1983) 158.

[10] J. L. van Hemmen, L. B. Ioffe, R. Kuhn and M. Vaas, Physica A 163 (1990) 386.

[11] K. Nokura, J. Phys. A 29 (1996) 3871.

Discussion of Reliability Criterion for US Dollar Classification by LVQ

Tosihisa Kosaka[+], Norikazu Taketani[+], Sigeru Omatu[++], and Kunihiro Ryo[+]
[+]Glory LTD, Himeji, Hyougo 670-8567 Japan
[+]Phone:+81-792-92-8445,Fax:+81-792-94-9603, Email: kosaka@tec.glory.co.jp
[++]Osaka Prefecture University, Sakai, Osaka 599-8531, Japan
[++]Phone:+81-722-54-9278,Fax:+81-722-57-1788,Email:omatu@cs.osakafu-u.ac.jp

Abstract- A bill money classification has become automated and it is important that the classifier must have higher accuracy. Generally, accuracy of classification is represented as recognition rate of sample data. However, classifying a bill money, we must evaluate the accuracy more strictly. In the pattern recognition, a Neural Network (NN) is studied and its ability is highly estimated. Among NNs a competitive NN has a simple structure and can be analyzed of the relation between the inputs and the outputs more easily than a layered NN based on the back-propagation method. Because of the reason, we use a competitive NN for the bill money classification and use the Learning Vector Quantization (LVQ) method for training the NN. We propose a reliability criterion based on a probability distribution for the classification by the LVQ method. Then we classify the US dollar by the LVQ and apply the reliability criterion to the classification. We show that the proposed method of bill money classification has higher accuracy.

I. Introduction

Automation of a bill money classification has become popular and it is required that the classifier must have higher accuracy. Generally, accuracy of classification is represented as a recognition rate of sample data. However, classifying bill money, we must evaluate the accuracy more strictly. In the pattern recognition, a neural network has been adopted since it has learning ability of pattern classification. Until now we have developed the layered NN for bill money classification and obtained high performance of classification[1]-[2]. But it takes much time for learning and many local minimum of error function. Therefore, we have also developed another type NN by using a competitive NN which has a simple structure and shows the relation between the inputs and the outputs more clearly than a layered NN and the meaning of the clustering of the input data. In this paper, we adopt the learning vector quantization (LVQ) method among competitive NNs.

II. Competitive Neural Networks

We will explain the competitive neural networks that are used to classify the bill money. The structure of a LVQ competitive network is shown in Fig. 1. The input for the LVQ is bill money data where an original image consists of 128x64 pixels and the input data to the network is compressed as 64x15 pixels to decrease the computational load. The output of the network consists of the US dollars, \$1, \$,2, \$5, \$10, \$20, \$50, \$100. US\$ to the transaction machine has four directions A,B,C, and D where A and B mean the normal direction and the upside down direction and C and D mean the reverse version of A and B.

In the input layer the original bill money data are applied and all the units at the input layer are connected to all the neurons at the output layer with connection weight W_{ij}. W_{ij} denotes the connection weight from the unit j in the input layer to unit i in the output layer. The connection weights W_{ij} are set by the random number at the beginning. Then the following learning algorithm of the connection weight vector is used in the LVQ method[3].

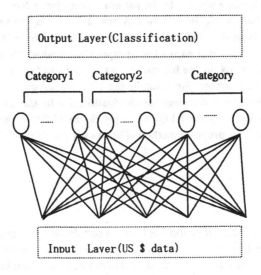

Fig. 1. Structure of the LVQ networks.

LVQ algorithm

Step 1. Find the unit c at the output layer which has the minimum distance from the input data x (t)

$$\|x(t) - W_c\| = \min_i \|x(t) - W_i\|$$

where $\| \ \|$ denotes the Euclidean norm and t denotes the iteration time.

Step 2. If the input x (t) belongs to Category c, then

$$w_c(t+1) = w_c(t) + \alpha(t)(x(t) - w_c(t))$$
$$w_i(t+1) = w_i(t), \quad i \neq c$$

and if the input x (t) belongs to the other Category j (j ≠ c), then

$$w_c(t+1) = w_c(t) - \alpha(t)(x(t) - w_c(t))$$

$$w_i(t+1) = w_i(t), \quad i \neq c$$

where $\alpha(t)$ is a positive function and denotes learning rate.

In the the usual LVQ $\alpha(t)$ is given by

$$\alpha(t) = \alpha_0(1 - \frac{t}{T})$$

where $(0 < \alpha_0 < 1)$ is a positive and T is a total number of learning iterations.

The above algorithm for selection of new weight vector Wc(t+1) can be explained graphically as Fig. 2 .

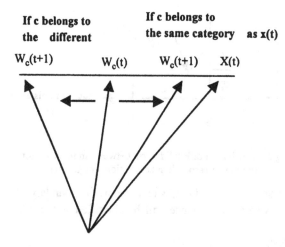

Fig.2. Principle of the LVQ algorithm where the right hand side shows the same category case of **x**(t) and Category c and the left hand side denotes the different category case.

In the above LVQ algorithm, the learning rate $\alpha(t)$ plays an important role for convergence. To adjust the parameter, Kohonen has proposed an optimization method without proof as follows:

$$\alpha_c(t) = \frac{\alpha_c(t-1)}{1 + s(t-1)\alpha_c(t-1)}$$

where s(t) =1 if **x**(t) belongs to the same Category c and s(t)=-1 if **x**(t) does not belong to the same Category c. Here, $\alpha_c(t)$ denotes the learning rate for the pattern of Category C.

Using the above OLVQ1 algorithm, we will classify the Italian bills in the following section.

III. Threshold for Firing

Bill money classification machines may get the foreign currency or irregular bills which were not trained before. From the LVQ algorithm, the proposed classification algorithm will fire the unit at the output layer which has minimum distance connection weight vector to the input vector. Thus, it may classify an undesirable bill as the desirable one. To prevent from classifying incorrectly, we set a threshold within which the input bill will be classified as objective bill.

Let S_i be a population of the firing data of cluster i at the output layer and let T_i be a sample space of the distances between the input and the correspondent weight vector for fired unit among S_i. Assume that the probability distribution of T_i is Gauusian with mean m_i and variance σ_i^2 and denote it by $N_i(m_i, \sigma_i^2)$. The distribution is shown in Fig.3.

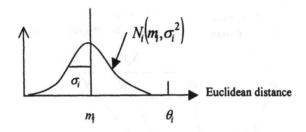

Fig.3. Distribution of Euclid norm between an input vector
and the corresponding connection weight vector.

From this distribution, the threshold θ_i which means the boundary of classification and the input data within this distance will be classified as the cluster i is set as follows:

$$\theta_i = m_i + n_i \sigma_i$$

n_i is selected empirically from [4.5, 6.5] . If $n_i = 4.5$, the probability that unit i does not fire for the data S_i becomes 3.4×10^{-6} and if $n_i = 6.5$, it becomes 4.0×10^{-11} from Gaussian distribution theory. The latter case means that the not fired bills are a few among 100,000,000,000 bills.

IV. Reliability

Using the Gaussian distribution of the Euclidean measure stated in Section III, we will propose the evaluation method of bill money classification.

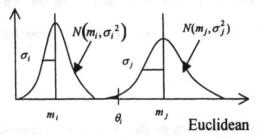

Fig. 4. Error distributions of two categories by the LVQ where θ_i denotes a

threshold value given by $\theta_i = m_i + n_i\sigma_i$.

To classify the bill we must determine the classification boundary(threshold). In this paper we denote it by θ_i . Then we can adopt the measure RM defined in the following relation as a misclassification rate:

$$RM = \int_{-\infty}^{\theta_-} \frac{1}{\sqrt{2\pi}\sigma_j} \exp(-\frac{(x-x_j)^2}{2\sigma_j^2}) dx \ .$$

This probability means that the bill in the cluster j might occupy the region of the cluster i. Thus, RM is called reliability measure and smaller value of RM is more reliable classification. Generally, this value is requested less than 10^{-10} in the vending machine market.

Using this RM, we will determine the threshold value θ_i in the following. In the beginning we set m_i by 4.5, nd then find RM from the Gaussian Table after classification. If the value of RM is less that 10^{-10} then we increase the value of RM as large as possible while it must be stopped when m_i becomes 6.5 since smaller value has no important meaning statistically. If we cannot achieve such a value of RM when we complete the learning, we increase the number of units in the cluster whose RM is larger than 10^{-10}. Repeating this procedure, we can achieve sufficiently small RM for any kind of US dollars for training data of 75 peaces for each bill. In what follows, we will apply this procedure to classify US dollars by using real data obtained from a transaction machine developed by us.

V. Bill Money

In this paper, we consider 8 kinds of US Dollars such as $1, $2, $5, $10, $20, $50, $100(old), and $100(new). The transaction directions of the bill to the transaction machine are four as shown in Fig.5. Number of the used data in this paper is 150 peaces for each category and as a total it is 2,400 peaces.

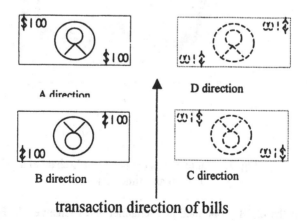

transaction direction of bills

Fig.5. Four directions of bill money.

Figure 6 shows the sensor locations at L-Channel and R-Channel and Fig.7 shows

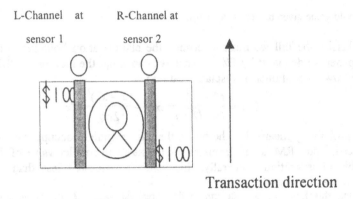

L-Channel at   R-Channel at

sensor 1   sensor 2

Transaction direction

Fig.6.   Sensors location at right and left parts.

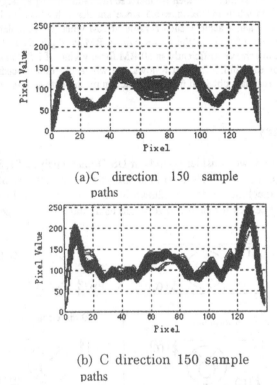

(a)C   direction   150   sample
paths

(b) C direction 150 sample
paths

Fig. 7.   Time Series data of $1 at L-channel and R-Channel are
merged as a time series.

the sample paths of $1   obtained at L-channel and r-Channel. In Fig.7 we show
measurement   data of 120 samples at each channel   and the total data from both
channel. Then some preprocessing has been performed to eliminate the noise and

decrease the variation of samples. The data of 120 will be compressed to 16 by 1/8 to make the size of the neural network smaller. After these preprocessing, we can get the results shown in Fig.8 where C direction data of $1 is illustrated.

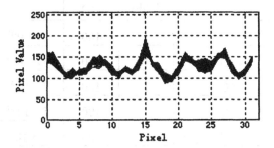

Fig. 8. C direction data of $1 after preprocessing.

These data will be used to classify the US dollar bills by using LVQ algorithm stated in Section II. Hence, the input to the competitive neural network is 32 and the number of the cluster at the output layer is 8 which denotes $1, $2, $5, $10, $20, $50, $100(old), and $100(new). In this case, we must specify the number of units in each cluster. Here, we have used the increasing method to determine the number of units such that if the minimum distance is not decreased, then the unit number is increased until perfect classification and reliability measure with 10^{-10} are achieved for training data of 75 samples.

Table 1 shows the classification results before learning. This means the results by the pattern matching since the initial connection weight vectors are set as the mean vector of the training data. Here, total means the average of RR and the minimum of RM. We define the recognition rate denoted by RR as

$$RR = \frac{correct\ classification\ number}{total\ number}.$$

Except for $ 50 the classification results are good but the RM is not sufficient. In Table1 we cannot perform the perfect classification in the case of $5 D direction which was classified into $50 D. To see the reason why such a misclassification might occur, we pick up the sample paths both of them. Figure 9 shows the sample paths for $5 D direction where the solid line shows the sample path for misclassified bill. Figure 10 shows the sample paths for $50 D direction which have similar form to those of $5 D direction. To make the misclassification we train the neural network based on the LVQ method. The results is shown in Table 2 (a) where perfect classification   has been performed and excellent RM values have been obtained for training data.

Finally, we have checked the classification ability for test data which were not trained at all.   Here, we have checked for 75 test data for each US $ of 8 kinds. The classification results are shown in Table 2 (b).

Table 1. Classification accuracy before learning.

$ directions	No of units	RR (%)	RM
$1 C	1	100	2×10^{-16}
$1 D	1	100	5×10^{-15}
$2 C	1	100	7×10^{-31}
$2 D	1	100	5×10^{-37}
$5 C	1	100	1×10^{-3}
$5 D	1	89.3	5×10^{-2}
$10 C	1	100	6×10^{-11}
$10 D	1	100	1×10^{-4}
$20 C	1	100	1×10^{-0}
$20 D	1	100	3×10^{-8}
$50 C	1	100	6×10^{-9}
$50 D	1	100	1×10^{-1}
old$100 C	1	100	2×10^{-2}
old$100D	1	100	2×10^{-9}
new$10C	1	100	4×10^{-27}
new$10D	1	100	1×10^{-20}
Total	16	99.83	1×10^{-0}

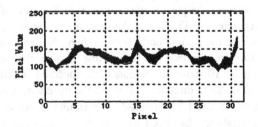

$5 D direction data where bold line shows the misclassified sample path.

Fig.9 The sample paths for $ 5 after preprocessing
where a bold line shows misclassified sample
by using the pattern matching of Table1.

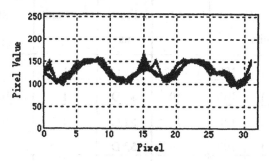

$50 D direction data which are similar

figures to $5.

Fig. 10. The sample paths for $50 which are similar pattern of the solid line of Fig.9.

For the test data the perfect classification has been achieved although the RM values are worse compared with those of Table 2. But the results show the real applicability of the proposed method.

VI. Conclusions

We have proposed a new classification method of Italian Liras by using the OLVQ1 algorithm. The experimental results show the effectiveness of the proposed algorithm compared with the conventional pattern matching method.

References

[1] S. Fukuda, T. Kosaka, and S. Omatu: Bill Money Classification of Japanese Yen Using Time Series Data, Trans. Of IEE of Japan, Vol.115-C, No.3, pp.354-360, 9995(in Japanese).

[2] J. Dayhoff: Neural Network Architectures: An Introduction, International Thompson Computer Press, New York, 1990.

[3] T. Kohonen: Self-Organizing Maps, Springer, Berlin, 1995.

Table 2. Classification accuracy :(a)After number optimization of units.
(b)Classification accuracy for test data of 75

	(a)				(b)		
\$ directions	No of units	RR (%)	RM	\$ directions	No of units	RR (%)	RM
\$1 C	1	100	1×10^{-14}	\$1 C	1	100	4×10^{-7}
\$1 D	1	100	2×10^{-15}	\$1 D	1	100	2×10^{-11}
\$2 C	1	100	5×10^{-32}	\$2 C	1	100	1×10^{-17}
\$2 D	1	100	2×10^{-42}	\$2 D	1	100	2×10^{-20}
\$5 C	2	100	9×10^{-20}	\$5 C	2	100	2×10^{-7}
\$5 D	4	100	2×10^{-12}	\$5 D	4	98.67	3×10^{-8}
\$10 C	1	100	9×10^{-23}	\$10 C	1	100	1×10^{-14}
\$10 D	2	100	2×10^{-18}	\$10 D	2	100	1×10^{-33}
\$20 C	4	100	4×10^{-10}	\$20 C	4	100	2×10^{-8}
\$20 D	2	100	1×10^{-13}	\$20 D	2	100	4×10^{-9}
\$50 C	2	100	2×10^{-12}	\$50 C	2	100	2×10^{-8}
\$50 D	6	100	1×10^{-12}	\$50 D	6	100	9×10^{-12}
old\$100 C	4	100	1×10^{-10}	old\$100 C	4	100	1×10^{-5}
old\$100 D	2	100	1×10^{-12}	old\$100 D	2	100	3×10^{-15}
new\$100 C	1	100	4×10^{-27}	new\$100 C	1	100	5×10^{-13}
new\$100 D	1	100	3×10^{-17}	new\$100 D	1	100	2×10^{-21}
Total	35	100	4×10^{-10}	Total	35	99.92	1×10^{-5}

Facility Location Using Neural Networks

F. Guerrero[1], S. Lozano[1], K.A. Smith[2], and I. Eguia[1]

[1]Escuela Superior de Ingenieros, University of Seville
C/ Camino de los Descubrimientos, s.n. 41092. Seville. Spain

[2]School of Business Systems, Monash University
3168. Clayton. Victoria. Australia

Abstract: Facility location problems occur whenever more than one facility need to be assigned to an equal number of locations at a minimal cost. The quadratic assignment problem is an example within this class of problems. This paper presents a new self-organizing approach to solve quadratic assignment problems. Our neural approach uses neuron normalization as well as a conscience mechanism to consistently find good feasible solutions. To test our neural approach, a set of test problems from the literature has been used. Further research avenues are suggested.

Keywords: facility location, neural networks, self-organizing neural network.

1. Introduction

The Quadratic Assignment Problem (QAP) is an important example of a broad class of combinatorial optimization problems that are known as permutation problems. There are two basic neural approaches for solving combinatorial optimization problems. The first one is the method of Hopfield and Tank [1], which uses a Hopfield neural network to minimize an energy function representation of an optimization problem. The other basic approach is the use of self-organizing neural approaches, which are based upon the work of Kohonen [2].

The major drawbacks of these neural approaches are that, while the solution quality offered by the Hopfield approach tend not to be comparable with those offered by other optimization techniques, the self-organizing neural approaches have been traditionally restricted to solving Euclidean problems by their nature [3].

Recently, however, a new Self-Organizing Neural Network (SONN) to solve combinatorial optimization problems with generalized assignment constraints has been proposed [4]. This SONN operates in a combinatorial way and employs a Hopfield network to enforce feasibility. This SONN has been successfully applied to solving practical optimization problems such as car sequencing problem [5], frequency assignment problem [6], and manufacturing cell formation problem [7].

In this paper, a new neural approach, which is based upon the previous SONN is designed to solving the QAP. In this new neural approach, constraints are adaptively enforced using neuron normalization as well as a penalty function that is computed through a conscience mechanism [8], rather than the Hopfield network previously used. These modifications mean that the new SONN is more elegant and faster than the previous SONN.

The paper is organized as follows. In section 2 the mathematical formulation of the QAP is introduced. Section 3 describes the new SONN. In section 4, first computational results are presented. Finally, Section 5 gives the main conclusions of this research as well as avenues for further study.

2. The Quadratic Assignment Problem

The QAP may be stated as:

Given n locations and n facilities, find an assignment of facilities to locations that minimizes the cumulative product of flow between every two facilities and distance between every two locations.

Let

- d_{kl} be the distance between locations k and l

- f_{ij} be the flow between facilities i and j

- $x_{ik} = \begin{cases} 1 & \text{if facility } i \text{ is assigned to location } k \\ 0 & \text{otherwise} \end{cases}$

The QAP can be modelled as:

$$\text{minimize } G(x) = \sum_{i=1}^{n} \sum_{j=1}^{n} f_{ij} \sum_{k=1}^{n} x_{ik} \sum_{l=1}^{n} d_{kl} x_{jl} \tag{1}$$

subject to:

$$\sum_{j=1}^{n} x_{ij} = 1 \quad \forall i \tag{2}$$

$$\sum_{i=1}^{n} x_{ij} = 1 \quad \forall j \tag{3}$$

$$x_{ij} = 0,1 \quad \forall i,j \tag{4}$$

The objective function (1) represents the total sum of product of flow between every two facilities and distance between every two locations. Since it is quadratic and non-convex, a number of local optima exist. Constraints (2) ensure that each facility is assigned to only one location. Constraints (3) force that each location is occupied by only one facility. In (4) the integrality constraints on the decision variables are imposed. The feasible solutions to (1)-(4) are composed of n! distinct permutations.

This problem can be solved exactly for very small sizes ($n \leq 25$). In real situations, however, a large number of problems lead to QAP instances of considerable size that can not be solved exactly [9]. As a consequence, heuristic approaches have to be designed to address these difficult problems.

3. Self-Organizing Neural Network

The architecture of the net is shown in Figure 1. It is a feedforward neural network with an input layer of n nodes (locations), and an output layer of n nodes (facilities).

The weight connecting input node k and an output node i is given by W_{ik} and represents the continuous relaxation of the decision variable x_{ik} in (1)-(4). Thus, W_{ik} is a measure of the degree of assignment of facility i to location k.

The training set consists of one input pattern for each location. The input pattern corresponding to location k^* is a vector of n components, where the only component different to '0' is the k^*-th component that is equal to '1'.

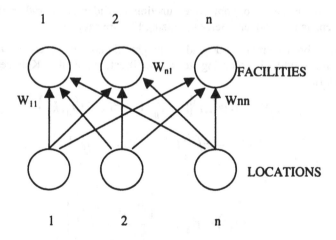

Figure 1. Architecture of the SONN

When the input pattern corresponding to location k^* is presented, for each node i of the output layer, we calculate the potential V_{ik^*} which is a linear combination of the cost of assigning facility i to location k^* and the degree of violation of constraints (3) (i.e., the same location is occupied by different facilities):

$$V_{ik^*} = \sum_j f_{ij} \sum_l d_{k^*l} W_{jl} + \beta \sum_{l \neq i} W_{lk^*} \qquad (5)$$

where β is a penalty parameter. Note that the first term of the potential can be alternatively seen as the partial derivative of $G(W)$ with respect to W.

The competition between the output nodes gives the winning node i_0 which is the facility with the minimum potential:

$$i_0 = \arg \min_i V_{ik^*} \qquad (6)$$

Furthermore, ranking the output nodes according to its potential (closest neighbor to farthest neighbor) as:

$$N(i_0, k^*) = \{i_0, i_1, i_2, \ldots, i_\eta\} \qquad (7)$$

where

$$V_{i_0k^*} \leq V_{i_1k^*} \leq V_{i_2k^*} \leq \ldots \leq V_{i_\eta k^*} \leq \ldots \leq V_{i_nk^*} \qquad (8)$$

and $\eta \geq 1$ is the neighborhood size. It should be noted that this definition of neighborhood is based on objective function considerations, rather than spatial arrangement as in Kohonen's Self-Organizing Feature Map [2].

Once both winning node and its neighborhood have been determined, the weights are modified according to a modification of the Kohonen's weight adaptation rule:

$$\Delta W_{ik^*} = \alpha(i,t)[1 - W_{ik^*}] \quad \forall i \in N(i_0, k^*) \qquad (9)$$

$$\Delta W_{ik^*} = -\beta(t) W_{ik^*} \quad \forall i \notin N(i_0, k^*) \qquad (10)$$

where

$$\alpha(i,t) = \beta(t) \exp\left[\frac{|V_{i_0k^*} - V_{ik^*}|}{|V_{i_0k^*} - V_{i_nk^*}|}\right] \qquad (11)$$

Learning rule (9) means that the degree of assignment of location k^* to every facility in the neighborhood of i_0 is increased. However, the increase is not uniform but depends on the facility potential: the higher the potential, the higher the increase. Learning rule (10) means that degree of assignment of location k^* to the non-neighboring facilities is decreased.

After that, and in order to explicitly enforce the constraints (2), the weights for each output node i are normalized as follows:

$$W_{ik} = \frac{\exp\left(-\left(1-W_{ik}\right)\middle/T\right)}{\sum_k \exp\left(-\left(1-W_{ik}\right)\middle/T\right)} \tag{12}$$

where T is a parameter (referred as temperature), which is lowered as the learning process proceeds. The normalization operation guarantees that when convergence is complete, each facility is assigned to only one location. A similar normalization procedure was used in [10].

During the learning process, the neighborhood size, the magnitude of the weight adaptations, and the temperature are gradually decreased. The complete algorithm follows:

1.   Randomly initialize the weights:

$$0 \le W_{ik} \le 1 \quad \forall i,k \tag{13}$$

2.   Randomly choose a location k^* and present its corresponding input pattern.

3.   Compute the potential V_{ik^*} for each output node i according to (5).

4.   Determine the winning node, i_0, as well as its neighboring nodes according to (6)-(8).

5.   Update weights, W_{ik^*}, connecting input node k^* with every output node i according to (9)-(10).

The updated weights are:

$$W_{ik^*} \leftarrow W_{ik^*} + \Delta W_{ik^*} \tag{14}$$

6.   Normalize weights to enforce the constraints (2) according to (12).

7.   Repeat from Step 2 until all locations have been selected as input patterns. This is a training epoch t. Repeat from Step 2 until $\left|\Delta W_{ik}\right| \approx 0 \; \forall i,k$.

8.   Anneal T to encourage integrality constraints (4). Decrease the neighborhood size η, and $\beta(t)$.

In the algorithm described above, constraints (2) and (3) are enforced in different ways. Thus, constraints (2) are encouraged by using neuron normalization, which converges to 0-1 solutions as the learning process proceeds. On the contrary, constraints (3) are enforced by a weighted penalty function (second term of the definition of potential).

Attempts to search out the correct penalty parameter β for the QAP by trial and error are computationally expensive. In this paper, a flexible way to automating the choice of the penalty parameter, which is based on a conscience mechanism, has been designed:

$$\beta(i) = \frac{G(W)}{n} \, win(i) \tag{15}$$

where $win(i)$ is the number of wins for the output node i on the current pass through the input patterns (locations).

The idea behind (15) is to penalize more those output nodes which are winning more times than the others. The inclusion of $\beta(i)$ in the second term of the definition of potential (5) is used to help enforce that each input pattern (location) is claimed by a unique output node (facility) and helps to prevent instability and oscillation.

4. Computational Results

The new neural approach described in the previous section was implemented in C. The computational experiments were performed on a Intel Pentium 166 Mhz PC. The problem instances given by Nugent et al [11] were used. The problem size ranges from 5 to 30 locations (or facilities). Both data sets and best knwon solutions can be obtained from the QAPLIB [12]. For the 8 problem instances except that of size equal to 30, the best known solutions are optimal solutions.

Due to the randomness in the order of presentations of the training patterns, it is appropiate to solve each problem more than once. In this line, each problem has been solved 5 times. Table 1 shows both best and worst solutions obtained by using the new neural approach for each one of the 8 problem instances. For each case, deviations to the best known solutions are provided. Furthermore, it should be said that for the 40 runs of the algorithm (5 runs * problem instances), feasible solutions were always obtained.

Table 1. Results of experiments: objective function

Problem instance		Best solution		Worst solution	
Size (n)	Best known obj. Function value	Obj. Funct. value achieved	Deviation (%)	Obj. Funct. value achieved	Deviation (%)
5	50	50	0	50	0
6	86	86	0	86	0
7	148	148	0	148	0
8	214	214	0	214	0
12	578	590	2.08	596	3.2
15	1150	1190	3.48	1204	4.7
20	2570	2670	3.89	2678	4.2
30	6128	6335	3.38	6367	3.9

These results seem to indicate that the new neural approach is effective in its ability to find feasible solutions. Also, the performance of the neural approach in terms of solution quality remains strong for the different problem instances. Another interesting result is that both best and worst solutions are quite similar, which shows the robustness of the algorithm under different runs of a problem instance.

Table 2 shows the CPU requirements for both best and worst solutions corresponding to each one of the eight problem instances. Note that, as a consequence of these small CPU times, the new SONN can be applied when near-optimal solutions are required.

Table 2. Results of experiments: CPU requirements

Problem instance	Best solution	Worst solution
Size (n)	CPU time (seconds)	CPU time (seconds)
5	0.95	1.09
6	1.21	1.28
7	2.54	2.25
8	2.65	3.01
12	4.21	4.53
15	7.55	8.27
20	8.23	7.95
30	10.01	12.78

5. Conclusions

In this paper, we have proposed a new neural approach for the Quadratic Assignment Problem. The net is an unsupervised Self-Organizing Neural Network,which uses neuron normalization and a penalty term to enforce feasibility in a adaptive way. In order to compute the penalty term, a conscience mechanism has been introduced.

We believe this neural approach presents attractive characteristics:

- Constraints are enforced within the learning process, i.e. without any kind of aditional tuning or second stage [4].

- Good feasible solutions are consistently found.

- This new neural approach has been designed as an extension of an existing hybrid neural approach and is opened to extension to other neural features.

Further experimentation will address issues susch as performance on larger problem instances, and parameter tuning and sensitivity analysis. Another possible extension is the inclusion of chaotic elements within the algorithm.

References

1. Hopfield J J, Tank D W 1985 "Neural" computation of decisions in optimization problems. *Biological Cybernetics* 52: 141-152

2. Kohonen T 1982 Self-organized formation of topologically correct feature maps. *Biological Cybernetics* 43: 59-69

3. Smith K 1999 Neural Networks for Combinatorial Optimisation: A review of more than a decade of research. *INFORMS Journal on Computing* 11: 15-34

4. Smith K, Palaniswami M, Krishnamoorthy M 1996 A hybrid neural approach to combinatorial optimization. *Computers and Operations Research* 23: 597-610

5. Smith K, Palaniswami M, Krishnamoorthy M 1998 Neural Techniques for Combinatorial Optimisation with applications. *IEEE Transactions on Neural Networks* 9: 1301-1318

6. Smith K, Palaniswami M 1997 Static and Dynamic Channel Assignment using Neural Networks. *IEEE Journal on Selected Areas in Communications* 15: 238-249

7. Lozano S, Guerrero F, Eguia I, Canca D, Smith K Cell formation using two neural networks in series. In: Dagli CH et al (eds) 1998 *Intelligent Engineering Systems Through Artificial Neural Networks Vol 8.* ASME Press, pp 341-346

8. Burke L I, Damany P 1992 The Guilty Net for the Traveling Salesman Problem. *Computers and Operations Research* 19: 255-265

9. Taillard E D 1991 Robust taboo search for the quadratic assignment problem. *Parallel Computing* 17: 443-455

10. Van de Bout D E, Miller III T K 1989 Improving the performance of the Hopfield-Tank neural network through normalization and annealing. *Biological Cybernetics* 62: 129-139

11. Nugent C E, Vollmann T E, Ruml J 1968 An experimental comparison of techniques for the assignment of facilities to locations. *Operations Research* 16: 150-173

12. Burkard R E, Karisch S E, Rendl F 1991 QAPLIB – A quadratic assignment problem library. *European Journal of Operational Research* 55: 115-119

Building Maps of Workspace for Autonomous Mobile Robots Using Self-Organizing Neural Networks

Katsuhiro Hori[1], Yukio Hashimoto[1], Shouyu Wang[2] and Takeshi Tsuchiya[3]

[1] Muroran Institute of Technology, Muroran 050-8585, Japan
[2] Kochi University of Technology, Kochi 782-8502, Japan
[3] Hokkaido University, Sapporo 060-0813, Japan

Abstract: This paper presents a new method of building maps of workspace for autonomous mobile robots using self-organizing neural networks. By this method, the topological maps of the workspace can be self-organized from the relative distance data between a robot and walls on the workspace only using ultrasonic distance sensors. However, when the shape of the workspace is complicated, an unsuitable map with dead nodes or dead links may be generated. In this paper, we consider the cause of the problem, and we propose a new building maps algorithm which consists of two learning stages.

1 Introduction

Building maps of workspace is an important task for workspace recognition of autonomous mobile robots. The method of building maps by learning using neural network has been proposed [1]-[3]. By this method, the topological maps of the workspace can be self-organized from the relative distance data between robot and wall on the workspace The relative distance data are collected only using ultrasonic distance sensors.

This method uses self-organizing neural networks called self-organizing feature map [4] or self-created and organizing neural network [5] to learn maps of workspace. The inputs of the neural networks are the relative distance data between robot and wall at many places of the workspace. After a sufficient learning, a topological map of the workspace can be built on the self-organizing layer of these neural networks The topological map consists of nodes and links. The nodes on the map are the representative positions of the workspace and the links on the map are the relations of the representative positions. Thus, the topological map can roughly express the workspace.

However, when the shape of the workspace is complicated, some dead nodes and dead links may be generated on the maps. In this paper, in order to cope with the problem, we consider the cause of the problem, and we propose a new building maps algorithm which consists of two learning stages: the nodes learning and the links learning.

2 Conditions of Robots and Workspace

In this study, the following three conditions are assumed about the workspace and the autonomous mobile robot. (1) The workspace is a closed space with obstacles. (2) The robot has no information about the workspace beforehand. (3) The robot has two or more ultrasonic distance sensors arranged in the uniform direction as in Figure 1.

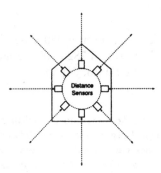

Figure 1: Autonomous mobile robot.

And it is specified that this autonomous mobile robot behaves in the following procedure shown in Figure 2.

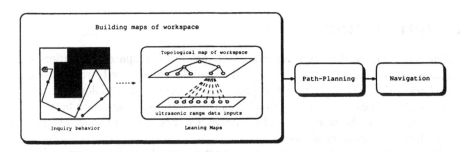

Figure 2: Behavior procedure of autonomous mobile robot.

At the first step, the robot behaves in inquiry on the unknown workspace, and collects the relative distance data between the robot and the wall at many places on the workspace for every fixed distance moving. This inquiry behavior consists of the combination of the straight-line moving and the random direction conversion carried out when the distance value from the wall becomes smaller than a fixed value. By inputting the collected relative distance data to a self-creating and organizing neural network and computing the learning algorithm described in the following chapter, a topological map which consists of nodes and links of the workspace can be generated on the self-organizing layer of the network. At the second step, the robot work out path-planning between the given destination with its present position using the map. At the third step, the given task (that is, moving to the destination) is achieved by moving along the planned path.

In this paper, we mainly discuss the first step: building maps of workspace.

3 Self-Organizing Neural Networks

3.1 Self-Creating and Organizing Neural Network (SCONN)

In order to implement the maps of the workspace, we use the self-creating and organizing neural network(SCONN) with the structure shown in Figure 3.

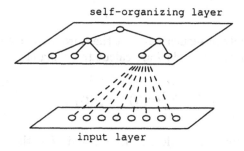

Figure 3: The structure of self-creating and organizing network (SCONN).

This SCONN consists of two layers. One layer is an input layer and another is a self-organizing layer.

Each neuron on the input layer and on the self-organizing layer has joined mutually. The learning of this network is carried out based on the competitive learning algorithm. Therefore, a neuron on the self-organizing layer which has the weight vector with the minimum distance from the input vector is selected as the winner neuron. At the initial state, there is only one neuron on the self-organizing layer. And the neurons on the network is self-created and organized, according to the change of the feature of input vectors, and then the topological map with a tree structure is generated on the self-organizing layer. Since the neurons on the self-organizing layer express the nodes on the topological map, 'neuron' and 'node' are equivalent. Therefore, we describe uniformly 'neuron' as 'node' after this.

3.2 Learning Algorithm

Figure 4 shows a block diagram of the learning algorithm for the SCONN, and the detailed steps of the learning algorithm are as follows:

At the first step, there is only one node on the self-organizing layer with small random weight at the primitive stage and its activation level is set large enough to respond to any input stimuli. At the second step, new input vector

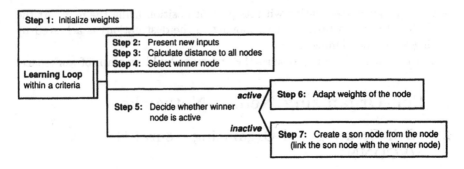

Figure 4: A block diagram of the learning algorithm for the SCONN.

is presented randomly or sequentially. At the third step, distances d_j between the input and each output node j are calculated using (1).

$$d_j{}^2 = \sum_{i=1}^{N} \{x_i(t) - w_{ij}(t)\}^2 \tag{1}$$

where $x_i(t)$ is the input to node i at time t and N is the demension of the input and $w_{ij}(t)$ is the weight from input node i to output node j at time t. At the fourth step, an output node with the minimum distance is selected as the winner node. At the fifth step, it is decided using (2) whether the winner node is active or inactive.

$$y_{wj} = \begin{cases} \text{is active,} & \text{if } d_{wj} < \theta(t) \\ \text{is inactive,} & \text{otherwise} \end{cases} \tag{2}$$

where y_{wj} is the output of the winner node, d_{wj} is the distance between the inputs and the winner node, and $\theta(t)$ is an activation level that is sufficiently wide at a primitive stage and decreases with time. In this study, we use (3) as the activation level.

$$\theta(t) = c_1 \exp(-c_2 t) + c_3 \tag{3}$$

where c_1, c_2, c_3 are constant. At the sixth step, the weights of an active winner node is adapted using (4).

$$w_{i,wj}(t+1) = w_{i,wj}(t) + \alpha(t)\{x_i(t) - w_{i,wj}(t)\} \tag{4}$$

where $w_{i,wj}(t)$ are the weights from the inputs to an active winner node and $\alpha(t)$ is the gain term that can be constant or decrease with time. At the seventh step, a son node is created from a mother node (an inactive winner node) using (5) and (6), and the son node is linked the mother node.

$$sj \ = \ sj + 1 \tag{5}$$

$$w_{i,sj}(t+1) \ = \ w_{i,sj}(t)$$
$$+ \ \beta(t)\{x_i(t) - w_{i,wj}(t)\} \tag{6}$$

where sj is the current number of total output nodes, $w_{i,sj}(t)$ are the weights from the inputs to a son node created from a mother node, and $\beta(t)$ is the resemblance factor that varies from 0 to 1.

In this algorithm, there can be three criteria to stop the program. Those criteria are iterations t, number of output nodes sj and activation level $\theta(t)$.

4 Building Maps of Workspace Using Self-Organizing Neural Networks

There are some problems on the maps of workspace built using directly the above-mentioned learning algorithm of SCONN. In this section, through a simulation case study, we consider the problems and propose a new improved learning algorithm.

4.1 Simulation Case Study

We use the workspace shown in Figure 5 as the unknown workspace in this simulation case study. The robot behaves in inquiry on the unknown workspace and measures the relative distance data from walls as shown in Figure 6, and then the topological map of the workspace is built by the learning algorithm of SCONN.

Figure 7 shows an example of simulation results of an autonomous mobile robot with eight sonar sensors($N = 8$) by the conventional learning algorithm of SCONN. This map is projected the topological map on the real (2D) space. Simulation parameters are $\alpha = 0.085, \beta = 0.85, c_1 = 450000, c_2 = 0.001, c_3 = 50000$. It is confirmed by Figure 7 that the workspace has been divided roughly.

However, there are two problems on this map. The first problem is that the links have the thin tree structure and there is no link between the nodes that can actually move on the workspace, that is, there are some dead nodes on the map. The second problem is that there are some dead links that can not actually move on the workspace.

4.2 Discussion about the Problems

In the first place, we try to solve the former problem that there are few links on the map by changing the structure of the map into graph structure from tree structure. And if the distance between the inputs and the winner node is smaller than the activation level (that is, if $d_{wj} < \theta(t)$), we not only adapt the

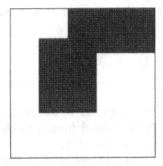

Figure 5: The shape of the unknown workspace used in the simulation case study.

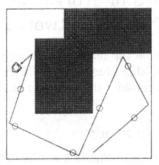

Figure 6: Inquiry behavior of autonomous mobile robot.

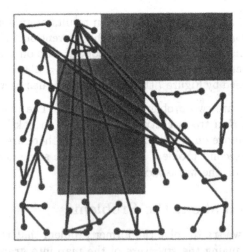

Figure 7: The map generated by the conventional learning algorithm of SCONN.

weights of the network but link to the winner node and the previous winner node.

The total number of links could increase by these improvements, but number of dead links also increase on the other hand. A simulation result by this improved learning algorithm is shown in Figure 8. Therefore, the algorithm which decreases the number of dead links is needed.

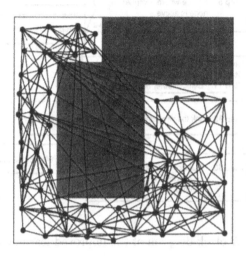

Figure 8: The map with graph structure.

Then, considering a generation factor of a dead link, we found out that the link which generated at the beginning of learning becomes a dead link in many cases. The basic function of the SCONN is generating representative nodes from input vectors and determining weights of the representative nodes. As the weights is not stabilized at the initial learning stage, the links which is generated at the time will extend by movement of the nodes and tend to grow into the links over an obstacle of the workspace as shown in Figure 9.

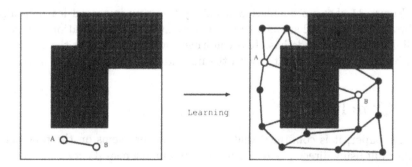

Figure 9: A generation factor of a dead link.

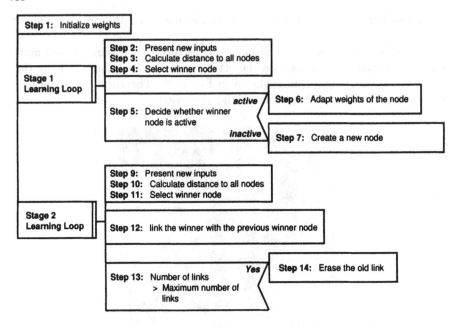

Figure 10: A block diagram of the improved learning algorithm.

4.3   Improved Learning Algorithm

Based on the above-mentioned consideration, we propose a new improved algorithm as follows. In this algorithm, at the beginning of leaning, only determination of representative nodes of the workspace (so-called vector quantization) is carried out. That is, at the first leaning stage, the movement of the representative nodes can converge and the rough division of the workspace can be done. And after this, at the second leaning stage, we generate the links on the map in the same way of the above-mentioned method. A block diagram of this new improved learning algorithm is shown in Figure 10.

Figure 11 shows a simulation result by this new improved learning algorithm. And Figure 12 shows a simulation result with the limitation of number of links which one node has. It is confirmed by Figure 12 that the dead links can be completely eliminated with the limitation of the number of links.

5   Conclusions

In this paper, it is confirmed that there are two problems on the topological maps built using directly self-creating and organizing neural network, when the shape of the workspace is complicated. In order to cope with these problems, we propose a new learning algorithm divided into the two learning stages.

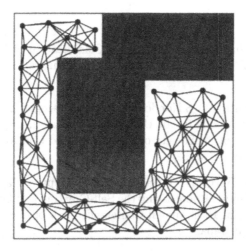

Figure 11: The map generated by the improved learning algorithm.

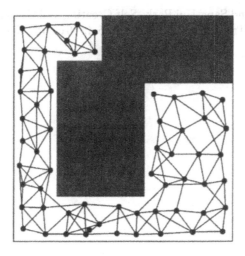

Figure 12: The map generated by the improved learning algorithm with the limitation of the number of links which one node has.

At the first stage, only representative nodes are learned, and at the second stage, links between the nodes are generated. As a result, the proper maps are built in these two stages. This algorithm was tested by the simulation for an autonomous mobile robot with eight ultrasonic distance sensors, and it was demonstrated that the algorithm is useful for the purpose.

References

[1] A.Kurz: Building Maps on a Learned Classification of Ultrasonic Range Data, Proc. IFAC Intelligent Autonomous Vehicles, Southampton, UK, 191/196, 1993.

[2] T.Ohisi, K.Furuta and S.Kondo: Workspace Recognition and Navigation of Autonomous Mobile Robot Using Self-Creating and Organizing Neural Network: Trans. of the Society of Instrument and Control Engineers, 33-3, 203/208, 1997(in Japanese).

[3] S.Takahashi, K.Hori, and Y.Hashimoto: Workspace Map Generation for Autonomous Mobile Robot Using the Self-Creating and Organizing Neural Network: Proc. of 16th Annual Conference of Robotics Society of Japan, 293/294, 1998(in Japanese).

[4] T.Kohonen: Self-Organization and Associative Memory, Springer-Verlag, 1984.

[5] Doo-II Choi and Sang-Hui Park: Self-Creating and Organizing Neural Networks, IEEE Trans. Neural Networks, 5-4, 561/575, 1994.

Neural Network Parameter Estimation and Dimensionality Reduction in Power System Voltage Stability Assessment

Sami Repo

Tampere university of technology, PO box 692, 33101 Tampere, Finland, sami.repo@cc.tut.fi

Abstract: The need for electric power system real-time security assessment has increased due to open system, increase in the number of power wheeling transactions and environmental concerns. In the paper special attention is focused on neural network parameter estimation and dimensionality reduction in order to improve neural network generalisation in large scale problem. The results prove the capability of neural networks to model the most critical voltage stability margin in a electric power system. The proposed approach is tested with IEEE 118-bus test network. The generalisation and training time of a neural network model can be improved significantly using proposed methods.

1. Introduction

Operators of an electric power system must define how much power can be safely transferred across the system. The security of a power system must be guaranteed in all conditions continuously. Management of power systems is becoming more difficult due to (i) increased power transfers in the open electricity markets, (ii) environmental constraints which restrict the expansion of the transmission network and power generation near load centres and (iii) requirements to provide the same services with existing power systems at lower costs. To guarantee power system security in all conditions, a number of contingencies (line or generator tripping) are studied in typical operation points. The number of different contingencies and operation points cannot be large, due to the massive computation burden, and that is why the computation of a power system security assessment is done in off-line.

Power systems are used in a conservative manner because the operation point is changing continuously and it may not correspond to points used in the off-line computation. This leads to a situation where power transmission could be increased in some situations without fear of insecurity if more accurate and up-to-date transmission limits were used. The interest in real-time security assessment has increased due to willingness to use the power system maximum capability in all conditions.

This paper presents a real-time voltage stability assessment approach based on neural networks (NNs). It is similar to earlier studies [1,2,3], except that NN generalisation capability has improved and NN training has been made faster. Feature extraction methods are used to compress the information presented for NN. This is very important in power system applications, because measurements are very redundant. Both NN training time and generalisation can be improved in this way. NN parameter estimation (training) is also very critical for generalisation. NN can memorise the training data if the number of parameters is too large or training data is presented too many times. The choice of NN training algorithm will affect the behaviour of the training process.

2. Voltage Stability Assessment

2.1 Voltage Stability

A power system at a current operation point is voltage stable if, following any disturbance, voltage near loads are close to the pre-disturbance values. Voltage instability is the absence of voltage stability and results in uncontrollable voltage decrease. Voltage stability is also called load stability due to power system limited capability to transmit power to the loads. According to the maximum power transfer theorem [4], operation point has a maximum load demand which can be delivered without instability. The main factor causing voltage instability is the inability of the power system to meet the demand for reactive power in heavily stressed systems. Other factors contributing to voltage stability are the generator reactive power limits, load characteristics, characteristics of reactive compensation devices and the action of voltage control devices.

Voltage stability can be studied by dynamic simulations or by using static voltage collapse point calculation methods [4]. Load flow based static analysis methods are often useful for fast approximate analysis of long-term voltage stability. Load flow equations diverge at an unstable operation point because equations have no solution. Special voltage collapse point calculation methods should be used to avoid numerical problems if accurate results are needed.

The power system should be operated with a sufficient voltage stability margin. The voltage stability margin is determined as a difference between the operation and voltage collapse point according to a key parameter (loading, line flows etc.). The security of the power system is determined by a post-contingency margin which needs knowledge of the most critical contingency (Figure 1). The computation of voltage stability assessment is very time consuming because many contingencies and post-contingency margins should be studied. The pre-contingency margin describes power system loadability and is not very interesting from the security point of view.

2.2 Neural Network Based Voltage Stability Assessment

The NN based approach is created for real-time voltage stability assessment for a transmission grid control centre. In this approach NN inputs are active and reactive line flows and bus voltages in a normal operation point. These are measured commonly in most power systems from almost all lines and buses. The output is the voltage stability margin for the most critical contingency. Inputs will be transmitted via SCADA system and the application will be part of the power system management program.

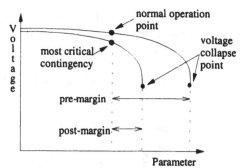

Fig. 1. Voltage stability margins.

Data needed for NN training is calculated with a power system load flow program [5]. There are two reasons why calculated data is used in this paper: formulation of good quality data from measured values for NN training may be difficult and a suitable power system model is, however, needed to calculate the voltage stability margins. Normal operation points are calculated by varying loads and production. The voltage stability margin is calculated by increasing the key parameter in the chosen direction in a post-contingency situation. The calculation of voltage collapse point is done by repeating load flows step by step in the calculation direction for chosen contingencies [3,6].

Problems in large scale power system applications are related to NN model generalisation. These problems are due to a large number of different operation points and variables. Operation points should include all possible network topology, production and loading situations. Especially the topology of the transmission network in the normal operation point will strongly affect the value of the post-contingency voltage stability margin. Network topologies are divided into topology classes. Each topology class includes similar network topologies according to behaviour and state of power system normal operation point. Each topology class will have its own NN model in order to improve model generalisation capability. NN model training will also be faster than training of one huge NN model. Only the most interesting and probable network topologies are included in topology classes to reduce the data calculation burden.

The number of variables can be extremely high in power system applications. Voltage stability is a system wide problem which cannot be modelled only with a part of the power system. It is extremely difficult to determine which variables are the best ones for NN models. The selection of NN inputs can be made with a knowledge of the voltage stability problem. The extraction of NN inputs is then done by classifying or transforming selected inputs. Furthermore, each NN (i.e. topology class) may have different inputs. When the number of NN inputs is reduced, the NN model training can be made faster and the generalisation capability can be improved.

Figure 2 presents the scheme of a real-time NN based voltage stability assessment approach. In real-time use, data is supplied through a state estimator (part of the power system management program) in order to minimise the number of missing and bad measurements. The number of NN inputs is reduced in a pre-processing stage. It is also very important to remember the "data boundaries", because the NN model has not the capability to extrapolate. "Data boundaries" are mainly characterised by power system topology. If the training data includes cases only from a healthy network, it cannot model voltage stability margin during important transmission line outage. The transmission line and generator outage list is used to decide which NN model is most appropriate for the current operation point.

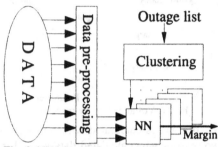

Fig. 2. NN based voltage stability assessment.

3. Neural Network Parameter Estimation and Generalisation

The multilayer perceptron (MLP) NN is the best and most widely used NN for function approximation tasks for large problems [7]. MLP NN parameter estimation is an unconstrained non-linear optimisation problem.

3.1 Parameter Estimation

The standard back-propagation training algorithm is based on the gradient descent (GD) method. The convergence of GD is very slow due to its tendency to zigzag on the flat error surface [7,8]. A momentum term allows GD to respond not only to the local gradient, but also to recent trends in the error surface [8]. It allows small features in the error surface to be ignored and prevents getting stuck in the local minimum. The performance of GD is also very sensitive to the setting of the learning rate. It may become unstable if the learning rate is too large and might take an enormously long time to converge if the learning rate is set at a very small value. The performance of the GD algorithm can be improved with an adaptive learning rate which tries to keep the learning rate at a maximum while keeping learning stable [8].

The slow convergence of GD can be improved with a resilient back-propagation (RPROP) algorithm [8]. RPROP uses only the sign of the derivative to determine the direction of parameter update. If the parameter continues to change in the same direction for several iterations, then the size of parameter change will be increased. Whenever the parameters are oscillating, the size of parameter change will be reduced. RPROP accelerates training in flat regions of the error function, although it seems a very imprecise method.

Although the error function decreases most rapidly in the direction of a gradient, this does not ensure the fastest convergence. Second order optimisation algorithms use the gradient and the Hessian matrix of error function to find the optimum solution. Second order methods are in principle no more likely to find a global minimum than are first order methods. However, their behaviour is better than that of first order methods.

The conjugate gradient (CG) algorithms are usually much faster than GD with a momentum term and adaptive learning rate, but are sometimes slower than RPROP [8]. In CG a search is made along conjugate gradient direction to determine step size which will minimise the performance function along that line. Scaled conjugate gradient (SCG) reduce NN training time compared to other CG methods because it can avoid the time consuming line search. The CG method needs only a little more computer memory than GD or RPROP. These are used for NN parameter estimation when there is a large number of parameters. The CG methods avoid completely the estimation and storage of the Hessian matrix [9].

Newton's method is an alternative to the CG methods for fast optimisation [9]. Its convergence is faster than CG methods. Unfortunately it is expensive to compute the Hessian matrix for MLP NN and there is no guarantee that the Hessian matrix of MLP NN is always non-singular. Quasi-Newton (QN) methods do not require calculation of second derivatives and a approximate Hessian matrix is always non-singular. QN methods are like Newton's method with line search, except that Hessian matrix is approximated by a symmetric positive definite matrix which is updated at each iteration. Broyden, Fletcher, Goldfarb and Shanno (BFGS) QN update has been most successful in published studies [10]. The BFGS needs more computation in each iteration and more computer memory than CG methods, but converges usually faster [8]. The additional computer memory is needed to store an approximate Hessian matrix.

The Newton's method becomes the Gauss-Newton (GN) method when the objective function is a sum of squares (non-linear least squares problem) and the approximate Hessian matrix is linear approximation of the residuals [9]. In the GN method using the information required to determine the first derivative vector, it is possible to approximate the second derivative matrix. The convergence of GN method can be more rapid than the convergence of QN methods. One negative aspect of the GN method is so-called large residual problem. In this problem GN method can fail or can converge slowly. This problem can be avoided in Levenberg-Marquardt (LM) method which is a restricted step algorithm [9,10]. The difficulty caused by non-positive definite Hessian matrix in Newton's method can also be avoided in restricted step algorithms. In non-linear least squares problem the LM method use the L_2 in norm. The LM method is the standard method for non-linear least squares problems due to its fast convergence and robustness [10]. The LM method starts with GD method with small step size. The LM method becomes more like the Newton's method using approximate Hessian matrix near an error minimum due to its faster convergence and better accuracy. The LM algorithm is the fastest method for NN parameter estimation for moderate size NNs, especially when accurate results are needed [8]. The main disadvantage of the LM method is the quite large computer memory requirement.

3.2 Generalisation

The NN is said to generalise when the function between input and output is computed "correctly" for the test data. The test data is not used in NN training or in NN design. It should, however, belong to the same population as training and validation data. When NN learns too many input-output relations or the number of NN parameters is too great, the NN may memorise the training data and therefore be less able to generalise [7]. The best way to avoid the overfitting problem is to use ample training data and limit the number of parameters as strictly as possible. Use of inadequate amount of training data will lead to an ill-conditioned parameter estimation problem. The overfitting problem appears when the training error is small but the test error is large. The choice of NN architecture and training algorithm affect for the overfitting problem.

The NN parameter estimation performance function can be changed by adding a regularisation term. It is a function which have a large value for smooth mapping functions (simple models) and a small value otherwise (complex models). Thus the regularisation encourage less complex models i.e. it is used to improve the generalisation [7,8]. If a term that consists of the mean square error (MSE) of NN parameters (weight decay) is added to the original performance function, some of the parameters will have smaller values. In this case the NN response will be less likely to overfit.

From the whole set of available data for training, some part is put aside for the purpose of validation and not used in training. In the early stopping method the generalisation capability of NN (validation error) is calculated during the training process after each training iteration [8]. When the NN begins to overfit, the validation error will rise although the training error decrease. If the validation error continues rising for a specified number of iterations, the training process is stopped and the NN parameters at the minimum validation error point are returned. In this method NN is usually overparameterized. Despite of that NN can generalise good enough.

The trial and error method is usually used in the NN design, but more advantageous methods like cascade correlation and NN pruning may also be used [7]. The goal of the NN design is to find the best NN architecture. A NN model with a lot of parameters generally requires a lot of training data or else it will memorise well. On the other hand, a NN model with too few parameters compared with the size and the complexity of the training data may not have the capability to generalise at all. The trial and error method is limited to test purposes because the number of studied architectures cannot be large. However, it produce usually sufficiently good results but does not guarantee the best architecture.

4. Dimensionality Reduction and Feature Extraction

The performance of the NN with respect to the dimension of the pattern space have a peak and increasing the dimension further may decreases it. The phenomena of performance decreasing as dimensionality of pattern space increases is known as the course of dimensionality. The process of extracting the useful information and translating the pattern into a vector is called feature extraction. Usually the raw data is not feed directly into the NN but rather processed before. The most important forms of preprocessing are dimensionality reduction and feature extraction. These are usually used together by dropping some vector components automatically and keeping more "feature rich" vector components. Dimensionality reduction and feature extraction methods will always lead to some loss of information [11]. The improvement in NN model accuracy may overcome the information loss if the right balance is found between the information loss and the NN processing capability [7].

Power system measurements are very redundant and the number of measurements is very large. Redundant variables do not bring new information into the model, because the same information is got by many variables. When the NN contains redundant variables, there may be linear dependency among the columns of the NN training algorithm's Jacobian matrix and hence singularity of that matrix. The information given by redundant variables should be compressed to improve NN generalisation and to reduce NN training time. The data presented to the NN model should include a minimum amount of information which guarantees good generalisation.

4.1 Principal Component Analysis

The total system variability can be presented by a smaller number of principal components, because there is almost as much information in the first p principal components as there is in the original variables [11,12]. Principal components are linear combinations of original variables. They represent a new co-ordinate system where the new axes represent the directions with maximum variability. The principal component analysis (PCA) is based on input data X covariance matrix Σ eigenvalue decomposition. The covariance matrix can be represented by eigenvalues λ_i and eigenvectors $\underline{p}_i$. The number of original variables is m.

$$\Sigma = X^T X = \lambda_1 \underline{p}_1 \underline{p}_1^T + ... + \lambda_n \underline{p}_m \underline{p}_m^T \tag{1}$$

The columns of PCA transformation matrix P are eigenvectors $\underline{p}_i$. The new transformed input matrix T, variances of new variables and covariances of new variables are presented in (2). The new transformed input matrix can be calculated by multiplying the original input matrix by the reduced transformation matrix P_{red} (p is the number of chosen principal components, $p<m$). The variances of new variables are the eigenvalues of the input data covariance matrix. The most important new variables are those whose variances, i.e. covariance matrix eigenvalues, are greatest. The first p principal components explain the variability of original input data almost totally. In addition, the new variables do not correlate with each other, because they are arranged in an orthogonal base.

$$T = XP_{red} = X\left[\underline{p}_1^T, \underline{p}_2^T, ..., \underline{p}_p^T\right]$$

$$\text{var}(\underline{t}_i) = \underline{p}_i^T \Sigma \underline{p}_i = \lambda_i \tag{2}$$

$$\text{cov}(\underline{t}_i) = \underline{p}_i^T \Sigma \underline{p}_j = 0, \quad i \neq j$$

The PCA is very good at data compression while maintaining as much information on the original variables as possible. However, it should be used very carefully in real-life approaches, because outliers with high noise level can make it useless. One possible solution for this problem is to feed input data through a filtering function or state estimator before PCA. Sometimes linear transformation should not be used, because essential information about input data may be lost. Then non-linear transformation should be tried [12].

4.2 K-means Clustering

The K-means clustering is based on the concept of input vector classification by distance functions. K-means clustering is well known, because it is very simple and it can manage large data sets [11]. It is based on the minimisation of the sum of squared distances from all points in a cluster domain to the cluster centre. It has the following steps:

1. Choose K initial cluster centres $z_1(1)$, $z_2(1)$, ..., $z_K(1)$. These are usually the first K samples of the data set.
2. At the kth iteration distribute the input vectors x among the K cluster domains, using the relation

$$x \in S_j(k) \quad if \; \|x - z_j(k)\| < \|x - z_i(k)\|,$$
$$i = 1,2,\dots K, i \neq j \tag{3}$$

   where $S_j(k)$ denotes the set of input variables whose cluster centre is $z_j(k)$.
3. Compute the new cluster centres $z_j(k+1)$ such that the sum of the squared distances from all points in $S_j(k)$ to the new cluster centre is minimised. This is simply the input variables' mean of $S_j(k)$. Therefore, the new cluster centre is

$$z_j(k+1) = \frac{1}{m_j} \sum_{x \in S_j(k)} x, \; j = 1,2,\dots K \tag{4}$$

   where m_j is the number of input variables in $S_j(k)$.
4. If $z_j(k+1) = z_j(k)$ for $j=1,2,\dots K$, the algorithm has converged and should be ended. All input variables have found their clusters and they will not change from one cluster to other. Otherwise go to 2.

The new transformed input matrix is formed from cluster centre mean vectors $z_j(k+1)$. In this way the number of new input variables is K. The behaviour of the K-means clustering is influenced by the number of cluster centres, the choice of initial centres, the order in which the input vectors are taken and properties of the data [11]. The algorithm will produce acceptable results when the data contain characteristic regions which are relatively far from each other. In practice, the use of K-means clustering will require experimenting with various number of clusters.

5. Test Results

The test results were calculated with IEEE 118-bus test network. The contingency list included four contingencies. The number of training, validation and test cases was 4370, 772 and 100, respectively. Modelling was done with MLP NN. Two feature selection methods and five NN parameter estimation algorithms [8,13] were studied to find the best combination. Each algorithm was studied with a different number of hidden layer nodes with the trial and error method. Furthermore, parameters were initialised five times to attain as good a local minimum as possible. The best combination was then further tested with early-stopping, regularisation and pruning methods to improve NN generalisation capability. Studies were made with PC (Pentium 90 MHz and 64 Mb ROM) and Matlab.

5.1 Testing of PCA

The number of NN model inputs were reduced from 490 to 10 with PCA. The first principal component explains 50 % of variability of all variables. The criterion for the number of principal components was that they should explain 90 % of total variability. The results of the parameter estimation algorithm study are presented in Figure 3a. LM is the best algorithm according to validation error. BFGS is almost as good as LM. Other algorithms are in the following order: SCG, RPROP and GD with adaptive learning rate and momentum term. The output error and the number of parameters can be reduced to a very low level with PCA and LM. The minimum NN model output error was reached with the LM algorithm and with nine hidden layer nodes. Figure 3b presents training times for all algorithms with respect to number of hidden layer nodes. The number of iterations used for different algorithms was: LM 100, BFGS, SCG and RPROP 1000 and GD 10000 iterations. The LM algorithm is again the best one. The LM algorithm is obviously the best choice when the number of NN parameters is not great.

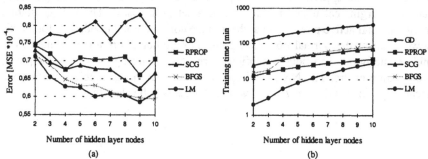

Fig. 3. PCA: (a) Validation errors and (b) training times with different training algorithms.

5.2 Testing of K-means

The selection of the number of NN inputs is not so straightforward with the K-means method as with PCA. In this study several numbers of cluster centres were tested with the LM training algorithm to choose the number of cluster centres for each measurement group. The number of cluster centres for active line flow, reactive line flow and voltage were 10, 15 and 15, respectively. The results of the NN training algorithm study are presented in Figure 4a. The LM training algorithm is again the best one and BFGS is almost as good as LM. The performance of SCG, RPROP and GD is reduced significantly compared with the PCA study. The minimum NN output error (LM with eight hidden layer nodes) is less than with PCA. However, the total number of NN parameters is much greater with K-means than with PCA, because the number of inputs is greater. Figure 4b presents NN training times for all algorithms with respect to the number of hidden layer nodes. The training time of all training algorithms is longer than that with PCA. This is due to the greater number of NN parameters. When the number of NN parameters is increased, the time needed for LM training increases very rapidly. According to the training time requirement, the BFGS training algorithm is much better than LM. Unfortunately, fast training algorithms like RPROP and SCG do not work well enough in parameter estimation.

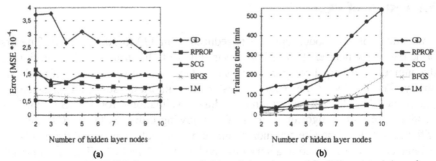

Fig. 4. K-means: (a) Validation errors and (b) training times with different training algorithms.

5.3 Generalisation

NN model (PCA and nine hidden layer nodes) generalisation capability is presented in Figure 5. The uppermost subfigure presents correct (x) and estimated (o) voltage stability margins calculated for the test data set (100 cases). The voltage stability margin is total active power load at the voltage collapse point divided by total active power load at the normal operation point. NN output errors are presented in the middle subfigure. The bottom subfigure presents the histogram of NN output approximation errors. The NN model approximates the voltage stability margin very accurately. The maximum error is less than 0.022, which is less than 2%.

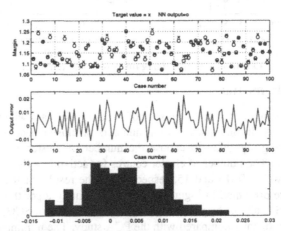

Fig. 5. NN generalisation test.

The early stopping method was tested with PCA and different NN training algorithms. It was found that generalisation did not improve, but the number of iterations was reduced in all cases. The generalisation capability of the GD training algorithm was reduced, because training was stopped too early. The generalisation capability of other training algorithms was about the same as without early stopping. The average number of iterations was 65 (LM), 150 (BFGS), 145 (SCG), 425 (RPROP) and 340 (GD). According to this study early stopping can be used to reduce the NN training time.

Regularisation was tested with PCA and the LM training algorithm. The NN model generalisation capability was only slightly better than that of the best NN model given by the trial and error method. The use of regularisation allows larger NNs to be used without loss of generalisation capability. In this way the NN design time can be reduced, because it is not necessary to study many different kinds of NN structures.

NN pruning was tested with the best NN structure. The starting point of pruning requires a well generalised NN. This means that NN pruning is suitable only for fine tuning of an NN model, not for NN design. NN pruning is very effective in the reduction of NN parameters. Without a significant decrement in the NN generalisation capability, about half of NN parameters could be removed. This may indicate that there are still too many input variables. The NN pruning technique used was optimal brain surgeon.

6. Conclusions

The NN model can approximate the post-contingency voltage stability margin very accurately and in real-time. This approach substitutes for a number of voltage collapse point calculations required to be done in traditional security assessment to ensure secure power system operation. However, this approach is not ready to be put into operation. There is still needed more research in the fields of NN modelling with large power systems, NN model updating when a power system has changed significantly and calculation of a huge number of training cases.

In this paper has been compared two dimensionality reduction and feature extraction methods, five NN training algorithms and three NN generalisation improvement methods. The number of NN parameters must be reduced to avoid problems from redundant variables, to improve NN generalisation and to reduce NN training and design time. When PCA is used in dimensionality reduction and feature extraction, less input variables can be used to guarantee good generalisation than when using the K-means method. The use of PCA is also more straightforward and easier than using the K-means method. The NN generalisation capability was found to be at the same level with both methods. When the size of data is huge, computation of PCA needs special algorithms for eigenvalue decomposition.

The LM training algorithm was found to be the best. When the number of NN parameters increases, the LM training algorithm needs a lot of computer memory and training time will be huge. In that case the BFGS or SCG training algorithm should be used. According to this study, the use of GD is not recommended.

The best way to improve NN generalisation is to limit the number of NN parameters so that NN has not the capability to overfit. The use of early stopping or regularisation reduces the NN design time. If the trial and error method is used in the NN design, the results are about the same as with early stopping or regularisation, but more computation is needed. The use of NN pruning is limited to special cases.

References

[1]   M. La Scala et al, "A neural network-based method for voltage security monitoring," IEEE Trans. Power Systems, Vol.11, No.3, 1996.

[2]   L. Wehenkel, "Contingency severity assessment for voltage security using nonparametric regression techniques," IEEE Trans. Power Systems, Vol.11, No.1, 1996.

[3]   S. Repo, "Real-time transmission capacity calculation in voltage stability limited power system," Bulk power system dynamics and control IV, Greece, 1998.

[4]   T. Van Cutsem, C. Vournas, "Voltage stability of electric power systems," Kluwer academic publishers, 1998.

[5]   S. Repo, J. Bastman, "Neural network based static voltage security and stability assessment," IASTED High technology in the power industry, USA, 1997.

[6]   S. Repo, P. Järventausta, "Contingency analysis for a large number of voltage stability studies," IEEE Power Tech, Hungary, 1999.

[7]   S. Haykin, "Neural networks, a comprehensive foundation," Prentice Hall, 1994.

[8]   The Math Works Inc., "Neural network toolbox user's guide," MATLAB, 1998.

[9]   R. Fletcher, "Practical methods of optimization," John Wiley & Sons, 1987.

[10] T.R. Cuthbert, "Optimization using personal computers with applications to electrical networks," John Wiley & Sons, 1987.

[11] J.T. Tou and R.C. Gonzalez, "Pattern recognition principles," Addison-Wesley, 1974.

[12] H.C. Andrews, "Introduction to mathematical techniques in pattern recognition," John Wiley & Sons, 1972.

[13] M. Nørgaard, "Neural network based system identification toolbox," Technical report 97-E-851, Technical University of Denmark, 1997.

Neural Networks-based Friction Compensation with Application in Servo Motor Systems

X. Z. Gao and S. J. Ovaska

Institute of Intelligent Power Electronics
Helsinki University of Technology
Otakaari 5 A, FIN-02150 Espoo, Finland
E-mail: gao@csc.fi, Seppo.Ovaska@hut.fi

Abstract: Compensation of negative effects caused by friction in high precision servo control systems is an important and challenging problem. Conventional compensation methods often rely on an explicit friction model, which is difficult to acquire accurately in practice. In this paper, we propose a neural network-based compensation scheme to cope with this problem. The visible disturbance resulting from friction is first identified by a BP (Back-Propagation) neural network. The friction compensator is constructed by cascading this neural identifier with the inverse model of the motor system. It is shown that our approach has the advantages of simplicity and generality. Moreover, no prior information concerning the friction is needed. Simulations are carried out to demonstrate the efficiency of the proposed method in compensating for deterministic as well as non-linear friction.

1. Introduction

With the rapid development of electric motor and drive industry, high resolution servo performance plays a pivotal role in electrical machines and robot arms [1]. Besides the internal nonlinear characteristics, external friction is the main factor that degrades the servo performance. In fact, it reduces the tracking accuracy drastically, if not carefully compensated. Many configurations, therefore, have been proposed to compensate for the harmful friction. For instance, adaptive control technology is employed to overcome the influences of friction in DC drives [2]. Other typical compensation methods include finding ways to increase the servo rigidity and utilizing a *feedforward* compensation scheme. However, these schemes may easily result in large amplifier gains, which cause saturation and affect the achievable system properties. In addition, although the compensation results of the available methods are satisfactory, certain assumptions have to be made before applying them. For example, in the feedforward compensation structure, it is usually required that the friction itself should be directly measurable. Otherwise, the friction has to be identified by necessary observers that can lead to fairly complex implementations [3]. Most of all, from the design point of view, these methods are inherently restricted within specific models of friction, and thus cannot be considered as general purpose solutions.

Due to their common capability of approximating nonlinear functions by learning from input/output samples pairs, artificial neural networks attract research attention from different communities [4], and become popular also in the control field [5]. One promising application of neural networks in this area is the related work on disturbance compensation in motor drive control systems [6], [7], [8]. In this paper, we propose a straightforward neural network-based friction compensation scheme for DC servo motor systems. The basic idea lies on applying a neural network to predict the dynamical effects of friction on the system performance. To eliminate these negative consequences, appropriate control actions are, hence, obtained by passing the prediction output of the neural network through an inverse model of the motor drive system. Simulations have been carried out to verify the effectiveness of the proposed scheme. The adaptation feature of neural networks gives our method a unique advantage over conventional approaches. Additionally, there is no strict constraint on the type of friction to be tackled in our method, except that it must be highly deterministic and repetitive.

This paper is organized as follows. In Section 2, we briefly discuss the structure of a representative DC motor drive control system. The simplified models for the motor together with the friction are also provided. Background knowledge of the BP neural network is presented in Section 3. We introduce our BP neural network-based friction compensation scheme in Section 4. In Section 5, based on the above description, verifying simulation experiments are considered, and comparative results are discussed. Finally, we draw some conclusions in Section 6.

2. DC Motor Drive Control Systems

DC motor is commonly applied in servo control systems as an effective actuator. It can provide both rotary and transitional motion. The electric circuit of the armature and a diagram of the rotor of a DC motor are shown in Fig. 1. In Fig. 1, input $u(t)$ is the source voltage and output $\dot{\omega}(t)$ is the rotation speed. The dynamical characteristics of a DC servo motor can be modeled by the following transfer function

$$G(s) = \frac{\dot{\omega}(s)}{u(s)} = \frac{K}{(Js+b)(Ls+R)+K^2}.$$ (1)

The primary symbols and their values are explained as follows:
electromotive force constant $K = 0.01$ Nm/A;
moment of inertia of the rotor $J = 0.01$ kgm²/s²;
damping ratio of the mechanical system $b = 0.1$ Nms;
electrical inductance $L = 0.5$ H;
electrical resistance $R = 1\ \Omega$.

We point out that small time constants of the servo system are neglected here, because they are assumed to have minor effects on the dynamical response of the overall system. Additionally, the rotor and shaft of the motor are supposed to be rigid.

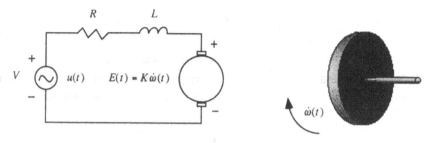

Fig. 1. Electric circuit of the armature and diagram of the rotor [9].

The aforementioned DC servo motor is widely used in industrial applications. In some particular cases, such as robotics manipulator control, accurate position or velocity control with the motor is desired. However, there are always external noise and disturbances existing in these motor drive systems, which deteriorate the resulting servo accuracy. From the control point of view, the measurement noise is considered stochastic in nature. Disturbances from friction, mechanical nonlinearity, and uneven mass distribution are, to great extent, deterministic. Consequently, the dynamical property of friction can often be described by time-invariant mathematical models. Although friction is neither random nor time-varying, it is sometimes highly nonlinear. This increases the difficulty of motor controller design [7]. The structure of a DC servo motor control system containing friction is illustrated in Fig. 2.

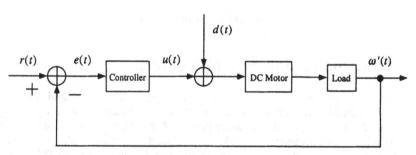

Fig. 2. Servo motor control system.

Here, we denote $d(t)$ as the external friction. Basically, friction is a complex phenomenon that must be interpreted by many elements such as stiction, hysteresis, and velocity dependence. It can cause large overshoots and steady tracking error especially at low velocities. Conventional friction compensation methods are designed based on explicit friction models. There are several popular models of friction, e.g., the Coulomb and exponential models [2]. Detailed discussion of these models is beyond the scope of our paper. As a matter of fact, friction is always velocity dependant. Next, we show a simplified relationship between friction $d(t)$ and the output velocity of the system $\omega'(t)$ in Fig. 3. The corresponding mathematics representation of the friction is written as:

$$d(t) = \begin{cases} a_1\omega'(t) + b_1 & \text{if } \omega'(t) > 0 \\ 0 & \text{if } \omega'(t) = 0 \\ a_2\omega'(t) + b_2 & \text{if } \omega'(t) < 0 \end{cases} . \tag{2}$$

Usually, $a_1 = a_2$ and $b_1 = -b_2$ because of the reasonable assumption of systematic characteristics. This simple friction model is used in the following simulations in Section 5.

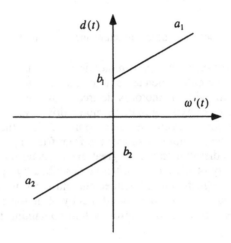

Fig. 3. Simplified friction model [2].

3. Back-Propagation Neural Network

Artificial neural networks [4], which are identically referred to as 'neural networks,' were first studied by the motivation to understand and imitate the working principle of human brain. Comprehensive descriptions of the foundation, structures, and learning rules of current neural networks can be found in the available literature. An interesting overview on the advances in neural control field is given in [5]. The BP neural network (also called *feedforward* or *multi-layer perceptron* network) is the most widely applied network model, due to its simple structure and efficient training algorithm. Moreover, it has been proved that a BP neural network with sufficient number of hidden nodes can approximate any nonlinear function arbitrarily well [10], [11]. A three layer BP neural network is illustrated in Fig. 4.

In brief, adjustable weights w connecting adjacent layers are updated during the training procedure to make the BP neural network realize nonlinear function mappings from the multi-dimensional input space to the output space. The learning algorithm of the BP neural network utilizes the gradient descent method:

$$\mathbf{w}^{new} = \mathbf{w}^{old} + \Delta\mathbf{w}, \tag{3}$$

where

$$\Delta\mathbf{w} \propto -\frac{\partial H}{\partial \mathbf{w}}. \tag{4}$$

H is the Sum Squared Error (SSE) of the BP neural network, which is usually defined as a quadratic function:

$$H = \frac{1}{2} \sum_{k=1}^{L} \sum_{i=1}^{N} (Y_i^{(k)} - y_i^{(k)})^2 . \tag{5}$$

In (5), $Y_i^{(k)}$ and $y_i^{(k)}$ are the reference and actual outputs of the BP neural network, respectively. L and N are the numbers of training pairs and network outputs. The approximation capability of the BP neural network has been extensively explored in the literature [4]. In addition to accurate reproduction of training data, the BP neural network can generate desired outputs for those fresh input patterns that are not present in the training procedure. Consequently, the BP neural network is well suited to be employed in our friction compensation scheme to identify and predict the effects of nonlinear friction.

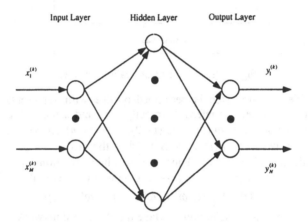

Fig. 4. Structure of BP neural network.

4. Friction Compensation Using Neural Network

As mentioned above, the essential idea of classic friction compensation methods is based on real-time acquisition of the friction values. This can be accomplished by measuring the friction directly or using an observer [12]. However, acquiring the friction accurately is difficult if not impossible in practical motor control systems. Inspired by the powerful nonlinear approximation capability of neural networks, we propose a neural network-based compensation solution. Our friction compensation scheme is divided into two steps. The first step is to predict the velocity error resulting from friction. The corresponding neural network-based identification scheme is shown in Fig. 5. Without principal difficulty, a virtually *friction-free* servo motor system can be constructed in a laboratory. The servo error $E(t)$ stemming from friction is defined as follows:

$$E(t) = \omega(t) - \omega'(t). \tag{6}$$

It is clearly visible that $E(t)$ is the difference between the output of the ideal system model (without $d(t)$) and the practical system (with $d(t)$). Our friction compensation goal is to minimize this error.

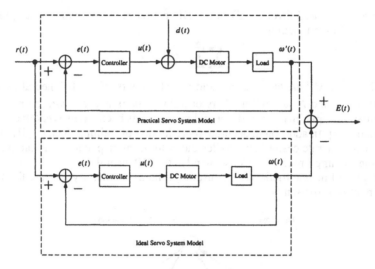

Fig. 5. Servo error identification scheme.

The identification step is performed off-line with representative training data collected in advance. The reference velocity $r(t)$ and actual velocity $\omega'(t)$ together with its history data $\omega'(t-1), \omega'(t-2), ..., \omega'(t-n)$ are selected as the inputs of the neural network. $E(t)$ is considered as the desired output, and n is the order of the identification model. Thus, the nonlinear relationship $\Psi[\cdot]$ between $E(t)$, $r(t)$, and $\omega'(t), \omega'(t-1), ..., \omega'(t-n)$ is set up by the neural network as:

$$E(t) = \Psi\big[r(t), \omega'(t), \omega'(t-1), ..., \omega'(t-n)\big]. \tag{7}$$

To simplify the presentation, we take the BP neural network as an illustrative example throughout this paper. The BP neural network training phase for servo error identification is revealed in Fig. 6. $E'(t)$ is the prediction output of the BP neural network. A Tapped Delay Line (TDL) is used to store the past velocity samples. However, there is one key problem in designing such a BP neural network-based identification scheme, i.e., how to choose an appropriate n. This problem is actually related to the choice of the neural identification model order, and can often be solved by *trial and error* [13]. As we know, both the DC servo motor system and velocity controller (usually a PID controller) are low order systems. Therefore, the identification model order is selected to be low, for example, $n = 2$.

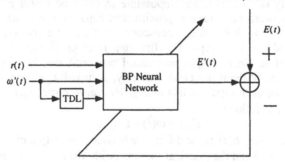

Fig. 6. BP neural network-based servo error identification structure.

After the BP neural network has been trained to give a satisfactory prediction of the 'control error,' it is applied on-line at the second step to compensate for the source of this error—friction, as shown in Fig. 7. Our friction compensator consists of two components: one is the trained BP neural network, and the other is the *inverse* model of the servo system. This inverse model can be obtained using the available knowledge of the DC motor and load. In Fig. 7, $u_N(t)$ is the compensation output. Let $\Phi^{-1}[\cdot]$ be the inverse model of the servo system, we get

$$u_N(t) = \Phi^{-1}[E'(t)]. \tag{8}$$

$u'(t)$ is the final control command. The control error resulting from the friction can thus be compensated in this way.

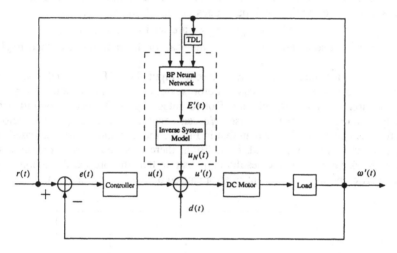

Fig. 7. BP neural network-based friction compensation
in a servo motor system.

Comparing with conventional friction compensation approaches, the proposed neural network-based scheme has several distinguishing features. First, it does not require that the friction must be measurable. The servo error is identified by using only the available outputs of the practical and ideal systems. Second, the applied BP neural network can effectively cope with the inherent nonlinearity in the friction. Besides, there is no restraint on the type of the friction except that it has to be deterministic and repetitive. Third, the collection of identification data for the servo error has no effect on the normal operation of the servo control system. We will demonstrate the effectiveness of our method with numerical simulations in the following section.

5. Simulations

Simulation results are given to provide an illustrative verification of the presented method in this section. The motor drive is simplified as a second order model discussed in Section 2 with representative coefficients:

$$G(s) = \frac{0.01}{0.005s^2 + 0.06s + 0.1001}. \tag{9}$$

The feedforward controller is a classical PID controller shown below:

$$G_{PID}(s) = K_P + \frac{K_I}{s} + K_D s. \tag{10}$$

The corresponding parameters for our PID controller are $K_P = 200$, $K_I = 10$, and $K_D = 100$, respectively. Note that these three values are not carefully fine-tuned, because the PID controller is used here only for the purpose of validating the neural network-based compensation scheme. We do not intend to compensate for the friction by optimizing the PID controller alone in our simulation. The BP neural network is constructed with four input nodes ($r(t), \omega'(t), \omega'(t-1), \omega'(t-2)$) and five hidden nodes. $a_1 = a_2 = 35$ and $b_1 = -b_2 = 1$ are set for the friction model presented in (2). It should also be emphasized that the load of the motor is assumed negligible.

Given a typical square wave reference input signal, Figs. 8 and 9 show the velocity responses of the DC servo motor system with and without our friction compensation scheme, respectively. Dotted lines represent desired output while solid lines represent actual response. We can conclude that in Fig. 8 the velocity waveform is apparently better than that in Fig. 9; the overshoots are decreased, and the oscillation is greatly reduced. In addition, there is an obvious improvement in terms of tracking time. These results demonstrate that the proposed method is potentially efficient in reducing control error caused by the external friction.

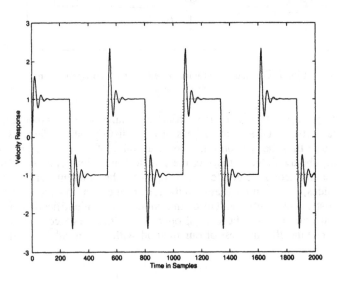

Fig. 8. Velocity output with friction compensation.

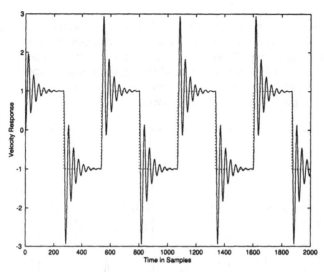

Fig. 9. Velocity output without friction compensation.

We point out that the compensation results demonstrated are acquired under the circumstance of fixed PID controller parameters. However, if the PID controller is re-tuned, the neural identifier should naturally be *retrained* to adapt to this change as well. Otherwise, the compensation performance will be deteriorated. Figs. 10, 11, and 12 give an illustration to support this argument. Suppose there is a 50% reduction in the PID controller parameters, i.e., the old PID controller is replaced by $K_p = 100$, $K_I = 5$, and $K_D = 50$. The velocity response of this closed loop servo system with the new PID controller is shown in Fig. 10, which behaves differently from that in Fig. 9.

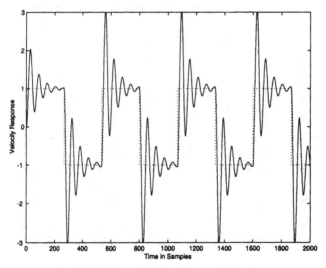

Fig. 10. Velocity output with the new PID controller
(without friction compensation).

Now, the compensation results with our original and retrained friction compensation schemes are shown in Figs. 11 and 12, respectively. The compensation result in Fig. 11 with unchanged neural friction compensator is worse than that in Fig. 8. Comparing Fig. 11 with Fig. 12, we can observe that the response performance of the latter is better than that of the former neural compensator with respect to overshoots. Therefore, our neural network-based friction compensation scheme can be regarded as controller dependent.

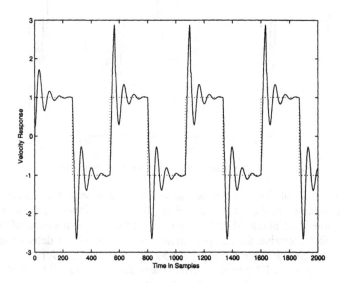

Fig. 11. Velocity output with the new PID controller
(with the original friction compensator).

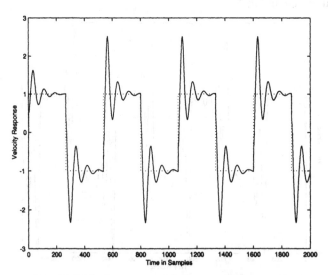

Fig. 12. Velocity output with the new PID controller
(with the retrained neural friction compensator).

One important issue related to the identification of servo error using the BP neural network is the *over-training* problem [14]. Generally, when the BP neural network is trained to give very accurate approximation of the training data, it will be less robust and more sensitive to the difference between the training data and the verification data. This over-training problem may become even serious in our friction compensation case. Since the effect of the friction cannot be compensated completely in practice, the preceding residual error will easily disturb the following compensation output in the presence of an over-trained BP neural network. Fig. 13 illustrates such a representative example, in which the over-trained BP neural network greatly interferes the system response, and leads to unstable behavior. A possible solution to the above problem is using the *early-stopping* technique [15]. It should be pointed out that the high-frequency oscillations demonstrated in Fig. 13 actually can be damped by a practical load, which is not taken into consideration in our simulation.

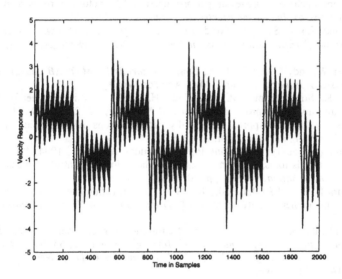

Fig. 13. Velocity output with an over-trained friction compensation scheme.

6. Conclusions

A neural network-based disturbance compensation scheme is introduced in this paper. Its validity has been verified by simulations. We emphasize that our method is not only suitable for friction compensation, but it can be generalized to compensate for other forms of deterministic disturbances such as mechanical nonlinearities. Nevertheless, the main drawback of this scheme is that identification of the disturbance effect is done off-line. Thus, it cannot be applied to tackle with time-varying disturbances. Future investigations may concentrate on studying new compensation algorithms with on-line identification capability.

References

[1] Kazmierkowski M. and Tunia H., 1994, *Automatic Control of Converter-Fed Drives*. Amsterdam, The Netherlands: Elsevier.

[2] Canudas C., Åstrom K. J., and Braun K., 1987, "Adaptive friction compensation in DC-motor drives," *IEEE Journal of Robotics Automation*, vol. 3, pp. 681-685.

[3] Saif M., 1993, "A disturbance accommodating estimator for bilinear systems," in *Proc. American Control Conference*, San Francisco, CA, pp. 945-949.

[4] Haykin S., 1994, *Neural Networks: A Comprehensive Foundation*. New York, NY: Macmillan College Publishing Company.

[5] Hunt K. J., Sbarbaro D., Zbikowski R. et al, 1992, "Neural networks for control systems—A survey," *Automatica*, vol. 28, pp. 1083-1112.

[6] Low T.-S., Lee T.-H., and Lim H.-K., 1993, "A methodology for neural network training for control of drives with nonlinearitis," *IEEE Transactions on Industrial Electronics*, vol. 39, pp. 243-249.

[7] Teeter J. T., Chow M.-Y., and Brickley J. J., Jr, 1996, "A novel fuzzy friction compensation approach to improve the performance of a DC motor control system," *IEEE Transactions on Industrial Electronics*, vol. 43, pp. 113-120.

[8] Du H. and Nair S. S., 1999, "Modeling and compensation of low-velocity friction with bounds," *IEEE Transactions on Control Systems Technology*, vol. 7, pp. 110-121.

[9] Messner W. and Tilbury D., 1998, *Control Tutorials for MATLAB and Simulink: A Web-based Approach*. Reading, MA: Addison-Wesley.

[10] Hornik K., Stinchcommbe M., and White H., 1989, "Multilayer feedforward networks are universal approximators," *Neural Networks*, vol. 2, pp. 359-366.

[11] Cybenko G., 1989, "Approximations by superpositions of a sigmoidal function," *Math. Contr. Signals. Syst.*, vol. 2, pp. 303-314.

[12] Armstrong-Helouvry B., Dupont P., and Canudas de wit C., 1994, "A survey of models, analysis tools, and compensation methods for the control of machines with friction," *Automatica*, vol. 30, pp. 1083-1138.

[13] Narendra K. S. and Parthasarathy K., 1990, "Identification and control of dynamical systems using neural networks," *IEEE Transactions on Neural Networks*, vol. 1, pp. 4-27.

[14] Hanson L. K. and Solamon P., 1990, "Neural network ensembles," *IEEE Transactions on Pattern Analysis and Machine Intelligence*, vol. 12, pp. 993-1001.

[15] Nelson M. C. and Illingworth W. T., 1991, *A Practical Guide to Neural Nets*. Reading, MA: Addison-Wesley.

Linear and neural dynamical models for energy flows prediction in facility systems

Lucia Frosini[1] and Giovanni Petrecca[1]

[1]Department of Electrical Engineering - University of Pavia
Via Ferrata 1, Pavia, I-27100 Italy
e-mail: lucia@unipv.it, petrecca@unipv.it

Abstract: A procedure for the short–term prediction of the thermal energy consumption of an hospital is shown in this paper. At first, linear ARX models are built to get information on the influence of the input variables on the output of the system. Therefore, non-linear models based on feedforward neural networks (NNARX) are built using the information provided by the linear estimate. The results obtained from the ARX and NNARX models are compared, concluding that NNARX models provide better results than ARX models, but the analysis of ARX models is necessary to obtain guidelines in the choice of the best regression vector as input for neural models.

1. Introduction

The optimal planning and management of energy flows in an industrial plant or in a tertiary site (hospitals, government and university buildings, malls, etc.) requires:
i)   the forecasting of the electric, thermal and cooling requirements;
ii) the optimal plant management;
iii) the choice of the optimal layout.
The optimal planning and management is based on both technical and economic coefficients becoming very important in the forthcoming European energy market situation, which will surely include provision of diversified tariffs from different utilities.
A methodology for the development of an economic optimization model for managing a facility system, based on the combination of mixed integer linear programming, dynamic programming and neural networks system identification, has been reported in a previous work [1].
Every process can be schematized by means of a flow chart where the two steps of the proposed methodology are described: firstly, the forecasting technique predicts the energy end user requirements on the basis of previous data, secondly, the dynamic programming manages continuous time spans in a discrete manner and the mixed integer linear programming gives the best variables value, according with the objective function and the technical and physical constraints.

In this paper we pay attention to the stage of prediction based on neural networks and we show in detail a case study concerning the prediction of the thermal energy consumption of an hospital.

2. Prediction of energy flows

In order to have a prediction of energy flows, such as electric energy consumption E, thermal energy consumption T, electric or thermal energy consumption for cooling C, the use of NNARX predictors is suggested (Fig. 1). The predicted output $\hat{y}(t\,|\,\theta)$ can depend on the past values $y(t-k)$, where $k = 1, 2, \ldots n_a$, and possibly on the past values of other exogenous variables $u(t-k)$, where $k = n_k, n_k-1,$ ... n_k-1+n_b:

$$\hat{y}(t\,|\,\theta) = g(y(t-1),...,y(t-n_a),u(t-n_k),u(t-n_k-1)...\,u(t-n_k-n_b+1)) \qquad (1)$$

The notation $\hat{y}(t\,|\,\theta)$ indicates that the predicted output is a "guess" for $y(t)$ given a particular parameter value θ. In particular, a Multi-Layer Perceptron (MLP) feedforward neural network can be chosen as g function. The exogenous variables, in the case of the air-conditioning of a building, can be the internal and external temperature and relative humidity and the occupation, which have to be measured with the same frequency as the dependent variables.

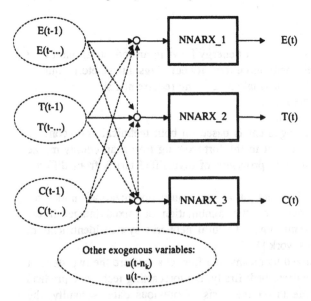

Fig. 1. Prediction of energy flows

Treatment of the problem requires knowledge of the data of thermal and electric energy consumption regarding a significant period which can be measured periodically with a frequency depending on the period of reference ("step") for which it is intended to make the short-term forecast (e.g. 2 hours, 6 hours, 1 day). This technique can be an effective support in matching optimization algorithms applied to facility systems.

3. Case study

The case study is aimed at the prediction of the thermal energy consumption of an hospital.

The thermal energy is used for:
*   heating and air conditioning of the buildings;
*   production of hot water for bathroom, kitchen and cleaning;
*   production of steam for sterilization and laundry.

The objective was to find a relation among the prediction of thermal energy consumption, the past values of the same variable and the past values of other variables, if necessary.

Such analysis will be able to optimize the planning and management of the plant through the prediction of the energy consumption of the subsequent hours, and to justify the considered too high energy consumption, i.e. to find the causes of energy wastes.

The available data are:
*   natural gas consumption (Sm^3) collected every two hours as difference between two successive records of the meter;
*   cold and hot water consumption (m^3) collected every two hours as difference between two successive records of the meter;
*   external temperature (°C) collected every half an hour by data logger;
*   internal temperatures (°C) collected every half an hour by data loggers, in four different zones of the hospital.

All the data are recorded in the period from 12 a.m. of the day 31/03/98 to 10 p.m. of the day 10/04/98.

In order to use a sampling interval of two hours for all the variables, we have calculated the mean of the values of the temperatures every two hours as mean of the previous four values.

We have divided the available data (126 for every variable) in two sets:
a) a training (or identification) set, from the 1st to the 86th value for each variable;
b) a test (or validation) set, from the 87th to the 126th value for each variable.

Each training set is scaled to zero mean and variance one and then each test set is scaled with the same constants.

4. An approach to prediction with neural networks

A main principle in identification is the rule "try simple things first". The idea is to start with the simplest model which has a possibility to describe the system and only to continue to more complex ones if the simple model does not provide reliable results in the validation stage [2]. When a new more complex model is investigated, the results with the simpler model give some guidelines how the structural parameters should be chosen in the new model.

Our approach to build prediction models based on neural networks has been the following:

• construction of ARX models to get information on the influence of the input variables on the output of the system, quickly and without danger of falling in local minima;

• use of these information for the construction of NNARX models based on feedforward neural networks;

• comparison between the best NNARX model and the best ARX model and validation of the best model at all.

5. Construction of the ARX models

An ARX (Auto Regressive eXogenus) model is described by the following equation:

$$A(q^{-1})y(t) = B(q^{-1})u(t) + e(t) \tag{2}$$

where y is the output of the dynamic model, u the input, e the disturbance or noise and q^{-1} the shift operator.

The corresponding predictor:

$$\hat{y}(t \mid \theta) = -a_1 y(t-1) - ... - a_{n_a} y(t-n_a) + b_1 u(t-n_k) +$$
$$... + b_{n_b} u(t-n_k - n_b + 1) \tag{3}$$

is thus based on the regression vector:

$$\varphi(t) = [y(t-1) \ ... \ y(t-n_a) u(t-n_k) \ ... \ u(t-n_k - n_b + 1)] \tag{4}$$

where n_a is equal to the number of poles, $n_b - 1$ is the number of zeros and n_k is the pure time-delay of the system [3].

The considered variables are: natural gas consumption, cold and hot water consumption, internal temperatures (recorded in four different zones) and external temperature.

Combining these variables we have analyzed the possible regression vectors. Moreover, we have considered the models that, having the same variables in the regression vector, are different for the number of past values of each variable.

We have processed only a significant part of the total number of the applicable models. This has been feasible because we have taken into account the information provided by the simpler models for the construction of the more complex models.

For each model, we have fixed $n_k = 1$ and we have made n_a and n_b vary from 2 to 8.

In order to analyze the linear models according to (2), we have supposed that $e(t)$ is a white noise.

Using ARX model, the least-squares procedure automatically makes the correlation between $e(t)$ and $u(t-k)$ zero for $k = n_k, n_k+1, ... n_k+n_b-1$, for the data used for the estimation.

We have selected the best linear models through the cross validation, i.e. we have identified the models by the Least Mean Squares (LMS) estimate using the training data set and hence we have chosen the models that provided the lowest values of the Mean of Squared Residuals (8) during the validation stage (MSRv). For this choice, we have considered also the value of the Mean of Squared Residuals during the identification stage (MSRi) that should not be too far from the value of MSRv. In fact, in most cases, the lower value of MSRv corresponds to a quite high value of MSRi (e.g. the model with the external temperature as input variable, with $n_a = 2$ and $n_b = 7$ gives MSRv = 0,3902 and MSRi = 0,6676).

In Table 1 are showed the best ARX models for every considered combination of input variables, where Tex is the external temperature, T1, T2, T3 and T4 the internal temperatures and W the water consumption.

Table 1. The best ARX models for every considered combination of input variables

u_1	u_2	u_3	u_4	MSRi	MSRv	n_a	n_{b1}	n_{b2}	n_{b3}	n_{b4}
Tex	-	-	-	0,4420	0,4935	8	8	-	-	-
T1	-	-	-	0,8926	0,9128	2	2	-	-	-
T2	-	-	-	0,9290	1,1286	3	3	-	-	-
T3	-	-	-	0,8882	0,9825	2	8	-	-	-
T4	-	-	-	0,8580	0,8465	3	4	-	-	-
W	-	-	-	0,6534	1,0246	6	6	-	-	-
Tex	T1	-	-	0,5958	0,4272	6	6	2	-	-
Tex	T4	-	-	0,4715	0,4664	8	8	2	-	-
Tex	W	-	-	0,8024	0,7432	3	3	3	-	-
Tex	T1	T4	-	0,3854	0,3845	8	8	8	8	-
Tex	T1	W	-	0,3522	0,4923	8	8	8	8	-
Tex	T4	W	-	0,4236	0,3983	8	8	2	3	-
Tex	T1	T4	W	0,4643	0,3827	8	8	2	2	2

6. Construction of the NNARX models

6.1 Feedforward networks

Neural network can be classified as feedforward networks and recurrent networks. In feedforward networks, the neurons are connected in such a way that all signals flow in one direction from input units to output units.

In recurrent network, there are both feedforward and feedback connections along which signals can propagate in opposite directions.

Feedforward networks have been applied to dynamic system identification with success. Because a feedforward network does not have dynamic memory, the tapped-delay-line method is usually adopted to enable it to represent a dynamic system. The method employs the current and past inputs and outputs of the system to be modelled as the inputs to the network. The next output of the system is used as a teaching signal [4].

Using feedforward neural networks, one can build non-linear models corresponding to ARX models.

6.2 Multi-layer Perceptron (MLP)

On the basis of the best ARX models, we built NNARX models using a Multi-Layer Perceptron (MLP) network.

The choice of this network was based also on the experience of other authors [4, 5] for its ability to model simple as well as very complex functional relationships.

We have considered a MLP with three layers: an input layer, an output layer and an intermediate or hidden layer.

For the Cybenko universal approximation theorem [6], a MLP with a single hidden layer is sufficient to compute a uniform ε approximation (where $\varepsilon > 0$) of any continuous function, under the condition to take a sufficient number of neurons.

Each neuron j in the hidden layer sums up its input signals x_i after weighting them with the strengths of the respective connections w_{ji} from the input layer and computes its output g_j as a function f of the sum:

$$g_j = f(\sum_{i=1}^{m} w_{ji} x_i) \tag{5}$$

where the function f can be linear, threshold, sigmoid, hyperbolic tangent or radial basis. The output of neurons in the output layer is computed similarly.

We have considered hyperbolic tangent functions or linear functions for the neurons in the hidden layer and linear functions for the neurons in the output layer.

The output $\hat{y}_1$ of the MLP is:

$$\hat{y}_1(w, W) = F_1(\sum_{j=1}^{q} W_{1j} f_j(\sum_{i=1}^{m} w_{ji} x_i + w_{j0}) + W_{10} \tag{6}$$

where:

m = number of inputs

q = number of hidden neurons

w_{ji} = weight between input x_i and hidden neuron j

W_{lj} = weight between hidden neuron j and output l.

The weights w and W (vector θ) are the adjustable parameters of the network and they are determined through the training process.

The training data are a set of inputs u(t) and corresponding desired outputs y(t):

$$Z^N = \left\{ \left[u(t), y(t) \right] \middle| \, t = 1, ..., N \right\} \tag{7}$$

The objective of training is to determine a mapping from the training data set to the set of possible weights $Z^N \rightarrow \hat{\theta}$, so that the network will produce predictions $\hat{y}(t)$ which are close to the true outputs y(t).

The prediction error approach is based on the introduction of a measure of closeness in term of a Mean of Squared Residuals (MSR):

$$V_N(\theta, Z^N) = \frac{1}{N} \sum_{t=1}^{N} (y(t) - \hat{y}(t \mid \theta))^T (y(t) - \hat{y}(t \mid \theta)) \tag{8}$$

The weights are found as:

$$\theta = \arg_\theta \min V_N(\theta, Z^N) \tag{9}$$

by an iterative minimization scheme:

$$\theta^{(i+1)} = \theta^{(i)} + \mu^{(i)} h^{(i)} \tag{10}$$

where $h^{(i)}$ is the search direction and $\mu^{(i)}$ the step size. A large number of training algorithms exist, each of which is characterized by the way in which search direction $h^{(i)}$ and step size $\mu^{(i)}$ are selected.

We have used the Levenberg-Marquardt method, due to its rapid convergence properties and robustness [5].

VII. Validation of the best NNARX model

For the choice of the number of the hidden neurons we used a "trial and error" method: we start from the simplest model (one neuron with linear function) and we increase the complexity of the network (one hyperbolic neuron, two linear neurons, one hyperbolic neuron and one linear neuron, two hyperbolic neurons, etc.). The method stops when the performance of the network, in term of MSRv and MRSi, becomes worse.

Moreover, once found the optimal number of neurons, we used the Optimal Brain Surgeon (OBS) method in order to obtain the best architecture of the network "pruning", if necessary, some interconnection weights [5].

Because of the random initialization of the weights, it's necessary to process the identification software of each NNARX model several times in order to obtain reliable values of MSR.

The best NNARX model has been this with only one input variable (the external temperature), $n_a = n_b = 8$, $n_k = 1$, two hyperbolic and one linear hidden neurons and totally connected network. This model has given the following results: MSRi = 0,3702, MSRv = 0,3268.

These results are better in respect of all the analyzed ARX models.

We observe that NNARX models with more than one input variable have given worse results compared to the corresponding linear models.

In our opinion, this failure is due to few data available for the identification of models with a too large number of parameters (the interconnection weights of the network).

Summarizing, the results have proven that a neural model with only one input variable allows to obtain better results in respect of linear models with more input variables and, therefore with higher costs for the collection of data.

8. Other approaches to prediction with neural networks

The prediction with neural networks has been implemented in several fields. In particular, we have considered the following researches:

1. In [7] a methodology for the development of an economic optimization model of the UCLA cogeneration plant is shown. A neural network model is developed for prediction of the future weather conditions upon which the expected future campus demands and plant component performance are dependent. The solution consists of predicting future values for the weather variables (temperature, humidity, etc.) and campus demand of steam, chilled water and electricity. This information is used to identify plant states which can satisfy the predicted campus demand. Two neural MLP predictors are used, one for the weather forecast, the other for the prediction of the energy demand.

2. In [8] a constrained multivariable control strategy along with its application in thermal power plant control is presented. In order to identify a system with three outputs and seven inputs, the authors build three second-order 7 inputs - 1 output NNARX models using MLP networks with one hidden layer and tansigmoid neurons. In a previous work [9], the same authors have implemented a second-order 4 inputs - 4 outputs model, using a MLP network with one hidden layer, to identify a simulation model of a boiler-tubogenerator unit. They have obtained good results in predicting the plant output, not only for a single step ahead, but also over a long prediction horizon.

3. In [4] are discussed the pros and cons of feedforward and recurrent networks for the identification of dynamical models. The authors point out the slow computation of the feedforward networks due to the large number of units in the input layer. This is because, if the order of the system to be identified is unknown, it must be over-estimated. The large number of units in the input layer also makes the identifier highly susceptible to external noise. In our opinion, this over-estimate can be avoid through the previous identification of the order of the system by linear models.

The authors [4] show that recurrent networks do not suffer from the above drawbacks. Among the available recurrent networks, the Elman network is one of the simplest types that can be trained using the standard backpropagation algorithm.

In Elman network, in addition to the inputs units, hidden units and outputs units, there are also context units. The context units memorize some past states of the hidden units, and so the outputs of the network depend on an aggregate of the previous states and the current input; the context units can be considered to function as one-step time delays. It is because of the property that partially recurrent networks possess the characteristic of a dynamic memory.

When the Elman network is used to model single-input single-output systems, only one unit is needed in the input layer and the output layer:

$$y(t) = -a_1 y(t-1) - a_2 y(t-2) - ... - a_n y(t-n) +$$
$$+ b_1 u(t-1) + b_2 u(t-2) + ... + b_n u(t-n) \tag{11}$$

To model the nth-order dynamic system represented by equation (11), 2n input units would be needed if a feedforward network is used. For an Elman network, the number inputs units is one, or n+1 if the context units are regarded as input units. Thus an Elman network will be significantly smaller in structure than a feedforward network when n is large.

Therefore recurrent networks probably could provide good results for energy flows prediction. However, a neural model based on recurrent networks does not correspond to an ARX model; it is rather equivalent to OE (Output Error) or ARMAX (Auto Regressive Moving Average eXogenous) models which have simulated or predicted outputs within their regressors.

This might lead to instability in certain areas of the network's operating range and it can be very difficult to determine whether or not the predictor is stable [2, 5].

9. Conclusions

A procedure for the short–term prediction of the thermal energy consumption of an hospital has been shown in this paper. This prediction can be employed for the optimal planning of the thermal plants of this facility system.

Non-linear models based on MLP feedforward neural networks have been implemented as second step of the procedure, only after the linear estimate of the best regression vectors. This procedure allows to evaluate a greater number of models compared to a procedure starting with the neural models estimate.

The results obtained from the ARX and NNARX models are compared, concluding that NNARX models provide better results than ARX models, but the analysis of ARX models is necessary to obtain guidelines in the choice of the best regression vector as input for neural models.

Further improvements could be brought using a larger number of data that allows to estimate more complex models.

Acknowledgment

The authors thank professor De Nicolao of the Department of Computer Engineering and Systems Science of the University of Pavia for his helpful advice, and Mr. Mascheroni of Honeywell S.p.A. for the provision of the data analyzed.

References

[1] Anglani N., Frosini L. and Petrecca G., "Forecasting and optimizing techniques for energy flows management in facility systems", *The 3rd Annual International Conference on Industrial Engineering Theories, Applications and Practice*, Hong Kong, Dec. 1998.

[2] Sjöberg J., Hjalmerson H. and Ljung L., "Neural Networks in System Identification", *10th IFAC symposium on SYSID*, Copenhagen, 1994.

[3] Ljung L., 1995, *System Identification Toolbox for use with MATLAB®*. The MathWorks.

[4] Pham D. T. and Liu X., 1995, *Neural Networks for Identification, Prediction and Control*. Springer Verlag.

[5] Nørgaard M., 1997, *Neural Network Based System Identification Toolbox*. Technical Report, Department of Automation, Technical University of Denmark.

[6] Haykin S., 1994, *Neural Networks – A comprehensive foundation*. Macmillan.

[7] Jones S., Fengler W., Hatch T. et al, "Performance Modeling for Economic Optimizing of Central Plant Operations", *International District Energy Association, 11th Annual College/University Conference*, Redondo Beach, Feb. 1998.

[8] Prasad G., Swidenbank E. and Hogg B. W., "A Neural Net Model-based Multivariable Long-range Predictive Control Strategy applied in thermal power plant control", *IEEE Trans. Energy Conversion*, Vol. 13, No. 2, Jun. 1998.

[9] Irwin G., Brown M., Hogg B. W. et al, "Neural Network modelling of a 200 MW boiler system", *IEEE Proc. Control Theory Appl.*, Vol. 142, No. 6, Nov. 1995.

An Optimal VQ Codebook Design Using the Co-adaptation of Learning and Evolution

Daijin Kim[1] and Sunha Ahn[2]

[1] Department of Computer Science and Engineering, POSTECH,
   San 31, Hyoja Dong, Nam Gu, Pohang, 790-784, Korea
   e-mail : dkim@postech.ac.kr
[2] Department of Computer Engineering, DongA University,
   840, Hadan Dong, Saha Gu, Pusan, 604-714, Korea
   e-mail : shahn@vlsi.donga.ac.kr

Abstract. This paper proposes a design method of an optimal VQ (Vector Quantization) codebook using the co-adaptation of self-organizing maps that attempts to incorporates the Kohonen's learning into the GA evolution. The Kohonen's learning rule used for vector quantization of images is sensitive to the choice of its initial parameters and the resultant codebook does not guarantee a minimum distortion. We alleviate these problems by co-adapting the codebooks by evolution and learning in a way that the evolution performs the global search and makes inter-codebook adjustments by altering the codebook structures while the learning performs the local search and makes intra-codebook adjustments by making each codebook's distortion small. Simulation results show that the evolution guided by a local learning provides the fast convergence, the co-adapted codebook produces better reconstruction image quality than the non-learned equivalent, and Lamarckian co-adaptation turns out more appropriate for the VQ problem.

1 Introduction

Vector quantization [1] is a simple and effective data compression technique that uses vectors instead of scalars to achieve low bit-rate and better image reconstruction quality [2]. Mathematically, a vector quantizer Q of dimension k and size N is a mapping from an arbitrary vector $X = \{x_1, x_2, \cdots, x_k\} \in \Re^k$ into a finite set C such that $Q : \Re^k \rightarrow C$, where C is called as a codebook that consists of N codevectors $C = (Y_1, Y_2, \cdots, Y_N)$. The vector quantizer is completely specified by determining N reproduction codevectors C and their corresponding nonoverlapping partitions $R = (R_1, R_2, \cdots, R_n)$ called Voronoi regions. A Voronoi region R_i is defined as

$$R_i = \{X \in \Re^k : \ \| X - Y_i \| \ \leq \ \| X - Y_j \| \quad i \neq j\} \tag{1}$$

and represents a subset of vectors of $\Re^k$, where the vectors are nearest to the codevector Y_i in the meaning of L^2 norm $\| \cdot \|$.

A design problem of good codebook is an important subject for an optimal vector quantization. Among the commonly used VQ design methods, the Linde-Buzo-Gray algorithm [3] using a clustering approach has been the most well-known one. As an alternative VQ design method, an unsupervised neural network called the self-organizing map (SOM) [4] that is adapted by Kohonen's learning rule has been widely applied to codebook design for vector quantization [5,6]. These algorithms are considered as the suboptimal ones at best because they are based on the local gradient descent method for which a prescribed distortion measure is decreased monotonically by updating the codebook entries iteratively. Furthermore, due to highly nonlinear nature of VQ problem, the distortion function is non-convex and may exhibit many local minima. Therefore, the resultant codebook does not guarantee the minimum distortion and its reconstruction performance is sensitive to the choice of initial parameters.

Recently, the ideas of combining evolution and learning based on the concepts inspired from the biological systems have attracted many attentions increasingly. Evolution changes the genotype of organism at the population level while learning changes the behavior of organism at the individual level. Change by evolution is made from reproducing organism selectively and performing genetic operations (recombination and/or mutation) in order to maintain inter-individual variability. The change due to evolution is generationally cumulative in a manner that some changes at a particular generation is superimposed upon other changes in the previous generations. Change by learning is made from interacting with a specific environment during its lifetime and incorporating aspects of the environment through its experience in its internal structure. The change due to learning is individually cumulative in a manner that some changes in an individual at a particular time of its lifetime are influenced by other changes at preceding times.

There are two hypotheses on how learning makes influences on evolution. One is the Lamarckian hypothesis [7] that phenotypic traits acquired by an organism during its lifetime is transcripted into the heritable genotype and can be passed on to the organism's offspring directly. D. Ackley and M. Littman [8] showed that Lamarckian evolution could be both easy to implement and potentially far more effective for optimization problems due to its fast convergence. Another is the Baldwin effect [9] that acquired characteristics are not inherited to its offspring directly but adaptive learning can guide the course of evolution indirectly in a way that learning alters the shape of search space and thereby provides good evolutionary paths towards sets of co-adapted individuals. G. Hinton and S. Nowlan [10] demonstrated that Baldwin effect allowed the learning organisms to evolve faster than their nonlearning equivalents even though the characteristics acquired by the phenotypes are not directly communicated to their genotypes. D. Parisi and S. Nolfi [11] also indicated that performing a learning task that was indirectly related to an organism's fitness could still be used to guide evolution. They used a crude estimate of the direction towards the local minimum as indirect learning. According to the result of their experiment, coarse approximation to the gradient can provide enough directionality to allow learning to successfully guide evolution.

This paper is organized as follows. A VQ codebook design based on Kohonen's learning rule is presented in Section II. An evolutionary VQ codebook design based on the genetic algorithms is explained in Section III. A new codebook design by co-adaptation that incorporates learning into evolution is proposed in Section IV. Simulation results are performed to demonstrate faster convergence and better image quality in terms of the number of generations and the root mean squared distortion error, respectively, in Section V. Finally, a conclusion is drawn.

2 Codebook Design by Kohonen's Learning Rule

The Kohonen's learning rule is an iterative procedure for training a sheet-like artificial neural network called SOM. The learning procedure is unsupervised or self-organized and is used for tuning the nodes to input vector distributions. The training of the SOM is initialized by assigning small random values to the weight vectors $C = \{Y_1, Y_2, \cdots, Y_N\}$ of the units in the network. Each iteration in the learning process consists of three steps: the presentation of an input vector to the network, the evaluation of the network, and an update of the weight vectors. The update of weight vectors is performed in the following manner.

After the presentation of an input vector, the Euclidean distance between the input vector and the weight vector is computed for all units in the network. The unit with the smallest distance is selected as a winner marked as unit w as

$$\| X(t) - Y_w(t) \| = \underset{i}{min}(\| X(t) - Y_i(t) \|). \tag{2}$$

Then, all units within a certain spatial neighborhood N_w around the unit w are updated as

$$Y_i(t+1) = \begin{cases} Y_i(t) + \alpha_{w,i}(t)\,[X(t) - Y_i(t)] & \text{if } i \in N_c(t) \\ Y_i(t) & \text{if } i \notin N_c(t) \end{cases} \tag{3}$$

Here, $N_c(t)$ is the the size of the neighborhood and is shrunken gradually with the lapse of time and $\alpha_{w,i}(t)$ is the step size of the adaptation of the weights, which is represented by a Gaussian weighting function as

$$\alpha_{w,i}(t) = \alpha_o(t)exp(-\frac{\| r_i - r_w \|^2}{\sigma(t)^2}) \tag{4}$$

where $\alpha_o(t)$ and $\sigma(t)$ are decreased with the lapse of time t. The distance term $\| r_i - r_w \|^2$ is replaced by the distance between two weight vectors $\| Y_i - Y_w \|^2$ because the topological distance is no longer meaningful in the problem of VQ codebook design.

3 Codebook Design by evolution

GAs [12] are the population-based searching and optimization techniques that encode a potential solution to a specific problem on a simple chromosome-like data structure and apply the genetic operators (selection, recombination, and mutation) to the chromosomes in order to achieve more good solution. It is believed that the GAs can provide more possibility of obtaining the global optimal solution because their trials are allocated in an exponentially differentiated way to a large number of subsets based on implicit competitions. GAs have been proven to be powerful methods in search, optimization and machine learning [13]. They encode a potential solution to a specific problem on a simple chromosome-like data structure and apply recombination operators to these structures to achieve optimization. When GA is applied to find an optimal VQ codebook, we need to consider the following problems.

3.1 Chromosome Representation

Assume that (1) the codebook size is constant and predetermined with N codevectors, (2) each codevector consists of k components, and (3) each component is a real number. Then, each chromosome (codebook) consists of $N \times k$ consecutive real numbers, where the first k real numbers are corresponding to the first codevector, the next k real numbers to the second codevector, etc. We adopt a real number representation of chromosomes [14] since each component of codevector is a real number.

3.2 Initial Population Generation

The initial population is critical for determining the convergence speed of GAs and the distortion of the resultant codebook. The initial lth components of M codevectors in the ith codebook are obtained from the uniform distributed randomly generated values as

$$Y_i(l) = [U_1[a_l, b_l], U_2[a_l, b_l], \cdots, U_M[a_l, b_l]]^T$$
$$i = 1, 2, \cdots, N, \quad and \quad l = 1, 2, \cdots k, \tag{5}$$

where N and M are the number of codebooks and the size of codebook, respectively, and $U_j[a_l, b_l](j = 1, 2, \cdots, M)$ is the lth component of the jth codevector that is obtained from the uniform distributed random number generator, where a_l and b_l are the lower and upper bound of the lth component of codevector, respectively. It is necessary to repeat the above random number generation $N \times M \times k$ times for the initial population.

3.3 Fitness Function

Each chromosome (codebook) produces the reconstructed vector $\hat{X}$ with respect to the original vector X. Then, the average distortion D over the training vectors $T = \{X_1, X_2, \cdots, X_{N_T}\}$ is represented as

$$D = \frac{1}{N_T} \sum_{t=1}^{N_T} d^2(X_t, \hat{X}_t) = \frac{1}{N_T} \sum_{t=1}^{N_T} \sum_{l=1}^{k} (x_{t,l} - \hat{x}_{t,l})^2, \tag{6}$$

where $x_{t,l}$ and $\hat{x}_{t,l}$ are the lth component of the tth training vector and reconstructed vector, respectively, N_T and k are the number of training vectors and the size of training vector, respectively. The fitness function F_i^G of the ith chromosome used for our codebook design is related to the ith codebook's average distortion D_i as $F_i^G = \frac{1}{1+D_i}$.

From this relation, it is noted that the small average distortion means the larger fitness function, which is more familiar with our understanding. Further, it is remarkable that the fitness function seems to satisfy the necessary condition [16] for the encoder to be optimal under the given codebook in that the chromosomes with higher fitness values will form good partitions of vectors since they produce the smaller average distortions.

3.4 Genetic Operations

The initial population of codebooks is then evolved by appropriate genetic operations in order to find an optimal codebook for the given images. The detailed explanation about the genetic operations used for an optimal codebook design is given as follows.

Reproduction We use a mixture of selection methods for reproducing the chromosomes. The first selection method is an elitism that the best chromosome with the highest fitness value is passed in the new population. The second selection method is a modified k-tournament method [15]. In this method, a chromosome having the best fitness value among the k chromosomes randomly selected from the upper class of chromosomes according to their fitness values is chosen for the reproduction. Two codebook chromosomes C' and C'' obtained by repeating the above procedure consecutively create a new chromosome C by applying the crossover and mutation operations explained later. New chromosome C is replaced by a certain chromosome $\tilde{C}$ that has the worst fitness value among the k chromosomes randomly selected from the lower class of chromosomes according to their fitness values. The above reproduction procedure is repeated as many times as *pselect* $\times |P|$, where $|P|$ is the population size. Finally, the remaining portion of the population set is filled by copying the population set in the order of magnitude of fitness values. Fig. 1 shows a hybrid reproduction method based on a mixture of three different reproduction methods.

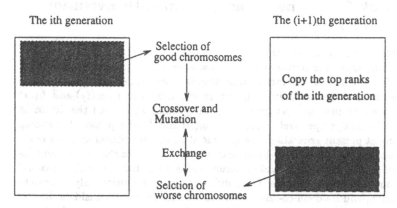

The ith generation

The (i+1)th generation

Selection of good chromosomes

Copy the top ranks of the ith generation

Crossover and Mutation

Exchange

Selction of worse chromosomes

Fig. 1. An illustration of the proposed hybrid reproduction scheme.

Crossover Basically, the crossover between two selected parent chromosomes C' and C'' is performed as follows. First, the codevectors of two selected chromosomes are matched by associating one vector in C' with the most similar (close) one in C''. where N_i' and N_i'' are the number of training vectors encompassed by the ith codevectors Y' and Y'', respectively. Second, the ith codevector Y_i of the new chromosome C that is obtained from the ith codevectors Y_i' and Y_i'' of two selected chromosomes C' and C'' is computed by the weighted centroid as

$$Y_i = \frac{N_i'Y_i' + N_i''Y_i''}{N_i' + N_i''}, \tag{7}$$

where N_i' and N_i'' are the number of training vectors encompassed by the ith codevectors Y' and Y'', respectively. This new codevector minimizes the merging errors ME_i defined by

$$ME_i = \frac{N_i'N_i''}{N_i' + N_i''} \| Y_i' - Y_i'' \|^2. \tag{8}$$

This operation is applied to the overall codevectors of two selected codebooks in the crossover probability P_c. It is remarkable that the crossover operation satisfies the necessary condition [16] for the decoder to be optimal for a given partition in that new codevectors are just the arithmetic average weighted by the number of included training samples.

Mutation Mutation over new chromosome C is performed as follows. First, a codebook $\check{C}$ is selected randomly from the population in the mutation probability P_m. Second, some codevectors in the selected codebook are randomly selected and they are mutated by the following. Let one of mutated codevectors in the selected codebook be $\check{Y}_i^j$. Then, the lth component of $\check{Y}_i^j$ is changed as

$$\check{Y}_i^j(l) = \text{random}\left[\check{Y}_i^{j-1}(l), \check{Y}_i^{j+1}(l)\right] \tag{9}$$

where $\text{random}[a, b]$ implies a randomly generated number between a and b.

4 Codebook Design by Co-adaptation with Evolution and Learning

Before introducing a co-adaptive algorithm for an optimal codebook design, we illustrate how learning affect the course of evolution as shown in Fig. 2 graphically. Here, $t_s, g_i(t_s)$ and $f_i(t_s)$ represent the starting time at present generation, the genome at t_s, and the fitness at t_s of the ith individual, respectively. Similarly, $t_f, g_i(t_f)$ and $f_i(t_f)$ represent the starting time at next generation, the genome at t_f, and the fitness at t_f of the ith individual, respectively. Further, $t_l, g_i(t_l)$ and $f_i(t_l)$ represent the ending time of learning at present generation, the genome at t_l, and the fitness at t_l of the ith individual, respectively. Notice that the 'o' and 'x' symbols denote the genome and fitness at a specific time, respectively and the amount of learning time for each individual is different from each other. Consider three different cases : evolution only, Lamarckian co-adaptation, and Baldwin co-adaptation. In the first case that considers the GA only without learning, the ith genome $g_i(t_f)$ is reproduced by changing genetically the ith genome $g_i(t_s)$ selected by the fitness $f_i(t_s)$. In the second case that considers the evolution with Lamarckian learning, the ith genome $g_i(t_f)$ is reproduced by changing genetically the ith genome $g_i(t_l)$ selected by the fitness $f_i(t_l)$. In the third case that considers the evolution with Baldwin learning, the ith genome $g_i(t_f)$ is reproduced by changing genetically the ith genome $g_i(t_s)$ selected by the fitness $f_i(t_l)$.

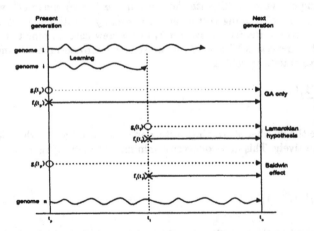

Fig. 2. Three different ways how learning influences on evolution.

Fig. 3 represents the co-adaptation scheme for an optimal codebook design. In this work, we adopt three different codebook design methods that are based on the evolution only, Lamarckian co-adaptation, and Baldwin co-adaptation, respectively. Among the many possible fitness functions, we choose the fitness function as

$$F_i^G = \frac{k_1}{1 + D_i'} \quad (Evolution\ only), \tag{10}$$

$$F_i^L = \frac{k_1}{1 + |t_l - t_s|} \times \frac{k_2}{1 + D_i'} \quad (Lamarckian\ learning), \tag{11}$$

$$F_i^B = \frac{k_1}{1 + |t_l - t_s|} \times \frac{k_2 \cdot |D_i' - D_i|}{D_i} \quad (Baldwin\ learning), \tag{12}$$

where t_s and t_l are the starting and ending time of learning, respectively, and D_i and D_i' are the ith codebook's distortion measure before and after learning, respectively. The first terms in the F_i^L and F_i^B reflect how rapidly learning is performed and the second terms in the F_i^L and F_i^B reflect the error measure after learning (Lamarckian co-adaptation) and the error difference between before and after learning (Baldwin co-adaptation).

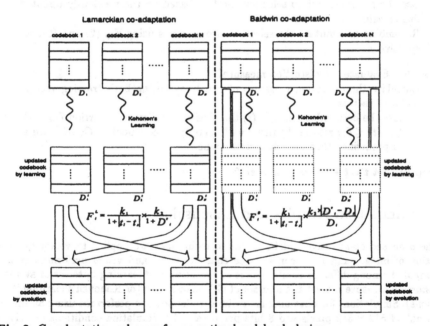

Fig. 3. Co-adaptation schemes for an optimal codebook design.

From the above figures, we introduce a new co-adaptation algorithm that incorporates evolution with learning for an optimal codebook design as follows.

Step 1: Initialization.
Prepare a training set $T = \{e_1, e_2, \cdots, e_{N_T}\}$, and set a time index $t = 0$.
Given a fixed number N, form a pool of N chromosomes $C(t) = \{C_1(t), C_2(t), \cdots, C_N(t)\}$, where each chromosome is corresponding to individual codebook.
Since each codebook consists of M codevectors and each codevector has k components, it needs to generate $M \times k$ real numbers uniformly at random

from the range $[x_{min}, x_{max}]$, where x_{min} and x_{max} are the lower and upper bound of one component of codevectors.

Step 2: Train each codebook.

Train each codebook using the Kohonen's learning rule (Eq. 3) until the resultant codevectors are not changed any more. The learning time for each codebook to be converged will be different because the learning environment (i.e., learning parameter and/or current codevectors) of each codebook is distinct each other.

Step 3: Calculate fitness value of each codebook.

The calculation of fitness value is depending on which type of learning is considered.

Use Eq. (10), Eq. (11) and Eq. (12) when considering no learning, Lamarckian co-adaptation, and Baldwin co-adaptation, respectively.

Step 4: Update the codevectors in the codebooks using the genetic algorithms.

Use the proposed hybrid selection method based on the previously updated fitness value.

Recombine and mutate the selected chromosomes using Eq. (7) and (9), respectively.

Step 5: Evaluate the distortion measure.

Calculate the distortion measures of codebooks using both training vectors and test vectors.

Choose the best codebook $C_{best}(t)$ that has the minimum distortion $D_{min}(t)$. If $D_{min}(t) \leq \delta$, halt the algorithm with an optimal codebook as $C_{best}(t)$. Here, δ is a predefined threshold distortion value.

Step 6: Set $t = t + 1$ and go to Step 2.

5 Simulation Results and Discussion

The proposed co-adaptive codebook design techniques are applied to vector quantization of images and their convergence characteristics and visual performances are compared among different combinations of evolution and learning in terms of average distortion measure against the number of GA generations (or Kohonen's learning iterations). Simulations for obtaining an optimal codebook are performed using an image "Lena" of 512×512 pixels and 8 bits per pixel. The simulation conditions are given as follows. The overall bpp (bit per pixel) R is assumed 0.25 in all simulations.

Recently, wavelets have been widely introduced as an effective tool for the analysis and representation of signals [17]. The wavelet transform (WT) produces a time-scale representation that is sensitive in time at small scales, and sensitive in frequency at large scales. It has been widely applied to image coding (compression) based on multiresolution signal decomposition since it does not suffer from the block distortion in contrast to transform image coding, and the quantization noise generated in a particular subband is limited such that it is not allowed to spread to other subbands. So, we choose the wavelet transform as the first level of signal decomposition. In our simulations, the images are decomposed into three subband levels (10 subbands) by the orthogonal wavelet transform [17,18] using the filter coefficients shown in Table 1 [19].

Table 1. The LPF's coefficients for decomposition and reconstruction.

Decomposition $h_0(n)$		Reconstruction $h_1(n)$	
$h_0(-2)$	-0.125		
$h_0(-1)$	0.25	$h_1(-1)$	0.25
$h_0(0)$	0.75	$h_1(0)$	0.5
$h_0(1)$	0.25	$h_1(1)$	0.25
$h_0(2)$	-0.125		

The subbands decomposed by biorthogonal wavelet transform are denoted by $I_d(i,j)$, where an index d represents the subband level and an index pair (i,j) represents the frequency band. The three subbands in the third level $I_2(0,1), I_2(1,0)$, and $I_2(1,1)$ are not processed since they are visually-insignificant. This elimination also reduces the computational time without significant degradation of visual performance of the reconstructed image. The three subbands in the second subband level $I_1(0,1), I_1(1,0)$, and $I_1(1,1)$ and the other three subbands in the first level $I_0(0,1), I_0(1,0)$, and $I_0(1,1)$ are vector quantized (vector size $k=16$ and $k=4$, respectively) and their codebooks are made by the conventional generalized Llyod algorithm (GLA) method [3]. The lowest frequency subband $I_0(0,0)$ is vector quantized with vector size $k=2$, and this subband is processed in a different way with the upper subbands because it is the most visually significant subband. In this work, four different methods of designing the codebook for the subband $I_0(0,0)$ are considered as Simulation 1: Kohonen's learning only, Simulation 2: evolution (GA) only, Simulation 3: evolution + learning (Lamarckian co-adaptation), and Simulation 4: evolution + learning (Baldwin co-adaptation). Fig. 4 illustrates an overall encoding and decoding procedure that has been used in our simulation.

Many sets of training vectors are necessary for determining the codebooks by the proposed co-adaptive methods. We select sets of training vectors such that they can reflect the coherent characteristic among the wavelet coefficients in each block. For example, the training vectors in the frequency bands of (0,1) are made by taking several adjacent pixels located in the direction of horizontal axis since they have many horizontal edge components. Similarly, the training vectors in the frequency bands of (1,0) are made by taking several adjacent pixels located in the direction of vertical axes since they have many vertical edge components. Also, the training vectors in the frequency bands of (1,1) are made by taking several adjacent pixels located in the direction of both horizontal and vertical axis since they have many diagonal edge components. Fig. 4 also illustrates three different scanning patterns for making the training vectors in three distinct blocks.

5.1 Convergence Characteristics

First, the convergence characteristics of the proposed codebook design methods are compared in terms of the distortion measure (See. Eq. (6)) in accordance with the number of GA evolutions or Kohonen's learning among four different design methods such as Kohonen's learning only, GA evolution only, evolution + leaning (Lamarckian co-adaptation), and evolution + learning (Baldwin co-adaptation). Table 2 shows the execution parameters of evolution and/or learning used in our simulations, where these parameters are experimentally determined from many trials. If a codebook design

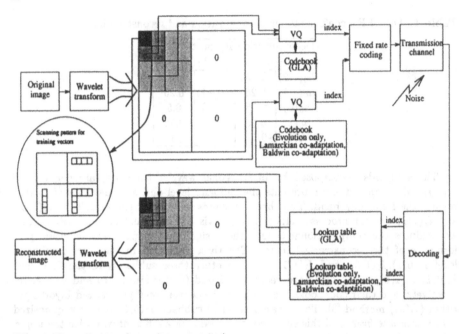

Fig. 4. Overall procedure of our simulations.

method does not depend on a certain execution parameter, it is denoted by a symbol '-' in a corresponding entry in the table.

Table 2. Execution parameters used for evolution and/or learning.

	Kohonen's Learning only	GA Evolution only	Evolution+Learning (Lamarckian co-adaptation)	Evolution+Learning (Baldwin co-adaptation)
No. of generations	-	200	200	200
Population size	1	100	100	100
Chromosome size	$2 \cdot 2^4$	$2 \cdot 2^4$	$2 \cdot 2^4$	$2 \cdot 2^4$
No. of k-tournament	-	3	3	3
pselect	-	0.3	0.3	0.3
P_c	-	0.7	0.7	0.7
P_m	-	0.3	0.3	0.3
$N_c(t)$	$100 \to 0.1$	-	$100 \to 0.1$	$100 \to 0.1$
$\alpha_c(t)$	$5 \times 10^{-3} \to 10^{-6}$	-	$5 \times 10^{-3} \to 10^{-6}$	$5 \times 10^{-3} \to 10^{-6}$
$\sigma(t)$	$100 \to 0.1$	-	$100 \to 0.1$	$100 \to 0.1$

Fig. 5 shows the evolution curves of distortion measure over GA generations while the codebooks for $I_0(0,0)$ subband are training. Each evolution curve represents the distortion measure D_i of the best codebook within the population at a specific GA's generation, where the distortion measure D_i is the mean squared error between the training vector and the vector recovered from the best codebook at the present generation (See Eq. (6)). The curves are obtained from three different types of co-adaptation such as evolution only, evolution + learning (Lamarckian co-adaptation), and evolution + learning (Baldwin co-adaptation), respectively. The convergence speed among three different co-adaptive codebook designs is compared by investigating the number of GA generations whose distortion measure reaches at 90% of its final value. It is found that the numbers of GA generations reaching at 90% of the final distortion measure are 168, 124, and 52 in the case of evolution only, evolution + learning (Baldwin co-adaptation), and evolution + learning (Lamarckian co-adaptation), respectively. From this result, it is noted that (1) learning of codebooks allows to evolve much faster than their nonlearning equivalents, and (2) Lamarckian co-adaptation of learning is more effective than Baldwin co-adaptation of learning in the case of codebook design problem.

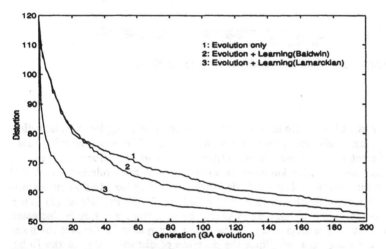

Fig. 5. Evolution curves of different codebook design methods.

Fig. 6 shows the learning curves of distortion measure in the very beginning of Kohonen's learning iterations while the codebooks for $I_0(0,0)$ subband are training. Each learning curve represents the distortion measure D_i of the best codebook within the population of each different co-adaptation method at each Kohonen's learning, where the distortion measure D_i is the mean squared error between the training vector and the vector recovered from the best codebook at the present iteration (See Eq. (6)). The curves are obtained from represents three different learning methods such as Kohonen's learning only, evolution + learning (Lamarckian co-adaptation), and evolution + learning (Baldwin co-adaptation), respectively. The parameters for the Kohonen's learning were experimentally chosen from many trials by the following. The neighborhood width $N_c(t)$ is decreased $100 \to 0.1$ with the lapse of time. Similarly, the magnitude parameter the $\alpha_o(t)$ and the width parameter $\sigma(t)$ of Eq. (4) are decreased $5 \times 10^{-3} \to 1 \times 10^{-6}$ and $100 \to 0.1$, respectively, with the lapse of time.

The convergence characteristics among different types of learning are compared by investigating how the learning curves are updated. From this simulation, we have obtained some valuable results by the following. (1) The learning curve does not converge well when the Kohonen's learning is used alone. When the fitness value is not changed any more by Kohonen's learning, we restart the Kohonen's learning with different initial values of the neighborhood parameter $N_c(t)$, and other parameters $\alpha_o(t)$

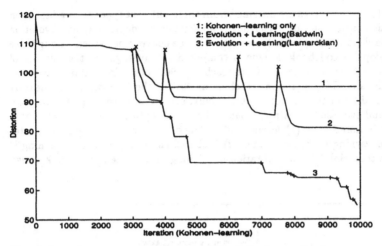

Fig. 6. Learning curves of different codebook design methods.

and $\sigma(t)$. This restart is effective in the beginning stage of learning but it is less effective as the learning is advanced progressively. We tried the Kohonen's learning many times with different values of parameters, $N_c(t)$, $\alpha_o(t)$, and $\sigma(t)$. However, we could not find any case that the distortion measure is converged within a tolerable number of Kohonen's learning iterations. It is concluded that it was very hard to find an optimal codebook with large codebook size (2×2^6) by Kohonen's learning alone. (2) When the Baldwin co-adaptation of learning is used, the distortion measure is decreased gradually as the learning is progressing. In the figure, the symbol 'x' denotes the point where the GA has been occurred. Since the decrease of distortion due to the Kohonen's learning does not change directly the genotype of chromosome in the case of the Baldwin co-adaptation of learning, i.e., evolutionary change is depending to the distortion measure before learning, the distortion measure at the point of GA occurrence is increased suddenly. However, the distortion measure at the occurrence of GA tends to decrease gradually. (3) When the Lamarckian co-adaptation of learning is used, the distortion measure is also decreased gradually as the learning is progressing. In the figure, the symbol '+' denotes the point where the GA has been occurred. Since the decrease of distortion due to the Kohonen's learning changes directly the genotype of chromosome in the case of the Lamarckian co-adaptation of learning, i.e., evolutionary change is depending to the distortion measure after learning, the distortion measure at the point of GA occurrence is decreased continuously. This ends up to accelerate the speed of evolution and therefore, the Lamarckian co-adaptation of learning is the most effective in the application of an optimal codebook design problem. Notice that there seems to be no learning between two consecutive GA evolutions in a certain learning interval. Kohonen's learning is existed between these two GA evolutions in reality, but the learning degrades the fitness value on the contrary. So, we ignore the learning effect between these two GA evolutions intentionally and perform the GA evolution based on the distortion measure before learning. (4) This distortion behavior of each learning method is appeared consistently, i.e., the distortion curves look similar with Fig. 6 even though some evolution and learning parameters have been changed.

5.2 Visual Performance

The visual quality of the reconstructed image is compared among different combinations of evolution and learning : Kohonen's learning only, GA evolution only, evolution + learning (Lamarckian co-adaptation), and evolution + learning (Baldwin co-adaptation). In this work, we do not consider the case of Kohonen's learning only because the method does not provide an optimal codebook within a tolerable computation time. To make the comparison fair, each codebook is taken from the best one at the 65th GA generation of each different co-adaptive codebook design method. The selected codebooks are used to reconstruct the lowest frequency band $I_0(0,0)$ and the wavelet coefficients of other frequency bands are equal among different co-adaptive codebook design methods.

Fig. 7. Reconstructed images from different codebook design methods.

Fig. 7 shows the reconstructed images obtained from three different codebooks as Upper left : original image, Upper right : evolution + learning (Lamarckian co-adaptation), Lower left : evolution + learning (Baldwin co-adaptation), and Lower right : GA evolution only. From the subjective visual test, it is judged that the reconstructed image is ranked good in the order of the Lamarckian co-adaptation, the Baldwin co-adaptation, and GA evolution alone. This decision is supported by the distortion measures at a specific generation (=60th generation) whose values are 57.2, 65.4, and 70.2 in the order of the Lamarckian co-adaptation, the Baldwin co-adaptation, and GA evolution only, respectively, as shown in Fig. 8.

6 Conclusion

We built an appropriate co-adaptation computational model that effectively applied these biologically inspired co-adaptation schemes to an optimal codebook design prob-

238

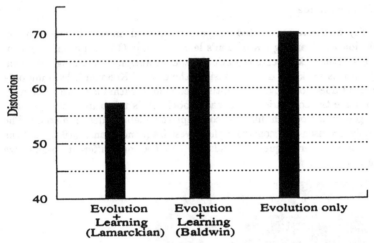

Fig. 8. Distortion measures at the 65th generation of different codebook design methods.

lem and evaluated the convergence speed to reach the satisfactory codebooks and the visual performance of resultant codebooks obtained from the different co-adaptation schemes. We found that co-adaptation schemes that combined evolution and learning provided faster convergence than nonlearning equivalents and Lamarckian co-adaptation was more effective method for the optimal codebook design problem than Baldwin co-adaptation. Further, the resultant codebook from the co-adaptation schemes provided better visual performance than nonlearning equivalents after the same amount of GA generations. We also showed that the co-adaptation codebook design methods satisfied two necessary but not sufficient conditions for the existence of an optimal vector quantizer in the sense of minimum MSE such that (1) the fitness function satisfies the necessary condition for the encoder to be optimal under the given codebook because the chromosomes with higher fitness values will form good partitions of vectors because they produce the smaller average distortions and (2) the crossover in the genetic operations satisfies the necessary condition for the decoder to be optimal for a given partition because new code vectors are just the arithmetic average weighted by the number of included training samples.

References

1. Gersho A., Gray R. M. (1992) *Vector quantization and signal compression*, Kluwer Academic Publishers
2. Nasrabadi N. M., King R. A. (1988) Image coding using vector quantization: A Review, *IEEE Transaction on Communication*, **36**, 2166-2173
3. Linde Y., Buzo A., Gary R. M. (1980) An algorithm for vector quantizer design, *IEEE Transaction on Communication*, **28**, 84-95
4. Kohonen T. (1989) *Self-Organization and Associative Memory*, Spring-Verlag, Berlin
5. Nasrabadi N. M., Feng Y. (1988) Vector quantization of images based upon Kohonen's self-organizing feature map, *IEEE Int. Conf. on Neural Networks*, **1**, 101-108

6. Karayiannis N. B., Pai P. (1996) Fuzzy algorithms for learning vector quantization, *IEEE Trans. on Neural Networks*, **7**, 1196-1211
7. Lamarck J. B. (1914) Of the influence of the environment on the activities and habits of animals, *Zoological Philosophy*, **1**, 106-127
8. Ackley D. E., Littman M. L. (1994) A case for Lamarckian evolution, In: C. G. Langton (ed) *Artificial Life III*, Addison-Wesley, 3-10
9. Baldwin J. M. (1896) A new factor in evolution, *The American Naturalist*, **30**, 441-451
10. Hinton G. E., Nowlan S. J. (1996) How learning can guide evolution, In: Belew R. K., Mitchell M. (eds), *Adaptive Individuals in Evolving Populations : Models and Algorithms*, Addison Wesley, 447-454
11. Parisi D., Nolfi S. (1996) The influence of learning on evolution, In: Belew R. K., Mitchell M. (eds), *Adaptive Individuals in Evolving Populations : Models and Algorithms*, Addison Wesley, 419-430
12. Holland J. H. (1975) *Adaptation in natural and artificial systems*, University of Michigan Press
13. Goldberg D. E. (1989) *Genetic Algorithms in Search, Optimization and Machine Learning*, Addison Wesley Press
14. Wright A. H. (1991) Genetic algorithms for real parameter optimization, In: Rawlins G. (ed), *Foundations of Genetic Algorithms*, Morgan Kaufmann Publishers, 250-220
15. Kim Daijin, Ahn Sunha (1999) A MS-GS VQ codebook design for wireless image communication using genetic algorithms, *IEEE Trans. on Evolutionary Computation*, **3**, 35-52
16. Lloyd S. P. (1982) Least squared quantization in PCM, *IEEE Transaction on Information Theory*, **28**, 127-135
17. Mallat S. G. (1989) A theory of multiresolution signal decomposition: The wavelet representation, *IEEE Transactions on Pattern Analysis and Machine Intelligence*, **11**, 674-693
18. Antonini M., Barlaud M., Mathieu P., Daubechies I. (1994) Image coding using wavelet transform, *IEEE Transactions on Image Processing*, **3**, 367-381
19. Cheong C. K., Aizawa K., Saito T., Hatori M. (1992) Subband image coding with biorthogonal wavlets, *IEICE Trans. Fundamentals*, **75**, 871-881
20. Li W., Zhang Y. (1994) A study of vector transform coding of subband-decomposed images, *IEEE Transaction on Circuits and Systems for Video Technology*, **4**, 383-391

Chapter 3: Evolutionary Computation

Papers:

An Emergence of Coordinated Communication in Populations of Agents with Evolution Simulated by Genetic Algorithm
V. Kvasnicka

Migration and Population Dynamics in Distributed Coevolutionary Algorithm
J. Pospichal

Royal Road Encodings and Schema Propagation in Selective Crossover
K. Vekaria and C. Clack

An Evolutionary Approach for the Design of Natural Language Parser
O. Unold

GA-Based Identification of Unknown Structured Mechatronics System
M. Iwasaki, M. Miwa, and N. Matsui

Integrating Genetic Algorithms and Interactive Simulations for Airbase Logistics Planning
N. L. Schneider, S. Narayanan, and C. Patel

Evaluation of Virtual Cities Generated by Using a Genetic Algorithm
N. Kato, H. Kanoh, and S. Nishihara

An Emergence of Coordinated Communication in Populations of Agents with Evolution Simulated by Genetic Algorithm

Vladimir Kvasnicka

Department of Mathematics
Slovak Technical University
812 37 Bratislava, Slovakia
email: kvasnic@cvt.stuba.sk

Abstract: The purpose of this communication is to demonstrate that in a population composed of agents that are capable of simple cognitive activities spontaneously emerges coordinated communication between agents. Each agent is characterized by meaning vectors (internal states) represented by n-dimensional binary vectors. Cognitive activities of agents are performed by simple formal device represented by mappings that map binary vectors onto symbol strings (signals) and conversely. An elementary communication act consists in (1) a random selection of two agents, where one of them is declared as speaker and the other one as listener, (2) the speaker codes its randomly selected meaning vector into a sequence of symbols and sends it to the listener as a message, and finally, (3) the listener decodes this received message into a meaning vector. A Darwinian evolution of population is simulated by simple version of genetic algorithms, where agent mappings are considered as chromosomes and their fitness is evaluated on the basis of distances between speaker meaning vector and listener meaning vector constructed from the received messages. If these distances are small, then for both agents, speaker and listener, fitness is increased. It is demonstrated that in the course of evolution agents gradually improve decoding of the received messages (they are closer and closer to meaning vectors of speakers) and all agents gradually start to use tightly related cognitive devices, i.e. all agents start to use the same vocabulary for common communication. Moreover, if agent meaning vectors contain regularities, then these regularities are manifested also in messages created by agents – speaker, i.e. similar parts of meaning vectors are coded by similar symbol substrings. This observation is considered as a manifestation of an emergence of a grammar system in the common coordinated communication.

Key words: genetic algorithms, coordinated communication, emergence, agent, Darwinian evolution.

1. Introduction

Human language [1-3] makes it possible to express a huge number of quite different meanings by token sequences composed of a small number of simple elements, and to interpret such sequences by the meanings that they contain. A standard meaning of the term "grammar" refers to the systematic regularities between meanings and their representation by token sequences in a language. These structural regularities of a language constitute a basis for expressions of novel meaning combinations. The hearer can from the regularities accurately interpret the received sequences as involving those familiar structures and relations, even though their specific combinations may have never been used before. This means that a communication system endowed by a grammar can be used to express new unusual meanings relevant to specific situations. The ability to communicate by a system composed of structural regularities represents an important achievement of a species, for which coordinated social activities are vital to survive, where the accurate communication between two individuals from the same species represents the definite selection advantage. Having this benefit it is natural to explain language communication as the result of Darwinian natural selection [4].

The purpose of this paper is to study a hypothesis that coordinated communication together with grammar regularities are results of an evolutionary process running in a population composed of individuals – agents. These agents are endowed by an ability to perform simple cognitive activities, and it is presumed that at the beginning of the evolution the coordinated communication between agents doesn't exist. Under the term *coordinated communication* between population of agents we understand such an exchange of messages between agents that is unified (with the same semantic contents) for the whole population. For an elementary communication act the term "coordinated" implies that both speaker and listener understand each message in the same way. This requirement is formally expressed by a sequence of elementary steps in the communication act, the speaker's internal state (e.g. corresponding to an internal representation of the environment) is coded into a signal - message (token sequence) received by the listener, who decodes this message in a form of internal state:

> A selection of speaker's internal state,
> ⇒ transformation of this internal state into a message represented by a token sequence,
> ⇒ the speaker sends the message to the listener,
> ⇒ the received message is decoded into a form of internal state.

From this simple communication scheme immediately follows that agents should be capable of simple *cognitive activities* that consist in coding of internal states into messages represented by token sequences (direct cognitive activity) and also in decoding of these received messages into internal states (inverse cognitive activity). Since our goal is to study an emergence of the coordinated communication for a population of agents capable of pairwise elementary communications, we have to introduce into the whole system also a learning process. The listener compares internal states constructed from the received messages with original internal states of speakers. If these internal states, original and decoded, are different, then listener (often also speaker) modifies parameters of his (or her) cognitive device so that this type of the difference between internal states in the forthcoming history of both agents is minimized. We presume that the speaker's internal states correspond to some external surrounding reality, which can be determined also by a listener. Therefore we postulate that a listener can find out in some other way the internal state of the speaker.

The creation of coordinated communication between agents can be interpreted as an emergence of new phenomenon in population of agents that are capable of simple pairwise communication accompanied by a learning process. Starting from the general ideas of *Darwin evolutionary theory*, an origin of coordinated communication in a population of agents may be sought in a fact that an ability of coordinated communication substantially increases agent fitness, i.e. there exists *a selection pressure for a spontaneous emergence of coordinated communication*. Individuals that are incapable of coordinated communication are strongly handicapped. Their evaluation by fitness is substantially smaller than an evaluation of those agents that are capable of coordinated communication. Frequency of appearance of individuals that are not able of coordinated communication decreases in the course of evolution.

Metaphor of Darwinian evolution [5] is applied to the population of agents with elementary pairwise communication as follows: A couple of agents – parents is quasirandomly selected from the population, where the probability of their selection is proportional to their fitness. Applying a reproduction process to this couple of agents, another couple of agents – offspring is created, which replaces in the population a couple of quasirandomly selected agents with small fitness. In this approach the term fitness expresses not only an exactness of prediction of internal states on the basis of received messages of speakers but also their mean lengths. Such agents are preferred, whose cognitive devices capable of unambiguous mapping of internal states onto signals (token sequences). These signals are restricted by a requirement that their length (number of tokens) is as short as possible. This second requirement of a minimal length of a created message supports a spontaneous emergence of coordinated communication that uses not only a simple "vocabulary" composed of short signals. The emerged cognitive devices must have built in a "grammar" that maps similar internal states (meaning) onto similar signals (i.e. grammar saves structural regularities appeared in internal states such that they appear also in signals). For instance, let us assume that we have a set of such internal states that have the same common "main" part and a "side" part, which is slightly varied from state to state. Then we expect that agents start to produce in the course of evolution such signals that reflect structural regularities of internal states, i.e. all the signals have the same common part and their side parts are varied in dependence on the corresponding side parts of meaning vectors. This ability of agent's cognitive devices to save structural regularities of internal states is considered as an indication of the emergence of grammar in the evolution of coordinated communication. In the opposite case, if the length of produced signals is not restricted, the evolution of coordinated communication is "directed "with high probability to such areas of fitness landscape where each internal state is coded by a different signal with minimal (or almost nonexistent) exploitation of structural regularities. Loosely speaking, an emerged language would be similar to a Chinese like language with many thousands different symbols and with very simple grammar (if any).

The problem of emergence of coordinated communication, where cognitive activities of agents are performed by neural networks or by other formal devices, is very intensively studied in current literature in evolutionary linguistics [6,7] and as well as in artificial life [10-16]. The present paper is based on an idea that agent cognitive devices are represented by simple mappings that are composed of rewriting rules. These rewriting rules are applied for mapping of internal states onto token strings (signals) and vice versa. Moreover, each rewriting rule is evaluated by a counter (represented by a nonnegative integer), it allows to some extent an inclusion of learning processes by comparing the resulting listener's internal state with the original speaker's internal state.

2. Specification of Agents

Let $P = \{A_1, A_2, ..., A_p\}$ be a population of agents (see Fig. 1). Agents mutually interact such that between a couple of agents runs an asymmetric interaction - communication, where the first agent is declared as a speaker and the second agent is declared as a listener. We will postulate the following four properties of agents (see Fig. 2)

1. Each agent is capable to send signals to another agent, where signals are strings of symbols $\{a, b, ...\}$.

2. Internal states of agents are described by binary meaning vectors $\alpha = (\alpha_1, \alpha_2, ..., \alpha_n) \in \{0, 1\}^n$.

3. A "brain" of agents is modeled by a mapping (composed of rewriting rules) that is capable of the following two mutually inverse activities:
   - maps received signals into internal states and
   - maps internal states into signals (in inverse task to the previous direct activity).

The transformation process of a binary meaning vector into a string composed of symbols $\{a, b, c, ..\}$ is formally determined as follows

$$G : \{0,1\}^n \rightarrow \{a,b,c,...\}^*$$ (1a)

$$\beta = G(\alpha)$$ (1b)

where $\{a,b,c,...\}^*$ is a set composed of all possible strings of q symbols $a,b,c,...,$
$\{a,b,c,...\}^* = \{a,b,...,aa,ab,...,aab,...,baac,...\}$. An inverse transformation H with respect to the transformation (1a-b) maps symbol strings into binary meaning vectors

$$G^{-1} = H : \{a,b,c,...\}^* \rightarrow \{0,1\}^n$$ (2a)

$$\alpha = G^{-1}(\beta) = H(\beta)$$ (2b)

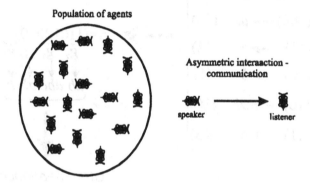

Figure 1. Population $P = \{A_1, A_2, ..., A_p\}$ is composed of a finite number of agents. In the process of communication {an asymmetric interaction) between two agents, the first agent - speaker sends a message to the second agent - listener.

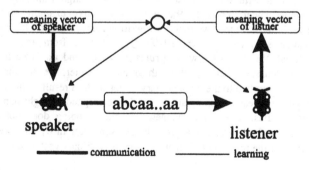

Figure 2. Schematic representation of an elementary asymmetric act of communication between two agents. Agent speaker maps its internal state (represented by a binary meaning vector) onto a signal (represented by symbol strings). An agent listener receives this signal and it transforms the received signal in a form of binary meaning vector. In the process of learning both speaker and listener modify their cognitive devices in order to decrease a difference between meaning vector of speaker and listener.

The mappings G and H may be realized by many different ways, e.g. by recurrent neural networks which were used in recent Batali paper [11]. The purpose of the present communication is to apply a different approach for the realization of these mappings. They are determined by sets of simple rewriting rules accompanied by integer counters and priority rules of their application (see Figs. 3 and 4).

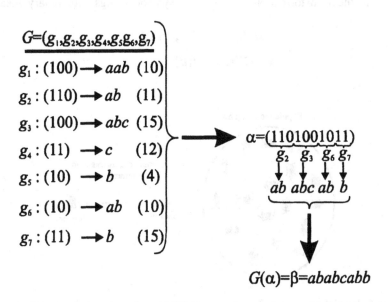

Figure 3. Illustrative example of an application of the mapping G (composed of seven rewriting rules) to a binary vector $\alpha=(1101001011)$. In the first step we look for such rewriting rules g that perfectly match the initial part of the binary vector α. In this case three rewriting rules g_2, g_4, and g_7 exist. If we get more than one such rule, then we select quasirandomly a rule with the greatest counter. This is algorithmically performed by Goldberg's roulette wheel approach [9], in our illustrative example we selected the rewriting rule g_2. In the next step we look for rewriting rules with such left-hand sides that perfectly match a subvector in $\alpha=(1101001011)$ starting from the fourth position (underlined entries). We get four rewriting rules g_1, g_3, g_5, and g_6. From these we select quasirandomly by roulette wheel a rule with greatest counter, we get the rule g_3. The method is repeatedly applied to the whole binary vector α. As a result of the used technique is a string of symbols created step-by-step from the corresponding right-hand sides of the selected rewriting rules. There may happen that the mapping does not contain such rewriting rule that perfectly matches a given binary subvector in α, then we say that the current application of the mapping G is inapplicable to the binary vector α.

Let the mapping G be specified as an ordered sequence of m rewriting rules

$$G = (g_1, g_2, ..., g_m) \tag{3}$$

A rewriting rule g_i is specified by the left-hand side determined by a binary subvector α_i and by the right-hand side specified by a symbol substring β_i

$$g_i : \alpha_i \rightarrow \beta_i \quad (c) \tag{4}$$

where c is the so-called counter represented by a nonnegative integer, its meaning will be explained in our forthcoming discussion on an application of the mapping to a binary vector.

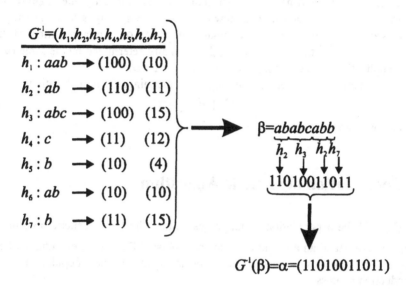

Figure 4. Illustrative example of applying an inverse rule H to a symbol string β.

Let us assume that a binary vector α is mapped by a mapping G onto symbol string $\beta \in \{a,b,c,...\}^*$ and that the first i components of α are already mapped. Its means that in the mapping G we look for rewriting rules such that their left-hand sides perfectly match a subvector in α starting from its $i+1$ component. A proper rewriting rule is selected by making use the following stochastic selection method based on Goldberg's roulette wheel approach: Let us assume that we have r rewriting rules $g_{i1}, g_{i2}, ..., g_{ir}$, with counters $c_{i1}, c_{i2},...,$ and c_{ir}, respectively. We calculate probabilities from these counters as follows

$$prob_k = \frac{c_{ik}}{\sum\limits_{l=1}^{r} c_{il}} \tag{5a}$$

Then we select a rule g_{ik} stochastically with a probability $prob_k$. This selection is realized after Goldberg [9] in such a way that for a random number $0 \leq random \leq 1$ we look for such an index $1 \leq k \leq r$ that

$$\sum\limits_{i=1}^{k-1} prob_k \leq random < \sum\limits_{i=1}^{k} prob_k \quad (for \quad k = 1,2,...,r) \tag{5b}$$

It means that k-th rule g_{ik} was quasirandomly selected.

The inverse mapping $H=G^{-1}$ is specified as a sequence of m rewriting rules that are inverse ones with respect to the rewriting rules of the mapping G

$$G^{-1} = (h_1, h_2, ..., h_m) \qquad (6a)$$

$$h_i : \beta_i \rightarrow \alpha_i \quad (c) \qquad (6b)$$

where h_i is an inverse rewriting rule created from the corresponding rewriting rule g_i in such a way that left-hand and right-hand sides are mutually interchanged, the counter c remains untouched. This new rewriting rule h_i assigns to a substring β_i (composed of symbols $a,b,c,...$) a binary subvector α_i. An application of the inverse mapping G^{-1} to a symbol string β is formally determined by the same rules as the application of the mapping G to a binary vector α.

A mapping G is called unambiguous if mappings G and G^{-1} satisfy

$$G^{-1}[G(\alpha)] = \alpha \qquad (7)$$

for each binary vector α from the set A.

4. Formulation of Genetic Algorithm

Let $\mathcal{G} = \{G\}$ be a population of mappings. Under the term "genetic algorithm" [8,9] we understand an application of the metaphor of Darwinian evolution, where a new population is created recurrently from the former population by a reproduction process

$$\mathcal{G} \rightarrow \mathcal{G}' := O_{repro}(\mathcal{G}) \rightarrow \mathcal{G} := \mathcal{G}' \rightarrow \mathcal{G}' := O_{repro}(\mathcal{G}) \rightarrow \mathcal{G} := \mathcal{G}' \rightarrow .... \qquad (8)$$

where the arrow means an evolutionary step. Let us specify details of the genetic algorithm. The first very important operation is a selection of mappings from the population such that the selection probability is directly proportional to mapping fitness. Formally, this selection operator is expressed by

$$G = O_{select}(\mathcal{G}) \qquad (9)$$

From a pair of parental mappings we create a pair of offspring by making use an operator of reproduction

$$(G'_1, G'_2) = O_{repro}(G_1, G_2) \qquad (10)$$

where created offspring are determined as follows

$$(G'_1, G'_2) = \begin{cases} (\tilde{G}_1, \tilde{G}_2) & (\text{if } random < P_{cross}) \\ (G_1, G_2) & (\text{if } random > P_{cross}) \end{cases} \qquad (11a)$$

$$(\hat{G}_1, \hat{G}_2) = O_{cross}(G_1, G_2) \qquad (11b)$$

$$\tilde{G}_1 = O_{mut}(\hat{G}_1) \quad \text{a} \quad \tilde{G}_2 = O_{mut}(\hat{G}_2) \qquad (11c)$$

```
𝒢:=randomly generated population of mappings;
time:=0;
while time<time_max do
begin time:=time+1;
      fitness evaluation;
      𝒢':=∅;
      while |𝒢'|<|𝒢|do
      begin G₁:=O_select(𝒢);  G₂:=O_select(𝒢);
            (G₁',G₂'):=O_repro(G₁,G₂);
            𝒢':=𝒢'∪{G₁',G₂'};
      end;
      𝒢:=𝒢';
end;
```

Algorithm 1. A pseudo Pascal code of genetic algorithm. The algorithm is initiated by a random generation of population $𝒢$ of mappings. Integer variable `time` is a counter of epochs of genetic algorithm. Outer cycle is repeated by `time_max` – times. In the bulk of this cycle a new population $𝒢'$ is generated by making use the stochastic operator of reproduction, see Fig. 5.

This means that the reproduction operation has a stochastic character (it is applied with the probability P_{cross}), in the opposite case reproduction operator consists in simple copying parental mappings G_1 a G_2 to new population. Applying the introduced formalism we may simply implement the method in a form of pseudo Pascal Algorithm 1 (see Fig. 5).

For an application of genetic algorithm to a simulation of evolution of coordinated communication in agent populations there must be specified two basic operations for mappings (chromosomes): (1) Mutation that stochastically assigns to a chromosome another chromosome

$$O_{mut} : G \rightarrow G' \tag{12}$$

and (2) crossover that stochastically assigns to a pair of chromosomes another pair of chromosomes

$$O_{cross} : (G_1, G_2) \rightarrow (G_1', G_2') \tag{13}$$

The mutation is performed with a probability $P_{mut}^{(select)}$ in two different ways, a transposition of two randomly selected rewriting rules or a stochastic modification of a randomly selected (with probability P_{mut}) rewriting rule, see Fig. 6. Formally, the mutation is expressed as follows

$$G' = O_{mut}(G) \tag{14}$$

where the mappings G and G′ are determined by

$$G = (g_1, g_2, ..., g_m) \text{ and } G' = (g_1', g_2', ..., g_m') \tag{15a}$$

$$O_{mut}(g_1, ... g_i, ..., g_m) = G' = (g_1', ... g_i', ..., g_m') \tag{15b}$$

$$g_i' = \begin{cases} (\tilde{\alpha}_i, \tilde{\beta}_i, \tilde{c}) & \left(\text{if } random < P_{mut}\right) \\ (\alpha_i, \beta_i, c) & \left(\text{if } random > P_{mut}\right) \end{cases} \tag{15c}$$

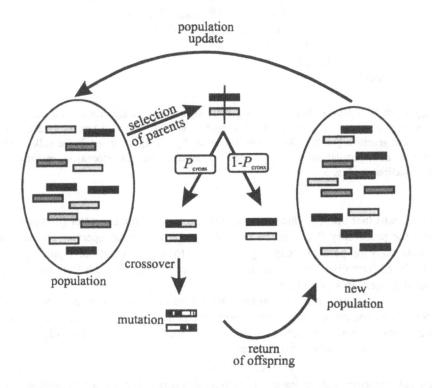

Figure 5. Diagrammatic representation of genetic algorithm, see Algorithm 1. A couple of mappings is quasirandomly selected from the population such that probabilities of their selection are proportional to their fitness. It means that the greater fitness implies the greater probability of selection to the process of reproduction. The selected couple of parental mappings undergoes the crossover operation with the probability P_{cross}. In the opposite case (with probability $1-P_{cross}$) the selected couple of mappings is not participating in the reproduction process, mappings are simply copied directly into the new population. If the new population has the same number of mappings (offspring) as the original parental population, then the reproduction process is stopped and the new population updates the original population. This process is repeated by a prescribed number of epochs or it is stopped when the population is sufficiently homogenous (i.e. almost all mappings from the population are the same).

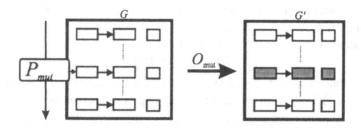

Figure 6. Diagrammatic visualization of the mutation applied to a mapping G, its application to a rewriting rule g_i is selected randomly with a probability P_{mut}. The used type of mutation consists in a modification of randomly selected rewriting rule, where either the left-hand side or the right-hand side of the rule is stochastically modified in such a way that an entry is omitted, introduced or modified.

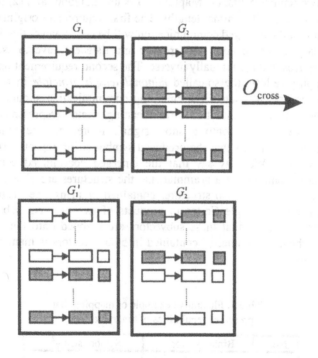

Figure 7. Diagrammatic visualization of the crossover applied to two mappings G_1 a G_2, where two new mappings G_1' a G_2' are created. The crossover consists in a random selection of the crossover point (a horizontal line on the left-hand side), behind this point mappings mutually interchange rewriting rules.

In the last formula (13c) an expression $\left(\tilde{\alpha}_i, \tilde{\beta}_i, \tilde{c}\right)$ means that in the original rewriting rule $g_i = (\alpha_i, \beta_i, c)$ either the left-hand side α_i or right-hand side β_i was stochastically modified that either one element is omitted, or one element is inserted, or finally one element is changed. In all these possible cases new counter $\tilde{c}$ is set to one, $\tilde{c} = 1$.

The crossover id formally expressed as follows

$$(G_1', G_2') = O_{cross}(G_1, G_2) \qquad (16)$$

where a couple of mappings G_1 a G_2 is stochastically changed onto another couple of mappings G_1' a G_2', see Fig. 7, These mappings are specified by

$$G_1 = \left(g_1^{(1)}, ..., g_i^{(1)}, g_{i+1}^{(1)}, ..., g_m^{(1)}\right) \text{ and } G_2 = \left(g_1^{(2)}, ..., g_i^{(2)}, g_{i+1}^{(2)}, ..., g_m^{(2)}\right) \qquad (17a)$$

$$G_1' = \left(g_1^{(1)}, ..., g_i^{(1)}, g_{i+1}^{(2)}, ..., g_m^{(2)}\right) \text{ and } G_2' = \left(g_1^{(2)}, ..., g_i^{(2)}, g_{i+1}^{(1)}, ..., g_m^{(1)}\right) \qquad (17b)$$

where the "crossover point" i is randomly selected.

Let $\mathcal{G} = \{G\}$ be a population of mappings G that are restricted by the following two requirements: (1) Mappings G is unambiguous and (2) the produced symbol strings β are of minimal lengths. The first requirement originates from our assumption that the emerged communication will be coordinated, that both sides in an elementary communication act (speaker and listener) have to use equivalent cognitive devices that are mutually inverse. The second requirement on minimality of the mapping reflects the so-called *minimal length principle* in a decoding of signals. In the present context the emerging communication should be based on unambiguous mappings that transform meaning vector (represented by binary vectors of fixed length) onto a short signal. Moreover, the requirement of minimality usually means that the produced symbol string repeats structures built in binary vectors. We may say that the emerged coordinated communication reflects also to some extent a grammar, i.e. the structures are binary vectors that are mapped onto other signal structures contained in strings, see Table 1. In this simple illustrative example Cartesian products of two different subvectors form binary vectors. We see that these subvectors are mapped onto the same symbol substrings, where a "symmetry" contained in binary vectors is manifested also by symbol strings.

Table 1. Illustrative example of mapping that preserves regularities of binary vectors.

No.	Binary vector[1]	Symbol string[2]
1	(11|01)	a|bc
2	(11|10)	a|cb
3	(00|01)	b|bc
4	(00|10)	b|cb

[1] Set of binary vectors is determined by the relation:
A={(11), (00)}×{(01),(10)}.

[2] Mapping G is determined by four rewriting rules:
$(11) \to a$, $(00) \to b$, $(01) \to bc$, $(10) \to cb$.

Let us study two agents $\mathcal{A}_S$ and $\mathcal{A}_L$, where the first one is declared as speaker and other one is listener. Their cognitive devices represented by mappings are denoted by G_S and G_L. Let us assume that the speaker's internal state is described by the binary meaning vector $\alpha \in \{0,1\}^n$, then applying its cognitive device represented by the mapping G_S to the meaning vector we get a symbol signal $\beta = G(\alpha) \in \{a,b,c,...\}^*$, this signal is sent to the listener $\mathcal{A}_L$. On the opposite side of the communication channel, speaker received the signal represented by the string β. By application of his inverse cognitive device H_L^{-1} the string is decoded in a form of binary vector $\alpha' \in \{0,1\}^*$ (in general, its length may be quite different from the length of speaker's meaning vector $\alpha \in \{0,1\}^n$). Let us define a distance between meaning vectors α and α' by

$$d(\alpha,\alpha') = \frac{1}{min(n,n') + |n-n'|}\left(\sum_{i=1}^{min(n,n')}|\alpha_i - \alpha'_i| + \frac{1}{2}|n-n'| \right) \qquad (18)$$

where n and n' are lengths of meaning vectors α and α', respectively. This definition of distance is reduced to an analogue of the standard L_1 (Hamming) distance, if both vectors have the same dimension, i.e. $n=n'$. If meaning vectors α and α' were randomly generated, then distance between them is with a high probability equal to 0.5. For the coordinated communication between speaker and listener the above distance should be vanishing, i.e. both meaning vectors α and α' become identical. Let us define a fitness increment as follows

$$\Delta f = \frac{1}{d(\alpha,\alpha') + \omega_{length}|\beta|} \qquad (19)$$

where ω_{length} is a positive penalization constants. The increment is larger if the distance $d(\alpha,\alpha')$ is smaller and the length of the message β is smaller. The second term in (19) reflects our requirement that emerged communication should be realized by shortest possible messages. With these requirements introduced we expect the emergence of an analogue of grammar in the coordinated communication. A successfulness of the elementary communication act may be characterized by the increment Δf, its greater value indicates a greater "coordination" in the mutual communication, i.e. the fitness of both agents is modified by

$$(f_S \leftarrow f_S + \Delta f) \text{ and } (f_L \leftarrow f_L + \Delta f) \qquad (20a)$$

Moreover, the increment Δf is used for an activation of simple learning process; whenever the increment is greater than a threshold value ϑ, then the counters of rewriting rules that were active in the current elementary communication act are increased

$$\Delta f > \vartheta \quad \Rightarrow \quad (c_i \leftarrow c_i + 1, \quad \text{for all active rules } g_i) \qquad (20b)$$

If this condition is satisfied, then for both the agents their cognitive devices are modified so that the counters of rewriting rules of the speaker mapping G_S are increased by one whenever they are applied to transformation process of the given meaning vector into a string of symbols. The same conclusion is true also for the mapping H_L whenever applied to received strings of symbols. The value of threshold ϑ may be used for change of intensity of learning process, an increasing of the threshold ϑ means that the learning process is weaker and weaker; in a limit case, if the threshold ϑ is a great positive number, then the learning process is fully turned off.

```
for i:=1 to p do
begin f[i]:=0;
         for j:=1 to m do c[i,j]:=0;
end;
for k:=1 to Nele-comm do
begin select randomly agent-speaker AH;
         select randomly agent-listener AL;
         α:=randomly selected meaning vector;
         β:=GS(α);
         α':=HL(β);
         Δf:=1/(d(α, α')+ωlength*|β|);
         f[S]:=f[S]+Δf;
         f[L]:=f[L]+Δf;
         if Δf>ϑ then modify counters of GS and GL;
end;
```

Algorithm 2. Pseudo Pascal implementation of fitness calculation, see fifth row in Algorithm 1. At the beginning of algorithm all counters and fitness are set to zero. The fitness evaluation consists in a repeated application of $N_{ele\text{-}comm}$ times elementary pairwise communications. Each elementary communication act is composed of random selection of two agents, these agents are declared as speaker and listener: The speaker maps a randomly selected meaning vector into a message, the listener receives the message and decodes it into a meaning vector. If a fitness increment Δf is greater than a threshold ϑ, then the fitness of the speaker and the listener are increased and also the counter of corresponding mapping is adapted.

The above modification of counters of selected rewriting rules may be understood as an adaptation process that modifies mappings – cognitive devices so that successful elementary communication acts give rise to an agent's memory that reflect to some extent the previous successful experiences of agents. A pseudo Pascal implementation of the above considerations on the fitness calculation are outlined in Algorithm 2. It specifies more deeply a statement "fitness evaluation" in Algorithm 1.

5. Simulation Calculations

The theory outlined in the preceding part of this communication will be applied to simulation calculations of emergence of coordinated communication. First of all, we introduce meaning vectors that are structured so that we may expect that the emerged coordinate communication will be endowed by the similar structure regularities as that one in the meaning vectors. This is considered as a manifestation of the grammar in the emerged coordinated communication. Let meaning vectors be determined as binary vectors created by a Cartesian product of three subsets of binary vectors

$$A = \tilde{A} \times \tilde{A} \times \tilde{A}, \quad \text{where} \quad \tilde{A} = \left\{ \underbrace{(011)}_{a}, \underbrace{(101)}_{b}, \underbrace{(110)}_{c} \right\} \tag{21}$$

It means that each 9-dimensional meaning vector is composed of three parts (that can be denoted by a, b, and c) that originate from the subset $\tilde{A}$ (see Table 2).

Table 2. Set of meaning vectors

No.	Meaning vector	Symb.[1]	No.	Meaning vector	Symb.[1]	No.	Meaning vector	Symb.[1]
1	(011011011)	aaa*	10	(101011011)	baa	19	(110011011)	caa*
2	(011011101)	aab	11	(101011101)	bab*	20	(110011101)	cab
3	(011011110)	aac	12	(101011110)	bac	21	(110011110)	cac
4	(011101011)	aba*	13	(101101011)	bba	22	(110101011)	cba
5	(011101101)	abb	14	(101101101)	bbb*	23	(110101101)	cbb*
6	(011101110)	abc	15	(101101110)	bbc	24	(110101110)	cbc
7	(011110011)	aca*	16	(101110011)	bca	25	(110110011)	cca
8	(011110101)	acb	17	(101110101)	bcb*	26	(110110101)	ccb
9	(011110110)	acc	18	(101110110)	bcc	27	(110110110)	ccc*

[1]Meaning vectors labeled by "star" symbol form the so-called test set.

Moreover, the set A is divided into disjoint subsets A_{train} and A_{test}, $A = A_{\text{train}} \cup A_{\text{test}}$, where the training subset A_{train} is composed of all meaning vectors from Table 2 that are not labeled by star symbol, consequently, the test subset A_{test} is composed of nine starred meaning vectors. The evolution simulation will be performed with respect to the training set of meaning vectors, and we will expect that the emerging coordinated communication is capable of correct mapping of testing meaning vectors into token strings so that the same grammar is used as for training meaning vectors.

Table 3. Rewriting rules of the fittest mapping after 300 epochs for $\vartheta = 1.5$

No.	rewriting rule	c	No.	rewriting rule	c	No.	rewriting rule	c
1	(11) → (e)	2	10	(10) → (a)	1	18	(001) → (bc)	26
2	(111) → (b)	5	11	(00) → (a)	2	19	(1) → (bed)	3
3	(00) → (d)	12	12	(001) → (eb)	27	20	(01) → (cb)	29
4	(101) → (a)	10087	13	(00) → (cbc)	3	21	(010) → (abe)	1
5	(1) → (be)	4	14	(00) → (aae)	1	22	(0) → (cde)	48
6	(11) → (e)	31052	15	(1) → (b)	184	23	(00) → (db)	1346
7	(0) → (b)	27951	16	(001) → (cbd)	1	24	(01) → (d)	717
8	(01) → (d)	12	17	(0) → (abc)	4	25	(11) → (e)	1
9	(11) → (cae)	2						

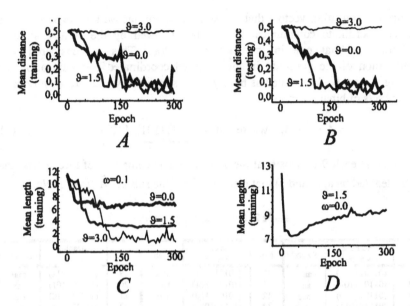

Figure 8. Four different plots of results obtained by the present approach performed for meaning vectors listed in Table 2. The first three plots were done for three different values of threshold parameter ϑ=0.0, 1.5, and 3.0 and for the penalization constant ω=0.1, whereas the last fourth plot D was done for the single value of threshold parameter ϑ=1.5 and for zero penalization constant ω=0.0. Plot A corresponds to the mean distances that are calculated for speaker and listener training meaning vectors. The plot B is the same as the plot A but for testing meaning vectors. The most important conclusion from these plots is that the mean distance for training as well as testing meaning vectors is decreased. In other words, a coordinated communication spontaneously emerged in the course of evolution. On the other hand, we see from plots A and B that if the threshold ϑ has great value (in our other calculations we put ϑ=0), i.e. when the learning was almost suppressed, then the emergence of coordinated communication is stopped. Plot C shows that the mean length of messages that are calculated for different values of the threshold ϑ is monotonously decreasing to a value that is constant in the rest of evolution. The last plot D shows the mean length of messages that are produced in a population without penalization of lengthy messages (we put ω=0.0).

The parameters of our simulation calculations were set as follows:

(1) n, meaning vector length ($n=9$)
(2) p, population size ($p=100$)
(3) P_{cross}, crossover probability ($P_{cross}=0.5$)
(4) P_{mut}, mutation probability ($P_{mut}=0.01$)
(5) t_{max}, number of epochs ($t_{max}=300$)
(6) $N_{ele\text{-}comm}$, number of elementary communications ($N_{ele\text{-}comm}=1000$)
(7) m, number of rewriting rules ($m=25$)
(8) ω_{length}, penalization coefficient of lengthy messages ($\omega_{length}=0.1$)
(9) q, number of symbols that form messages ($q=5$)

(10) ϑ, threshold for fitness modification (ϑ=0.0, 1.0, and 2.0)

(11) $length_{left}$, $length_{right}$, maximal length of the left-hand and right-hand sides of rewriting rules ($length_{left}$=3, $length_{right}$=3)

(12) Initial population of mappings is randomly generated, all rewriting rule counters are set to one.

Table 4. Mapping of meaning vectors by rewriting rules shown in Table $2^{1,2}$.

No.	α	β	α_{inv}
1	(011011011)*	(bebebe)	(011011011)
2	(011011101)	(bebea)	(011011101)
3	(011011110)	(bebeeb)	(011011110)
4	(011101011)*	(beabe)	(011101011)
5	(011101101)	(beaa)	(011101101)
6	(011101110)	(beaeb)	(011101110)
7	(011110011)*	(beebbe)	(011110011)
8	(011110101)	(beeba)	(011110101)
9	(011110110)	(beebeb)	(011110110)

[1] Only the first nine meaning vectors are displayed

[2] Meaning vectors labeled by the star belong to the testing set.

Results are summarized in Fig. 8, plots A to B correspond to the mean distance calculated for training and testing meaning vectors. An emergence of the coordinated communication is nicely indicated by plot A where it is shown (see plots for thresholds ϑ=0 and ϑ=1.5) that in the course of evolution the mean distance between speaker and listener meaning vectors is vanishing. Consequently, we may say that the emerged communication is coordinated, that is all agents at the end of evolution start to map training meaning vectors unambiguously into symbol strings and all agents are able to decode uniquely these strings into the corresponding meaning vectors. On the other hand, if the threshold ϑ is set to a sufficiently great number (see plot A, where ϑ=3.0), then the learning process is turned off and an emergence of the coordinated communication is fully suppressed. The same observations hold (see plot B) when in the course of evolution a coordination communication is verified by testing meaning vectors (we have to emphasize that the whole population evolution is realized over the training set A_{train} of meaning vectors). We see that the testing vectors are in the course of evolution interpreted on the same level of exactness as the training vectors. The plot C corresponds to the mean length of messages, we see that this important parameter almost monotonously decreases to values that are constant for the rest of evolution. This feature of the emerged coordinated communication is caused by the fact, that lengthy messages are penalized (see eq. (19)). Finally, the plot D shows a dependence of the mean length of messages when the penalization is turned off (i.e. we put $\omega = 0.0$). The resulting rewriting rules are summarized in Table 4. We see that only six of twenty-five rules have counter substantially different from number one, i.e. that have sufficiently high level of probability to be applied in the process of mapping meaning vectors into symbol strings and vice versa. In Table 4 there are listed first nine meaning vectors (for their full list see Table 2), we see that rewriting rule listed in Table 3 map these meaning vectors onto such symbol strings that are mapped again into the same meaning vectors. This means that this resulting mapping (cognitive device) is unambiguous so that the requirement (7) is satisfied, $G^{-1}\big[G(\alpha)\big] = \alpha$, for all meaning vectors. What is very interesting, the resulting coordinated communication (based on the mapping listed in table 3) correctly classifies also testing vectors, i.e. it is capable of a generalization outside the training set. The testing meaning vectors are also unambiguously mapped into symbol strings and vice versa. As one may see from table 4, symmetry of meaning vectors (we remember that they were formed by Cartesian products), their symmetry is also reflected by the resulting mapping so that similar pieces of meaning vectors are mapped onto similar pieces of symbol strings. For example, three meaning subvectors listed in Table 2 are mapped onto the following substrings: *be*, *a*, and *eb*, respectively. Then an arbitrary meaning vector listed in Table 2 is coded by a proper combination of these three substrings. This nice property of the merged mapping may be considered as an outline of a grammar in our coordinated communication, similar pieces of meaning vectors are mapped into similar pieces of symbol strings.

6. Summary and Discussion

Suppose that we look for an answer to a question, how human languages (or in general, coordinated communication in population of agents that are endowed by specific cognitive activities) are evolutionary acquired. Within classic linguistic or classic cognitive sciences we expect an answer of very general character without detailed mechanisms of their emergence of languages. Substantial breakthrough happens in this field if we start to consider recent human being as a result of Darwinian evolutionary process, as something what has itself evolutionary trajectory from the long past to present. Then we can accept an idea that human language communication is gradually evolved from its simplest forms to current highly developed coordinated communication formally expressed by generative grammars. Simultaneously with evolution of language there must appear also an evolution of cognitive device (brain) that makes possible necessary cognitive activities of mapping of meaning states into linearly structured communication signals and vice versa.

The present computer science is equipped by algorithmic devices based on the metaphors coming from the life sciences (in particular a metaphor of human brain and a metaphor of Darwinian evolution) that make possible computer simulations of processes of the evolution of language as coordinated communication. These simulations are based on an assumption that those individuals of population that are endowed by this property have a substantial selection advantage with respect to those ones that are unable of coordinated communication. The cognitive device of agents was realized in our model calculations by simple concept of a mapping composed of rewriting rules that are simultaneously evaluated by counters reflecting their application frequencies in the previous history. This simple formal tool enables to map symbol strings of variable lengths into meaning binary vectors of the fixed length. The concept of positive integer counters enables an involvement of the learning processes that are vitally important in our model of elementary communication acts (see Fig. 2). On the other hand, Darwinian evolution is simulated by simple genetic algorithm based on a linear representation of genotype by binary vectors of the fixed dimension. Applying both these approaches of the modern computer science we can model an evolution (emergence) of the coordinated communication in populations of agents that are endowed by specific cognitive activities.

Let us return to the question formulated at the beginning of this conclusion Section. From our results we may give an answer to this question. When the evolution of coordinated communication runs in a Darwinian manner, its detailed mechanisms may be simulated computationally; the obtained results are summarized as follows:

1. *The evolution (emergence) of language may be in general considered as an evolution of a complex dynamic system composed of many elementary components-agents that mutually interact through simple elementary communication acts (see Fig. 2). Objects of this evolution are mappings that map meanings into symbol signals and vice versa. An integral part of the elementary communication between two agents is the process of learning, where both participating agents change their cognitive device so that possible discrepancies between speaker and listener meanings are minimized.*

2. *At the beginning of evolution agent cognitive devices are randomly generated, that is they do not manifest any coordinated activities. Loosely speaking, elementary communication acts run mainly stochastically in this starting stage of evolution, when a measure of understanding between speaker and listener is almost vanishing. Agent – speaker randomly maps meaning vector into a symbol string, and similarly, agent – listener also randomly maps the received symbol string into a meaning vector.*

3. *In the course of evolution coordination of mutual communication between agents is substantially increased. This means that meaning vectors of speakers are gradually closer and closer to meaning vectors of listeners, i.e. listeners correctly decoded received messages sent by speakers.*

Loosely speaking, agents start to use a common dictionary in direct coding of meaning vectors into messages and also in inverse decoding of messages into meaning vectors. An architecture of cognitive device is evolved in such a way that produced cognitive activities within elementary communication acts are gradually more and more effective.

4. *Our model calculations show that an inclusion of learning process within elementary communication between speaker and listener is very important for the speed of evolution of coordinated communication. If the learning is turned off (e.g. by increasing of the threshold ϑ to great values), then the speed of emergence of coordinated communication is substantially slower than in the case that the learning is included (the threshold ϑ is a small positive number).*

5. *The emerged coordinated communication preserves regularities of meaning vectors that are manifested in signals so that similar meaning subvectors are mapped onto similar parts of signals. This property may be considered as an indication of simple grammar built in the emerged cognitive devices. An emergence of grammar is supported by our fitness calculation requirement that shorter signals are preferred (or in other words lengthier signals are penalized)*

We conclude this communication by a claim that a design of scenarios that specify detailed mechanisms of emergence and evolution of different cognitive activities (like the scenario presented above) represents main contribution of the modern computer science for cognitive sciences. Cognitive sciences are thus becoming transformed into a science where hypothesis can be also verified computationally.

Acknowledgments

This work was supported by the grants # 1/4209/97 and # 1/5229/98 of the Scientific Grant Agency of Slovak Republic.

References

[1] F. de Saussure: Course in General Linguistics. Duckworth, Translated by R. Harris, London 1983 (original French edition was published in 1916).

[2] N. Chomsky: *Syntactic Structures.* Mounton & Co., The Hague 1957.

[3] N. Chomsky: *Knowledge of Language: Its Nature, Origin, and Use.* Praeger, New York 1987.

[4] S. Pinker and P. Bloom: Natural language and natural selection. *Behavioral and Brain Sciences,* **13**(1990), 707. This paper is available on ftp://ftp.princeton.edu/pub/harnad/ BBS/WWW/bbs.pinker.html.

[5] C. Darwin: *The Origin of Species.* Penguin, London (1985) (original edition was published in 1859).

[6]   *Approaches to the Evolution of Language: social and cognitive bases.* Edited by J. R. Hurford, M. Studdert-Kennedy and C. Knight. Cambridge University Press, Cambridge 1998. An information about the book is available on http://www.ling.ed. ac.uk/~jim/evoconf.html.

[7]   J. R. Hurford: The Evolution of Language and Languages". In *The Evolution of Culture*, edited by R. Dunbar, C. Knight and C. Power, Edinburgh University Press, Edinburgh 1999. This paper is available on http://www.ling.ed.ac.uk/~jim/dunbar. knight.power.s.ps

[8]   J.H. Holland: *Adaptation in Natural and Artificial Systems.* University of Michigan Press (1975).

[9]   D.E. Goldberg: *Genetic Algorithms in Search, Optimization, and Machine Learning.* Addison-Wesley (1989).

[10]   J. Batali: Computational Simulations of the Emergence of Grammar. In [6]. This paper is available on http://cogsci.ucsd.edu/~batali/papers/ grammar.html.

[11]   J. Batali: The Negotiation and Asquistion of Recursive Grammars as a Result of Competition Among Exemplar, to be published. This paper is available on http://cogsci. ucsd.edu/~batali/rephrase.ps.gz.

[12]   J. Batali: Innate Biases and Critical Periods: Combining Evolution and Learning in the Acquisition of Syntax. In *Proceedings of the Fourth Artificial Life Workshop*, edited by R. Brooks and P. Maes. The MIT Press, 1994, pp.160-171. This paper is available on http://cogsci.ucsd.edu/~batali/papers/alife-syntax.ps.

[13]   L. Steels: Synthesising the origins of language and meaning using co-evolution, self-organisation and level formation. In [6]. This paper is available on http://arti.vub.ac. be/steels/edin.ps.

[14]   L. Steels: A self-organizing spatial vocabulary. *Artificial Life Journal* 2(1966). This paper is available on http://arti.vub.ac.be/steels/space.ps.

[15]   B.J. MacLennan and G. M. Burghardt: Synthetic Ethology and the Evolution of Cooperative Communication. *Adaptive Behavior* 2(1994), 161-187. This paper is available on http://www.cs.utk.edu/~mclennan/anon-ftp/SEECC.ps.Z.

[16]   G. W. Werner and M.G. Dyer: Evolution of Communication in Artificial Intelligence. In Artificial Intelligence II. Edited by C. G. Langton, J. D. Farmer, and S. Rasmusen, Addison Wesley 1991.

Migration and Population Dynamics in Distributed Coevolutionary Algorithm

Jiri Pospichal

Department of Mathematics, Slovak Technical University, 812 37 Bratislava, Slovakia
email: pospich@cvt.stuba.sk

Abstract: An optimization in both stable as well as in a changing environment is studied by a modification of a genetic algorithm, where two kinds of chromosomes are optimized - candidate solutions to a given problem and test cases. Similar problems without time-dependent function optimization were already well studied as the host-parasite co-evolutionary genetic algorithms. Solutions are tested by test cases, either randomly selected, or from the vicinity of solutions in case of a space distribution. During this process solutions gather fitness according to their ability to pass correctly the test cases and test cases gather fitness, when a solution fails them. Three modifications of such an algorithm are compared. The first one is most closely related to a genetic algorithm, with fixed sizes of both solution and test cases populations, and no space distribution of chromosomes. The second algorithm tries to modify the size of solution-population and number of used test cases in the inverse proportion to the success of the other species. It means, that when the solutions fail test cases, there is no need to increase the number of test cases, but the number of newly generated solutions should increase. When solutions are good, the opposite strategy is applied. The third algorithm uses spatial distribution of chromosomes, but instead of a standardly used two-dimensional grid, algorithm uses populations situated on two circles, which rotate in opposite directions. The investigated problem uses sorting networks as candidate solutions and permutations as test cases. The changing environment is simulated by disabling some of possibilities for an exchange of couples of entries, changing thus the fitness function of sorting networks.

1. Introduction

The attempts to model a coevolutionary behavior and its application has been started successfully by Hillis [1], who used a model of host-parasite coevolution. Coevolution in other meanings, which we shall not explore here, is a cooperative behavior, where two species must cooperate to achieve high fitness [2-5] or a co-evolution, when fitness of one member depends on the rest of population and not on a fixed fitness function [6].

In the following work we shall compare three different types of algorithms solving the classical model of Hillis, who evolved sorting networks representing host populations against the test cases in the forms of sequences of numbers to be sorted representing the parasite population. All three algorithms will have the same type of mutations of sorting networks. No crossover will be used, since from our numerical experiences follows, that a crossover of two sorting networks works most probably as a giant mutation instead of an exchange of reasonable parts of information. This observation corresponds to the claims by Fogel [7]. The selection pressure of the evolution will be mostly applied on sorting networks, while test cases will be evolved with only a very slight selection pressure. This also follows from our numerical experience, where the number of correct sorting networks is negligible, while the number of test cases, for which a randomly created sorting network fails, is enormous.

The first algorithm will be similar to the classical genetic algorithm such, that both the host as well as the parasite populations have a stable size. The selection from the host population uses nonlinear ranking, while the only selective process applied to the parasite population is a replacement of the parasites - test cases, which were correctly sorted by all the host networks tested by them. Every network is tested by 5 randomly chosen test cases, and its fitness is in the inverse proportion to the number of cases, where in the test case filtered through the network still remains a bigger number before a smaller one.

The second algorithm tries to apply autocorrective mechanism controlling both the number of test cases used for evaluation of a sorting network as well as the number of sorting networks. The total number of tests determined by the multiplication of "number of sorting networks" times "number of test cases for each network" should remain the same. When there is a lot of sorting networks evaluated by the same value, which is the best achieved value in the current generation, one should try to distinguish between those networks by testing their descendants each against more test cases, where the number of descendants is smaller. On the other hand, when the number of best evaluated sorting network is small, one could generate more of their descendants, using smaller number of test cases against them. A similar idea was already successfully tested in optimization [8]. More theoretically oriented work was studied in [9,10].

The third algorithm applies spatial distribution for the populations of hosts and parasites. However, unlike the typical two-dimensional grid placed on a toroid [1,11-14], we have placed the populations in one dimension, i.e. each population forms a linear sequence, where the end is "connected" to the beginning, thus creating a circle to avoid a boundary problem. More exactly, the hosts (i.e. sorting networks) are "placed" on one circle, while the parasites (test cases) are placed on another circle. While the hosts are tested against a few corresponding parasites on the other circle, the circle turns around, so at the next time step the host will be tested by another subset of parasites. The selection of the sorting networks into next generation is based on a tournament between the networks placed in a neighborhood (up to k networks to the left and up to k networks to the right of the current position). Such a selection does not require a complicated and lengthy selection procedure used in the classical genetic algorithm. When a favorable solution appears in some position, it does not immediately occupy the whole population, but it spreads to both sides around the circle. The velocity of this dispersion/migration depends on a size (length) of a neighborhood, for which the descendant is selected by a tournament. The bigger the neighborhood, the faster the favorable solution spreads around the circle.

Finally, to test the algorithms in a changing environment, we suggest a mechanism that during the course of evolution disables exchange of numbers in two positions for certain couples of positions. A network containing such an exchange would still partially work, only the couple of numbers in the incriminate positions would not be swapped, if the first number is bigger than the second one.

2. General Framework of Algorithms

The goal of the algorithms is to find a correct sorting network to sort a sequence with a given number of entries [15], i.e. for any given input sequence of numbers it results in a correctly ordered output sequence. The way, how such a network works, is illustrated in fig. 1. Numbers of an input sequence (4,1,3,2) enter the network from the left. The horizontal lines depict the positions in a sequence and vertical lines connecting always two horizontal lines represent comparison-swap operations (so called comparators). A comparator compares two numbers from the positions of horizontal lines, which it connects, and if they are in a wrong order, swaps them each to the other position. If we want a sequence in an increasing order, and the upper number is larger than the lower one, we would switch them. Every comparator can be described by a couple of numbers describing the lines, which it

connects. For sorting n numbers $\binom{n}{2}$ comparators can be used, which can be each

indexed by i. Such a comparator comparing numbers in positions j, k can be then described as $[j{:}k]_i$. For example, for 4-number sequence we can use comparators $[1{:}2]_1,[1{:}3]_2,[1{:}4]_3,[2{:}3]_4,[2{:}4]_5,[3{:}4]_6$. The sequence of comparators $[1{:}2],[3{:}4],[1{:}3],[2{:}4],[2{:}3]$, used in the figure 1, can be then described as a sequence of their indexes (1,6,2,5,4).

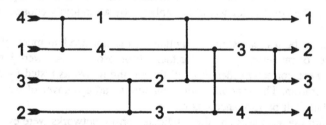

Figure 1: Example of a sorting network for a test case sequence (4,1,3,2), which correctly sorts the sequence into a resulting sequence (1,2,3,4). The network can be described as a sequence of comparators $[1{:}2],[3{:}4],[1{:}3],[2{:}4],[2{:}3]$, where the numbers in brackets are index numbers of connected lines, where the switch takes place.

For a given number of n to be sorted we firstly generate and index all possible comparators. The sizes of shortest sorting networks (i.e. with the smallest possible number of comparators) are already known for small numbers of n sorted entries (for greater n the shortest possible sizes are not proved to be exact). Therefore in our approach a candidate sorting network has always the smallest known number of comparators. For $n=6$, 7, 8 the used numbers of comparators are $m=12$, 16, and 19, respectively. The candidate sorting networks are therefore defined as sequences of a length m of indexes of comparators, $\left\{1,...,\binom{n}{2}\right\}^m$. Such sequences are defined as host chromosomes in our tested versions of an evolutionary algorithm. An ultimate test of the sorting network was to order correctly all $n!$ permutations of sequences comprised of numbers $1...n$. When such a network was found, it was called a correct sorting network and an algorithm was stopped. However, such a test was performed for assessment of the algorithm only, and it was not used to influence otherwise the run of the algorithm.

Two basic operations of genetic algorithms are crossover and mutation. We had a bad computational experience with an application of a crossover (we used a classical one-point crossover). It substantially prolonged an average number of generations to reach a correct sorting network. It might be due to the fact, that the crossover in this case works more like a giant mutation than an exchange of valid pieces of information [7]. Therefore the crossover was removed from our versions of coevolutionary algorithm. Candidate sorting networks will be denoted $(x_1,...,x_n)$, where x_i means the index of ith comparator. We have used two kinds of mutations, exchange and point mutation, described by equations 1a and 1b, each with 50% probability:

$$\left(x_1,...,x_j,...,x_i,...,x_m\right)_{mutated} := \left(x_1,...,x_i,...,x_j,...,x_m\right) \tag{1a}$$

$$\left(x_1,...,\tilde{x}_i,...,x_m\right)_{mutated} := \left(x_1,...,x_i,...,x_m\right), \text{ where } \tilde{x}_i = random(\left\{1,...,\binom{n}{2}\right\}) \tag{1b}$$

Indexes i,j of the switched comparators in eq. (1a) are randomly selected. Point mutation goes through all m possibilities of an index i, each time deciding with a $2/m$ probability, whether the point mutation will be used to replace the ith comparator. The comparator would be replaced by a randomly chosen comparator, as in eq. (1b).

Since the test cases, i.e. permutations, did not need so much of evolution, no crossover or mutation was used for their chromosomes, they were just replaced, when they were correctly ordered by all the sorting networks tested by them in the current generation. The versions with a mutations and crossover of test cases were also tried, but their performance was worse.

Both the test cases and the candidate sorting networks were evaluated by a function, which counted the number of cases, when the test case sequence filtered through the candidate network had a bigger number before a smaller one (not only adjacent cases were counted)

$$f(x_1,...,x_m) = \text{number of cases}\left(x_i > x_j \text{ for } i < j\right) \tag{2}$$

Because of the great number of possible test cases, which growth is factorial with n, only a few test cases were used for an evaluation of each sorting network in a population during one generation. For the purposes of the assessment of the algorithms the candidate sorting networks were applied also against the whole set of permutations, calculating, how big a percentage of them was filtered incorrectly.

3. Description of Algorithms

3.1 The "Classical" Genetic Algorithm

To asses the effectiveness of the used algorithms, it should be compared with some more standard approach. Therefore we used a genetic algorithm with a crossover and mutation applied for both sorting networks. The fitness function was based on an inverse of the result of eq. (2) summed for several test cases with added number one to prevent division by zero. Of course, fitness of test cases was evaluated directly by that sum, because the fittest test case is that, which filters incorrectly through a tested sorting network. This algorithm failed to perform adequately even in comparison with an Olsson's algorithm [11] based on a distributed population, when the number of evaluations needed to achieve a valid sorting network was compared. By further elaboration with a fitness evaluation and a selection mechanism it was determined, that best results were achieved when a selection of the sorting networks was most selective, while the selection of test cases was most benevolent. Both populations had a stable size of 200 chromosomes.

```
t:=0;
Initialize population of S of sorting networks, |S|:=200;
Initialize population of T of test cases, |T|:=200;
while t<5000 do
begin
    Evaluate S and T by a function f by testing each s∈S
    against 5 randomly chosen test cases from T;
    Determine and normalize fitness of S;
    Replace test cases from T evaluated by zero by randomly
    generated test cases;
    t:=t+1; S':=∅;
    while |S'|<200 do
    begin
        Select pseudorandomly s∈S;
        x:=s; Mutate x;
        S':=S'∪x;
    end;
    S:=S';
end;
```

Algorithm 1. The genetic algorithm without crossover, with a fixed sizes of populations of candidate sorting networks and test cases, where the selection of sorting networks is done by nonlinear rank selection.

The selection from the sorting networks used nonlinear ranking, where the fitness of the network was evaluated by

$$\textit{fitness of sorting network } i = \left(\textit{rank}(i)\right)^8 \tag{3}$$

when we order the sorting networks by their value determined by (2) and *rank(i)* equals to the number of networks in population (200 in our case) for the most fit network, while it equals number one for the least fit network. In fact, it practically means the selection of the best. Since there were many cases, when the value of function (2) was the same for several networks, in such instance their rank was calculated as a medium of ranks of these networks. The only selective process applied to the test cases was their replacement if they were correctly sorted by all the candidate sorting networks tested by them. Every network was tested by 5 randomly chosen test cases.

3.2 The Genetic Algorithm with a Variable Size Population

The second algorithm we tried to explore was based on an idea that when there are only a few best chromosomes evaluated by the same fitness, the evolutionary optimization should be directed toward more global search, while when there are many chromosomes evaluated as the best ones, one should direct the search more locally. To keep the size of population within limits, the upper bound of the population size was set to 1000, while the lower bound to 5. Since we wanted the resources (computational time) for each generation to be kept fixed, and as the main demand on those resources was considered the number of testing of sorting networks against test cases, we decided to keep this number stable. Only the best sorting networks were used as parents for the next generation.

```
t:=0;
Initialize population of S of sorting networks, |S|=200;
Initialize population of T of test cases, |T|=200; trial=5;
while t<5000 do
begin
    Evaluate S and T by a function f by testing each s∈S
    against trial of randomly chosen test cases from T;
    Determine fitness of S; S'':= cases of S with the best
    fitness; best:=|S''|;
    Replace test cases from T evaluated by zero by randomly
    generated test cases; t:=t+1; S':=∅;
    if best>5 then trial:=min(200, trial+1)
    else if best<5 then trial:=max(1, trial-1);
    while |S'|<⌊1000/trial⌋ do
    begin
        Select randomly s∈S'';
        x:=s''; Mutate x; S'=S'∪x;
    end;
    S:=S';
end;
```

Algorithm 2. The genetic algorithm with variable size populations of candidate sorting networks evaluated against a variable number of test cases. No crossover was used, and the only the sorting networks with best fitness were selected as parents for the next generation.

Control of the global/local character of the search was translated into control of the number of best cases. It was empirically decided, that this number should be equal to five, and the total number of the tests single sorting network against a single test case should be about 1000. The population of sorting networks started with 200 sorting networks and each of them was tested by 5 randomly chosen test cases. When there was less then 5 best sorting networks, which served as the parents for the next generation, the number of evaluation of sorting networks against test cases was reduced by 1 (but at least one test case for each sorting network must have been tested). Since the total number of tests should be 1000 or less, the number of sorting networks created in the next generation was determined by $g=1000/$(number of test cases for each network), which number was rounded to the greatest integer less than or equal to g. On the other hand, when there was more than 5 best sorting networks, the number of testing of sorting networks against test cases was increased by 1 (with a theoretical upper bound 200) and the number of sorting networks in the next generation was calculated in the same way as in the previous case.

3.3 The Genetic Algorithm with Host and Parasite Populations Distributed on Two Circles

The genetic algorithm with host and parasite populations distributed randomly on a 2-dimensional grid has a disadvantage, that many moves do not result in a test of the chromosomes and can be therefore viewed as inefficient [11]. The basic idea behind the current version of the algorithm was to preserve a spatial distribution of chromosomes and its influence on the evolution, while removing the redundant random wandering of chromosomes on a grid. Therefore the sorting networks were situated in a one-dimensional array of 200 cells connected in a circle, resulting in a periodic boundary conditions. The test cases were placed into another one-dimensional array of 200 cells connected into a circle. The arrays were placed over themselves, and each sorting network was evaluated by 5 closest test cases in the opposite array, which means, that the sorting network placed in position i was tested by test cases from positions $i-2,i-1,i,i+1,i+2$. The circle of test cases rotated by 5 entries each generation, so that a sorting network, which would remain in the same position i, would be tested by another 5 test cases, which were before placed in positions $i+3$, $i+4$, $i+5$, $i+6$, and $i+7$. As in previous algorithms, the test case, which was correctly filtered by all five opposite sorting networks, was replaced in the next generation by a new randomly generated test case.

The selection of the sorting network, which would be mutated and placed onto current position in the next generation was done in a form of tournament between sorting networks occupying the current position and its neighborhood. The larger the neighborhood would be, the more the algorithm would be driven towards a local optimum instead of the search for the global optimum. In out computation we used a neighborhood size of 41, it means the current position and 20 positions to the left and to the right. We searched the neighborhood from the left to the right placed positions and selected the first sorting network with the best evaluation.

Evaluation procedure

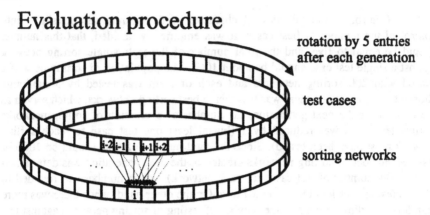

Figure 2: Illustration of the evaluation procedure of the algorithm 3, where sorting networks are placed each in an entry of the lower array connected to the circle, and test cases are placed each in an entry of the upper array connected to the circle. A sorting network i from the lower array is tested by 5 test cases from the upper array indexed $i-2,i-1,i,i+1,i+2$. After each generation the upper circle is rotated by 5 entries, so that the corresponding network i would be tested by next 5 test cases.

Selection procedure

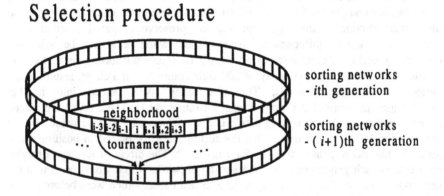

Figure 3: Illustration of the evaluation procedure of the algorithm 3, where the upper circle array represents the ith generation of sorting networks, and the lower circle array represents the next $(i+1)$th generation of the sorting networks. The selection procedure to the next generation uses a tournament within the neighborhood of the entry to be filled, where in the case of more best entries with the same fitness there is used the first one from the left. All the entries in the new generation are then mutated.

```
t:=0;
Initialize population of S of sorting networks, |S|:=200;
Initialize population of T of test cases, |T|:=200;
trial=5; neighborhood:=41;
while t<5000 do
begin
    for i:=1 to 200 do for j=1 to trial do
    evaluate s_i and test_{(i+j-(trial-1)/2+5*t+|T|) mod |T|} ∈ T
    by a function f;
    Replace test cases from T evaluated by zero by randomly
    generated test cases; t:=t+1; S':=∅; i:=0;
    while i<200 do
    begin
        i:=i+1; best:=i;
        for i:=2 to neighborhood do
        if evaluation(s_best)
            > evaluation(s_{(i+j-(neighborhood-1)/2+5*t+|S|) mod |S|})
        then best:=(i+j-(neighborhood-1)/2+5*t+|S|) mod |S|;
        x:=s_best; Mutate x; S':=S'∪x;
    end;
    S:=S';
end;
```

Algorithm 3. The genetic algorithm with sorting network and populations distributed on two circles without crossover, but with fixed sizes of populations of candidate sorting networks and test cases, where the selection of a sorting networks for a current position is always done by a tournament within the sorting networks from the neighborhood of the position.

4. Results

Results for three different sizes of test cases for all three presented algorithms and for the algorithm described in [11] considered here for a comparison are presented in Table 1. The suggested algorithms showed a better performance in all but one case, where for a test size 8 only the Algorithm 3 performed substantially better than the algorithm from [11]. The labels of algorithms correspond to the algorithms in this paper, except for the algorithm 4, which represents results taken from [11] for a comparison. There is a slight discrepancy between a maximum number of evaluations used in the calculations. Limits for algorithms here was $5 * 10^6$ of evaluations of a sorting network against a test case, while for n=8 in the [11] there was used about $3 * 10^3$ networks times $2 * 10^4$ generations; however, in [11] only part of networks was tested during each generation.

Meaning of the single entries of the table is following:

- length n of test means the length of test sequences, i.e. sizes of permutations used as test cases
- length m of nets means the length of sorting networks, i.e. the number of comparators in the network

- algorithms means algorithm number in the paper, whereas no. 4 belongs to algorithm described in [11]
- correctly sorted means average from 10 runs of percentage of test cases correctly sorted by the best network
- valid runs of 10 means number of runs of total 10 runs which found a valid sorting network within the 5000 generations
- eval. $(*10^5)$ means an average number of evaluations from the 10 runs (must be multiplied by 10^5). In case that a valid network was found in less than 10 runs, the number of evaluations was calculated only from those runs

Table 2: Comparison of performance of the algorithms

length n of test	6				7			
length m of nets	12				16			
algorithms	1	2	3	4	1	2	3	4
correctly sorted	100%	100%	100%	100%	100%	100%	100%	99.7%
valid runs of 10	10	10	10	10	10	10	10	9
eval. $(*10^5)$	0.7	0.6	0.4	1.0	1.7	6.1	2.5	7.0

length n of test	8			
length m of nets	19			
algorithms	1	2	3	4
correctly sorted	93.1%	87.7	96.0%	91.0%
valid runs of 10	1	0	4	1
eval. $(*10^5)$	30.0	>50	17.9	13.7

The overall performance of the algorithms is shown in figure 4 for a test case size 7. It shows ordered distributions of numbers of evaluations of sorting networks against test cases in order to reach a valid sorting network. From the graph it can be seen, that e.g. if the algorithm 1 exceeds number of evaluation 300 000, it was probably trapped in some local minimum far from a global minimum, and therefore it would be more efficient to run the algorithm from the beginning than to wait, until the algorithm jumps over some local barrier in the search space. The curves of the distributions have basically the same shape, where at the end of the distributions there is a relatively small fraction of the runs which were probably caught in some local minimum, so it took them a long time to achieve a global minimum. The performances during single runs of the algorithm are shown in figure 5. One can see, that these sample runs correspond to the fact, that the algorithm 1 has a best performance for a test size 7, because the ratio of correctly sorted test cases from all possible permutations for the best sorting network in a current generation goes very quickly to 1. On the other hand, the performance of the algorithm 2 in this case is the worst one, the line of the plot is very ragged and touches ratio value 1 only in one case. Since none of the algorithms employs elitism, i.e. best solutions may not be saved in the population, the best solution can be unfavorably mutated, so that the best solution in the next generation can be worse.

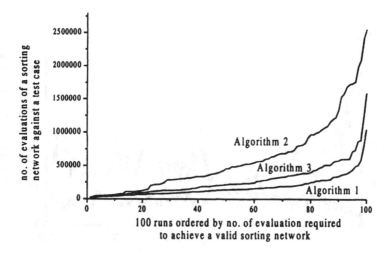

Figure 4: Comparison of results from 100 runs for all three algorithms described in this paper for networks sorting test cases of the length 7. At the end of the distributions there is a relatively small fraction of the runs which were probably caught in some local minimum.

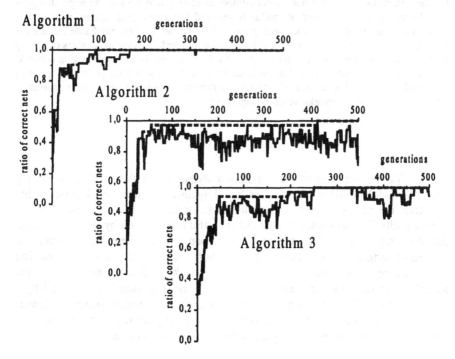

Figure 5: Comparison of a performance for the three described algorithms for a size 7 of the test problem. The vertical axis shows the percentage of correctly sorted permutations from all possible permutations. The full ragged line shows results for the best network in a current generation, the dotted nondecreasing line shows the performance of best network found so far during the current run.

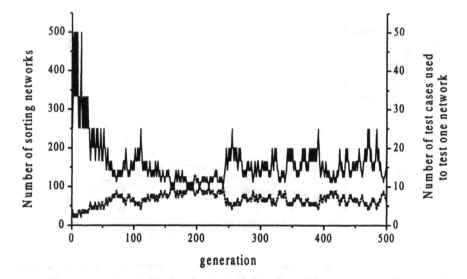

Figure 6: An example of population dynamics of a run of the algorithm 2 for the size of the test case equal to 7. The full line shows number of sorting networks in the population, the dotted line shows the number of randomly selected test cases used against each sorting network to evaluate it (the corresponding axis labels are at the right hand side). Since the total number of the tests in a generation is kept roughly fixed, increase in numbers of sorting networks automatically decreases the number of test cases used against each sorting network. The ideal condition for an optimization is set in advance to have 5 cases sorting networks evaluated by zero. If the number of such sorting networks is smaller, in the next generation the number of sorting networks would be increased.

The figure 6 shows a population dynamics of the number of sorting networks and the number of test cases tested against each network. The possibility of a total extinction of either sorting networks or test cases is prevented here by setting limits. The dynamics is also rather artificially affected by the requirement, that the total number of evaluation within a population is kept fixed, so that the number of sorting networks directly determines the total number of tested cases. The actual number of test cases in the population is kept fixed (it equals 200), but actually used test cases are selected randomly. The dynamics is also rigged by a preliminary specification of a number of sorting networks, which must not fail tests, so that when there is less of such networks, in the next generation we generate more networks and let them be tested by a smaller number of test cases.

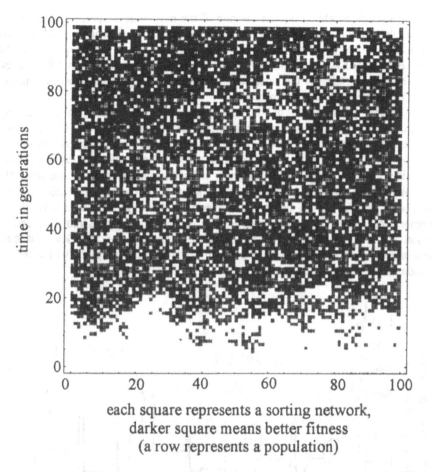

each square represents a sorting network,
darker square means better fitness
(a row represents a population)

Figure 7: The figure shows a spatial structure of evolution of sorting networks during one run of the Algorithm 3 for a dimension of test case equal to 6, for a neighborhood equal to 7 and for a size of the population of sorting networks equal to 100. Each square of the plot corresponds to one position in a population. The darkness of each square expresses quality of the corresponding network. One can see, that in the beginning the generated solutions are not very good, while in the later generations clusters of dark squares appear representing good solutions. Each horizontal line presents a whole population in one time step (with a periodic boundary condition), vertical line of squares represent a development of a sorting network in a particular position during generations.

The figure 7 shows a spatial structure of evolution of sorting networks during one run. The figure is very similar to that of cellular automata, in fact, the algorithm with its definition of the neighborhood is very close to one-dimensional cellular automaton. For every position the replacement of a current sorting network by a new sorting network in the next generation is decided by a tournament of all networks in the neighborhood. The main basic difference from a cellular automaton is a stochastic nature of random changes during a mutation, which follows the selection of a network into the next generation. When a good network is found, the size of the neighborhood defines the rate of the migration or spread in time of the favorable solution to neighboring positions. In case that we define a neighborhood in another way to take into account absolute positions of the sorting network in the one dimensional space, it is easy to get an evolutionary algorithm with niches. However, the instance of the algorithm we tried was very inferior in performance to the already presented three algorithms.

The performance of the algorithms was also tested in a changing environment, or in other words with a fitness function depending also on time. Such a function was created by "knocking out" functioning of some of the comparators in sorting networks. A fraction (3 in our case) from all possible comparators (21 for the size 7 of test cases) were randomly selected, and when a sorting networks contained such comparators, they did not swap a couple higher_number—lower number to their correct order, so the sorting network performed like if the comparator was deleted. This considerably distorted the fitness function, and to insert the time dependence, every 50 generations one of the "knocked out" comparators was brought back to perform correctly, and its place was taken by another randomly selected comparator.

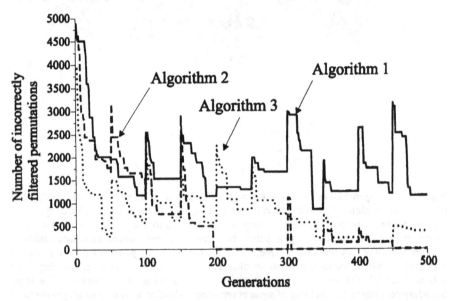

Figure 8: Performance of algorithms in a changing environment. Sorting networks for sequences of the length 7 are used, when three of possible 21 comparators are disabled, so that the sorting network can include them, but they do not swap the numbers. After each 50 generations one of the disabled comparators switches a place with a randomly chosen working comparator, so that the fitness function evolves through the time.

The figure 8 shows, that unlike the performance on a fixed optimization problem, situation for a time-dependent fitness function is substantially different. The algorithm 1, which performed best for the used size of the problem, becomes the worst one, and the worst second algorithm performs relatively well. The best algorithm is the third one, which suggests, that space distribution, even if only in one dimension, can help to sustain diversity in a population enough to quickly adapt to a changing environment. These test were repeated for 10 runs, each for 1000 generation, with averages of incorrectly filtered permutations from the set of all permutations equal for algorithms 1, 2, and 3 to 1378, 1297, and 1205, respectively.

5. Conclusions

The discussion from the previous section shows that comparing results from the presented algorithms can be seriously influenced not only by setting of various control parameters and by the type of optimization problem, but it can vary even for slightly different sizes of the same problem. We also planed to describe an algorithm, where the populations would be divided into subpopulations distributed in niches, where each niche is cohabited by a subpopulation of solutions and test cases and a more specified migration would occur. Unfortunately, the performance of such a setup proved for a given problem rather inefficient, therefore this algorithm was removed from the presented work.

As an example of influence of the size of a problem onto performance can be used Algorithm 1, which performed worst of the three introduced algorithms for the test cases of the size 6 (see Table 1), while performing best for the size of the value 7. Such a difference can be explained by a fact, that while the size of the test cases increased only by one, the actual size of the search space was increased from $number\_of\_possible\_comparators^{size\_of\_network} = 15^{12} = 129746337890625$ to $21^{16} = 1430568690241985328321$, which is a more than 10^7 times more.

While the Algorithm 2, which was designed to autoadjust its tendency to explore the search space intensively/extensively (i.e. to look for a local/global optimum), was the least efficient for higher dimension of the sorting network search, it performed quite well in the changing environment.

The Algorithm 3, designed to keep a diversity of the search by a one dimensional spatial distribution of chromosomes, performed very well for the sorting network search (in fact it was the best one for the highest tested size 8 of test cases), it was also the best one in the changing environment. It does not require sorting of the fitness in the whole population like the Algorithms 1 and 2, and its intensity for a local/global search preferences can be controlled by the size of the neighborhood for a tournament. Therefore we believe it is suitable for further investigation.

Acknowledgments

This work was supported by the grants 1/4209/97 and 1/5229/98 of the Scientific Grant Agency of Slovak Republic.

References

[1] Hillis D. W., 1991, Co-evolving parasites improve simulated evolution as an optimization procedure. In Langton C. G., Taylor C., Farmer J. D., and Rasmunssen S. (eds), *Artificial Life II, SFI Studies in the Sciences of Complexity*, Volume 10, pp. 313-324, Addison-Wesley.

[2] Husbands P., Mill F., 1991, Simulated co-evolution as the mechanism planning and scheduling. In Belew R. K., Booker L. B. (eds), *Proceedings of the Fourth International Conference on Genetic Algorithms*, pp. 264-270, Morgan Kaufmann Publishers.

[3] Paredis J., 1995, The symbiotic evolution of solutions and their representations. In Eshelman L. (ed), *Proceedings of the Sixth International Conference on Genetic Algorithms*, pp. 359-365, Morgan-Kaufmann

[4] Paredis J., 1994, Co-evolutionary constraint satisfaction. In Davidor Y., Schwefel H.-P., and Männer R. (eds), *Proc. 3rd Parallel Problem Solving from Nature*, pp. 46-55, Springer, Berlin.

[5] Potter M. A., 1997, *The design and analysis of a computational model of cooperative coevolution*. PhD Thesis, George Mason University, Fairfax, Virginia.

[6] Pollack J. B. and Blair A. D., 1996, Coevolution of a backgammon player. *Artificial Life V: Proc. 5th Int. Workshop on the Synthesis and Simulation of Living Systems*. MIT Press, Cambridge, MA.

[7] Fogel D. B., 1995, *Evolutionary Computation: Toward a New Philosophy of Machine Intelligence*. The IEEE Press, New York.

[8] Pospichal J., Kvasnička V., 1998, Altruism in Evolutionary Optimization. In *Mendel '98*, ISBN 80-214-1199-6, pp. 105-110, PC-DIR, Brno.

[9] Kephart J. O., 1994, How Topology Affects Population Dynamics. In Langton C. G. (ed), *Artificial Life III, SFI Studies in the Sciences of Complexity*, Proc. Vol. XVII, Addison-Wesley, Reading, MA.

[10] Forst C. V., Reidys C., 1997, On Evolutionary Dynamics. In Langdon C. G., Shimohara K (ed), *Artificial Life V: Proc. 5th Int. Workshop on the Synthesis and Simulation of Living Systems*, MIT Press, Cambridge, MA.

[11] Olsson B., 1996, Optimization Using A Host-Parasite Model with Variable-Size Distributed Populations. In *Proceedings of the 1996 IEEE 3rd International Conference on Evolutionary Computation (ICEC'96)*, IEEE Press.

[12] Eriksson R., Olsson B., 1997, Cooperative Coevolution in Inventory Control Optimisation. In Smith G. D., Steele N. C., and Albrecht R. F. (eds), *Artificial Neural Networks and Genetic Algorithms: Proceedings of ICANNGA 97*, pp. 583-587, Springer.

[13] Olsson B., 1998, Evaluation of a Simple Host-Parasite Genetic Algorithm. In Porto V. W., Saravanan N., Waagen D., and Eiben A. E. (eds), *Evolutionary Programming VII: Proceedings of the Seventh Annual Conference on Evolutionary Programming*, pp. 53-62, Springer.

[14] Olsson B., 1998, A Host-Parasite Genetic Algorithm for Asymmetric Tasks. In Nedellec C., Rouveirol C. (eds), *Machine Learning: ECML98*, pp 346-351, Springer.

[15] Knuth D. E., 1998, *The Art of Computer Programming*, Vol. 3. Addison-Wesley, Reading, MA.

Royal Road Encodings and Schema Propagation in Selective Crossover

Kanta Vekaria and Chris Clack

Department of Computer Science, University College London, Gower Street, London, UK
{K.Vekaria, C.Clack}@cs.ucl.ac.uk

Keywords: Genetic Algorithms, Adaptive Recombination, Crossover, Schema Propagation

Abstract

Recombination operators with high positional bias are less disruptive against adjacent genes. Therefore, it is ideal for the encoding to position epistatic genes adjacent to each other and aid GA search through genetic linkage. To produce an encoding that facilitates genetic linkage is problematic. This study focuses on selective crossover, which is an adaptive recombination operator. We propose three alternative encodings for the Royal Road problem. We use these encodings to analyse the performance of selective crossover with respect to different encodings. This study shows that the performance of selective crossover is consistent and is not affected by alternative encodings of a problem, unlike two-point crossover. The encodings are also used to understand the behaviour of selective crossover in terms of schema propagation. Experimental results indicate that selective crossover provides a better balance between exploration and exploitation than conventional recombination operators.

1 Introduction

Genetic algorithms (GAs) are a powerful general-purpose search method based on mechanisms drawn from natural evolution. The effectiveness of the genetic algorithm depends heavily on its parameters (mutation rate, crossover rate, selection method, and population replacement scheme) and the synergy of the encoding and recombination operators.

The encoding or representation is a central issue in genetic algorithms (Mitchell, 1996) as they directly manipulate the coded representation of the problem. Traditional encoding schemes use fixed length, binary representation schemes; however, selecting a representation scheme that facilitates solutions of the problem by a genetic algorithm is in itself an optimisation problem.

An example of a representation problem is the "linkage problem". Two loci are strongly linked when there is little or no recombination between them. Recombination is traditionally prevented (or restricted to very low rates) when the two loci are very close together on the chromosome. A lack of recombination between loci is advantageous when certain combinations of alleles at different loci result in high fitness, and alternate combinations have low fitness i.e. epistasis exists amongst these alleles. There is almost no crossing over because they are physically so close together on the chromosome. This kind of physical proximity is called "genetic linkage".

The "Building Block Hypothesis" (Goldberg, 1989) suggests that interacting bits that are relatively close are less likely to be disrupted in a canonical genetic algorithm (one that uses one point crossover). Thus, genes that are thought to be interacting should be positioned near each other on the chromosome. Further studies on gene positions and crossover by Eshelman, Caruana and Schaffer (1989) showed that crossover operators with high positional bias, such as one-point and two-point crossover, are less disruptive against adjacent genes.

When it is the combination of alleles at different loci that is important and the GA uses a crossover operator with a high positional bias, the encoding should reflect this linkage by placing these genes close to each other on the chromosome. As a result the occurrence of crossover between these alleles is reduced. To produce such an encoding is difficult without knowing ahead of time which genes are important and related to each other in useful schemas. Choosing such fixed representations without a priori knowledge of the problem can be problematic for the GA user; how is one to decide the best encoding for one's problem and one's GA?

Many techniques that adapt the representation have been proposed to overcome the linkage problem. Holland (1975) proposed inversion, a reordering operator. Inversion works by giving each allele an index indicating its actual position in the chromosome (for evaluation purposes). Two points are then chosen in the string and the bits between them are reversed to produce a new ordering. Crossover then occurs on this new ordering thereby, producing more orderings. The purpose of reordering was to find orderings in which beneficial schemas are more likely to survive under one-point crossover.

Goldberg, Korb & Deb (1989) proposed the messy GA that evolves the representation. The messy GA (mGA) uses a variable length representation where each allele has an index indicating its actual position in the chromosome, but all loci do not have to be specified in the chromosome and loci can be specified more than once. The mGA has two phases; a primordial phase – where building blocks of a particular order are generated and a juxtapositional phase – where building blocks are recombined using cut and splice operators that mimic one-point crossover. To use mGAs we are faced with the same problem of not knowing a priori knowledge of the problem. What is the useful schema order for the primordial phase? Also when evaluating strings where all loci are not specified; how can you compute the true fitness? If the individual contains loci that are interacting, missing alleles are crucial in defining fitnesses for individuals. Further extensions have been made to the mGA called the Gene Expression Messy GA (GEMGA) (Kargupta 1996) where each gene has a position, value, weight and a linkage set. GEMGA has better performance than the mGA but it also requires an optimal schema order in the initial population.

The Linkage Learning GA (LLGA) developed by Harik (1997) used alleles that were also tagged with their actual positions. An exchange operator similar to two-point was used for recombination. In his study he compared LLGA on problems constituted by a number of non-overlapping building blocks of a maximum size k and on uniformly-scaled problems (problems where all building blocks give the same contribution to the fitness i.e. the one-max problem). Unfortunately the LLGA did not work well for easy uniformly-scaled problems and the study was only limited to non-overlapping building blocks.

Genetic algorithms are generally used for large, complex and poorly understood search spaces where there is little or no a priori knowledge about the search space; hence GA practitioners use methods of trial and error to determine the representation and recombination operators to use. One can avoid such practices by using adaptive operators that have little or no positional bias.

In this study we describe three alternative encodings for the Royal Road problems. We decrease the genetic linkage in the encoding of the Royal Road so that the low-level schemas have high defining lengths. We empirically observe, using these Royal Road encodings, how schemas are propagated under an adaptive recombination operator - selective crossover (Vekaria & Clack, 1998). We compare the performance (in terms of the number of evaluations taken to find a solution) and schema survival rate of two-point, uniform and selective crossover. We show that the schema survival rate and performance of selective crossover is the same regardless of how the Royal Road is encoded in the chromosome (with or without linkage embedded in the encoding).

This paper is organised as follows: the next section introduces selective crossover. Section 3 gives a brief overview of Royal Road functions and describes three alternative encodings. It goes on to describe the methods used in this study. Section 4 discusses the results and Section 5 concludes.

2 Selective Crossover

Recombination, also known as crossover, is a commonly used operator in a GA. Traditionally, recombination has been considered as the primary operator of a GA and thought to be responsible for the generation and propagation of solutions. More recently, there have been many studies on the role played by traditional crossover operators, compared with mutation, in a GA (Schaffer & Eshelman 1991, Spears 1993 and Wu, Lindsay & Riolo 1997). Crossover operators have also been classified on their usefulness in terms of generating and propagating solutions (Eshelman & Schaffer 1995). There are now many different ways of implementing recombination (Spears 1997). Some forms of recombination are more suitable for certain problems than others and some proposed for general problems incorporate adaptive methods and thus are classed as adaptive recombination operators. Selective crossover (Vekaria & Clack, 1998) is one such adaptive recombination operator. For a survey on recombination operators the reader is referred to Spears (1997).

2.1 Representation

In selective crossover each individual has associated with it a real-valued vector, and thus each allele has an associated dominance value. Recombination uses two parents to create two children. During recombination two parents are selected and their fitness is recorded. The dominance value of each allele in both parents is compared linearly across the chromosome. The allele that has a higher dominance value contributes to Child 1 along with the dominance value. If both dominance values are equal then crossover does not occur at that position.

Consider an l bit representation and let $\Omega = \{0,1\}^l$ be the search space. Each individual in selective crossover has a gene vector G and a dominance vector D. The fitness function is $\phi(G)$.

$$G = (g_1,\ldots,g_i,\ldots,g_l) \quad where \quad g_i \in \{0,1\}$$
$$D = (d_1,\ldots,d_i,\ldots,d_l) \quad where \quad d_i \in \mathbf{R}^+$$

And therefore:

$$G^{A,k} = (g_1^{A,k},\ldots,g_i^{A,k},\ldots,g_l^{A,k}) \quad where \quad g_i^{A,k} \in \{0,1\}$$
$$D^{A,k} = (d_1^{A,k},\ldots,d_i^{A,k},\ldots,d_l^{A,k}) \quad where \quad d_i^{A,k} \in \mathbf{R}^+$$
$$A \in \{P,C\} and \ k \in \{1,2\}$$

Here P indicates a parent vector and C indicates a child vector. k refers to the parent or child number (either 1 or 2).

Two parents are chosen for crossover, Parent 1 ($P,1$) and Parent 2 ($P,2$), with gene vectors $G^{P,1}$ and $G^{P,2}$ respectively. Their corresponding dominance vectors are $D^{P,1}$ and $D^{P,2}$. A crossover can be represented by inheritance masks M^1 and M^2.

$$M^k = (m_1^k,\ldots,m_i^k,\ldots,m_l^k) \quad where \quad m_i^k \in \{0,1\} and \ k \in \{1,2\}.$$

Crossover will produce two children Child 1 ($C,1$) and Child 2 ($C,2$), with gene vectors $G^{C,1}$ and $G^{C,2}$ respectively. The inheritance mask for Child 1 is M^1 and is given in Equation 1. The inheritance mask is created by comparing the parent dominance vectors. In simple terms, each element m_i^1 in M^1 is 1 if the element $d_i^{P,1} > d_i^{P,2}$ in the parent dominance vectors. The inheritance mask M^2 of Child 2 is $M^2 = 1 - M^1$.

$$M^1 = S(S(D^{P,1} - D^{P,2}) + 1) \tag{1}$$

Where S is a sign function (see Appendix).

Using the inheritance mask the resulting gene vector for Child k (where $k \in \{1,2\}$), after recombination, is therefore:

$$G^{C,k'} = M^{k'}(G^{P,1} - G^{P,2}) + G^{P,2'} \tag{2}$$

Where X' represents a transposition of vector X.

The dominance vectors of the new Child k are updated to reflect fitness increase $\mathcal{FS}^k$ with respect to its parents. We only update the dominance values of those genes that were changed during crossover, so we take the hamming distance of the parent and child to update the appropriate dominance values in the dominance vector $D^{C,k}$.

The hamming distance is computed by the exclusive-OR operator $\oplus$. Therefore the resulting dominance vector for Child k is:

$$D^{c,k'} = M^{k'}\left(D^{p,1} - D^{p,2}\right) + D^{p,2'} + \mathcal{FI}^k\left(G^{c,k} \oplus G^{p,k}\right)'$$

(3)

Where:

$$\mathcal{FI}^k = \begin{cases} \phi\left(G^{c,k}\right) - \phi\left(G^{p,k}\right) & \text{, if } \phi\left(G^{c,k}\right) > \phi\left(G^{p,k}\right) \\ \max\left(0, \left(\phi\left(G^{c,k}\right) - \phi\left(G^{p,2}\right)\right)\right) & \text{, if } k = 1 \\ \max\left(0, \left(\phi\left(G^{c,k}\right) - \phi\left(G^{p,1}\right)\right)\right) & \text{, if } k = 2 \end{cases}$$

(4)

$\mathcal{FI}^k$ is the fitness increase, which is computed by taking the difference between fitness values of Child k and its corresponding Parents (as mentioned earlier $\phi\left(G^{\wedge,k}\right)$ is the fitness function). If there is no fitness increase then the dominance values are not updated (i.e. $\mathcal{FI}^k = 0$).

2.2 Example

Figure 1 gives an example of selective crossover: the shaded alleles have a higher dominance value than its competing allele. To keep diversity in the population Child2 inherits the non-dominant alleles.

Parent 1 – fitness = 0.36

0.4	0.3	0.01	0.9	0.1	0.2
1	0	0	1	0	0

Parent 2 – fitness = 0.30

0.01	0.2	0.4	0.2	0.9	0.3
0	1	1	1	1	0

Child 1

0.4	0.3	0.4	0.9	0.9	0.3
1	0	1	1	1	0

Child 2

0.01	0.2	0.01	0.2	0.1	0.2
0	1	0	1	0	0

Figure 1: Recombination with selective crossover

Child 1 – fitness = 0.46

0.4	0.3	0.4	0.9	0.9	0.3
1	0	1	1	1	0

Child 2 – fitness = 0.20

0.01	0.2	0.01	0.2	0.1	0.2
0	1	0	1	0	0

⇩ Increase dominance values

Child 1 – fitness = 0.46

0.4	0.3	0.5	0.9	1.0	0.3
1	0	1	1	1	0

Child 2 – fitness = 0.20

0.01	0.2	0.01	0.2	0.1	0.2
0	1	0	1	0	0

Figure 2: Updating dominance values.

After crossover the two new children are evaluated. If a single child's fitness is greater than the fitness of either parent, the dominance values (of those alleles that were exchanged during crossover) are increased proportionately to the fitness increase (Eq. 4). This is done to reflect the alleles' contribution to the fitness increase. Figure 2 gives an example of the mechanism. It follows on from the selective crossover example given in Figure 1. In Figure 2, only Child1 has an increase in fitness of 0.1 (compared with the fittest parent) hence its dominance values get updated. In Figure 1 the bit values of Parent1 and Parent2 at loci 1 and 2 did not get exchanged during crossover and the bit values at loci 4 are the same in both parents; this also applies to loci 6. Thus, after selective crossover, the alleles that caused a change in the chromosome are only those held at loci 3 and 5. Since the change of those alleles at loci 3 and 5 resulted in an increase in fitness, only their dominance values get increased in Child1 (shaded in Figure 2).

During initialisation of the population, the dominance values are initialised with a value of zero. Since all dominance values are equal, an inheritance vector (Eq. 1) that permits crossover cannot be created; therefore uniform crossover is used to create the inheritance masks in the first generation. Eq 2 and 3 still apply.

A study on biases in selective crossover (Vekaria & Clack 1999) showed that initialising the dominance values with zero further improved the performance of selective crossover.

3 Experimental Details

For this study we compare selective crossover with two-point and uniform crossover against Royal Road functions that have different encodings.

3.1 Royal Road Functions

Royal Road (RR) functions (Mitchell, Forrest, & Holland 1991; Forrest, & Mitchell, 1993) are a class of problems designed for the study of GA behaviour on landscapes that contain building blocks. Building blocks in this context are schemas of short order and short defining length that contribute to an individual's fitness. These schemas are hierarchically structured in the Royal Road functions and are pre-defined with corresponding fitness values. The shape of the landscape is very much like a staircase consisting of four levels (see Figure 3A). The lowest level (level 0) schemas are the shortest in order and in defining length. Level 1 schemas comprise of a combination of adjacent level 0 schemas. Level 2 schemas comprise of a combination of adjacent level 1 schemas and so on. The highest level schema is the solution and the fitness is calculated as a sum of all the schema fitnesses that exist in the individual. The landscape of RR functions can be altered by either removing or adding levels and by changing the steepness of the steps (by increasing or decreasing the fitness difference between levels)

The RR functions are synthetic problems specifically created to understand the behaviour of a canonical GA that uses one-point or two-point crossover. The predefined structure inherent in RR functions allows them to be a prime candidate to understand behaviours of other different recombination operators. In our study we use the RR function described in Figure 3(A). We change the traditional encoding to see what effect it has on the performances of two-point, uniform and selective crossover.

In these new encodings, the order of all schemas remains the same; however, the defining lengths of level 0 schemas are increased and if necessary other levels too.

We look at four different encodings of the RR function. The first is shown in Figure 3(A). This encoding contains genetic linkage embedded within the encoding. This linkage is gradually broken in B, C and D. In the second encoding (B) the defining lengths of level 0 schema is 14. In the third encoding the defining lengths of level 0 and level 1 schemas are 28 and 29 respectively. Notice how the level 1 schema is fragmented as the defining length of the level 0 schema is increased. This is more noticeable in (D). In (D) the defining length of level 0 schemas is 56.

3.2 Schema Propagation

The mechanics of selective crossover (Section 2) indicated that the dominance values are the key element that determines which allele will be passed onto which child (Eq. 1). The dominance values are also updated with respect to the fitness increase (Eq. 3 and 4). As fitness increase is determined by the fitness function itself, selective crossover cannot provide a universal understanding of schema propagation for all problems when it is being used for recombination. Therefore we look at RR functions to understand schema propagation in selective crossover.

To investigate whether selective crossover is successful in propagating schemas during recombination and to determine the survival rate of a schema in RR we count the number of level 0 schemas prior to recombination and after recombination.

We look at each level 0 schema of only those individuals that underwent recombination to see if:

a)   the schema survived recombination.
b)   the schema was constructed.

3.3 The Genetic Algorithm

A generational GA was used for this study. The parameters are as follows:

Population size: 128
Chromosome length: 64
Crossover rate: 0.6
Mutation rate: 0.01
Selection: SUS algorithm (Baker 1987)

50 independent runs were made for selective, two-point and uniform crossover. The GA was allowed to run until a solution was found.

4   Results and Analysis

For each recombination event we kept track of all level 0 schemas in the parents prior to crossover. The new children were then analysed to see how many were constructed under recombination. This analysis was carried out on the four encodings described in Figure 3.

Table 1 shows the performance of selective, uniform and two-point crossover in terms of the number of evaluations taken to find the solution. In the original encoding of the RR function DL7 (Figure 3A) two-point crossover is superior; taking on average 38287 evaluations whilst selective and uniform take 55857 and 74128 respectively. However, as the defining length of the RR increases, the performance of two-point crossover diminishes to 72077 evaluations. In contrast the performance of selective and uniform crossover remain consistent. Selective crossover takes approximately 57000 evaluations to find a solution with any encoding. Uniform crossover takes approximately 70000 evaluations to find a solution with any encoding.

The graphs in Figures 4, 5 and 6 shows the behaviour, in terms of schemas propagation, of selective, uniform and two-point crossover with the four different encodings described in Figure 3. As there are eight level 0 schema the survival rate is out of all eight schemas; hence the maximum value on the y-axis is 8.

	(A) RR DL7	(B) RR DL14
S1	1111111•••	1•1•1•1•1•1•1•••
S2	•••••••1111111••	•1•1•1•1•1•1•1••
S3	••••••••••••••1111111•••••••••••••••••••••••••••••••••••	••••••••••••••1•1•1•1•1•1•1•••••••••••••••••••••••••••••
S4	•••••••••••••••••••••1111111••••••••••••••••••••••••••••	•••••••••••••••1•1•1•1•1•1•1••••••••••••••••••••••••••••
S5	••••••••••••••••••••••••••••1111111•••••••••••••••••••••	••••••••••••••••••••••••••••1•1•1•1•1•1•1•••••••••••••••
S6	•••••••••••••••••••••••••••••••••••1111111••••••••••••••	•••••••••••••••••••••••••••••1•1•1•1•1•1•1••••••••••••••
S7	••1111111•••••••	••1•1•1•1•1•1•1•
S8	•••1111111	•••1•1•1•1•1•1•1
S9	11111111111111••	11111111111111••
S10	••••••••••••••11111111111111••••••••••••••••••••••••••••	••••••••••••••11111111111111••••••••••••••••••••••••••••
S11	••••••••••••••••••••••••••••11111111111111••••••••••••••	••••••••••••••••••••••••••••11111111111111••••••••••••••
S12	••11111111111111	••11111111111111
S13	1111111111111111111111111111••••••••••••••••••••••••••••	1111111111111111111111111111••••••••••••••••••••••••••••
S14	••••••••••••••••••••••••••••1111111111111111111111111111	••••••••••••••••••••••••••••1111111111111111111111111111
S15	11111111111111111111111111111111111111111111111111111111	11111111111111111111111111111111111111111111111111111111

	(C) RR DL28	(D) RR DL56
S1	1•••1•••1•••1•••1•••1•••1•••••••••••••••••••••••••••••••	1•••••••1•••••••1•••••••1•••••••1•••••••1•••••••1•••••••
S2	•1•••1•••1•••1•••1•••1•••1••••••••••••••••••••••••••••••	•1•••••••1•••••••1•••••••1•••••••1•••••••1•••••••1••••••
S3	••1•••1•••1•••1•••1•••1•••1•••••••••••••••••••••••••••••	••1•••••••1•••••••1•••••••1•••••••1•••••••1•••••••1•••••
S4	•••1•••1•••1•••1•••1•••1•••1••••••••••••••••••••••••••••	•••1•••••••1•••••••1•••••••1•••••••1•••••••1•••••••1••••
S5	••••••••••••••••••••••••••••1•••1•••1•••1•••1•••1•••1•••	••••1•••••••1•••••••1•••••••1•••••••1•••••••1•••••••1•••
S6	•••••••••••••••••••••••••••••1•••1•••1•••1•••1•••1•••1••	•••••1•••••••1•••••••1•••••••1•••••••1•••••••1•••••••1••
S7	••••••••••••••••••••••••••••••1•••1•••1•••1•••1•••1•••1•	••••••1•••••••1•••••••1•••••••1•••••••1•••••••1•••••••1•
S8	•••••••••••••••••••••••••••••••1•••1•••1•••1•••1•••1•••1	•••••••1•••••••1•••••••1•••••••1•••••••1•••••••1•••••••1
S9	11••11••11••11••11••11••11••••••••••••••••••••••••••••••	11••••••11••••••11••••••11••••••11••••••11••••••11••••••
S10	••11••11••11••11••11••11••11••••••••••••••••••••••••••••	••11••••••11••••••11••••••11••••••11••••••11••••••11••••
S11	••••••••••••••••••••••••••••11••11••11••11••11••11••11••	••••11••••••11••••••11••••••11••••••11••••••11••••••11••
S12	••••••••••••••••••••••••••••••11••11••11••11••11••11••11	••••••11••••••11••••••11••••••11••••••11••••••11••••••11
S13	1111111111111111111111111111••••••••••••••••••••••••••••	1111••••1111••••1111••••1111••••1111••••1111••••1111••••
S14	••••••••••••••••••••••••••••1111111111111111111111111111	••••1111••••1111••••1111••••1111••••1111••••1111••••1111
S15	11111111111111111111111111111111111111111111111111111111	11111111111111111111111111111111111111111111111111111111

Figure 3: The traditional encoding of a RR function is (A). Schemas S1 to S8 are level 0 schemas, with a fitness value 8, S8-S12 are level 1 schemas, with a fitness value of 16 and S13 to S14 are level 2 schemas, with a fitness value of 32. S15 the solution thus has a fitness value of 192. DL# is the defining length of level 0 schemas.

In Figure 6 two-point crossover is unable to effectively propagate schema as the defining length is increased (as expected from previous studies by Holland (1975) and Spears (1998)). Without genetic linkage embedded within the encoding two-point crossover is not a reliable recombination operator. Uniform crossover is more consistent in propagating schema with or without genetic linkage, but the payoff is performance; uniform crossover requires more evaluations. Selective crossover is also consistent in propagating schema with different encodings but its performance is superior to uniform crossover in all encodings and superior than two-point crossover when genetic linkage is not embedded in the encoding.

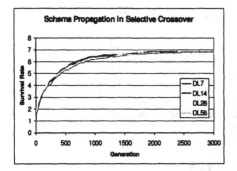

Figure 4: The survival rate of any level 0 schema under selective crossover with different RR encodings.

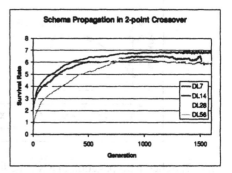

Figure 6: The survival rate of any level 0 schema under two-point crossover with different RR encodings.

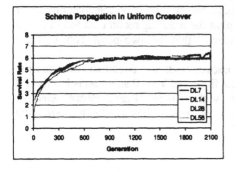

Figure 5: The survival rate of any level 0 schema under uniform crossover with different RR encodings.

	Selective	Uniform	Two-point
DL7	55857 (25749)	74128 (27469)	38287 (16894)
DL14	61007 (39576)	69525 (37003)	42671 (22462)
DL28	57320 (34162)	69376 (35672)	51351 (22172)
DL56	57807 (28367)	75834 (47465)	72077 (35568)

Table 1: Mean number of evaluations taken to find the solution for different RR encodings. The standard deviation is shown in parentheses.

From the graph in Figure 4 we can clearly state that selective crossover has no positional bias. Selective crossover will perform equally as well on any encoding of the problem. No a priori knowledge is required to 'tune' the encoding in the hope of further increasing performance. Selective crossover consistently shows a steady growth in schema propagation. Clearly two-point crossover has an advantage when genetic linkage is embedded in the encodings; however, as mentioned before 'how is one to decide the best encoding for one's problem and one's GA?'

To understand the growth in schema propagation in selective crossover we compared the survival rate of all three recombination operators. Due to the limitations of this paper we only present graphs for DL28 shown in Figure 7.

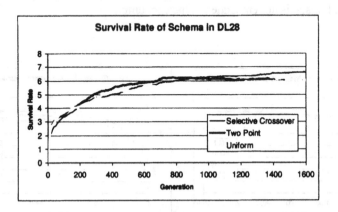

Figure 7: Schema propagation in RR encoding DL28

In Figure 7 the growth in schema propagation of selective crossover is quite slow in early generations but then in future generations exceeds uniform and two-point crossover. This behaviour is a result of large amounts of exploration in early generations. This is necessary whilst the population is still diverse and to reduce the chances of premature convergence. Having explored the landscape selective crossover is able to determine good schemas and to exploit them during recombination in future generations. This exploitatative power exceeds that of two-point and uniform crossover.

This study also showed that selective crossover finds it difficult to *construct* the level 0 schemas. This is due to the fact that the fitness of level 0 schemas is 8 only if it contains all 1's leaving all other 2^8-1 schemas with a fitness of zero. Hence, selective crossover cannot exploit the 1's in the schema until it has randomly found the complete set of 8 1's.

5 Conclusions

Recombination operators with high positional bias are less disruptive against adjacent genes; therefore, to exploit this behaviour the encoding should reflect this genetic linkage by placing epistatic genes close to each other on the chromosome. This then reduces disruption amongst these genes. To produce such an encoding a great deal of knowledge is required about the landscape being searched and in most GA optimisation problems such knowledge is difficult to acquire.

This paper showed that selective crossover is an adaptive recombination operator that guarantees behaviour on any encoding of the Royal Road problem. Three alternative encodings for the Royal Road problem were described and used to understand the effect an encoding has on performance of selective crossover. Selective crossover showed consistent performance regardless of the encoding, unlike two-point crossover.

From this study we can conclude that given any random encoding of the Royal Road function, selective crossover will take approximately 57000 evaluations to find a solution. No a priori knowledge is required to tune the encoding for guarantees on performance.

Schema propagation was studied to understand the behaviour of selective crossover in terms of schema survival rates. This showed that selective crossover, like uniform crossover, has no positional bias. It can successfully propagate schema, during recombination, at equal rates independent of the defining length.

The growth in the schema survival rate was compared with two-point and uniform crossover. This showed that selective crossover is very explorative in early generations, showing a lower schema survival rate, which is ideal whilst diversity still exists in the population. However in subsequent generations it is very exploitative with the highest schema survival rate compared with two-point and uniform crossover. Selective crossover provides a better balance between exploration and exploitation than conventional recombination operators.

To conclude, the performance of selective crossover is not affected by the encoding of the problem. Selective crossover is very much like uniform crossover in that it has no positional bias. In contrast, it does not have the high disruptive qualities of uniform crossover. Selective crossover uses directional and credit biases (Vekaria & Clack 1999) that exploit schema, thereby providing a better balance between exploration and exploitation.

Appendix: Sign Function

Given vectors M and N of length l where:

$$M = (m_1, \ldots, m_i, \ldots m_l) \quad where \quad m_i \in \mathbf{R}$$
$$N = (n_1, \ldots, n_i, \ldots n_l) \quad where \quad n_i \in \{0,1\}$$

For any $i \in [1,l]$

$$S(M) \to N \mid n_i = s(m_i) = \begin{cases} 1, & if \quad m_i > 0 \\ 0, & if \quad m_i = 0 \\ -1, & otherwise \end{cases}$$

Acknowledgements

This research was supported by the Engineering and Physical Sciences Research Council. The first author would like to thank Dr Richard Chandler for his invaluable help, Jungwon Kim and Manlio Valdivieso-Casique for their constructive feedback.

References

Baker, J. E. (1987) Reducing Bias and Inefficiency in the Selection Algorithm. In J.J Grefenstette, editor, *Proceedings of the 2nd International Conference on Genetic Algorithms*, 14-21. Lawrence Erlbaum Associates.

Eshelman, L. J.,Caruana, R. A., & Schaffer J. D. (1989) Biases in the Crossover Landscape. In J. David Schaffer, (editor), *Proceedings of the Third International Conference on Genetic Algorithms*, 10-19. Morgan Kaufmann.

Forrest, S. & Mitchell, M. (1993) Relative Building-Block Fitness and the Building Block Hypothesis. In L. D. Whitley, editor, *Foundations of Genetic Algorithms 2*, 109-126. San Francisco, CA: Morgan Kaufmann.

Goldberg, D. E. (1989*) Genetic Algorithms in search, optimization and machine learning.* Addison-Wesley.

Goldberg, D. E., Korb, B. & Deb, K. (1989) Messy Genetic Algorithms: Motivation, Analysis, and First Results. In *Complex Systems*, Vol. 3. 493-530.

Harik, G. R. (1997) *Learning gene linkage to efficiently solve problems of bounded difficulty using genetic algorithms.* Doctoral dissertation, University of Michigan, Ann Harbor.

Holland, J. H. (1975) *Adaptation in Natural and Artificial Systems.* MIT Press.

Kargupta, H. (1996) The Gene Expression Messy Genetic Algorithm. In *Proceedings of the IEEE International Conference on Evolutionary Computation,* 814-819 IEEE Press.

Mitchell, M. & Forrest, S. & Holland, John H. (1991) The Royal Road for Genetic Algorithms: Fitness Landscapes and GA Performance. In F. J. Verala & P. Bourgine (eds.), *Toward a Practice of Autonomous Systems: Proceedings of the First European Conference on artificial Life,* 245-254. Cambridge, MA:MIT Press.

Mitchell, M. (1996) *An Introduction to Genetic Algorithms.* MIT Press.

Schaffer, J. D. & Eshelman, L. J. (1991) On Crossover as an Evolutionary Viable Strategy. In In R. Belew and L. Booker (eds.), *Proceedings of the Fourth International Conference on Genetic Algorithms,* 61-68. Morgan Kaufmann.

Spears, W. M. (1993). Crossover or Mutation? In L. Darrell Whitley, editor, *Proceedings of Foundations of Genetic Algorithms 2,* 221-237. Morgan Kaufmann.

Spears, W. M. (1997), Recombination Parameters. In T. Bäck, D. Fogel and Z. Michalewicz (ed.), *The Handbook of Evolutionary Computation,* Oxford University Press.

Spears, W. M. (1998) *The Role of Mutation and Recombination in Evolutionary Algorithms.* Doctoral dissertation, George Mason University, Virginia.

Vekaria K. & Clack C. (1998). Selective Crossover in Genetic Algorithms: An Empirical Study. In Eiben et al. (eds.). *Proceedings of the 5th Conference on Parallel Problem Solving from Nature,* 438-447. Springer-Verlag.

Vekaria K. & Clack C. (1999). Biases Introduced by Adaptive Recombination Operators. In Banzhaf et al. (editors.) *Proceedings of the Genetic and Evolutionary Computation Conference,* CA: Morgan Kaufmann.

Wu, S., Lindsay, R. K. & Riolo, R. L. (1997) Empirical Observations on the Role of Crossover and Mutation. In Thomas Bäck (editor), *Proceedings of the Seventh International Conference on Genetic Algorithms,* 362-369. Morgan Kaufmann.

An evolutionary approach for the design of natural language parser

Olgierd UNOLD

Institute of Engineering Cybernetics
Wroclaw University of Technology
Wyb. Wyspianskiego 27, 50-370 Wroclaw, Poland
e-mail:unold@ci.pwr.wroc.pl

Abstract: In this paper, we provide theoretical bases for the use of two classes of evolutionary computation, that is evolutionary programming and genetic programming, that support automated inference of fuzzy automaton-driven parser of natural language. This parser, called fPDAMS (fuzzy nondeterministic pushdown automaton with associative memory access), works in the stratificational knowledge representation system (SKRS system), which is an attempt at formalizing the multi-layer structure of natural language. In the evolutionary programming (EP) each chromosome is a representation of a transition graph of fPDAMS. The key mechanism used in this approach is asexual mutation. Genetic programing (GP) can be applied to natural language parser if a mapping is established between the point-labeled tree used in GP and the transition graph of the automaton. The method of mapping is based on the encoding technique called "Cellular Encoding" and proposed by F.Gruau.

1. Introduction

Man thinks and communicates in natural language (NL). NL is most important media and method of knowledge representation. Natural language processing systems (NLP systems), the applied artificial intelligence, support natural front-end for human being, that is input and output in NL. To make possible wide use of NLP systems, we have to first equip the NLP system with the adaptable analyzer, which can be taught while working.

This paper sketches an evolutionary approach to inference the architecture of natural language parser. The developed parser is based on fuzzy automaton, which works in the stratificational knowledge representation system (SKRS).

2. Fuzzy automaton-driven parser of NLP system

In [1-4] we have proposed a fuzzy parser of natural language texts (so-called fPDAMS - fuzzy nondeterministic pushdown automaton with associative memory access) which works in the stratificational knowledge representation system [5].

Stratificational knowledge representation system is an attempt at formalizing the multilayer structure of natural language. The SKRS is based on a multistratum semantic network whose nodes contain particular linguistic units (single words, the so-called structural-semantic components, sentences). The node-to-node links correspond to relations between particular linguistic units.

If a natural language text representing the acquired knowledge is to be mapped onto the stratificational semantic network, the particular linguistic units must be isolated in the analysed text, such as word groups and sentences. If we decide to limit our input texts to isolated clauses, we can limit the sentence decomposition task to two subtasks: one of word group decomposition into the particular words and one of sentence decomposition into word groups. In order to increase both knowledge integrity in the knowledge base and the efficiency of the search algorithm, an assumption has been made that it is not phrases, but the structural-semantic components (the components) that are the objective of the sentence decomposition, the components being word groups constituting semantic units which can no longer be decomposed, each of which has its own syntax characteristics. Sentence decomposition into structural-semantic components in the SKRS is performed by a fuzzy automaton-driven parser fPDAMS. An automaton decomposition model in the SKRS can be used due to the reduction of the process of sentence decomposition into structural-semantic components to the process isolating appropriate substrings of symbols in the string of symbols representing lexical categories.

The construction of the proposed analyzer is based on a fuzzy automaton [6]. The recent application of finite-state approach in NLP shows the usefulness of automata in this area of AI [7]. The basic advantage of applying a fuzzy automaton to process natural language lies in the existence of an established algorithm for the next state selection based on values of the characteristic functions, allowing the characteristic functions value to change with time. In that case fuzzy automaton-driven parser „gets accustomed" to the input syntax constructions, imitating human behavior.

Figure 1 shows a graph of a subautomaton $fA(l_1)$ that isolates from the string of the lexical categories of the words a substring having a syntactical category of the nominal group. There is another subautomaton, $fA(l_3)$, incorporated into the $fA(l_1)$ subautomaton structure, which accepts a substring having a syntactical category of the adjectival group. The subautomaton $fA(l_3)$ is represented by the subgraph $fG(l_3)$ in the transition graph. The label $w_p i_s$ means that a transition along the edge is possible when w_p (standing for lexical categorie of the word from the analysed sentence) is the input symbol that is being read. During the transition the instruction i_s is performed. The label $w_p i_s q_t l_u$ means that during the transition, the instruction i_s is performed with the parameters q_t l_u. The symbol r_k denotes a specific set of type w symbols; an edge marked with a label that contains the r_k symbol represents a bundle of edges each of which is marked with one of the symbols belonging the r_k set.

The transition from the q_i to the q_j state at the input signal w_p has some degree of membership (usually denoted as $\mu_{ij} \in [0,1]$). There is the following role for the characteristic function μ_{ij} which describes the transition relation of the fPDAMS automaton: for a given state q_i and input w_p of the automaton, the next state q_j of the automaton is determined so that to maximize the characteristic function value corresponding to the transition. By the way, note that the fuzzy automaton is similar to the stochastic automaton. Both fuzzy and stochastic automata are examples of the acceptors of weighted languages. However, in stochastic automata the transitions are applied according to a probability distribution, there exists no uncertainty about the generated string. In fuzzy automata all applicable transitions are executed to some degree and transtions weights, in contrary to stochastic automaton, do not need to sum to 1, if there exist several alternative transitions at any state.

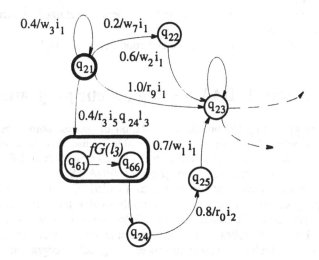

Fig. 1. A fragment of the transition subgraph $fG(l_1)$ of the fuzzy subautomaton $fA(l_1)$

The graph transition represents an analysis of a fragment of the input string of the lexical categories, the analysis result being stored on one of the six stacks of the fPDAMS automaton. The automaton operation is determined by a finite number of instructions. These instructions constitute the transition rules to allow the automaton to transit from one configuration to another. Limited number of instructions, straight rules of activity permit to use fPDAMS as a kind of linguistic tool.

3. Developing fPDAMS using evolutionary programming

Evolutionary programming [8] can operate on fPDAMS's as follows:
1) Initially a population of parent fPDAMS's is randomly or by hand constructed. The fPDAMS is represented by its transition graph.
2) The parents are tested in the environment, that is for each parent fPDAMS the collection of positive and negative examples of sentences is offered. The fitness of the automaton is then measured on the basis of the fraction of correctly analyzed sentences.
3) Offsprings fPDAMS's are created by mutating each parent automaton. Mutation can change both the topology and the connection weights. The topology-modifying operators can change the topology of the fPDAMS by: adding new state to the transition graph or removing the existing state. The operators changing the connection weigths operate on the terms $\mu w_{p_i s} q_{r_u}^l$ modifing the individual elements.
4) The offsprings are evaluated over the existing environment in the same manner as their parents.
5) Those fPDAMS's that provide the best fitness are retained to become parents of the next generation.
6) Steps 3)-5) are repeated until the end condition is reached.

4. Developing fPDAMS using genetic programming

Genetic programming [9] breeds a population of rooted, point-labeled trees with ordered branches. From here also, to be able to use this class of evolutionary computation in evaluating the architecture of parser, one should first find suitable method of mapping the transition graph of fPDAMS to tree structure. In [10] Frederic Gruau proposed interesting technics called "Cellular Encoding", in which genetic programming is used to concurerently evolve the architecture of a neural network. We apply "Cellural Encoding" to the evolution of the fPDAMS's. "Cellular Encoding" relies in fact on operating on indirect structures, so-called grammar-tree, which are subject to process of genetic programming. Every such structure represents graph (of neural networks [11], of electric circuit [12], or of the transition graph of the fPDAMS). The grammar-tree is the genotype and the fPDAMS constructed in accordance with the tree's instruction is the phenotype.

   The population of grammar-trees representing the fPDAMS's is subject to the standard scheme of evolutionary programming except that genetic programming allows sexual, genetic crossover operation. The recombination operator crossover is realized by exchanging the subtrees of two parent individuals. The mutation operator brings some variations to the genome and is realized by inclusion of randomly initialized grammar-trees.

5. Conclusions

We have proposed the concept of use the evolutionary approach to inference of natural language parser fPDAMS. The theoretical bases for the use of evolutionary programming and genetic programming that supports automated designed of the architecture of the fPDAMS were provided. Fuzzification of the parser and its evolutionary developing are the first steps toward the self-learning, "smart" analyzer of the natural language texts. The presented adaptable sentence analyzer, which can be taught while working, can be part of various NLP systems.

References

[1]   Unold O., *Automatic Analysis of Natural Language Texts in Man-Machine Communication* [in:] Wojtkowski G. at al (ed.), Systems Development Methods for the Next Century, Plenum Publishing Corp., New York, 1997, pp. 185-193.

[2]   Unold O., *A Fuzzy Automaton Approach to Dialog Systems*, Proc. of the IASTED International Conference-ASC'98, Cancun, Mexico, May 1998, pp. 215-218.

[3]   Unold O., *Application of Fuzzy Sets in Natural Language Processing*, Proc. of the 6th Congress on Intelligent Techniques and Soft Computing EUFIT'98, Aachen, Germany, September 1998, pp.1262-1266.

[4]   Unold O., *Toward fuzziness in natural language processing,* [in:] Advances in Soft Computing – Engineering Design and Manufacturing, Springer-Verlag, London, 1999, 554-567.

[5]   Unold O.: *A Stratificational Knowledge Representation System*, [in:] Vetulani Z., Abramowicz W. (ed.) Language and Technology, Academic Printing House PLJ, Warszawa 1996, pp. 177-181 (in Polish).

[6]   Kandel A., Lee S.C., *Fuzzy Switching and Automata: Theory and Applications*, Crane Russak, New York, 1979.

[7]   Roche E., Schabes Y.: *Finite-State Language Processing*, A Bradford Book, The MIT Press, Cambridge, Massachusetts, 1997.

[8]   Fogel L.J., Owens A.J., Walsh M.J., *Artificial Intelligence through Simulated Evolution*, J.Wiley, Chichester, 1996.

[9]   Koza J., *Genetic Programming*, MIT Press, Cambridge, MA, 1992.

[10]   Gruau F., *Cellular encoding of genetic neural networks*, Technical report 92-91, Ecole Normale Superieure de Lyon, Institut IMAG, 1992.

[11]   Friedrich C.M, Moraga C., *An Evolutionary Method to Find Good Building-Blocks for Architectures of Artificial Neural Networks*, Proc. of the Sixth Int. Conf. on Information Processing and Management of Uncertainty in Knowledge-Based Systems IPMU'96, Granada, Spain, 1996, pp.951-956.

[12]   Andre D., Bennet III F.H., Koza J., Keane M., *On the Theory of Designing Circuits using Genetic Programming and a Minimum of Domain Knowledge*, Proc. of the 1998 IEEE Congress on Computional Intelligence WCCI'98, Anchorge, Alaska, 1998, pp. 130-135.

GA-Based Identification of Unknown Structured Mechatronics System

Makoto Iwasaki, Masanobu Miwa, and Nobuyuki Matsui

Dept. of Elec. and Comp. Eng., Nagoya Institute of Technology
Gokiso, Showa, Nagoya 4668555, Japan, iwasaki@elcom.nitech.ac.jp

Abstract: Soft computing techniques, e.g., Neural Networks, Fuzzy inference, evolutionary computation, and chaos theory, have been applying to a wide variety of control systems in industry because of their control capability and flexibility. They are also powerful to handle the complicated mechatronics systems with various nonlinearities which are difficult to be modeled by mathematical formulas. This paper presents a novel autonomous algorithm for the identification of unknown structured motion control systems using Genetic Algorithms (GA), where the optimal order of a system polynomial and the optimal set of its coefficients can be determined by means of the optimization ability of the GA. The effectiveness of the proposed identification can be verified by experiments using the typical mechanical systems with velocity controller.

1 Introduction

Most of motion control strategies are based on the control theory, where the compensators are generally designed using CAD systems on the basis of mathematical system models and are computationally processed using high performance CPUs. In such strategies, the motion control accuracy is affected by both the structured and unstructured uncertainties [1] in the models, where, generally speaking, the structured uncertainties are caused by errors in the parameterization for exact mathematical models, and the unstructured uncertainties are due to the structural modeling errors. Although the structured uncertainties can be compensated for by system identification algorithms and the unstructured uncertainties can be also compensated for by robust control algorithms, these approaches are very computationally intensive and require the higher processing capability of CPUs. Recent digital signal processors (DSPs) and/or reduced instruction set computers (RISCs) have the capability performing such complicated control processing on line, by means of their optimal hardware architecture and instruction sets [2]. However, as the control specifications become severer, various nonlinearities in controller elements and mechanisms, e.g., the saturation of compensator outputs, the nonlinear friction, and the lost motion (backlash + torsional property) of gears, should be handled.

Recently, soft computing techniques have been an attracting interest in industrial application fields [3, 4]. Since the soft computing algorithms are

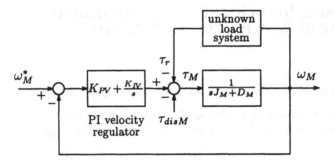

Figure 1: Block diagram of PI velocity control system.

bio-mimetic based strategies and can handle qualitative techniques with no mathematical model, they are easily applied to complex systems. In such diverse algorithms, the evolutionary computation techniques, e.g., Genetic Algorithms (GA) [5], Evolutionary Programming (EP) [6], Evolution Strategy (ES) [7], and Genetic Programming (GP) [8], are particularly applied as the optimization strategies [9]. For example, the GA can be actively applicable to identify unknown plant systems [10] and/or to design a variety of control systems [11, 12].

This paper presents an autonomous algorithm for the identification of unknown structured mechatronics systems to assist the compensator design of motion control systems. In this research, under the assumption that the transfer characteristics of the plant system can be modeled by a polynomial expression, the GA searches the optimal order of the polynomial and the optimal set of its coefficients by means of the optimization ability. The proposed scheme has a distinctive feature that the determination of the system structure and the identification of its parameters can be simultaneously attained, in contrast to the conventional identification algorithms, e.g., a linear regression technique under the given system structure. The effectiveness of the proposed identification is verified by experiments. In the experiments, a one-mass rigid system and a two-mass resonant system are selected as the typical mechanisms in motion control systems.

2 Identification of Mechatronics System

2.1 Motion Control System

Fig.1 shows a block diagram of a typical motion control system with proportional plus integral (PI) velocity control,

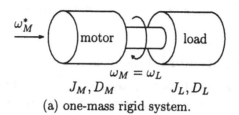

$$\omega_M = \omega_L$$

(a) one-mass rigid system.

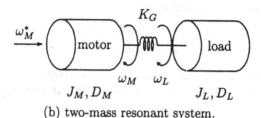

(b) two-mass resonant system.

Figure 2: Typical mechanical structure in motion control systems.

where ω_M^*; motor angular velocity command,
ω_M; motor angular velocity,
τ_M; motor torque,
τ_r; reaction torque from load system,
τ_{disM}; motor disturbance torque,
J_M; motor moment of inertia,
D_M; motor viscous damping coefficient,
K_{PV}, K_{IV}; velocity PI gains.

Here, the main objective of the research is to identify the mechanical system model based on a transfer characteristic of the motor velocity for the command. In the identification, it is assumed that the structure of load mechanical system is unknown. From the block diagram of Fig.1, the transfer function of ω_M for the command ω_M^* is given as follows under $\tau_{disM} = 0$:

$$\frac{\omega_M}{\omega_M^*} = \frac{sK_{PV} + K_{IV}}{s^2 J_M + s(D_M + K_{PV}) + K_{IV}} + \frac{1}{s^2 J_M + s(D_M + K_{PV}) + K_{IV}} \frac{\tau_r}{\omega_M^*}. \tag{1}$$

In eq.(1), the reaction torque τ_r is load mechanical structure dependent. The typical industrial mechanisms can be generally classified into two: one is a one-mass rigid system and the other is a two-mass resonant system with a low stiff component, as shown in Fig.2, where ω_L; load angular velocity, J_L; load moment of inertia, D_L; load viscous damping coefficient, and K_G; torsional constant. In Fig.2(a), no torque is reacted to the motor from load due to the rigidness of joint. In Fig.2(b), the existence of low stiff mechanisms in the load system induces the oscillatory τ_r, thus resulting in the resonant mechanical vibrations.

2.2 System polynomial expression

In the one-mass system of Fig.2(a), since the motor and load are rigidly coupled, i.e., $\omega_M = \omega_L$, and the apparent system inertia and damping can be handled as $J (= J_M + J_L)$ and $D (= D_M + D_L)$, the transfer function can be given in eq.(2) under $\tau_r = 0$ in eq.(1).

$$\frac{\omega_M}{\omega_M^*} = \frac{sK_{PV} + K_{IV}}{s^2 J + s(D + K_{PV}) + K_{IV}} \tag{2}$$

On the other hand, in the case of the two-mass system of Fig.2(b), the following 4th order transfer function can be obtained by substituting $\tau_r = K_G(\omega_M - \omega_L)/s$ in eq.(1):

$$\frac{\omega_M}{\omega_M^*} = \frac{b_1 s^3 + b_2 s^2 + b_3 s + b_4}{s^4 + a_1 s^3 + a_2 s^2 + a_3 s + a_4} \tag{3}$$

where
$$a_1 = \frac{J_M D_L + J_L(K_{PV} + D_M)}{J_M J_L},$$
$$a_2 = \frac{J_M K_G + D_L(K_{PV} + D_M) + J_L(K_{IV} + K_G)}{J_M J_L},$$
$$a_3 = \frac{K_G(K_{PV} + D_M + D_L) + K_{IV} D_L}{J_M J_L},$$
$$a_4 = \frac{K_{IV} K_G}{J_M J_L},$$
$$b_1 = \frac{K_{PV} J_L}{J_M J_L},$$
$$b_2 = \frac{K_{PV} D_L + K_{IV} J_L}{J_M J_L},$$
$$b_3 = \frac{K_{PV} K_G + K_{IV} D_L}{J_M J_L},$$
$$b_4 = \frac{K_{IV} K_G}{J_M J_L}.$$

In order to simulate the velocity responses by numerical calculations in processor-based controller, it is convenient to represent the transfer characteristic in the form of a recurrence formula. These transfer functions in eqs.(2) and (3) can be transformed to the following discrete polynomial with sampling number n, using the bilinear transformation:

$$\omega_M(n) = \sum_{i=0}^{N} A_i \omega_M^*(n - i) + \sum_{i=1}^{N} B_i \omega_M(n - i) \tag{4}$$

where the order of polynomial for each system is $N = 2$ in the one-mass system or $N = 4$ in the two-mass system, respectively. As a result, the determination of the order N and coefficients A_i and B_i in the polynomial results in the identification of the motion control system.

3 GA-Based System Identification

3.1 Principle of System Identification

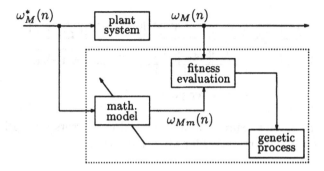

Figure 3: Identification system using GA.

Table 1: Specification of GA operation.

population size	50
tournament selection	
multi-point crossover	crossover rate: 0.9
bit mutation	mutation rate: 0.01

The proposed identification system is shown in Fig.3, where the "plant system" is an objective velocity control system to be determined and the "mathematical model" is given by the polynomial in eq.(4). In the proposed system, the parameters in the mathematical model are identified by the genetic process so that the actual velocity ω_M and the model velocity ω_{Mm} should coincide with each other. That is, the GA optimizes the combination of N, A_i, and B_i, which makes the fitness function F_{fit} in eq.(5) minimum, using the time basis velocity response data $\omega_M(n)$ and $\omega_{Mm}(n)$ obtained from the plant system and the mathematical model, respectively.

$$F_{fit} = \sum_{n=1}^{S} (\omega_M(n) - \omega_{Mm}(n))^2 + w_f N \qquad (5)$$

The second term in right hand side in eq.(5) is a penalty term with a weight w_f for the polynomial order. In this fitness function, the optimal parameter identification can be attained under the minimum order of model structure, since not only the integral of squared error but the order are evaluated considering the information criteria [13].

3.2 Genetic Operation

Specifications of GA operation are listed in Table 1. The order and each coefficient in eq.(4) are expressed in 2-bit and 10-bit binary chromosomes,

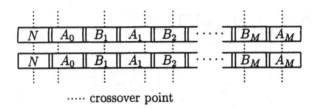

····· crossover point

Figure 4: Chromosome expression of order and coefficients for mathematical model, and multi-point crossover for every parameter.

respectively, and a set of the chromosomes composes an individual in the GA process as shown in Fig.4. Genetic operations are individually processed for each chromosome because the parameters in eq.(4) are independent each other. For example, the crossover is performed for every chromosome corresponding to each parameter as shown in Fig.4. In the following experimental studies, a 92-bit individual (2-bit + 10-bit × 9) is designed to handle the model whose order is $N \leq 4$. This word length allows each individual in population to express a mathematical model with arbitrary order, while only the part of A_i and B_i corresponding to N has effective information on the model, e.g., in the case of $N = 2$, $A_3 \sim A_4$ and $B_3 \sim B_4$ have no significant information. All identification processing indicated in the dotted box in Fig.3 is performed using a DSP-TMS320C30 (33 MHz), where the computational processing for one generation in GA can be done within 30 ms.

4 Experimental Verification

4.1 Configuration of Experimental Setup

Experimental studies are given for a robot arm with a flexible joint which is modeled by a two-mass resonant system as shown in Fig.5. In experiments, the one-mass rigid system can be also examined, by removing the load arm of this system. Specifications of the prototype is listed in Table 2, where the nominal mechanical parameters are given by mechanical specification sheets and the PI motor velocity control is digitally performed with a sampling period of T_s. In the identification process, a motor velocity command ω_M^* of pseudo white noise whose frequency range is of 0 to 40 Hz is given to achieve the generalization ability in the identified model. In order to evaluate the fitness, the motor velocity for the white noise command is stored for 0.1 s ($= S \times T_s$, the integral number S in eq.(5) is 250) to prepare the time basis data $\omega_M(n)$, and the weight w_f is selected as 2.0.

Fig.6 shows the motor velocity responses for step motor velocity command whose amplitude is 43.6 rad/s. In the figure, the plant response is detected from the real system and the model response is calculated using the mathematical model whose parameters are set by the nominal mechanical ones in

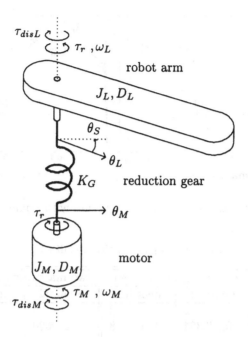

Figure 5: Model of two-mass resonant system for robot arm.

Table 2: Specifications of prototype.

nominal mechanical parameters			control parameters		
J_M	1.02×10^{-4}	[kgm^2]	K_{PV}	0.0387	[As]
D_M	0.003	[kgm^2/s]	K_{IV}	2.279	[A]
J_L	5.815×10^{-4}	[kgm^2]	T_s	0.444	[ms]
D_L	0.008	[kgm^2/s]			
K_G	16150.55	[Nm/rad]			

Table 2. Two responses do not show good agreement in both cases of one-and two-mass systems. This disagreement is due to existence of unstructured uncertainties, e.g., nonlinear friction in the plant system.

4.2 Experimental Results

Table 3 is the identification results by the proposed scheme and Fig.7 shows the averaged fitness-generation characteristics for 20 trials of the identification process, both for one- and two-mass systems. Notice here that no pre-knowledge of the plant structure is given before the identification, i.e., the initial population in GA system is composed of individuals with the arbitrary order of model ($N \leq 4$). From both results in the table, although the order of model can be determined correctly, the identified coefficients do not coincide with

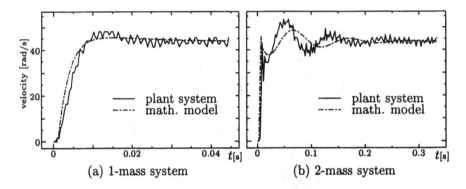

(a) 1-mass system   (b) 2-mass system

Figure 6: Velocity responses for step command with nominal parameters.

Table 3: Identification results.

| parameters | one-mass rigid system | | two-mass resonant system | |
	nominal	identified	nominal	identified
order N	2	2	4	4
A_0	0.078364	0.038318	0.220620	0.275171
A_1	0.002019	0.004165	−0.434021	−0.465493
A_2	−0.076345	−0.030342	−0.006987	−0.009169
A_3			0.434026	0.353079
A_4			−0.213628	−0.146334
B_1	1.829284	1.799902	3.532134	3.253958
B_2	−0.833322	−0.817302	−4.615719	−4.144086
B_3			2.634535	2.482893
B_4			−0.550959	−0.593743
fitness	1376.964	187.631	2842.167	1028.269

the corresponding nominal ones. However, it is apparent that the identified combinations make the fitness smaller. This means that the coefficients of the discrete polynomial have been determined to make the fitness in eq.(5) minimum by compensating for the effects of unstructured uncertainties on the actual plant velocity response.

Fig.8 shows the velocity responses of the plant system and the identified mathematical model for the pseudo white noise command used for the identification. As is apparent from the both, the model responses can be successfully fitted to the plant ones. Fig.9, on the other hand, shows the velocity responses for a stepwise command to verify the generalization ability in the identified models. In both cases, the plant and model responses show good agreement comparing with those in Fig.6 and, thus proving the effectiveness of the proposed identification.

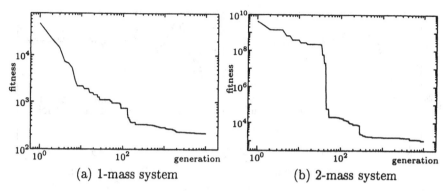

(a) 1-mass system (b) 2-mass system

Figure 7: Characteristic of fitness-generation in genetic operation.

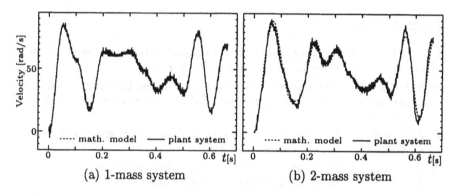

(a) 1-mass system (b) 2-mass system

Figure 8: Velocity responses for pseudo white noise command by identification results.

5 Conclusions

In this paper, a novel identification algorithm for motion control systems using GA is presented. The proposed scheme requires no pre-knowledge of the mechanical system structure unlike conventional identification techniques, thus allowing the system to be autonomous. Experimental results using a prototype verify the effectiveness of the proposed identification.

Acknowledgement

The part of this research was supported by eager discussions with members of "The project on development of emergent soft computers" in Nagoya Industrial Science Research Institute. The authors would like to express sincere gratitude to them.

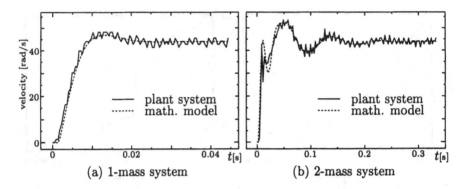

Figure 9: Velocity responses for step command by identification result.

References

[1] J. C. Doyle, B. A. Francis, and A. R. Tannenbaum, *"Feedback Control Theory,"* Macmillan Publishing Company, New York, 1992.

[2] N. Matsui, T. Takeshita, and M. Iwasaki, "DSP-Based High Performance Motor/Motion Control," *Proc. of International Federation of Automatic Control '96*, pp.393–398, 1996

[3] R. G. Hoft and Y. Dote, "Soft Computing in Power Electronics," *Proc. of Soft Computing in Industry '96*, pp.15–20, 1996.

[4] N. Matsui, "Applications of Soft Computing to Motion Control," *Proc. of 5th Int'l Workshop on Advanced Motion Control*, pp.272–281, 1998.

[5] D. E. Goldberg, *"GENETIC ALGORITHMS in Search, Optimization, and Machine Learning,"* Addison Wesley, 1989.

[6] L. J. Fogel et al., *"Artificial Intelligence through Simulatd Evolution,"* John Willey & Sons, 1966.

[7] I. Rechenberg, *"Evolutionsstrategie '94,"* frommann-hozboog, 1994.

[8] J. R. Koza, *"Genetic Programming,"* The MIT Press, 1992.

[9] K. F. Man, K. S. Tang, and S. Kwong, "Genetic Algorithm: Concepts and Applications," *IEEE Trans. on Industrial Electronics*, Vol.43, No.5, pp.519–534, 1996.

[10] T. Hatanaka, K. Uosaki, and Y. Yamada, "Evolutionary Approach to System Identification," *Preprints of 11th IFAC Symposium of System Identification*, pp.1383–1388, 1997.

[11] S. S. Ge, T. H. Lee, and G. Zhu, "Genetic Algorithm Tuning of Lyapunov-Based Controllers: An Application to a Single-Link Flexible Robot System," *IEEE Trans. on Industrial Electronics*, Vol.43, No.5 ,pp.567–574, 1996.

[12] K. S. Tang, K. F. Man, and D. W. Gu, "Structured Genetic Algorithm for Robust H∞ Control Systems Design," *IEEE Trans. on Industrial Electronics*, Vol.43, No.5 ,pp.575–582, 1996.

[13] H. Akaike, "An Objective Use of Bayesian Models," *Ann. Inst. Statist. Math.*, Vol.29, pp.9–29, 1977

Integrating genetic algorithms and interactive simulations for airbase logistics planning

N.L. Schneider, S. Narayanan, and Chetan Patel

Dept. of Biomedical, Industrial, & Human Factors Engineering, Wright State University

207 Russ Engineering Center, Dayton, OH 45435, USA. email: snarayan@cs.wright.edu

Abstract: There is a need for a modeling approach that can integrates simulation with human interaction and optimization to exploit the analyst's ability in addressing difficult-to-quantify issues while utilizing computer models and algorithms to perform complex computations in systems analysis. In this paper, we present an approach that integrates human interaction with simulations and genetic algorithms for the repair-time analysis problem in airbase logistics. We present the integrative approach along with the computer system implementation, and evaluation in the context of a realistic repair time analysis problem.

1. Introduction

An airbase consists of different entities, including aircraft, mechanics, support equipment, and flight schedulers, that work together to achieve the overall requirements of a mission. Aircraft are assigned to specific sorties listed in the mission log based on the abilities of the aircraft and the goals of the mission. The aircraft take off, perform special mission-oriented tasks, land, and undergo systematic maintenance checks of their subsystems. Each aircraft is comprised of many subsystems which, during operational use, may fail. These subsystems include the electronics and hydraulics systems as well as engines and ventilation. Subsystems can also experience different degrees of failure hence the maintenance needed may be as simple as adding additional cooling fluid or as involved as changing the turbine blades on an engine. If there are no maintenance problems, the aircraft wait for their next scheduled take-off. Otherwise, the aircraft is sent to maintenance before it is readied for its next flight. Skilled mechanics and specially designed equipment are required at an airbase to repair these failures. Depending on how a subsystem fails, hangar space, tools and spare parts may also be required.

All these resources must be available at the airbase. Logisticians must carefully consider the needs of completing a mission and allocate resources to fill those needs. Failure to accurately predict the correct amount of resources may lead to costly excess supplies and personnel, or to a dearth of supplies and personnel, which can leave an airbase dangerously unprepared for stressful or unexpected situations.

Simulation is a powerful tool often used to evaluate existing and potential systems. It can test the viability of resource combinations to successfully achieve a given mission or goal. Simulation models are used in experimenting with new designs or possible improvements to determine system performance under varying conditions [1].

While simulation is a flexible, powerful methodology for analysis of complex airbase systems [2], current simulation methods have several limitations. First, traditional discrete-event simulation accommodates very little human interaction [3]. Analysts are typically involved in tweaking the simulation parameters or performing what-if analysis of simple system configurations without full understanding of how the system is acting. This can lead to reduced trust in the simulation-generated solution. Second, simulation is a methodology focused on describing a system and there tends to be little support for prescriptive methods such as optimization. Simulation can be used prescriptively through what-if analysis, but manual tweaking of a single solution is inefficient.

Recent research has led to the integration of response surface equations to simulations [4], which augment descriptive simulations with optimization techniques. However, many of these techniques rely on purely quantitative formations of the problem and do not account for real-time constraints, uncertainty and qualitative factors. There is a need for a modeling approach that can integrate simulation with human interaction and with optimization to exploit the analyst's ability to address difficult-to-quantify issues while utilising computer models and algorithms to perform complex calculations in system analysis. This paper presents a novel approach that integrates interactive simulation with genetic algorithm (GA)-based optimization to solve the problem of resource allocation in the complex system of airbase logistics [5]. This paper first describes the approach, then discusses its implementation, and concludes with the presentation of the evaluation.

2. The Approach

Traditionally, simulation has been based on a descriptive model with the focus on evaluating performance and drawing conclusions about what happens in the system [6]. It allows analysts to study the results of their choices for the system but does not necessarily direct them towards changes that will improve the system. Through detailed understanding of the dynamics and interactions of the system components, an expert can study a simulation and, based on criteria for success, decide to modify it or leave it undisturbed. Mollaghasemi and Evans [4] argue that traditional simulation is not enough to effectively analyse and optimize large, complex systems. Prescriptive models, on the other hand, are used to generate decisions, not analysis, about system design [6]. In optimizing a system, it is important to know not only the current performance, but also what changes will lead to improved performance. Prescriptive models do this by working toward the system goals. Analysts need a prescriptive modeling tool to explore the solution space and find good, feasible designs and remove poor designs from consideration

[7]. Also, analysts need to interact with the simulated designs used a descriptive modeling method to study the effects of different scenarios and to develop trust in the designs' strength and flexibility. To meet this two-pronged objective of descriptive and prescriptive modeling methods, a framework that integrates genetic algorithms (GA) with interactive simulations is needed. When analysts only want to evaluate the response of a specific design configuration, descriptive simulation models are useful. However, if analysts wish to improve that response, descriptive methods are insufficient and must be accompanied by optimization [4]. Prescriptive modeling methods can provide decisions or guides about the number and type of resources required by a successful system [6].

An alternative to the traditional descriptive modeling approach, the Modeling-to-Generate Alternatives (MGA) technique [7] is designed for complex, incompletely defined problems. MGA is first based on the assumption that solutions derived from mathematical programming models, such as a linear programming, rarely represent sound solutions to real problems. This is due to simplification of the model and qualitative issues that cannot be adequately defined by the model's algorithms. MGA is also based on the assumption that the number of alternatives presented and the degree of uniqueness of each alternative affect the analyst's ability to integrate data received from the model with other information sources. A high degree of difference between alternatives helps the analyst to better choose among them. Corresponding to the limitations and variability of human decision-making (see 8, 9), MGA generates only a few different solutions for analysis and consideration. Analysts are better able to rank solutions when they are few in numbers and their differences are readily discernable. This is a joint cognitive systems approach that combines the human and computer into the decision-making system, enlarging the capability of each while neutralizing limitations. With MGA, many design alternatives are screened and poor designs are removed from the potential solution set. The remaining designs are known to perform well with respect to the system's quantitative goals and are noticeably different from each other. The number of alternatives presented to the analyst should be small so that the human does not become overwhelmed [7]. MGA is an improvement over traditional approaches. It does not constrain the range of design alternatives solely based on the solutions derived from mathematical models and presents the analyst with more than one solution for study. The framework described in this paper closely follows MGA and combines prescriptive and descriptive modeling methods. A prescriptive mathematical model employing a GA explores the solution space and generates alternatives while a descriptive, interactive simulation enables analysts to study the alternatives and employ qualitative constraints.

2.1 Overall Architecture

The framework, known as Java-based architecture for developing interactive simulation (JADIS) combines prescriptive modeling, in the solution explorer (SE) and implemented as a GA-based optimization program, and descriptive modeling, in the interactive analyzer (IA) and implemented as a graphical, interactive simulation program. Figure 1 depicts JADIS and shows the role of the analyst and the flow of information in moving from initial solution estimates to the final solution decision.

2.1.1 Solution Explorer

With the SE, analysts are able to rapidly study the solution space and optimize a set of initial design guesses. Analysts provide the initial designs, a quantitative fitness function (a mathematically expressed combination of goals and constraints) for the system, and the number of iterations for the solution search process. This information is input into the input specifier and the simulator that has been designed to model the system under study. The program then evaluates the designs' performances based on the fitness function and encodes the designs for use by the GA. The GA selects designs for modification and improvement. This new iteration of designs is decoded and inputted into the simulator and the process continues until the analyst-chosen number of iterations is reached. The GA was chosen for this project for several reasons. It has been successfully applied to optimization problems in the fields of communication networks, personal computer reliability, electromagnetic devices, and truss design [10]. The GA is a heuristic optimization method based on biology's survival of the fittest. Each set of designs is represented as a string, named "chromosome" to parallel terminology from biology. Each chromosome has a fitness value, corresponding to how well it meets the fitness function. For each iteration or "generation," the GA selects chromosomes to parent the next generation, showing preference for strings of greater fitness. Parent strings are chosen to produce a child string and their data are mixed in the process known as "crossover." Crossover is critical as it combines strings known to contain successful information. Additional random changes can occur to the child string through "mutation." Mutation permits the creation of otherwise impossible data combinations, expanding the search for successful results beyond the confines of the initial input sets. When a child is completed, it is tested for fitness and rated, making it ready to be considered for parenting the next generation of chromosomes. This process continues until the algorithm reaching the "terminating generation" or "terminating condition" which is preprogrammed stopping pint of the GA. The GA method performs a randomized, directed search. Calculus-based algorithms are very difficult to use in complex problems such because the system goal, expressed as a fitness function, is complex, time-dynamic, and discontinuous. Calculus-based algorithms are also weak in scope, only able to find a local minimum, and are tied to the existence of derivatives. Enumerative algorithms require calculations at every possible point in the domain's search space, making them grossly inefficient,. Random search algorithms perform similarly to enumerative approaches, only randomly selecting a set of pints to evaluate instead of employing a definitive method.

The GA is favored as a heuristic optimization tool because of its versatility and capability [10]. It is different from traditional optimization methods for four reasons. First, it does not operate directly on design parameters but rather with strings containing many parameters. Second, it optimizes entire populations of a design simultaneously instead of a single design. Third, it works with the fitness function, or goal, of the system and not it's derivative, which may be discontinuous. Finally, it uses probabilistic transition rules instead of deterministic rules to guide the search for improvement. We used the stochastic binary tournament for selecting the parent strings. In stochastic binary tournament selection, two parent strings are selected at random and the fitter of the two is reproduced in the next generation. This comparison process is repeated until the next generation reaches the current population size. The GA is able to consider the entire solution space without prejudice for local minima or the overall cost function or its derivative used in many optimization processes. By rating the fitness of its creations, it can quickly prune the search space and present the analyst with a selection of enhanced solutions for their final approval.

2.1.2 Interactive Analyzer

The interactive analyzer or IA uses data sets already selected as good solutions for a given system goal and allows the analyst to test the solutions under different and stressful conditions. The IA has the capability of adding realistic, random events to the running simulation such as the deployment of aircraft to another base or the temporary reduction of mechanics due to additional tasks and holidays. The analyst is free to alter the behavior of system resources in addition to adding and removing the amount of different resources, and can observe the consequences of those action through immediate graphical feedback on the display and through later statistical analysis based on data written to output files. This allows the analyst to consider many "what-ifs" associated with the uncertainty of the system [3]. This framework has the capability of automatically crating and evaluating alternatives based on the analysis's initial ideas and objectives for the system. The SE takes the initial solutions suggested by the analyst and presents an altered set of solutions for the analyst's detailed quantitative and qualitative study with the IA.

2.2 Application to Airbase Logistics

We implemented the entire architecture using the Java programming language by also enhancing the GaLib package downloaded from the publicly available MIT web site. We then applied the architecture to an airbase logistics system.

As described in the introduction, an airbase consists of different entities, including aircraft, mechanics, support equipment, and flight schedulers, that work together to achieve the overall requirements of a mission. Aircraft are assigned to specific sorties listed in the mission log based on the abilities of the aircraft and the goals of the mission. The aircraft take off, perform special mission-oriented tasks, land, and undergo systematic maintenance checks of their subsystems. Each aircraft is comprised of many subsystems which, during operational use, may fail. These subsystems include radios and navigational aids as well as the engines and ventilation system. Subsystems can also experience different degrees of failure hence the maintenance needed may be as simple as adding additional cooling fluid or as involved as installing a new engine. If there are no maintenance problems, the aircraft wait for their next scheduled take-off. Otherwise, the aircraft is sent to maintenance before it is readied for its next flight. Skilled mechanics and specially designed equipment are required at an airbase to repair these failures. Depending on how a subsystem fails, hangar space, tools and spare parts may also be required. All these resources must be available at the airbase. Logisticians must carefully consider the needs of completing a mission and allocate resources to fill those needs. Aircraft must be assigned to complete the mission requirements of the base. The base should have enough aircraft to avoid mission aborts caused by an insufficient number of aircraft. In addition, failures and routine maintenance visits must be dealt with efficiently and speedily. Excessive waits for maintenance work may because necessary aircraft to be unavailable for flight and lead to aborted missions. Failure to accurately predict the correct amount of aircraft and maintenance resources may lead to a costly excess of supplies and personnel, which drains the Air Force's budget unnecessarily, or to a dearth of supplies and personnel, which can leave an airbase dangerously unprepared for stressful or unexpected situations. Analysts must have the access to relevant data, organized in an intelligent manner, which will help them make decisions on how, or if, to modify a system. Analysts need access to statistics to gauge system performance based on the performance of important indicators (e.g., the number of aborted missions and cumulative maintenance wait time). With misleading or poorly organized data, analysts may reach the wrong conclusions about the system, leading to errors.

The particular scenario studied in this paper involves the missions of the FX-99, a notional aircraft. Mechanics, equipment, hangars and spare parts are used to repair failures and conduct routine maintenance on the FFX-99's meager collection of 23 subsystems. The schedule of flights, failures, and even the maintenance behaviors are based on data files that can be manipulated to test different possibilities during IA analysis. Both the solution explorer and the interactive analyzer were evaluated. Details of the individual evaluations are provided in Schneider [5]. Results are summarized below.

In the evaluation of the solution explorer, eight design solutions served as the initial population for the SE evaluation. For each generation, the designs were studied with different seed values and the generation mean was recorded. The Mann-Whitney test demonstrated an improvement in the population mean ($p > 0.002$) very rapidly. Subsequent improvements in the designs took more generations. Although there was convergence after six generations, it was not clear how close the converged solution was to the optimal.

In the evaluation of the interactive analyzer, three participants familiar with Air Force logistics each modified four design solutions (there was a fourth participant less familiar with the domain whose data turned out to be an outliner). Due to the convergence of the SE experiment, new designs were chosen for the IA experiment. Each participant modified the designs once, and the modified designs were recorded. All designs, original and subject-modified, were rerun many times. The performance of every subject-modified design, expressed as the fitness value, was compared with that of its original under the same random events, and the difference between fitness values was calculated. Results indicated that the interactive analysis phase resulted in statistically significant improvement 30% of the time and there was no significant changes the rest of the time. However, 50% of the time the results improved slightly and 20% of the time the results degraded slightly.

3. Conclusions:

The solution explorer saved initial search time for the analyst. It explores the design space quickly and efficiently, saving the analyst from this tedium, and presents a final solution set that outperforms the original. If the solution set had included more than one unique design, the analyst would have had the opportunity to conduct a detailed analysis on each solution to determine which best met the combination of quantitative and qualitative goals. Design convergence significantly impaired the SE. This occurred because a single, strong design was repeated with increasing frequency in later generations, weeding out other designs. Recall that the purpose of the JADIS-SE was to quickly search for a set of distinct solutions to present to the analyst for detailed, quantitative and qualitative study. While traditional methods presented the user with one design, selected by a mathematical model as superior to the rest, the JADIS framework tried to incorporate the concept of *choice* between good solutions.

The overall effectiveness of the interactive analyzer was good. However, comments made by participants during the experiments highlighted some shortcomings of the IA. The notional view of the airbase was unsatisfactory for determining what resource shortages existed. While it showed where aircraft were to be found, it did not explain which subsystem failure had sent an aircraft to maintenance or what resources were needed to repair it, and the work status of specific mechanics and equipment. In addition to the lack of detailed information provided to participants during the experiment, participants were also not given the opportunity to alter the behaviors of system entities. The simulation has the ability to change the flight schedule, and the time and type and amount of resources needed to repair a subsystem, but this remained untested. Future research will focus on incorporating additional complexities in the evaluation.

Our overall approach is philosophically similar to the interactive evolutionary computation approach [11]. In the interactive evolutionary computation approach, humans constantly provide subjective evaluation to the fitness function used by techniques such as genetic algorithms. This approach has been successfully applied to engineering, education, and artistic fields. In contrast, our approach decomposed the planning into solution explorer and interactive analyzer components and the human input was primarily used in the interactive analyzer phase. The fitness function used in the solution explorer only has implicit human input. In complex planning problems involving numerous interactive constraints it is difficult for the human to provide subjective fitness function values for every design. However, it will be interesting to explore dynamic fitness function values based on human input in future research.

References

1. Banks, J., and Carson, J. S. II., 1984, *Discrete-Event System Simulation*. New Jersey:Prentice-Hall International Series in Industrial and Systems Engineering.

2. Popken, D. A., 1992, An object-oriented simulation environment for airbase simulation. *Simulation*. 59(5), 328- 338.

3. Narayanan, S., Patel, C., Schneider, N. et al, 1997, Integrating object-oriented simulation and interactive   optimization methods for logistics systems analysis. (Report no. *AL/HR-TR-1997-0126*). Ohio: Armstrong Laboratories.

4. Mollaghasemi, M. and Evans G. W., 1994, Multicriteria design of manufacturing systems through simulation optimization. *IEEE Transactions on Systems, Man, and Cybernetics*. 24(9), 1407-1411.

5. Schneider, N. N., 1998, Integrating genetic algorithms with interactive simulation for repair-time analysis in air   force logistics. *Unpublished Masters Thesis*, Department of Biomedical & Human Factors Engineering, Wright State University, Dayton, Oh.

6. Dietrich, B., 1991, A taxonomy of discrete manufacturing systems. *Operations Research*. 39(6), 886-902.

7. Brill, E. D., Flach, J. M., Hopkins, L. D. et al, 1990, MGA: A decision support system for complex, incompletely defined problems. *IEEE Transactions on Systems, Man, and Cybernetics*. 20(4), 745-757.

8. Hogarth, R. M., 1981, Beyond discrete biases: Functional and dysfunctional aspects of judgmental heuristics. *Psychological Bulletin*, 90(2), 197-217.

9. Corbin, R. M., 1980, Decisions that might not get made. In T. S. Wallsten (Ed.), *Cognitive Processes in Choice and Decision Behavior* (pp. 47-67). New Jersey: Erlbaum.

10. Goldberg, D. E., 1989, *Genetic Algorithms in Search, Optimization, and Machine Learning*. Massachusetts: Addison-Wesley Publishing Company.

11. Takagi, H., 1998, Interactive evolutionary computation - Cooperation of computational intelligence and human   KANSEI. *Proceedings of the 5th International Conference on Soft Computing*, Izuka, Fukuoka,,Japan,October16-20.

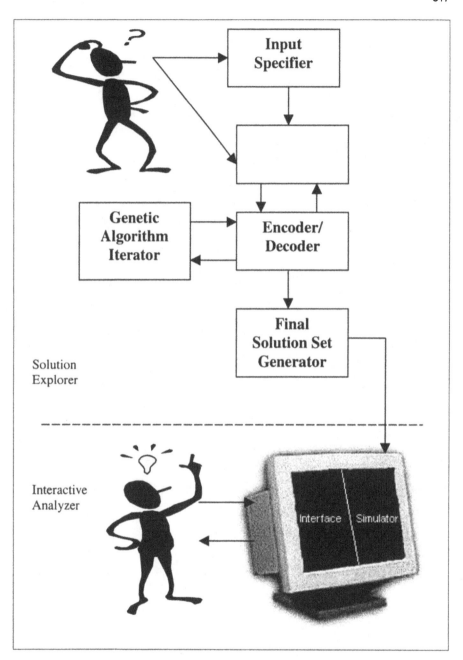

Figure1: Integration of solution explorer, interactive analyzer and humans in our approach.

Evaluation of Virtual Cities Generated by using a Genetic Algorithm

Nobuko Kato[1], Hitoshi Kanoh[2], and Seiichi Nishihara[2]

[1] Department of Information Science and Electronics,
Tsukuba College of Technology, Tsukuba, 305-0005, Japan ,
nobuko@a.tsukuba-tech.ac.jp
[2] Institute of Information Sciences and Electronics, University of Tsukuba,
Tsukuba, 305-8573, Japan , {kanoh,nishihar}@is.tsukuba.ac.jp

Abstract: A novel method for automatically modeling virtual cities is described and evaluated. A city is considered to be composed of several city blocks, and each block has feature values, such as average building height. The feature values are determined using a genetic algorithm. A chromosome represents the building layout of a virtual city and a fitness function is used to measure the similarity between neighboring blocks. Subjective evaluation showed that this method produces realistic virtual cities.

1 Introduction

A variety of virtual cities are widely used for World Wide Web sites and 3-D games. Modeling virtual cities requires a great deal of expense and effort. A method for modeling a virtual city based on a photograph of an actual city has been proposed [1], but new cities cannot be modeled this way. Using artificial life techniques has recently attracted much interest for producing various novel patterns [2]~[4]. We have developed a method that uses artificial life techniques to model original virtual cities that have the characteristics of actual cities[5].

A virtual city is usually composed of a road network that provides the basic structure of the urban area and a large number of buildings that are arranged in blocks. We developed a genetic algorithm (GA) for determining the values used for these layout features.

In the following section we discuss the factors that determine the landscape of a city and describe our basic modeling strategy. In section 3 we describe the coding method and the fitness function used to determine the feature values of city blocks. We present the results of our evaluation in section 4.

2 Strategy

2.1 City landscape

An actual city is composed of several types of areas: residential, commercial, high-rise, etc. Each area has its own characteristics, and the way these areas are distributed and shaped determines the landscape of the city. Each type of area has a thematic uniformity: that is, areas of the same type share similar characteristics including floor space, building height, distance between buildings and roads, building use, roofline, and roof pitch.

2.2 Parameters that define cities

Based on these characteristics, a city can be defined by using the following three parameters:

[area] number, shape and arrangement of areas,
[average] average of building height or floor space over the entire city,
[variance] variance in building height or floor space over the entire city.

The shapes of the buildings are also a significant parameter in defining the landscape of a city. We can easily control the average of the building height or floor space and the building shapes by changing the value range and building models, respectively. Therefore, we determine the areas and variances by using a GA. We can easily generate two types of cities: an orderly city in which the buildings are regularly arranged like a planned city, and a disorderly city in which the buildings are randomly arranged. Our goal is to generate a virtual city that is an appropriate mixture of both types.

2.3 Strategy

Our strategy for modeling a virtual city has five key components.

- We define "city block" to mean a region surrounded by roads, and we regard a city to be a set of city blocks.
- The city landscape is defined by how the features of the city blocks are distributed over the city.
- The distribution of feature values is obtained using a genetic algorithm.
- The fitness of each individual is defined as the composite similarity of feature values between neighboring blocks.
- The feature values are used to lay out the building objects, which are generated separately by a CAD system, on the road network.

GAs have been successfully used to solve many combinatorial optimization problems [7][8]. We also utilize GAs to find solutions, which have properties: orderly and disorderly. Because they perform a global search and give more than one semi-optimal solution, they can produce feature values for several cities with similar characteristics at one trial.

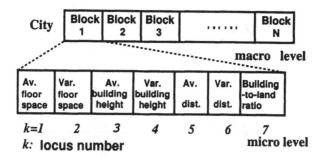

Av.: average
Var.: variance
dist.: distance between buildings and roads

Fig. 1. Coding method.

Table 1. Index of average floor space.

Index	1	2	3	4	5	6	7	8	9	10	.....	19	20
average floor space (m^2)	50	70	90	120	150	200	300	500	1000	1800	.....	9000	10000

3 Layout Method

3.1 Coding

We use four factors to define an area of a city:

- floor space
- building height
- distance between buildings and roads
- building-to-rand ratio

We use the average and variance of these factors as the feature values of city blocks. There can thus be several kinds of areas depending on the factor in which we are interested. The coding method we use to determine the building layout is shown in Fig. 1. A chromosome represents a city. We use a two-level hierarchical locus structure: the macro level corresponds to the blocks, and the micro level corresponds to the feature values. Each gene at the micro level is represented by an index. Table 1 shows, for example, an index of the average floor space.

3.2 Fitness function

The fitness value of each individual is calculated using

$$f = \frac{1}{2}(C_1 + C_2) \tag{1}$$

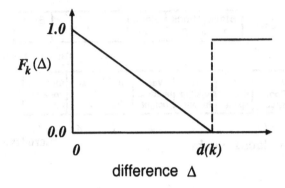

Fig. 2. Similarity function between neighboring blocks.

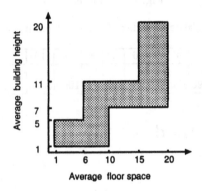

Fig. 3. Constraint between average building height and average floor space.

, where C_1 is the fitness of the similarity between neighboring blocks, C_2 is the fitness of the relationship between feature values inside a block.

[A] Fitness of similarity between neighboring blocks: C_1

Let $G_i(k)$ denote the gene at the k-th locus in the i-th block, and $F_k(\Delta)$ denote the similarity function between neighboring blocks (see Fig.2). C_1 is calculated using

$$C_1 = \sum_{i=1}^{N} \sum_{k=1}^{7} w(k) A_i(k) / N, \qquad (2)$$

where

$$A_i(k) = \sum_{j \in S_i} F_k(\Delta_{ij}(k)) / m_i$$

$$\Delta_{ij}(k) = |G_i(k) - G_j(k)|$$

Table 2. Parameters for GA.

Parameter	Values / Method
Crossover	Uniform
Mutation rate	0.001
Selection	Roulette
	Elite conservation
Upper limit of generation	200
Population size	500

Fig. 4. Example of virtual cities with basic parameters.

i, j: block number ,
N: number of blocks in a city ,
S_i: list of neighboring blocks to block i ,
m_i: number of blocks in S_i $(m_i = |S_i|)$,
$w(k)$: weighting factor.

[B] Fitness of relationship between feature values inside a block: C_2
We place a constraint on the relationship between the average building height and the average floor space so that the landscape is not unnatural. The shaded region in Fig. 3 illustrates this constraint. We define M_1 as the number of blocks without the shaded region and M_2 as the number of blocks that have no buildings, so that the number of empty blocks may decrease. C_2 is calculated using M_1 and M_2:

$$C_2 = 1 - \frac{(M_1 + M_2)}{2N} \tag{3}$$

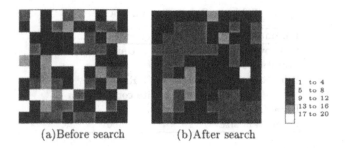

(a)Before search   (b)After search

1 to 4
5 to 8
9 to 12
13 to 16
17 to 20

Fig. 5. Distribution of $G_i(3)$.

4 Evaluation

4.1 Condition

For our evaluation experiments we used grid-type road networks, each with 10x10 blocks. The parameters used for the GA are shown in Table 2. The basic parameters for the fitness function were defined as follows: $(d(1), d(2), ..., d(7)) = (10, 6, 10, 6, 5, 6, 3)$, and $(w(1), w(2), ..., w(7)) = (1, 1, 1, 1, 1, 1, 1)$.

To evaluate the generated types of areas, we defined the area of a virtual city as follows. An area for locus k is defined as a cluster of city blocks having similar feature values, $G_i(k)$. An area size is the number of blocks in a cluster. The number of areas was calculated on the elite individual after quantizing $G_i(k)$ to five levels. The following data are the average of 20 trials.

4.2 Area generation

An example virtual city generated using the basic parameters is shown in Fig. 4. The distribution of the average building height, $G_i(3)$, before and after the search is shown in Fig. 5. The labels 1 to 4, for example, mean $1 \leq G_i(3) \leq 4$. Table 3 shows the number of areas for locus k=1,3,5, and 7 before and after searching. These results show that the generated virtual cities have several types of areas.

Table 3. Number of areas with basic parameters.

Features of area	Number of areas	
	Before search	After search
Average floor space ($k = 1$)	0.1	2.9
Average building height ($k = 3$)	0.0	2.2
Average distance between buildings and roads ($k = 5$)	0.1	1.1
Building-to-land ratio ($k = 7$)	0.1	1.1

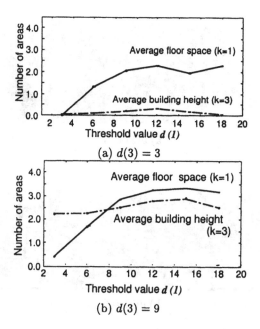

Fig. 6. Number of areas for locus $k = 1$ and locus $k = 3$ with $d(1)$.

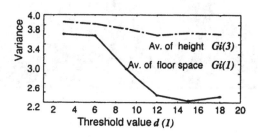

Fig. 7. Variances of $G_i(1)$, $G_i(3)$ with $d(1)$ $(d(3) = 3)$

4.3 Diversity of virtual cities

The number of areas for locus $k = 1, 3$ is shown as a function of the threshold value $d(1)$ in Fig.6. When $d(1)$ is high, areas for $k = 1$ are generated, while those for $k = 3$ are not, suggesting that we can make several types of areas for k when $d(k)$ is high.

Figure 7 shows the variances of $G_i(1)$ and $G_i(3)$ as a function of $d(1)$. As $d(1)$ increases, the variances of $G_i(1)$ decrease, where that of $G_i(3)$ remain high. This means we can generate virtual cities with equivalent building floor spaces but different building heights.

a) $d(3) = 3$

b) $d(3) = 9$

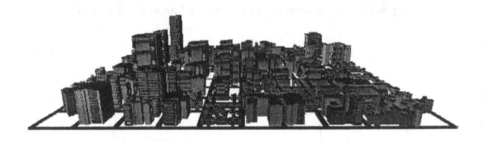

c) $d(3) = 12$

Fig. 8. Examples of virtual cities with $d(3) = 3, 9, 12$

We can therefore generate a various virtual cities with different types of areas, by using different parameter values.

4.4 Subjective evaluation

We used subjective evaluation to clarify the effectiveness of the generated virtual cities. The subjects were our university students, and the number of valid answers is 56. They evaluated image samples projected on a screen.

Effect of $d(k)$ value

First, we determined how the value of $d(3)$ affected the evaluation results. We used three values of d(3) to generate three virtual cities – Fig.8 a),b), and

Table 4. Effect of d(3) on evaluation

Figures	d(3)	Mean points	95% confidence interval
Fig.8 a)	3	3.3	2.9-3.6
Fig.8 b)	9	3.9	3.5-4.3
Fig.8 c)	12	4.7	4.4-5.0

Table 5. Effect of different building objects

Figures	Mean points	95% confidence interval
Fig.9 a)	2.1	1.8-2.5
Fig.9 b)	3.4	3.1-3.7
Fig.8 c)	5.5	5.2-5.7

c) – and defined a rating scale that ranged from 1 (disorderly) to 7 (orderly). An orderly city is one in which the buildings are regularly arranged, and a disorderly city is one in which they are randomly arranged. As shown in Table 4, the generated virtual cities were evaluated as being at about the middle of the scale. This means we can generate virtual cities that are somewhere between orderly and disorderly. By increasing the value of $d(3)$, we can increase the degree of orderliness. That is, our methods can be used to generate different types of virtual cities by controlling the value of $d(3)$.

Effect of different building objects

Next, we determined how using different tables (see Table 1) and building objects affected the evaluation. We used different indexes of average building height and building objects to generate three virtual cities –Fig.9 a),b) and Fig.8 c)–. We again defined a rating scale that ranged from 1 (suburban) to 7 (urban). As shown in Table 5, we can generate various virtual cities that give impressions ranging from urban to suburban.

5 Conclusion

We proposed a novel method for modeling virtual cities that uses a genetic algorithm to determine the values used to lay out the buildings. Subjective evaluation showed that it can generate a variety of virtual cities that have areas similar to those in actual cities.

Many factors are related to actual cities, and we are not able to put all of them into a fitness function. For example, we did not include the shape and

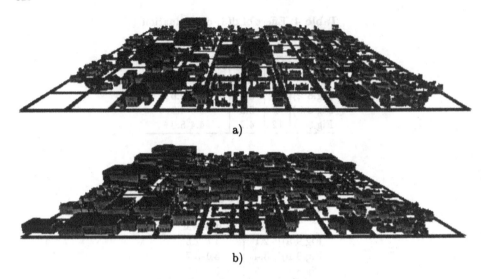

a)

b)

Fig. 9. Examples of virtual cities with different building objects

arrangement of the areas in our fitness function. Instead of determining other factors explicitly, one way to solve this problem is to use a method based on the user's selection of a favorite virtual city. GAs are appropriate technique since they give more than one semi-optimal solution as candidate individuals. We thus need to combine our method with an interactive one based on human evaluation.

References

[1] T.Endo, A.Katayama, H.Tamura, M.Hirose, T.Tanikawa and M.Saito: Image-based walk-through system for large-scale scenes. *Proc. International Conference on Virtual Systems and MultiMedia (VSMM'98)*,1998

[2] Christopher G. Langton: *Artificial Life.* Addison Wesley, 1989.

[3] Prusinkiewicz, P. and Lindenmayer, A.: The Algorithmic Beauty of Plants. (The Virtual Laboratory). Springer-Verlag,1986.

[4] Karl Sims: Evolving 3d morphology and behavior by competition. *Artificial Life IV*,pp.28–39,1994.

[5] N.Kato, T.Okuno, A.Okano, H.Kanoh, and S.Nishihara, "An alife approach to modeling virtual cities," *IEEE 1998 International Conference on System, Man, and Cybernetics*, pp. 1169–1174, 1998.

[6] N.Kato, A.Okano, T.Okuno, H.Kanoh, and S.Nishihara, "Modeling Virtual Cities Using Genetic Algorithms," *Proceedings of the 1999 IEEE Midnight-Sun Workshop on Soft Computing Methods in Industrial Applications*, pp. 134–139, 1999.

[7] D.E.Goldberg, *Genetic Algorithms in Search, Optimization and Machine Learinng.* , Addison Wesley, 1989.

[8] John S.Gero: *Genetic Engineering and Design Problems.* Evolutionary Algorithms in Engineering Applications, Springer-Verlag, pp.47-68, 1997.

Chapter 4: Probabilistic Computing

Papers:

Minimizing Real Functions by Scout
F. Abbattista and V. Carofiglio

Qualitative Similarity
Y. Y. Yao

Survival Probability for Uniform Model on Binary Tree: Critical Behavior and Scaling
A. Y. Tretyakov and N. Konno

Stochastic Modelling of Multifractal Exchange Rates
V. V. Anh, Q. M. Tieng, and Y. K. Tse

Bayesian State Space Modeling for Nonlinear Nonstationary Time Series
G. Kitagawa

Noise Induced Congestion in Coupled Map Optimal Velocity Model of Traffic Flow
S. Tadaki, M. Kikuchi, Y. Sugiyama, and S. Yukawa

Geometrical View on Mean-Field Approximation for Solving Optimization Problems
T. Tanaka

Probabilistic Computational Method in Image Restoration Based on Statistical-Mechanical Technique

K. Tanaka

Bayesian Neural Networks: Case Studies in Industrial Applications

A. Vehtari and J. Lampinen

Minimizing Real Functions by Scout

Fabio Abbattista, Valeria Carofiglio

Dipartimento di Informatica - Universita' di Bari
fabio@di.uniba.it

Abstract: In this paper the recently proposed optimization algorithm, the Scout algorithm, is applied to the minimization of some real functions, commonly used to test the performances of optimization algorithms. The Scout algorithm is inspired by the natural behavior of human or animal scout in exploring unknown geographical regions, and in their ability to exploit information coming from past experience. The experimental results are compared with that obtained by an efficient optimization technique called Differential Evolution.

1. Introduction

The Scout algorithm has been inspired by the human (and animal) scouts. They are able to find, in unknown areas, useful places such as rivers, food, refuges, enemies and preys. They exploit the trails leaded by other scouts or by the prey, and they remember the safe path visited in previous explorations.

As stated in [1], a *natural algorithms* should possess the following five distinctive characteristics:
1- inspired from a natural process;
2- easy to implement on a parallel computer (intrinsically parallel);
3- not deterministic;
4- based on the Darwinian concept of evolution and selection;
5- it should use the positive feedback.

Scout is a natural algorithm in which near all the five characteristics are present; as it will be clear in the following, Scout uses the selection concept but, in its current implementation, a weak form of evolution occurs.

The paper is organized as follows: section 2 describes the details of Scout; in section 3 we present the results of the application of Scout to the problem. Finally, conclusions and further work are presented in section 4.

2. The Scout algorithm

The Scout algorithm is a stochastic optimization technique on binary search spaces $(S = \{0,1\}^l)$.

The basic idea is that for each cycle (c), the components of the **k** generated solutions should be the more frequently ones contained into the best solutions of the preceding cycles.

The basic element is represented by a probability vector (P^c = $<p_i>$, i = 1...l), in which each element p_i gives the probability that the corresponding component of each solution (S^c_j = $<s_i>$, j = 1...k, i = 1...l) will be "1".

Scout is an iterative algorithm and, in each cycle, it performs the following operations:

1 - **Generation**. Generation of **k** solutions (S^c_j = $<s_i>$: j = 1...k , i = 1...l) by using the probability vector (P^c = $<p_i>$, i = 1...l);

2 - **Selection**. Evaluation of each solution by using the function (f : S --> R) to be optimized. Definition of the set of positive solutions ($\{Sp^c_j\}$ = { S^c_j | f (S^c_j) > $fmax^{c-1}$: fmax = $max_{1 \leq j \leq k}$ f (S^{c-1}_j) } where the cardinality is | { Sp^c_j } | = Num_S);

3- **Updating of P^c**. To compute the new P^{c+1} the positive solutions in which the i_th element is equal to "0" or to "1" have to be considered ({ $S_{j,i,v}$ } = { Sp^c_j | s_i = v : v = 0 or v = 1 } where the cardinality is | { $S_{j,i,v}$ } | = $N_{i,v}$). Next, it is necessary to compute the average function value on the sets { Sp^c_j } (namely $\bar{f}\{Sp^c_j\}$) and { $S_{j,i,v}$ } (namely $\bar{f}\{S_{j,i,v}\}$). Let

$$\lambda^v_i = \frac{\bar{f}\{S_{j,i,v}\}}{\bar{f}\{Sp^c_j\}}$$

then it follows:

$$P_i = \frac{P_i + \lambda^1_i}{1 + \lambda^0_i + \lambda^1_i}$$

These 3 steps are iterated until the stop condition occurs; in our experiments we used a fixed number of cycles (*Max_Cycles*).

Scout generates solutions corresponding to the binary configuration with higher bit probability value. On the basis of the performance of these solutions, the vector P^c is updated. The updating, in this situation, will increase p_i's having high values and it will decrease p_i's with low values. After a very few cycles, p_i's will have only values 1 or 0 and the solutions will tend to be equal each other (each solution will correspond to the bit configuration with the highest probability value), leading to the stagnation of the algorithm. To overcome this situation, the step 1 is also driven by a factor **E** measuring the tendency to explore alternative paths, similar to the mutation operator in evolutionary algorithms. When the probabilistic choosing has been performed, for each s_i, the algorithm generates a random number α. If this number α is less than the parameter value $\mathbf{E}$ (that is the event "To explore alternative path" occurs) the chosen s_i value is inverted. In such a way, the algorithm has the possibility to take unlikely decisions, and to explore new solutions.

The following algorithm synthesizes the preceding description of Scout. In this algorithm we assume a maximization problem. Only slight changes are needed in the minimization case.

- $c:=1$;
- $fmax^{c-1}:=-\infty$;
- Call Initialize_P^c;
- Repeat
    - $fmax^c:=-\infty$;
    - $ftot:=0$;
    - $\bar{f}\left\{Sp_j^c\right\}:=0$;
    - $Num\_S:=0$;
    - For $i=1...l$ and ($v=0$ or $v=1$)
        - $\Delta P_{i,v}:=0$;
        - $N_{i,v}:=0$;
    - End_for
    - For $j=1...k$
        - Call Generation;
        - Call Selection;
    - End_for
    - Call Update_P^c;
    - $c:=c+1$;
- until ($c>Max\_Cycles$).

where:
$ftot$ = the sum of the function values of all the positive solutions in a cycle;
$\Delta P_{i,v}$ = the increment of each p_i, not yet normalized, evaluated on the basis of the positive solutions in a cycle.
In the Appendix, the pseudo-code of the four called procedures is reported.

To conclude, it could be stated that despite its simple conceptual structure, Scout merges the advantages and the important features of natural groups or algorithms: the competition, the cooperation and the evolutionary character. The competition performed at each cycle among the different solutions is very similar to the selection operator of evolutionary algorithms. In these, selection operates on the basis of the fitness function, that is, the objective function evaluated on individuals (or a derived value). The cooperation is represented by the "parallel updating" of vector P performed by the positive solutions. This step is conceptually similar to the pheromone trace in the Ant System [1]. The evolutionary character is present, in a reduced form, by means of the parameter E, a form of mutation of the probabilistic behavior of the agents.

3. Experimental results

To analyze the behavior of Scout we use 8 of the five De Jong functions plus some additional functions[3]:

F1) First DeJong function (Sphere)

$$f(\bar{x}) = \sum_{j=0}^{2} x_j^2$$

where $x_j \in [-5.12, 5.12]$

The global minimum is $f(\bar{0}) = 0$

F2) Second DeJong Function (Rosenbrock's saddle)

$$f(\bar{x}) = 100 * (x_0^2 - x_1)^2 + (1 - x_0)^2$$

where $x_j \in [-2.048, 2.048]$

The global minimum is $f(\bar{1}) = 0$

F3) Third DeJong Function (Step)

$$f(\bar{x}) = 30 + \sum_{j=0}^{4} \lfloor x_j \rfloor$$

where $x_j \in [-5.12, 5.12]$

The global minimum is $f(\overline{-5 - \varepsilon}) = 0$ where $\varepsilon \in]0, 0.12]$

F4) Fourth DeJong Function (Quartic)

$$f(\bar{x}) = \sum_{j=0}^{29} (x_j^4 * (j+1) + \eta)$$

where $x_j \in [-1.28, 1.28]$, η is a random value uniformly distributed in the range $[0, 1[$.

The global minimum is $f(\bar{0}) \leq 30 * E[\eta] = 15$

F5) Fifth DeJong Function (Shekel's Foxholes)

$$f(\bar{x}) = \cfrac{1}{0.002 + \sum\limits_{i=0}^{24} \cfrac{1}{i+1+\sum\limits_{j=0}^{1}(x_j - a_{ij})^6}}$$

where $x_j \in [-65.536, 65.536]$, $a_{i0}=\{-32,-16,0,16,32\}$ or $i=0,..,4$ and $a_{i0}=a_{,\text{mod}(5,0)}$, $a_{i1}=\{-32,-16,0,16,32\}$ or $i=0,5,10,15,20$ and $a_{i1}=a_{i+k,1}$, for $k=1,..4$

The global minimum is $f(-32,32) \cong 0.998004$

F6) Corona's Parabola

$$f(\bar{x}) = \sum_{j=0}^{3} \begin{cases} 0.15*(z_j - 0.05*Sgn(z_j))^2 *d_j & \text{if } |x_j - z_j| < 0.05 \\ d_j * x_j^2 & \text{otherwise} \end{cases}$$

where $x_j \in [-1000,1000]$, $z_j = \left| \left| \dfrac{x_j}{0.2} \right| + 0.49999 \right| * Sgn(x_j) * 0.2$ and

$d_j = \{1,1000,10,100\}$

The global minimum is $f(\bar{x}) = 0$ with $|x_j| < 0.05$, for $j = 0,..3$

F7) Griewangk's Function

$$f(\bar{x}) = \sum_{j=0}^{9} \frac{x_j^2}{4000} - \prod_{j=0}^{9} \cos\left(\frac{x_j}{\sqrt{j+1}} \right) + 1$$

where $x_j \in [-400,400]$

The global minimum is $f(\bar{0}) = 0$

F8) Zimmermann's Problem

$$f(\bar{x}) = 9 - x_0 - x_1$$

where $x_j > 0$, for $j=0,1$, $(x_0-3)^2+(x_1-2)^2 \leq 16$ and $x_0*x_1 \leq 14$

The global minimum is $f(7,2) = 0$

The experimental results are compared with the results reported in [3], in which the Differential Evolution has been proposed. In [3], authors described two different versions, named DE1 and DE2.

The parameters directly influencing the behavior of the Scout algorithm are S (the number of scouts) and E (the tendency of the Scout algorithm to explore alternative solutions).

In all the experiments, the number of evaluations performed, that is the number of scouts times the number of cycles, has been kept constant (20,000). The values used for the two parameters were s=10 and $0.000 \leq E \leq 0.020$. All the tests were averaged over 10 trials. The analysis of the results was based on the two following measures:

- the number of evaluations needed to reach the global minimum;
- the number of trials (out of 10) in which the global minimum is found.

In the table 1, the comparison between Scout and DE versions is reported in terms of the number of evaluations needed to find the global minimum of the functions. The results are grouped in two classes: the first class includes experiments in which Scout and DE found the global minimum in all the 10 runs (Standard Deviation = 0.0); the second class is related to the experiments in which Scout did not find the global minimum in all the runs (standard deviation > 0.0), but it found the global minimum at least in 1 run.

As it can be seen from table 1, Scout performance are usually better than Differential Evolution, in the first class of results. In fact, Scout outperforms DE, both DE1 and DE2, for 5 of the functions (71,4%). For functions F2 and F8 DEs are superior. Moreover, for function F7 the Scout algorithm has not been able to reach the global minimum in all the 10 runs, but when it reached the minimum it has been faster than DE1 and DE2.

Table 1: Comparison of the performance of Scout vs. Differential Evolution methods.

Function	DE1	DE2	Scout
Standard Deviation = 0.0			
F1	490	392	360
F2	746	615	13730
F3	915	1300	190
F4	2378	2873	550
F5	735	828	220
F6	834	1125	750
F8	1559	1076	8550
Standard Deviation > 0.0			
F7	22167	12804	5140

4. Conclusions

The paper reports the results of the Scout algorithm applied to real function optimization. The experimental results are very promising. The Scout algorithm seems to be faster, in terms of number of evaluation, that the Differential Evolution method. It should be noticed that Differential Evolution scored third at the ICEC '96 contest [4]. We pllaned to test the Scout algorithm on a larger set of real function, including all of the functions of the ICEC'96 and ICEC'97 contests.

References

[1] M. Dorigo, V. Maniezzo and A. Colorni, 1994, Introduzione agli algoritmi naturali, *AICA Rivista di Informatica.* **XXIV, 3,** 179-197.

[2] F. Abbattista, F. Bellifemine, D. Dalbis, The Scout Algorithm applied to the Maximum Clique Problem, Advances in Soft Computing, R.Roy, T. Furuhashi and P.K. Chawdhry (eds), Springer, 1998.

[3] R. Storn, K. Price, Differential Evolution - A simple and efficient adaptive scheme for global optimization over continuous spaces, Tech. Rep., TR-95-012, March 1995.

[4] Storn, R. and Price, K., "Minimizing the real functions of the ICEC'96 contest by Differential Evolution" IEEE Conference on Evolutionary Computation, Nagoya, 1996.

Appendix

In this Appendix, we report the four procedure called by the Scout algorithm. Three of them, namely Generation, Selection and Updating_ P^c, have been described in section 2; the Initialize_P^c procedure simply sets all the elements of the vector P to 0.5 (that is the same probability for all p_i's).

Procedure Initialize_P^c;
- For i=1...l
    - P_i:=0.5;

End.

Procedure Generation;
- For i=1...l
    - s_i=1 whit probability p_i;
    - α:=Random();
    - IF ($\alpha<E$)
        - s_i:=1-s_i;

End.

Procedure Selection;
- f:=f(S_j^c);
- IF (f> fmax^c)
    - fmax^c :=f;

  End_if
- IF (f> fmax^{c-1})
    - Num_S:=Num_S+1;
    - ftot:= ftot+f;
    - For i=1...l
        - v= s_i;
        - $\Delta P_{i,v}$:= $\Delta P_{i,v}$+f;

- $N_{i,v} := N_{i,v}+1;$

    End_for
  End_if
End.

Procedure Update_P^c;
    • IF (Num_S>0)
        • $\bar{f}\{Sp_j^c\} := $ ftot/Num_S;
        • For i=1...l and (v=0 or v=1)
            • IF ($N_{i,v}$>0)
                • $\bar{f}\{s_{j,i,v}\} = \Delta P_{i,v}/N_{i,v};$
                • $\lambda_{i,v} = \bar{f}\{s_{j,i,v}\} / \bar{f}\{Sp_j^c\};$
           else
                • $\lambda_{i,v} = 0;$
           End_if
        End_for
        • For i=1...l

$$P_i = \frac{P_i + \lambda_i^{\ 1}}{1 + \lambda_i^{\ 0} + \lambda_i^{\ 1}};$$

        End_for
    End_if
End.

Qualitative Similarity

Y.Y. Yao

Department of Computer Science
University of Regina
Regina, Saskatchewan, Canada S4S 0A2
Email: yyao@cs.uregina.ca

Abstract: The notion of qualitative similarity is introduced and examined from the view point of measurement theory. Similarities among objects are described qualitatively by comparison of two pairs of elements. More precisely, qualitative judgements of similarity are interpreted through the statement that "a is more similar to b than c to d" for four elements a, b, c, and d. Relationships between commonly used approaches for modeling similarity and qualitative similarity are analyzed. Conditions on a qualitative similarity relation are identified so that it can be represented truthfully by a quantitative similarity measure or a fuzzy similarity relation.

1. Introduction

The concept of similarity plays an important role in many fields, such as data analysis and clustering [5], granular computing [10,11,13], approximate reasoning [1,2], and approximate information retrieval [7]. From the extensive existing studies on similarity, we may observe two classes of approaches. The notion of similarity is typically described and interpreted by either a binary relation or a quantitative similarity measure. There is a lack of study on qualitative interpretation of similarity characterized by the relation "more similar to". The main objective of this paper is therefore to introduce and examine such a notion of qualitative similarity.

The simplest way for describing similarity is the use of a reflexive and symmetric relation called a compatibility or tolerance relation. The reflexivity and symmetry reflect our intuitive understanding of similarity. An element should be similar to itself. If an element a is similar to another element b, then it is natural to infer that b is also similar to a. One may also use an equivalence relation to model similarity, although the added transitivity may not be entirely meaningful. The dichotomous interpretation of similarity is a major drawback of the binary relation based view. Two elements are viewed as being either similar or not similar without considering the degrees of similarities. A solution to this problem is the use of a quantitative similarity measure or a fuzzy similarity relation. However, it also introduces new problems. The requirement of numeric values for measuring similarity among objects may be too restrictive. It may be more desirable if one can discuss similarity in qualitative terms. Moreover, it is necessary to provide interpretations for the numeric similarity values and guidelines for their proper usage. All these problems have not been received enough attention.

The measurement theory provides useful tools for the study of interpretation, representation, and measurement of qualitative judgements [4,8]. It deals with qualitative and quantitative meaningful and meaningless statements, proper usage of numeric values, and connections between qualitative and quantitative judgements. One naturally expects that the above mentioned problems with quantitative similarity measures may be resolved in a measurement-theoretic setting. We will attempt to achieve this goal in the rest of the paper.

In Section 2, the concept of qualitative similarity relations is introduced. Existing approaches for modeling similarity are interpreted using qualitative similarity relations. In Section 3, representation theorems for measuring qualitative similarity are discussed by considering conditions on qualitative similarity relations. The results from this study may provide a new and different view for modeling similarity. In particular, the qualitative axioms on similarity may give more insights into this fundamental and important notion.

2. Qualitative Similarity Relations

Let U denote a finite and non-empty set called universe. We interpret similarity among elements of U by a binary relation $\succ$ on $U \times U$ or a quaternary relation on U. For four elements $a, b, c, d \in U$, we interpret $(a, b) \succ (c, d)$, or $ab \succ cd$, to be the statement that "a is more similar to b than c to d". This relation $\succ$ is called a qualitative similarity relation, indicating the intended physical interpretation of the relation. The qualitative judgements of similarity may be interpreted in terms of preference through the notion of exchanges [4]. Since $(a, b) \succ (c, d)$ suggests that a is more similar to b than c to d, it may be reasonable to assume that one would prefer to receive b in exchange for a than to receive d in exchange for c. This interpretation is more appealing if one element if fixed. For $(a, b) \succ (a, d)$, one would prefer to receive b than to receive d in exchange for a, as a is more similar to the former.

In the absence of strict preference, namely, both $\neg((a, b) \succ (c, d))$ and $\neg((c, d) \succ (a, b))$ hold, we say that (a, b) and (c, d) are indifferent. One may interpret indifference in several ways. It may happen that a is similar to b to the same degree that c is similar to d. In some situations, it may not be meaningful to compare (a, b) and (c, d) regarding their similarities. One may use the notion of indifference for such cases. A weak preference is defined as either strict preference or indifference. In this way, two additional binary relations are defined:

$$(a, b) \sim (c, d) \iff \neg((a, b) \succ (c, d)) \wedge \neg((c, d) \succ (a, b)),$$
$$(a, b) \succeq (c, d) \iff ((a, b) \succ (c, d)) \vee ((a, b) \sim (c, d)). \tag{1}$$

The indifference relation $\sim$ is a symmetric relation and the weak preference relation $\succeq$ is a reflexive relation. Furthermore, any two pairs of elements are comparable under $\succeq$. These properties are given by:

Symmetry of $\sim$:	$(a,b) \sim (c,d) \Longrightarrow (c,d) \sim (a,b),$
Reflexivity of $\succeq$:	$(a,b) \succeq (a,b),$
Comparability of $\succeq$:	for any $(a,b),(c,d) \in U \times U,$
	$(a,b) \succeq (c,d)$ or $(c,d) \succeq (a,b)$ or both hold.

Clearly, the reflexivity of $\succeq$ follows from the comparability of $\succeq$.

From a qualitative similarity relation $\succ$ on $U \times U$, we can define two qualitative similarity relations on U with respect to an element $a \in U$ as follows:

$$b \succ_{a\to} c \Longleftrightarrow (a,b) \succ (a,c),$$
$$b \succ_{\to a} c \Longleftrightarrow (b,a) \succ (c,a). \tag{2}$$

Both relations can be viewed as restrictions of $\succ$. The first relation orders elements of U based on a's similarities to them, while the second relation orders elements of U based on their similarities to a. The physical meaning of these two relations may be clearer than that of $\succ$. In general, $\succ_{a\to}$ and $\succ_{\to a}$ may not necessarily be the same relation. They become the same if we assume a kind of symmetry on similarity as expressed by $(a,b) \succ (c,d) \Longrightarrow (b,a) \succ (d,c)$.

To illustrate its richness and usefulness, we show that many existing approaches for modeling similarity can be conveniently expressed in terms of qualitative similarity relations.

An equivalence relation on a universe U is reflexive, symmetric, and transitive, which represents the simplest and well studied interpretation of similarity. Two elements are either equivalent or not equivalent, and equivalence classes form a partition of the universe. Let $E \subseteq U \times U$ denote an equivalence relation on U. A qualitative similarity relation can be defined by:

$$(a,b) \succ (c,d) \Longleftrightarrow aEb \wedge \neg(cEd). \tag{3}$$

The qualitative similarity $\succ$ indeed divides the set $U \times U$ into two classes, E and $E^c = U \times U - E$. It offers a qualitative interpretation of equivalence: two equivalent elements of a pair in E are more similar to each other than two non-equivalent elements of a pair in E^c. In terms of the derived relation $\succ_{a\to} = \succ_{\to a}$, we have:

$$b \succ_{a\to} c \Longleftrightarrow aEb \wedge \neg(aEc). \tag{4}$$

That is, a is more similar to its equivalent elements than to its non-equivalent elements. From the qualitative similarity relation $\succ$, the equivalence relation can be recovered by:

$$aEb \Longleftrightarrow (a,a) \sim (a,b)$$
$$\Longleftrightarrow a \sim_{a\to} b. \tag{5}$$

In general, one may use a weaker binary relation on U to describe similarities among elements of U. A compatibility/tolerance relation is only reflexive and symmetric. The use of compatibility relation is motivated by the observation that although a is similar to b and b is similar to c, a is not necessarily similar to c. The previous formulation is also applicable to compatibility relations, except that the qualitative similarity relation $\succ$ induced by an equivalence relation satisfies more properties than that induced by a compatibility relation.

Consider a similarity measure $S : U \times U \longrightarrow \Re^+$, where $\Re^+$ stands for the set of non-negative real numbers. The measure S reflects degrees of similarities among elements of U. A qualitative similarity relation $\succ$ on $U \times U$ is defined by: for $a, b, c, d \in U$,

$$(a, b) \succ (c, d) \iff S(a, b) > S(c, d). \tag{6}$$

More precisely, $\succ$ induces a partition of $U \times U$, which is ordered so that pairs in a higher level show greater similarity than pairs in a lower level, and pairs in the same level have the same degree of similarity. There does not exist a ono-to-one correspondence between quantitative similarity measures and qualitative similarity relations. From two different similarity measures, we may derive the same qualitative similarity relation.

3. Characterization of Qualitative Relations

In the last section, we have shown that the notion of qualitative similarity relations may be interpreted through other concepts. However, we will no longer pursue in this direction. Instead, qualitative similarity relations are used as a primitive notion in the development of a model for qualitative similarity. A fundamental issue is the characterization of qualitative similarity by stating properties or axioms that a qualitative similarity relation must obey. These properties reflect our perception of similarity, and we would expect to have them from a rational person in judging similarity. Moreover, under certain conditions one may use a quantitative similarity measure to truthfully represent a qualitative similarity relation.

For the quantitative representation of qualitative similarity, we consider two properties: for arbitrary $a, b, c, d, e, f \in U$,

Asymmetry :
$$(a, b) \succ (c, d) \implies \neg((c, d) \succ (a, b))$$
Negative transitivity :
$$\neg((a, b) \succ (c, d)) \wedge \neg((c, d) \succ (e, f)) \implies \neg((a, b) \succ (e, f)).$$

They are intuitively reasonable. One cannot state both a is more similar to b than c to d, and c is more similar to d than a to b. If one does *not* think that a is more similar to b than c is to d, *nor* thinks that c is more similar to d than e to f, then it should be expected that a is *not* more similar to b than *e* to *f*. A qualitative similarity relation satisfying these two axioms is called a *week order* [3]. A weak order has additional properties: for $a, b, c, d \in U$,

(i). exactly one of $(a, b) \succ (c, d), (c, d) \succ (a, b)$ and $(a, b) \sim (c, d)$ holds,

(ii). either $(a, b) \succeq (c, d)$ or $(c, d) \succeq (a, b)$ holds,

(iii). the relation $\succ$ is irreflexive,

(iv). both $\succ$ and $\succeq$ are transitive,

(v). the relation $\sim$ is an equivalence relation.

Recall that the relation $(a, b) \succeq (c, d)$ states that a is at least as much similar to b as c to d. If $\succ$ is a weak order, the indifference and weak preference relations are related by:

$$(a, b) \sim (c, d) \iff (a, b) \succeq (c, d) \wedge (c, d) \succeq (a, b),$$
$$(a, b) \succeq (c, d) \iff \neg((c, d) \succ (a, b)). \tag{7}$$

Thus, $(a, b) \succeq (c, d)$ may also be paraphrased as saying that it is not the case $(c, d) \succ (a, b)$.

One may use a quantitative measures for representing qualitative similarity, as shown by the following representation theorem from measurement theory [3,8].

Theorem 1. *Suppose U is a finite non-empty set and $\succ$ is a binary relation on $U \times U$. There exists a real-valued function $S : U \times U \longrightarrow \Re$ satisfying the condition,*

$$(a, b) \succ (c, d) \iff S(a, b) > S(c, d), \tag{8}$$

if and only if $\succ$ is a weak order. Moreover, S is uniquely defined up to a strictly monotonic increasing transformation.

In fact, a similarity function can be computed using the formula [8]:

$$S(a, b) = \text{the number of } (c, d) \text{ such that } (a, b) \succ (c, d). \tag{9}$$

That is, the similarity of a to b is determined by the number of other pairs that show less similarity. The function S satisfies condition (8). Moreover, S is a bounded and non-negative function.

If only the two axioms of weak order are required, qualitative similarity is measured based on an *ordinal scale* [3,4,8]. For an ordinal scale, it is meaningful to examine the order induced by the similarity function rather the actual values of the function. In other words, comparison is a valid operation. Other arithmetic operations, such as addition and subtraction, are not necessarily meaningful [4]. It is meaningless to make the statement that a is twice as much similar to b as c to d, although it may happen that $S(a, b) = 2S(c, d)$. The similarity measure S is not necessarily a symmetric function.

There are two ways to view these axioms [3,4]. The *prescriptive* or *normative* interpretation is concerned with the principles that one must follow to specify a qualitative similarity relation. The axioms are looked upon as conditions of rationality. That is, *rational* judgments must allow the measurement in terms of a quantitative similarity function. On the other hand, the *descriptive* interpretation treats the axioms as testable conditions. Whether one can measure a qualitative similarity relation depends on whether it is a weak order.

Asymmetry and negative transitivity are required by the quantitative representation of qualitative similarity. From properties (i)-(iv) and the discussion of last section, we can see that $\succ$ induces a partition of $U \times U$, which is ordered so that pairs in a higher level show greater similarity than pairs in a lower level, and pairs in the same level have the same degree of similarity. In order to describe more precisely our intuitive understanding of similarity, additional axioms must be studied.

Typically, similarity is perceived to be reflexive. An element must be similar to itself. For an element $a \in U$, no other elements should be more similar to a than a itself. This suggests the adoption of the axiom: for $a, b \in U$,

$$\text{Reflexivity of similarity}: \quad (a,a) \succeq (a,b).$$

Although this axiom is stated on the qualitative similarity relation, it reflects more on our intuitive understanding of similarity itself. It should not be confused with saying that $\succeq$ is reflexive. The reflexivity of similarity only states the relative order of two elements with respect to a fixed element.

For two elements $a, b \in U$, the reflexivity of similarity does not suggest anything about similarity of a to a in comparison with b to b. One may argue that a is similar to itself as much as any other element b is similar to itself, i.e., $(a,a) \sim (b,b)$, as there may not be any good reason for treating a and b differently. Following the same argument, we can say that a is similar to itself at least as much as an arbitrary element b similar to another arbitrary element c. In summary, we have the following strong version of reflexivity of similarity:

$$\text{Strong reflexivity of similarity}: \quad (a,a) \succeq (b,c).$$

The strong reflexivity of similarity implies $(a,a) \succeq (a,b)$, $(a,a) \succeq (b,a)$, and $(a,a) \sim (b,b)$. As a special case of the last one, we have $(a,a) \sim (a,a)$, which is implied by the asymmetry of $\succ$.

The next axiom deals with the symmetry of similarity. For two elements $a, b \in U$, it would be reasonable to assume that a is similar to b as much as b is similar to a. Furthermore, the exchange of positions of a and b will not affect their similarity in comparison with other pairs. If a is more similar to b than c to d, then b is also more similar to a than c to d. Similarly, if a is more similar to b than c to d, then a is more similar to b than d to c. Therefore, we consider three types of symmetry:

Symmetry of similarity :

(S1). $(a, b) \sim (b, a)$,
(S2). $(a, b) \succ (c, d) \implies (b, a) \succ (c, d)$,
(S3). $(a, b) \succ (c, d) \implies (a, b) \succ (d, c)$.

They imply other reasonable properties.

Lemma 1. *If a qualitative similarity relation $\succ$ on $U \times U$ satisfies (S2) and (S3), then it has the following properties:*

(S4). $(a, b) \sim (c, d) \implies (b, a) \sim (c, d)$,
(S5). $(a, b) \sim (c, d) \implies (a, b) \sim (d, c)$,
(S6). $(a, b) \sim (c, d) \implies (b, a) \sim (d, c)$,
(S7). $(a, b) \succ (c, d) \implies (b, a) \succ (d, c)$.

By the symmetry of the indifference relation $\sim$, one can see that (S4), (S5), and (S6) are indeed equivalent to each other. In general, the three axioms (S1), (S2), and (S3) are not equivalent to each other, nor can any one be derived from the others. If a qualitative similarity relation is a weak order, then they are equivalent.

Lemma 2. *If a qualitative similarity relation $\succ$ on $U \times U$ is a weak order, then (S1), (S2), and (S3) are pairwise equivalent.*

By summing up the results so far, we immediately arrive at a representation theorem, as a Corollary to Theorem 1, for measuring a qualitative similarity relation with a normalized and symmetric similarity function.

Corollary 1. *There exists a real-valued function $S : U \times U \longrightarrow [0, 1]$ satisfying the conditions,*

(1). $(a, b) \succ (c, d) \iff S(a, b) > S(c, d)$,
(2). $S(a, a) = 1$,
(3). $S(a, b) = S(b, a)$,

if and only if the qualitative similarity relation $\succ$ on $U \times U$ is a weak order and satisfies the strong reflexivity of similarity $(a, a) \succ (b, c)$ and the symmetry of similarity $(a, b) \sim (b, a)$.

A fuzzy similarity relation $\mathcal{R}$ is defined by a membership function $\mu_{\mathcal{R}} :$ $U \times U \longrightarrow [0, 1]$ satisfying the properties: for $a, b, c \in U$,

reflexivity : $\mu_{\mathcal{R}}(a, a) = 1$,

symmetry : $\mu_{\mathcal{R}}(a, b) = \mu_{\mathcal{R}}(b, a)$,

T transitivity : $\mu_{\mathcal{R}}(a, c) \geq T(\mu_{\mathcal{R}}(a, b), \mu_{\mathcal{R}}(b, c))$, where T is a binary operation on $[0,1]$ called triangular norm, or t-norm for short [7]. Examples of t-norms are min, product, and bounded difference defined by $\max(0, x + y - 1)$. The T-transitivity is also known and max-T transitivity, as it can be equivalent expressed as: for $a, c \in U$,

$$\mu_\mathcal{R}(a, c) \geq \max_{b \in U} T(\mu_\mathcal{R}(a, b), \mu_\mathcal{R}(b, c)). \tag{10}$$

If the min operation is used, one obtains the proposal by Zadeh [12]. In this case, we obtain the max-min transitivity: for $a, c \in U$,

$$\mu_\mathcal{R}(a, c) \geq \max_{b \in U} \min(\mu_\mathcal{R}(a, b), \mu_\mathcal{R}(b, c)). \tag{11}$$

The max-min transitivity can be reexpressed as: for $a, b, c \in U$,

$$\text{either } \mu_\mathcal{R}(a, c) \geq \mu_\mathcal{R}(a, b) \quad \text{or} \quad \mu_\mathcal{R}(a, c) \geq \mu_\mathcal{R}(b, c). \tag{12}$$

The max-min transitivity can be stated in a manner similar to the transitivity of a crisp binary relations: for all $\alpha \in [0, 1]$,

$$\mu_\mathcal{R}(a, b) \geq \alpha \wedge \mu_\mathcal{R}(b, c) \geq \alpha \Longrightarrow \mu_\mathcal{R}(a, c) \geq \alpha. \tag{13}$$

This property is commonly stated using the notion of α-cuts of a fuzzy relation [7,9]. A fuzzy binary relation satisfying reflexivity and symmetry is called a proximity relation [7]. Corollary 1 in fact identifies the class of qualitative similarity relations that can be represented truthfully by proximity relations. We now turn our attention to the class corresponding to fuzzy similarity relations.

Equations (12) and (13) suggest the following two forms of qualitative max-min transitivity: for $a, b, c, e, f \in U$,

Max–min transitivity of similarity :
    (T1).   either $(a, c) \succeq (a, b)$ or $(a, c) \succeq (b, c)$ holds,
    (T2). $(a, b) \succeq (e, f) \wedge (b, c) \succeq (e, f) \Longrightarrow (a, c) \succeq (e, f)$.

These two axioms are not equivalent.

Lemma 3. *For an arbitrary qualitative similarity relation $\succ$ on $U \times U$, (T2) implies (T1). If $\succ$ is a weak order, then (T2) and (T1) are equivalent.*

With the qualitative max-min transitivity, a representation theorem corresponding to fuzzy similarity relations is given below.

Corollary 2. *There exists a fuzzy similarity relation $\mu_\mathcal{R} : U \times U \longrightarrow [0, 1]$ satisfying the condition,*

$$(a, b) \succ (c, d) \Longleftrightarrow \mu_\mathcal{R}(a, b) > \mu_\mathcal{R}(c, d),$$

if and only if the qualitative similarity relation $\succ$ on $U \times U$ is a weak order and satisfies the strong reflexivity of similarity $(a, a) \succ (b, c)$, the symmetry of similarity $(a, b) \sim (b, a)$, and max-min transitivity (T2).

In this section, we have considered some basic properties of similarity by transforming quantitative properties into qualitative properties. In general, a quantitative axiom contains much more information than a qualitative axiom. The main advantages of using qualitative axioms is that they have a more transparent physical interpretation. More importantly, they clearly show the underlying assumptions that are implicitly made when using a quantitative similarity measure.

4. Conclusion

In this paper, we introduce and examine the concept of qualitative similarity. Some intuitively reasonable axioms are used for the measurement of qualitative similarity. Conditions on a qualitative similarity relation are identified so that it can be truthfully represented by a qualitative similarity measure and a fuzzy similarity relation. The main results presented in the paper follow from measurement theory. It should, perhaps, be emphasized that the present investigation of qualitative similarity is complementary to existing studies on similarity. It may lead to an intuitive or qualitative understanding of the notion of similarity, which may be more important than the formal results. This point, in fact, has been convincingly demonstrated by many authors in measurement theory [6,8].

The axioms studied in this paper only allow the measurement of similarity using an ordinal scale. This impose very restricted usages of the numeric similarity values. If we believe that these axioms are the *only* reasonable properties of similarity, we in fact only consider the relative order of elements regarding their similarities. Otherwise, additional axioms must be proposed and studied.

References

[1] Dubois, D. and Prade, H. Similarity-based approximate reasoning, in: *Computational Intelligence - Imitating Life*, Zurada, J.M., Marks II, J. and Robinson, C.J. (Eds.), IEEE Press, New York, pp. 69-80, 1994.

[2] Esteva, F., Garcia-Calvés, P. and Godo, L. Relating and extending semantical approaches to possibilistic reasoning, *International Journal of Approximate Reasoning*, 10, 311-344, 1994.

[3] Fishburn, F.C. *Utility Theory for Decision Making*, Wiley, New York, 1970.

[4] French, S. *Decision Theory - An Introduction to the Mathematics of Rationality*, Ellis Horwood Limited, Chichester, 1988.

[5] Jardine, N. and Sibson, R. *Mathematical Taxonomy*, Wiley, New York, 1971.

[6] Keeney, R.L. and Raiffa, H. *Decision with Multiple Objectives: Preferences and Value Tradeoffs*, Wiley, New York, 1976.

[7] G.J. Klir and B. Yuan, *Fuzzy Sets and Fuzzy Logic, Theory and Applications*, Prentice Hall, New Jersey, 1995.

[8] Roberts, F.S. *Measurement Theory with Applications to Decisionmaking, Utility, and the Social Sciences*, Addison-Wesley, Reading, 1979.

[9] Yao, Y.Y. Combination of rough and fuzzy sets based on α-level sets, in: *Rough Sets and Data Mining: Analysis for Imprecise Data*, Lin, T.Y. and Cercone, N. (Eds.), Kluwer Academic Publishers, Boston, pp. 301-321, 1997.

[10] Yao, Y.Y. Granular computing using neighborhood systems, in: *Advances in Soft Computing: Engineering Design and Manufacturing*, Roy, R., Furuhashi, T. and Chawdhry, P.K. (Eds), Springer-Verlag, London, pp. 539-553, 1999.

[11] Yao, Y.Y. and Zhong, N. Granular computing using information tables, manuscript, 1999.

[12] Zadeh, L.A. Similarity relations and fuzzy orderings, *Information Science*, **3**,177-200, 1971.

[13] Zadeh, L.A. Towards a theory of fuzzy information granulation and its centrality in human reasoning and fuzzy logic, *Fuzzy Sets and Systems*, **19**, 111-127, 1997.

Survival Probability for Uniform Model on Binary Tree: Critical Behavior and Scaling

Alex Yu. Tretyakov [1] and **Norio Konno** [2]

[1] Institute of Technology and Engineering, College of Sciences, Massey University, Private Bag 11 222, Palmerston North, New Zealand, atretiak@fractal.is.tohoku.ac.jp
[2] Department of Applied Mathematics, Faculty of Engineering, Yokohama National University, Tokiwadai 79-5, Yokohama, 240-8501, Japan, norio@mathlab.sci.ynu.ac.jp

Abstract: We use Monte Carlo simulations and scaling to study the uniform model on binary tree, introduced by Puha. Scaling is shown to hold, with critical exponents differing from the contact process on binary tree. Estimates for the static critical exponent obtained by two different approaches are slightly below the lower bound given by Puha, with the lower bound being within the uncertainty brackets.

1 Introduction

We consider the uniform model on the homogeneous tree $\mathbf{T}$, an infinite graph without cycles, in which each vertex has the same number of nearest neighbors, which we will denote by $d \geq 2$. Let o be a distinguished vertex of $\mathbf{T}$, which we call the origin. For $x, y \in \mathbf{T}$, the natural distance between x and y, $\mid x - y \mid$, is defined by the number of edges in the unique path of $\mathbf{T}$ joining x to y.

Here we introduce the uniform model, ξ_t, on $\mathbf{T}$, which is a continuous-time Markov process on $\mathbf{T}$. This model was first introduced by Puha [1]. The dynamics of the model are given by the following transition rates: for $x \in \xi$ with $\xi \subset \mathbf{T}$,
$\xi \to \xi \setminus \{x\}$ at rate 1, if $\sharp\{y \in \xi :\mid y - x \mid = 1\} \leq 1$,
$\xi \to \xi \cup \{y\}$ at rate β, if $y \notin \xi$ with $\mid y - x \mid = 1$,
where $\sharp A$ is the cardinality of A.

From now on we will restrict ourselves to the binary tree, i.e. $d = 2$. Let ξ_t^o denote the uniform model starting from the origin. Define the survival probability $s(\beta)$ by
$$P(\xi_t^o \neq \phi \text{ for all } t \geq 0).$$
The critical value of the survival probability can be defined by
$$\beta_c = \inf\{\beta \geq 0 : s(\beta) > 0\}.$$
Very recently she [2] proved that critical birth rate $\beta_c = 1/4$.

Next we consider the critical exponent γ on the survival probability $s(\beta)$ which is defined as follows if it exists:

$$\lim_{\beta \downarrow \beta_c} \frac{\log s(\beta)}{\log(\beta - \beta_c)^\gamma} = 1$$

As for the critical exponent, she [2] also proved that

$$\limsup_{\beta \downarrow \beta_c} \frac{s(\beta)}{(\beta - \beta_c)^{5/2}} > 0$$

and

$$\liminf_{\beta \downarrow \beta_c} \frac{s(\beta)}{(\beta - \beta_c)^{1+\sqrt{13}/2}} > 0$$

Puha's result implies that the critical exponent $\gamma \in [5/2, 1+\sqrt{13}/2]$ if it exists. So, a natural question is what is the rigorous value. This is the motivation of our paper, and our estimate by Monte Carlo simulation suggests the lower bound $5/2$ is correct as we will show later.

Two remarks should be made about her result and our estimate on critical exponent. First note that the results present a striking contrast to the critical exponent on the survival probability of the contact process on the binary tree. The dynamics of the contact process are given by the following transition rates: for $x \in \xi$ with $\xi \subset \mathbf{T}$,

$\xi \to \xi \setminus \{x\}$ at rate 1,

$\xi \to \xi \cup \{y\}$ at rate β, if $y \notin \xi$ with $\mid y - x \mid = 1$.

So the dynamics on birth of the contact process are the same as those of the uniform model. For further details concerning contact processes on $\mathbf{T}$, see Part 1 in Liggett [3] and Chapter 10 in Konno [4], for example. In a similar way, we can define the survival probability $s^{cp}(\beta)$, critical value β_c^{cp}, and critical exponent γ^{cp}. In this process, Tretyakov and Konno [5] gave the following estimate on this critical value:

$$\beta_c = 0.5420 \pm 0.0005.$$

Furthermore, recently Schonmann [6] showed that the critical exponent γ^{cp} is one. On the other hand, Puha's results imply that the critical exponent γ is not equal to one, and moreover that γ lies in the interval $[5/2, 1 + \sqrt{13}/2]$.

Secondly, the uniform model on the one - dimensional integer lattice, $\mathbf{Z}^1$, is equivalent to the biased voter model. As for the biased voter model, see Chapter 3 of Durrett [7] for details. In this case, the survival probability $s(\beta)$ can be obtained explicitly as follows:

$$s(\beta) = \begin{cases} (\beta - 1)/\beta, & \text{if } \beta \geq 1 ; \\ 0, & \text{if } 0 \leq \beta \leq 1. \end{cases}$$

So we see that critical value $\beta_c = 1$ and critical exponent $\gamma = 1$. This exponent is also different from that of the uniform model on the binary tree.

2 Simulation

For the purposes of computer simulation, we rewrite the definition of the uniform model on binary tree as follows: Particles occupy sites of a binary tree (each site of a binary tree is either empty or contains a single particle). A particle is taken at random. With probability p, it attempts to create a new particle on a randomly chosen nearest neighboring site. The attempt succeeds if the site is empty. In the alternative event, happening with probability $1-p$, the particle disappears leaving an empty site if the majority (at least two) of the nearest neighboring sites are empty.

One can show that parameter p relates to β as $p = 3\beta/(1 + 3\beta)$. Thus, critical point should be found as $p \equiv p_c = 3/7 \approx 0.43$ (corresponding to $\beta = 1/4$).

We start simulation from a single particle and a lattice of size one. The lattice is being created dynamically, as newly created particles require sites to be put on. This technique can be applied because of the absence of loops in a binary tree. The lattice size in the simulation is effectively infinite and there is no boundary condition, but simulation time in the survival region is limited by the amount of available computer memory.

One time step in the simulation corresponds to a number of elementary events equal to the number of particles in the system at the beginning of the step. After a fixed number of time steps, simulation is restarted with the same initial condition. Survival probability at a given time point can be estimated as the number of realizations with particles remaining, divided by the total number of realizations. Also, we observe the time dependence for the mean number of particles, with average taken over all realizations.

Fig 1. shows the mean number of particles over 1000 time steps for a number of values of the control parameter p. Average is taken over 20000 realizations. An abrupt change in behavior can be noted between $p = 0.42$ and $p = 0.44$. Scale invariance at the critical point can be observed as a tendency of the double logarithmic plot for $p = 0.43$ to fall on a straight line at intermediate time values. Simulation results are clearly in agreement with the exact value for the critical point provided by Puha.

Survival probability after 100 time steps is given in Fig. 2 (for 1000000 realizations, apart from 6 points with the biggest p, for which only 25000 realizations were taken; near the extinction point greater number of realizations is required to suppress random error). For an infinite time we expect $s(p) \sim (p-3/7)^\gamma$, and critical exponent γ should be given by the slope of $\lg(s)$ plotted versus $\lg(p-3/7)$. In Fig. 2 deviation from a straight line for p close to extinction point is due to finite time of the simulation. Still, power law can be confirmed for intermediate p. The slope appears to be slightly less than 2.5 (the dotted line in the figure is drawn with a slope exactly equal to 2.5). By taking slopes of line segments drawn between successive points, throwing away values that do not cluster together, averaging and calculating mean square deviation, we estimate γ as 2.3 ± 0.2.

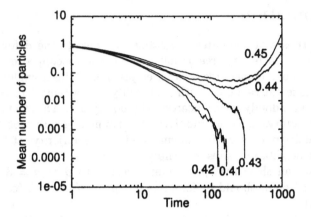

Figure 1: Mean number of particles as a function of time for different values of control parameter p.

To confirm this estimation independently, we also calculate the value of δ, the dynamical critical exponent for the survival probability, defined by

$$P(t) \sim t^{-\delta}$$

at the critical point [8]. By applying the standard local slopes technique [9] and by fitting the data directly, both for 1000 steps and for 100 steps simulation at the critical point, we conclude that the value of δ is close to 2 (2.0 ± 0.1).

The following scaling relationship is expected for $P(t)$ [8]:

$$P(t)t^\delta \sim \Psi((p - p_c)^\nu t).$$

Here, ν is the critical exponent for the correlation length. As it is evident from the preceding equation, ν can be estimated by plotting $P(t)t^\delta$ versus $(p - p_c)^\nu t$ and choosing the value of ν, for which plots for different $p - p_c$ fall on the same curve. The result of such procedure is given in Fig. 3. Under an assumption that $\delta = 2$, we have $\nu = 1.15 \pm 0.10$. γ can now be obtained from

$$\gamma = \nu\delta$$

[10]. The result is clearly consistent with $\gamma = 2.3 \pm 0.2$, obtained directly.

3 Conclusions

Two different Monte Carlo - based approaches give for γ a value slightly lower than the lower bound by Puha. This may indicate that the lower bound is indeed the exact value.

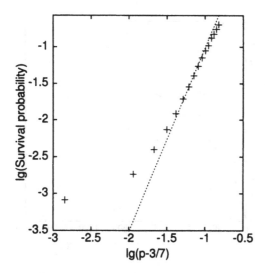

Figure 2: Survival probability versus distance from critical point.

For the uniform model, both static exponent γ and dynamic exponent δ for the survival probability differ from the ones for ordinary contact process on binary tree, for which γ and δ are both equal to 1 [5, 6].

References

[1] A. Puha, *J. Th. Probab.*, Vol. 12, p. 255, 1999.

[2] A. Puha, to appear in *Ann. Probab.*, 1999.

[3] T. M. Liggett, *Stochastic Interacting Systems*, Springer, 1999.

[4] N. Konno, *Phase Transitions of Interacting Particle Systems*, World Scientific, 1994.

[5] A. Yu. Tretyakov and N. Konno, *J. Phys. Soc. Jpn.*, Vol. 64, p. 4069, 1995.

[6] R. Schonmann, *J. Stat. Phys.* , Vol. 90, p. 1429, 1998.

[7] R. Durrett, *Lecture Notes on Particle Systems and Percolation*, Wadsworth, 1988.

[8] P. Grassberger, *J. Phys. A*, Vol. 22, p. 3673, 1989; P. Grassberger and A. de la Torre, *Ann. Phys. (NY)*, Vol. 122, p. 373, 1979; I. Jensen, *Phys. Rev. A*, Vol. 46, p. 7379, 1992.

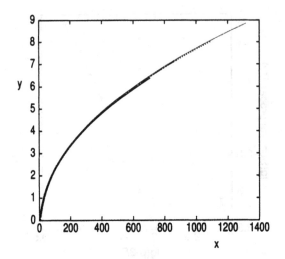

Figure 3: $y \equiv P(t)t^\delta$ versus $x \equiv (p - 3/7)^\nu t$ for $\delta = 2$ and $\nu = 1.15$. Data for $p = 0.43, 0.44, \cdots, 0.55$ are included (13 different sequences).

[9] I. Jensen, *Phys. Rev. Lett.* , Vol. 70, p. 1465, 1993.

[10] I. Jensen, *Phys. Rev. E* , Vol. 50, p. 3623, 1994.

Stochastic modelling of multifractal exchange rates

V. V. Anh[1], **Q. M. Tieng**[1] and **Y. K. Tse**[2]

[1]Centre in Statistical Science and Industrial Mathematics, Queensland University of Technology, GPO Box 2434, Brisbane, Q. 4001, Australia

[2]Department of Economics, National University of Singapore, Kent Ridge, Singapore 119260

Abstract: The existing concept of cointegration applies to integrated processes (in the Box-Jenkins ARIMA framework) or processes with long-range dependence. These processes are assumed to display a monoscaling behaviour (such as that of a fractional Brownian motion). On the other hand, many turbulent processes are known to be intermittent, hence possess multiscaling characteristics. This paper develops a concept of cointegration for these stochastic multifractals. A model is suggested for testing for cointegration and applied to the exchange rates of three major currencies.

1 Introduction

The concept of cointegration plays a key role in studying the causal relationships and prediction of nonstationary processes. The concept was introduced into time series and econometrics by Granger [16], and was discussed in a formal setting in Engle and Granger [12]. The underlying idea is based on the concept of stochastic equilibrium, that is, while the processes may display nonstationarity individually, a linear combination of them may behave as a system in equilibrium (such as a stationary process). The processes in the system are then said to be cointegrated. It is clear that an equilibrium cointegrated system is easier to handle than the nonstationary individuals; hence the concept is useful in the analysis and prediction of nonstationary processes.

In the literature, cointegration theories have been developed for integrated processes in the Box-Jenkins ARIMA framework (see Engle and Granger [12], Johansen [20], [21], Davidson [9], Banerjee *et al.* [3], Gonzalo and Granger [15]), and for processes with long-range dependence, yielding fractional cointegration (see Cheung and Lai [8], for example). These processes are typically monoscaling (i.e., their scaling behaviour can be characterised by a single parameter). An important example is fractional Brownian motion (fBm) with Hurst index H (see Samorodnitsky and Taqqu [29]). Their scaling behaviour can be described by

$$E\,|X\,(t) - X\,(t - r)|^q \sim r^{qH}, \; q \geq 0, \tag{1}$$

as $r \to 0$. Here, the symbol $\sim$ means that the ratio of the left-hand side to the right-hand side tends to a finite constant as $r \to 0$. It is noted that fBm exhibits long-range dependence (LRD) when $\frac{1}{2} < H < 1$ and the scaling exponent of (1), i.e., $\zeta\,(q) = qH$, is a linear function of q with a single parameter, namely, the Hurst index H.

It is commonly accepted in the studies on turbulence that the scaling of turbulent processes is more complex than (1). In fact, turbulent processes may possess a continuum of scales/singularity strengths, and the exponent $\zeta\,(q)$ will therefore be a nonlinear function of q, in which case, they are known as multifractals and display distinct intermittency (i.e. spiky appearance in their sample paths and bursty pattern in their increments) (see Mandelbrot [23], Schertzer and Lovejoy [30], Meneveau and Sreenivasan [27], Davis et al. [10], Frisch [13]).

This paper will concentrate on a class of nonstationary processes, which *in the second order* behave like a Brownian motion (i.e., $H = \frac{1}{2}$); but they are not assumed to be Gaussian. In fact, we shall assume that they are multifractal with a multiplicative cascade structure generated from a binomial distribution (defined in Section 2). These processes are useful to model financial data and geophysical data, for example. In particular, we shall demonstrate that major exchange rate series such as the Japanese yen, the Deutsche mark and the British pound possess the above characteristics. Existing theories of cointegration are not suitable for these multifractals. We shall suggest a new definition of cointegration and develop a method for its testing. The method will be applied to the above exchange rate series.

There has been strong empirical evidence in support of a unit root in the exchange rates of currencies that are free to float (see, for example, Meese and Singleton [26], Meese and Rogoff [25] and Baillie and Bollerslev [2]). In Baillie and Bollerslev [2] it was further argued that the exchange rates of seven major currencies are cointegrated. That is, they are tied together under a long run relationship. This finding, however, was later challenged by Sephton and Larsen [31] and Diebold, Gardeazabal and Yilmaz [11]. These authors argued that the Baillie-Bollerslev result is "fragile" (varies with the sample) and cannot be substantiated. In a rejoinder, Baillie and Bollerslev [1] put forward the possibility that the exchange rates may be fractionally cointegrated. They maintained that "a form of cointegration does exist between the exchange rates, so that they do not drift apart in the long run".

In this paper, we introduce the notion of stochastic multifractals to exchange rate data. This concept enriches the time series structure of the financial series. Furthermore, we extend the concept of fractional cointegration to multifractal series. Our empirical results show that some exchange rate series demonstrate fractional cointegration under the new definition.

2 Intermittency models

Let $\{X(t), t \in \mathbb{R}\}$ be a nonstationary process with stationary increments. This kind of processes has been found appropriate to represent data in geophysics and turbulence, for example (Davis et al. [10]). Define

$$Y(t) = \frac{|X(t) - X(t-1)|}{E|X(t) - X(t-1)|} \tag{2}$$

and

$$Y(t;r) = \frac{1}{r} \int_{t-\frac{r}{2}}^{t+\frac{r}{2}} Y(s)\,ds, \ r > 0. \tag{3}$$

It is noted that $Y(t) \geq 0 \ \forall t$ and $EY(t) = 1$. $Y(t;r)$ is a smoothing (coarse graining) of $Y(t)$ with smoothing window of size r. We shall assume that $Y(t;r)$ is an intermittent process (see Chapter 8 of Frisch [13]). The scaling behaviour of $Y(t;r)$ can be described by

$$\sum (Y(t;r))^q \sim r^{\tau(q)}, \ q \in \mathbb{R}, \tag{4}$$

as $r \to 0$, where the sum is taken over all disjoint intervals of length r. The generalised fractal dimension, D_q, of $Y(t)$ is then defined as

$$D_q = \frac{\tau(q)}{q-1}. \tag{5}$$

Hentschel and Procaccia [19] showed that D_0 is the fractal dimension of the support of $Y(t)$, D_1 and D_2 are the information dimension and the correlation dimension of $Y(t)$ respectively.

An intermittency model for $Y(t)$, and hence $X(t)$, is essentially a parametrisation of $\tau(q)$. There have been a large number of intermittency models suggested in the literature of turbulent processes. Of historical interest, the key models include Kolmogorov's lognormal model, the random curdling model (Mandelbrot [23]), the β-model (Frisch et al. [14]), the random β-model (Benzi et al. [4]), the α-stable model (Schertzer and Lovejoy [30]) and the binomial p-model (Meneveau and Sreenivasan [27]). In Borgas [6], a comparison of intermittency models was undertaken and it was concluded that the binomial p-model was the most satisfactory model and best represented the measurements. Along this line we shall assume that $Y(t)$ is generated by a multiplicative cascade with a binomial generator characterised by a probability p, $0 < p \leq \frac{1}{2}$. In other words, we consider an interval E of unit length and construct a Cantor set $F = \bigcap_{i=0}^{\infty} E_i$ on this interval, where $E_0 = E$, E_k contains 2^k subintervals of length 2^{-k} obtained by dividing each subinterval of E_{k-1} into two halves.

We next define a positive measure μ on F as follows. Let $0 < p \leq \frac{1}{2}$ be given and consider a unit mass on E_0 (i.e. $\mu_0 = 1$). Split this unit mass between the two intervals of E_1 so that one has mass p and the other has mass $1 - p$, the allocation being random. This defines μ_1, which has a constant value of $2p$ on one interval and a constant value of $2(1-p)$ on the other interval. Continue in this way, so that the mass on each interval of E_k is divided randomly into the proportions p and $1-p$ between its two subintervals in E_{k+1}. This defines a sequence $\{\mu_k\}$, which is a positive martingale; hence it converges almost surely to a limiting mass distribution μ on F (Kahane [22]).

Our basic assumption is that $Y(t)$ is generated by such an iterative process (resulting in a multiplicative cascade, as is known in turbulence), and its scaling behaviour/intermittency is described by (4). Let us now determine the function $\tau(q)$. Each generation of the cascade is defined by E_k and μ_k. For each $0 \le j \le k$, a number $\binom{k}{j}$ of the 2^k intervals of E_k have mass $p^k (1-p)^{k-j}$, where $\binom{k}{j} = \frac{k!}{j!(k-j)!}$. By the binomial theorem,

$$
\begin{aligned}
\sum (Y(k; 2^{-k}))^q &= \sum_{j=0}^{k} \binom{k}{j} p^{qj} (1-p)^{q(k-j)} \\
&= (p^q + (1-p)^q)^k.
\end{aligned}
\tag{6}
$$

Putting $2^{-k} = r$ (i.e., $k = -\frac{\log r}{\log 2}$), it follows from (6) that

$$
\begin{aligned}
\log \sum (Y(k; r))^q &= k \log (p^q + (1-p)^q) \\
&= -(\log r) \frac{\log (p^q + (1-p)^q)}{\log 2} \\
&= \log \left(r^{-\log_2 (p^q + (1-p)^q)} \right).
\end{aligned}
\tag{7}
$$

From (4) and (7) we get

$$
\begin{aligned}
\tau(q) &= \lim_{r \to 0} \frac{\log \sum (Y(k; r))^q}{\log r} \\
&= -\log_2 (p^q + (1-p)^q),
\end{aligned}
\tag{8}
$$

and hence

$$
D_q = -\frac{\log_2 (p^q + (1-p)^q)}{q-1}.
\tag{9}
$$

A related exponent, which is more convenient for fitting data, can be introduced by defining

$$
E(Y(t; r))^q \sim r^{1-q+\tau(q)}, \quad q \ge 0
\tag{10}
$$

(Monin and Yaglom [28], p. 534). Define

$$
K(q) = -\tau(q) + q - 1.
\tag{11}
$$

Then, for the binomial cascade described above,

$$
K(q) = \log_2 (p^q + (1-p)^q) + q - 1.
\tag{12}
$$

Remark 1 *It follows directly from (10) that $K(0) = 0$, and since $EY(t; r) = 1$ by definition, we also have $K(1) = 0$. Writing Y for $Y(t; r)$ and considering r sufficiently small, we get from (10) and (11) that*

$$K(q) = \frac{\log E(Y^q)}{\log r}. \tag{13}$$

Thus

$$K'(q) \log r = \frac{(E(Y^q \log Y)) \log Y}{E(Y^q)},$$

$$K''(q) \log r = \frac{E(Y^q)\left(E\left(Y^q (\log Y)^2\right)\right) \log Y - (E(Y^q \log Y))^2 \log Y}{(E(Y^q))^2}.$$

But

$$(E(Y^q \log Y))^2 = \left(E\left(Y^{\frac{q}{2}} Y^{\frac{q}{2}} \log Y\right)\right)^2$$
$$\leq E(Y^q) E\left(Y^q (\log Y)^2\right)$$

by Schwarz's inequality. Consequently, $K''(q) \geq 0$, so that $K(q)$ is a convex function. It also follows from (13) that $K(q) < 0$ iff $E(Y^q) < 1$, which holds only if $0 < q < 1$. These results are useful when the model (12) is fitted to empirical data.

Let us now get back to the positive measure Y generated by the binomial cascade algorithm as described above. For each $t \in E_0$, let

$$\alpha(t) = \lim_{r \to 0^+} \frac{\log Y(B(t; r))}{\log r}$$

be the local dimension of Y at t, where $B(t; r) = \{s; |s - t| < r\}$. Let $G(\alpha) = \{t; \alpha(t) = \alpha\}$ and let $f(\alpha)$ be the Hausdorff dimension of $G(\alpha)$. We call $f(\alpha)$ the *singularity spectrum* of Y, and we say that Y is a *multifractal measure* if $f(\alpha) \neq 0$ for a continuum of α. In this definition, monofractals therefore consist of singularities all of the same strength. As a result, the class of monofractals display scale invariance of the type

$$Y(\sigma t) \overset{d}{=} \sigma^\alpha Y(t), \tag{14}$$

where α is the scaling exponent of the monofractal, and $\overset{d}{=}$ means equality in distribution. That is, the probability density functions on both sides of (14) are the same yielding

$$E(Y(\sigma t))^q = \sigma^{q\alpha} E(Y(t))^q \tag{15}$$

and consequently, the monofractal is characterised by a single scaling exponent (i.e. $\alpha(q) = \alpha$, a constant, for all q, and the graph of $f(\alpha)$ consists of one point). An important example of a monofractal is the fractional Brownian motion with Hurst index H, in which case $\alpha(q) = H \; \forall q$, and $\alpha = \frac{1}{2}$ for Brownian motion. On the other hand, multifractals will display heterogeneous scaling, which is characterised by a nonlinear scaling function such as $\tau(q)$ or $K(q)$ as defined above.

Remark 2 *In order to examine directly whether $Y(t)$ is monofractal or multifractal, it is more convenient to consider another scaling exponent given by*

$$E\,|Y(t) - Y(t-r)|^q \sim r^{\zeta(q)}, \; q \geq 0,$$

as $r \to 0$. It is seen that $\zeta(0) = 0$, and no other exponent is known a priori (in contrast to $K(q)$, where there are two a priori exponents: $K(0) = 0$ and $K(1) = 0$). Following the same argument as for $K(q)$, it can be shown that $\zeta''(q) \leq 0$; hence $\zeta(q)$ is concave. Furthermore, under the condition that $Y(t)$ is bounded, the function $\zeta(q)$ is monotonically nondecreasing (Marshak et al. [24]). These results imply that, if $Y(t)$ is a monofractal, its scaling will be simply given by $\zeta(q) = q\alpha$, where α is a constant. In particular, for fractional Brownian motion with Hurst index H, we have $\zeta(q) = qH$. This result is a convenient tool to test if $Y(t)$ is a monofractal.

There is a relationship between $f(\alpha)$ and $\tau(q)$. In fact, let τ^* denote the concave conjugate of τ (also known as the Legendre transform of τ), i.e.,

$$\tau^*(\alpha) = \inf_{q \in \mathbb{R}} \{q\alpha - \tau(q)\}.$$

Hentschel and Procaccia [19], Halsey et al. [18] showed heuristically that, if the meaure Y is constructed from a cascade algorithm and if $\tau(q)$ and $f(\alpha)$ are smooth and concave, then $\tau^*(\alpha) = f(\alpha)$ and dually $f^*(q) = \tau(q)$.

This relationship is called the *multifractal formalism*, which is also known as the *thermodynamic formalism* because of the analogue of the Gibbs state, pressure and variational principle in thermodynamics (Bohr and Rand [5]). The multifractal formalism is a useful tool in applications. In fact, it yields that

$$\tau(q) = \sup_{0 < \alpha < \infty} \{f(\alpha) - q\alpha\}.$$

Suppose that the supremum is attained at $\alpha = \alpha(q) > 0$. Then

$$\frac{d}{d\alpha}(f(\alpha) - q\alpha) = 0,$$

which implies

$$q = \frac{df}{d\alpha}(\alpha(q)), \tag{16}$$

and

$$\begin{aligned}
\frac{d\tau}{dq} &= \frac{df}{d\alpha}\frac{d\alpha}{dq} - \alpha(q) - q\frac{d\alpha}{dq} \\
&= -\alpha(q).
\end{aligned} \tag{17}$$

Given a sample of $X(t)$, the mass exponent $\tau(q)$ can be computed from (2)-(4) (via log regression). The function $\alpha(q)$ is then given by (17) and the singularity spectrum $f(\alpha)$ is obtained from

$$\tau(q) = f(\alpha(q)) - q\alpha(q). \tag{18}$$

For the binomial cascade model, direct computations from (8) yields

$$\alpha(q) = -\frac{p^q \log_2 p + (1-p)^q \log_2(1-p)}{p^q + (1-p)^q}, \tag{19}$$

$$f(\alpha) = \log_2\left(p^q + (1-p)^q\right) - \frac{q\left(p^q \log_2 p + (1-p)^q \log_2(1-p)\right)}{p^q + (1-p)^q}. \tag{20}$$

The expression for $\alpha(q)$ in (19) can also be put in the following forms:

$$
\begin{aligned}
\alpha(q) &= -\log_2 p \frac{p^q + (1-p)^q \frac{\log_2(1-p)}{\log_2 p}}{p^q + (1-p)^q} \\
&= -\log_2(1-p) \frac{\frac{\log_2 p}{\log_2(1-p)} p^q + (1-p)^q}{p^q + (1-p)^q}.
\end{aligned}
$$

For $0 < p \le \frac{1}{2}$ we get $\frac{\log_2(1-p)}{\log_2 p} \le 1$. Thus,

$$-\log_2(1-p) \le \alpha(q) \le -\log_2 p. \tag{21}$$

Remark 3 *The strongest singularity of $Y(t)$ corresponds to $q \to \infty$. It is seen from (19) and (21) that, as $q \to \infty$,*

$$\alpha_{\min} = -\log_2(1-p). \tag{22}$$

Now,

$$\frac{d\alpha_{\min}}{dp} = \frac{1}{\log 2} \frac{1}{1-p} > 0.$$

Consequently $\alpha_{\min}$ is monotonically increasing as p increases from 0 to $\frac{1}{2}$. In other words, the strongest singularity, hence the degree of intermittency, of $Y(t)$ decreases as p increases. This result plays a key role in our definition of cointegration of multifractals given in the next section.

3 Cointegration of multifractals

The concept of cointegration was introduced in Granger [16], and organized in a formal setting in Engle and Granger [12], to study the long-run equilibrium relationship between a number of time series. A time series X_t is said to be integrated of order d, denoted by $X_t \in I(d)$, if it has an ARIMA (m, d, q) representation given by

$$(1 - \theta_1 B - \ldots - \theta_m B^m)(1 - B)^d X_t = (1 - \psi_1 B - \ldots - \psi_q B^q)\varepsilon_t, \tag{23}$$

where B is the lag operator $BX_t = X_{t-1}$, ε_t is white noise, and the roots of the AR and MA polynomials are assumed to lie outside the unit circle. Now, consider a vector $(X_{1t}, ..., X_{kt})$ of $I(d)$ time series. If a vector $\alpha = (\alpha_1, ..., \alpha_k)$ exists such that $Y_t = \alpha_1 X_{1t} + ... + \alpha_k X_{kt} \in I(d-b)$, where $b > 0$, the time series are said to be cointegrated, and the resulting cointegrated system has an error correction representation. Existing empirical works typically consider $d = b = 1$; that is, $X_1, ..., X_k \in I(1)$ and $Y \in I(0)$, which is the class of stationary time series. Thus this theory requires that the equilibrium error Y_t to be mean reverting, even though $X_1, ..., X_k$ are nonstationary (e.g. displaying stochastic trends).

As noted in Granger and Weiss [17] and formalised in Cheung and Lai [8], the above concept of cointegration can be extended to the case of d and b being fractional. As investigated in Cheung [7], for example (and the references therein), a number of exchange rate series were found to display LRD, in which case the value of d in (23) varies in the range $\left(1, \frac{3}{2}\right)$ (or $d \in \left(0, \frac{1}{2}\right)$ if the differenced series are considered). For d fractional, the fractional differencing operator is defined as

$$(1 - B)^d = \sum_{k=0}^{\infty} (-1)^k \binom{d}{k} B^k = \sum_{k=0}^{\infty} \frac{\Gamma(k-d)}{\Gamma(k+1)\Gamma(-d)} B^k,$$

where $\Gamma(\cdot)$ is the Gamma function. A representation of the form (23) (with d fractional) is then called an autoregressive fractionally integrated moving average (ARFIMA) process. Let $X_1, ..., X_k$ be ARFIMA processes with the same LRD exponent $d \in \left(1, \frac{3}{2}\right)$. If a vector $a = (a_1, ..., a_k)$ exists for which the resulting linear combination $Y = a_1 X_1 + ... + a_k X_k$ is an ARFIMA process with an LRD exponent d_1 strictly smaller than d, we say that $X_1, ..., X_k$ are fractionally cointegrated.

This paper will consider the case in which the processes $X_1(t), ..., X_k(t)$ are nonstationary with uncorrelated increments $X(t) - X(t-r)$, $r > 0$; but the processes are not assumed to be Gaussian. In other words, these processes will have a spectral density of the form

$$f(\lambda) = \frac{c}{\lambda^2}, \ c > 0, \ \lambda \in \mathbb{R} \tag{24}$$

(interpreted in the sense of the Schwartz distributions), but are not Brownian motions. In fact we shall allow these processes to be multifractal, and their multifractality/intermittency is characterised by a multiplicative cascade generated from a binomial distribution with probability p (see (2)-(4) and (8)). We shall demonstrate in the next section that some major exchange rate series (such as the Japanese yen, the Deutsche mark and the British pound) display such characteristics. For simplicity, we shall say these processes have generator p. It seems very unlikely that any linear combination of them would exist as a stationary process. Consequently, both definitions of cointegration mentioned above do not apply to this situation. On the other hand, the development of Section 2 and particularly Remark 3 suggests the following definition:

Definition 1 *Let* $X_1, ..., X_k$ *be nonstationary processes with generators* $0 < p_1, ..., p_k \leq \frac{1}{2}$ *respectively as described above. If a vector* $a = (a_1, ..., a_k)$ *exists such that the linear combination* $Y = a_1 X_1 + ... + a_k X_k$ *has a generator* p_Y *with*

$$p_Y > \max\{p_1, ..., p_k\},\qquad(25)$$

we say that the processes $X_1, ..., X_k$ *are cointegrated.*

Remark 4 *Similar to Brownian motions, the process* Y *is mean reverting. But this mean reversion depends on the value of* p_Y. *When* $p_Y \to \frac{1}{2}$, *the process* Y *becomes monoscaling; hence a linear combination of* $X_1, ..., X_k$ *(themselves nonstationary and multifractal) may exist as a Brownian motion. Hence, in a sense, our definition of cointegration means convergence to Brownian motion.*

Remark 5 *Given* $X_1, ..., X_k$, *we may consider the regression*

$$X_1 = a_2 X_2 + ... + a_k X_k + u,\qquad(26)$$

where u *is a white noise, and the coefficients* $a_2, ..., a_k$ *are obtained using least squares. Then, a linear combination for cointegration analysis can be obtained from*

$$Y = X_1 - \hat{a}_2 X_2 - ... - \hat{a}_k X_k.\qquad(27)$$

4   Empirical results

We apply the above theory to analyse the exchange rates of three major currencies. As pointed out by Baillie and Bollerslev [1], exchange rates may be tied together through a long-memory fractionally integrated type of process. The usual analysis of fractionally integrated time series, however, is restricted to monofractal processes. We shall show in this section that exchange rate data exhibit intermittency similar to turbulence data. A generalized concept of fractional cointegration is then examined.

We use the nominal spot exchange rates of the British pound (BP), Japanese yen (JY) and Deutsche mark (DM). These series are provided by the Commodity Systems Inc. They represent daily observations from January 1985 through July 1998, totalling 3421 observations.

From the plot of the normalised (by dividing by their means) series in Figure 1, we observe a fractional Brownian motion appearance in each series. Hence we assume that their spectral density takes the form

$$f(\lambda) = \frac{c}{|\lambda|^{2H+1}}, \ c > 0, \ 0 < H < 1, \ \lambda \in \mathbb{R},\qquad(28)$$

near frequency 0. The log periodograms $\hat{f}(\lambda)$ of BP, DM and JP against $\log \lambda$ are computed. The result is shown for DM in Figure 2 as an example. Their Hurst exponents H, estimated from the regression

$$\log \hat{f}(\lambda) = c - (2H+1)\log|\lambda| + u, \ u \sim WN,\qquad(29)$$

are 0.5241, 0.5190 and 0.5148 for BP, DM and JY, respectively.

The fitted model (the solid line) for DM is also given in Figure 2. These estimates indicate that all three time series appear to be Brownian motion ($H = \frac{1}{2}$). However, a closer look at their differenced series, such as that of JY shown in Figure 3, implies that a scaling structure more complex than that of Brownian motion exists; in fact the presence of intermittency is quite apparent in these differenced series. We then compute the exponents $\zeta(q)$ as a function of q as defined in Remark 2 using log regression. For comparison we also compute $\zeta(q)$ for Brownian motion (i.e., fBm with $H = 0.5$). The $\zeta(q)$ functions are plotted in Figure 4. It can be seen that $\zeta(q)$ is a linear function of q for fBm as expected (of a monoscaling process), but is clearly nonlinear for DM and JY, indicating that these latter series are multifractal. On the other hand, the $\zeta(q)$ curve for BP appears to be linear. The exponents $K(q)$ are next computed as in (10), again using log regression. These are shown in Figure 5. It is seen that they display the theoretical convex shape as demonstrated in Remark 1, with zeroes at $q = 0$ and $q = 1$; the $K(q)$ curve for JY is furthest from that of fBm, indicating that JY is more intermittent than DM and BP. The model (12) is then fitted using nonlinear least squares to these $K(q)$ curves and also to those obtained from the residuals $c_1 = BP + \hat{a}_1 DM$, $c_2 = BP + \hat{a}_2 DM + \hat{a}_3 JY$, $c_3 = DM + \hat{a}_4 JY$, where the coefficients $\hat{a}_1, ..., \hat{a}_4$ are least squares estimates. The estimates for the corresponding p are given in the table below.

Table 1: Estimates of the generator p from model (12).

$K(q)$	BP	DM	JY	c_1	c_2	c_3
p	0.325	0.320	0.275	0.320	0.300	0.340

Figure 6 demonstrates for DM that the fitted model describes the data very well; excellent fit is similarly obtained for JY and BP. A low value of p for JY confirms that this series is more intermittent than BP and DM. According to the criterion (25), the results of Table 1 indicate that DM and JY are cointegrated, while BP does not seem to be part of this cointegration. The computed functions $\zeta(q)$ and $K(q)$ in Figures 4 and 5 support very well this cointegration. In fact, the curves $\zeta(q)$ and $K(q)$ for c_3 are closer to that of fBm than the corresponding curves for DM and JY (see Figures 4 and 5).

In summary, the idea underpinning our definition of cointegration is that each individual process is nonstationary and non-Gaussian/multifractal, but a linear combination of them may approach Brownian motion. The above empirical results illustrate the practical relevance of this theory.

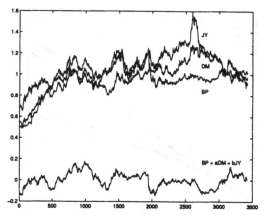

Figure 1: The time series JY, DM and BP normalised by dividing by their means respectively. At the bottom is the residual series $BP + \hat{a}DM + \hat{b}JY$, which has a Brownian motion appearance.

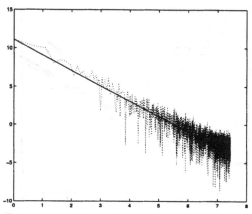

Figure 2: The log periodogram and fitted model (continuous line) of the DM series.

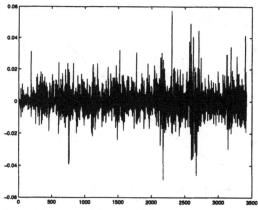

Figure 3: The differenced series $JY(t) - JY(t-1)$.

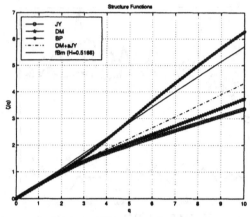

Figure 4: The exponents $\zeta(q)$ of BP, DM, JY, $DM + \hat{a}JY$ (i.e., c_3) and fBm with $H = 0.5$.

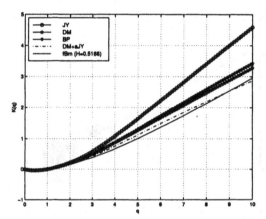

Figure 5: The exponents $K(q)$ of BP, DM, JY, $DM + \hat{a}JY$ and fBm.

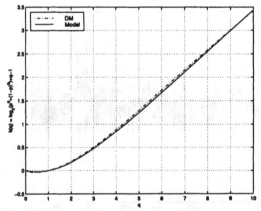

Figure 6: The exponents $K(q)$ and its fitted model for DM.

References

[1] R. T. Baillie and R. Bollerslev. Cointegration, fractional cointegration, and exchange rate dynamics. *Journal of Finance*, 49:737–745, 1994.

[2] R. T. Baillie and T. Bollerslev. Common stochastic trends in a system of exchange rates. *Journal of Finance*, 44:167–181, 1989.

[3] A. Banerjee, J. Dolado, J. W. Galbraith, and D. F. Hendry. *Co-Integration, Error-Correction, and the Econometric Analysis of Non-Stationary Data.* Oxford University Press, 1993.

[4] R. Benzi, G. Paladin, G. Parisi, and A. Vulpiani. On the multifractal nature of fully developed turbulence and chaotic systems. *J. Phys. A*, 17:3521–3531, 1984.

[5] T. Bohr and D. Rand. The entropy function for characteristic exponents. *Physica D*, 25:387–398, 1987.

[6] M. S. Borgas. A comparison of intermittency models in turbulence. *Phys. Fluids A*, 4(9):2055–2061, Sept. 1992.

[7] Y.-W. Cheung. Long memory in foreign-exchange rates. *Journal of Business & Economic Statistics*, 11(1):93–101, Jan. 1993.

[8] Y.-W. Cheung and K. S. Lai. A fractional cointegration analysis of purchasing power parity. *Journal of Business & Economic Statistics*, 11(1):103–112, Jan. 1993.

[9] J. Davidson. The cointegration properties of vector autoregression models. *Journal of Time Series Analysis*, 12(1):41–62, 1991.

[10] A. Davis, A. Marshak, W. Wiscombe, and R. Cahalan. Multifractal characterizations of nonstationarity and intermittency in geophysical fields: Observed, retrieved, or simulated. *Journal of Geophysical Research*, 99(D4):8055–8072, 20 Apr. 1994.

[11] F. X. Diebold, J. Gardeazabal, and K. Yilmaz. On cointegration and exchange rate dynamics. *Journal of Finance*, 49:727–735, 1994.

[12] R. F. Engle and C. W. J. Granger. Co-integration and error correction: Representation, estimation and testing. *Econometrica*, 55:251–276, 1987.

[13] U. Frisch. *Turbulence.* Cambridge University Press, 1995.

[14] U. Frisch, P. L. Sulem, and M. Nelkin. A simple dynamical model of intermittent fully developed turbulence. *J. Fluid Mech.*, 87:719–736, 1978.

[15] J. Gonzalo and C. Granger. Estimation of common long-memory components in cointegrated systems. *Journal of Business & Economic Statistics*, 13(1):27–35, Jan. 1995.

[16] C. W. J. Granger. Some properties of time series data and their use in econometric model specification. *Journal of Econometrics*, 16:121–130, 1981.

[17] C. W. J. Granger and A. A. Weiss. Time series analysis of error-correcting models. In S. Karlin, T. Amemiya, and L. A. Goodman, editors, *Studies in Econometrics, Time Series, and Multivariate Statistics*, pages 255–278. Academic Press, Inc., 1983.

[18] T. C. Halsey, M. H. Jensen, L. P. Kadanoff, I. Procaccia, and B. J. Shraiman. Fractal measures and their singularities: The characterization of strange sets. *Phys. Rev. A*, 33:1141–1151, 1986.

[19] H. G. E. Hentschel and I. Procaccia. The infinite number of generalized dimensions of fractals and strange attractors. *Physica D*, 8:435–444, 1983.

[20] S. Johansen. Statistical analysis of cointegration vectors. *Journal of Economic Dynamics and Control*, 12:231–254, 1988.

[21] S. Johansen. Estimation and hypothesis testing of cointegration vectors in Gaussian vector autoregressive models. *Econometrica*, 59(6):1551–1580, Nov. 1991.

[22] J.-P. Kahane. Produits de poids aléatoires indépendants et applications. In J. Bélair and S. Dubuc, editors, *Fractal Geometry and Analysis*, pages 277–324. Kluwer Academic Publishers, 1991.

[23] B. B. Mandelbrot. Intermittent turbulence in self-similar cascades: divergence of high moments and dimension of the carrier. *J. Fluid Mech.*, 62:331–358, 1974.

[24] A. Marshak, A. Davis, R. Cahalan, and W. Wiscombe. Bounded cascade models as non-stationary multifractals. *Phys. Rev. E*, 49:55–69, 1994.

[25] R. A. Meese and K. Rogoff. Empirical exchange rate models of the seventies: Do they fit out-of-sample? *Journal of International Economics*, 14:3–24, 1983.

[26] R. A. Meese and K. J. Singleton. On unit roots and the empirical modeling of exchange rates. *Journal of Finance*, 37:1029–1035, 1982.

[27] C. Meneveau and K. R. Sreenivasan. The multifractal nature of turbulent energy dissipation. *J. Fluid Mech.*, 224:429–484, 1991.

[28] A. S. Monin and A. M. Yaglom. *Statistical Fluid Mechanics*, volume II. MIT Press, Cambridge, 1975.

[29] G. Samorodnitsky and M. S. Taqqu. *Stable Non-Gaussian Random Processes*. Chapman and Hall, New York, 1994.

[30] D. Schertzer and S. Lovejoy. The dimension and intermittency of atmospheric dynamics. In L. J. S. Braddbury, F. Durst, B. Launder, F. W. Schmidt, and J. H. Whitelaw, editors, *Turbulence Shear Flow 4*, pages 7–33. Springer, Berlin, 1985.

[31] P. S. Sephton and H. K. Larson. Tests of exchange market efficiency: Fragile evidence from cointegration tests. *Journal of International Money and Finance*, 10:561–570, 1991.

Bayesian State Space Modeling for Nonlinear Nonstationary Time Series

Genshiro Kitagawa

The Institute of Statistical Mathematics
4-6-7 Minami-Azabu, Minato-ku, Tokyo 106-8569 Japan
email: kitagawa@ism.ac.jp

Abstract: Bayesian time series modeling can be achieved by general state space model. Recursive filtering and smoothing formula corresponding to the Kalman filter and smoother can be obtained for the general state space model, and can be realized by the implementations based on numerical integration or Monte Carlo approximation. Selection among the possible candidate models can be done by the information criterion AIC. When we need to use the Monte Carlo filter, we may have difficulty in parameter estimation, since the computed log-likelihood contains a sampling error. For such a situation, we also developed a self-adjusting state space model, with which the state and unknown parameters are simultaneously estimated. Numerical examples clearly shows the usefulness of the developed method.

1. Introduction

In statistical time series analysis, a model is built based on the data and prior knowledge on the object and the objective of the analysis. By using a proper model, it is possible to combine various knowledge on the object and the information from other and current data sets, and thus can extract essential information from the data. This is the main feature of statistical information processing procedures.

On the other hand, there is a danger of extracting biased result if we use an improper models. Therefore, in information processing based on a model, use of a proper model is crucial. Information criterion AIC is an objective criterion to evaluate the goodness of fit of statistical model and facilitates to compare various models quite freely.

In time series analysis, the prior knowledge on the components can usually be expressed by smoothness priors which can be combined into state space model form. Although, many important problems in time series analysis can be solved by using the ordinary state space model, it is sometimes required to use more general nonlinear model or non-Gaussian models. By the recent progress of computing ability, the use of nonlinear or non-Gaussian time series model is becoming realistic.

In this paper, the non-Gaussian filtering and smoothing method for the estimation of the state of the nonlinear non-Gaussian state space model are shown. A Monte Carlo filter and smoother introduced in this paper were developed to further extend the range of the applications of the models [7].

To mitigate the difficulty associated with the use of Monte Carlo filter and the non-Gaussian filter, a self-organizing state space model is also shown. In this approach, the state vector is augmented by unknown parameters of the model and the state and the parameters are estimated simultaneously by the recursive filter and smoother. This type of recursive estimation was earlier tried in the engineering literature (e.g., [10], [11]). Most of those works used the extended Kalman filter. Very likely the main reason for the difficulty in recursive parameter estimation was the lack of a practical nonlinear non-Gaussian smoothing algorithm. In the method shown in this paper, more accurate approximations to the marginal posterior densities of the state and the parameters are obtained by the Monte Carlo smoother.

2. General State Space Model and State Estimation

2.1 The Model and the State Estimation Problem

Consider a nonlinear non-Gaussian state space model for the time series y_n,

$$x_n = F_n(x_{n-1}, v_n) \tag{1}$$
$$y_n = H_n(x_n, w_n), \tag{2}$$

where x_n is an unknown state vector, v_n and w_n are the system noise and the observation noise with densities $q_n(v)$ and $r_n(w)$, respectively. (1) and (2) are called the system model and the observation model, respectively. The initial state x_0 is assumed to be distributed according to the density $p_0(x)$. $F_n(x, v)$ and $H_n(x, w)$ are possibly nonlinear functions of the state and the noise inputs.

This model is an extension of the well-known state space model

$$x_n = F_n x_{n-1} + G_n v_n \tag{3}$$
$$y_n = H_n x_n + w_n, \tag{4}$$

with Gaussian white noise sequences, $v_n \sim N(0, V_n)$ and $w_n \sim N(0, W_n)$.

The above nonlinear non-Gaussian state space model specifies the conditional density of the state given the previous state, $p(x_n|x_{n-1})$, and that of the observation given the state, $p(y_n|x_n)$. This is the essential features of the state space model, and it is sometimes convenient to express the model in this general form based on conditional distributions

$$x_n \sim Q_n(\,\cdot\,|x_{n-1}) \tag{5}$$

$$y_n \sim R_n(\,\cdot\,|x_n). \tag{6}$$

With this model, it is possible to treat the discrete process as well.

2.2 Non-Gaussian Filter and Smoother

The most important problem in state space modeling is the estimation of the state vector x_n from the observations, since many important problems in time series analysis can be solved by using the estimated state vector. The problem of state estimation can be formulated as the evaluation of the conditional density $p(x_n|Y_t)$, where Y_t is the set of observations $\{y_1, \ldots, y_t\}$. Corresponding to the three distinct cases, $n > t$, $n = t$ and $n < t$, the conditional distribution, $p(x_n|Y_t)$, is called the predictor, the filter and the smoother, respectively.

For the standard linear-Gaussian state space model, each density can be expressed by a Gaussian density and its mean vector and the variance-covariance matrix can be obtained by computationally efficient Kalman filter and smoothing algorithms ([2]).

For general state space models, however, the conditional distributions become non-Gaussian and their distributions cannot be completely specified by the mean vectors and the variance covariance matrices. Therefore, various types of approximations to or assumptions on the densities have been used to obtain recursive formulas for state estimation, e.g., the extended Kalman filter ([2]), the Gaussian-sum filter ([1]) and the dynamic generalized linear model ([12]).

However, the following non-Gaussian filter and smoother ([5]) can yield an arbitrarily precise posterior density.

[Non-Gaussian Filter]

$$p(x_n|Y_{n-1}) = \int p(x_n|x_{n-1})p(x_{n-1}|Y_{n-1})dx_{n-1}$$

$$p(x_n|Y_n) = \frac{p(y_n|x_n)p(x_n|Y_{n-1})}{p(y_n|Y_{n-1})}, \tag{7}$$

where $p(y_n|Y_{n-1})$ is the predictive distribution of y_n and is defined by $\int p(y_n|x_n)p(x_n|Y_{n-1})dx_n$.

[Non-Gaussian Smoother]

$$p(x_n|Y_N) = p(x_n|Y_n)\int\frac{p(x_{n+1}|x_n)p(x_{n+1}|Y_N)}{p(x_{n+1}|Y_n)}dx_{n+1}. \tag{8}$$

However, the direct implementation of the formula requires computationally very costly numerical integration and can be applied only to lower dimensional state space models.

2.3 Monte Carlo Filtering

In the Monte Carlo filtering ([4], [7]), we approximate each density function by many particles which can be considered as realizations from that distribution. Specifically, assume that each distribution is expressed by using m ($m=10,000$, say) particles as follows:

$$
\begin{array}{lll}
\{p_n^{(1)}, \ldots, p_n^{(m)}\} & \sim \ p(x_n|Y_{n-1}) & \text{Predictor} \\
\{f_n^{(1)}, \ldots, f_n^{(m)}\} & \sim \ p(x_n|Y_n) & \text{Filter} \\
\{s_{n|N}^{(1)}, \ldots, s_{n|N}^{(m)}\} & \sim \ p(x_n|Y_N) & \text{Smoother}
\end{array}
$$

Namely, we approximate the distributions by the empirical distributions determined by m particles. This means that $p(x_n|Y_{n-1})$ is approximated by the probability function

$$
\Pr(x_n = p_n^{(j)}|Y_{n-1}) = \frac{1}{m}, \qquad \text{for } j = 1, \cdots, m. \tag{9}
$$

Then it will be shown that a set of realizations expressing the one step ahead predictor $p(x_n|Y_{n-1})$ and the filter $p(x_n|Y_n)$ can be obtained recursively as follows.

[Monte Carlo Filter]

1. *Generate a random number $f_0^{(j)} \sim p_0(x)$ for $j = 1, \ldots, m$.*

2. *Repeat the following steps for $n = 1, \ldots, N$.*

    (a) *Generate a random number $v_n^{(j)} \sim q(v)$, for $j = 1, \ldots, m$.*

    (b) *Compute $p_n^{(j)} = F(f_{n-1}^{(j)}, v_n^{(j)})$, for $j = 1, \ldots, m$.*

    (c) *Compute $\alpha_n^{(j)} = p(y_n|p_n^{(j)})$ for $j = 1, \ldots, m$.*

    (d) *Generate $f_n^{(j)}$, $j = 1, \ldots, m$ by the resampling of $p_n^{(1)}, \ldots, p_n^{(m)}$.*

2.4 Monte Carlo Smoothing

The above algorithm for Monte Carlo filtering can be extended to smoothing by a simple modification. The details of the derivation of the algorithm is shown in [7].

An algorithm for smoothing is obtained by replacing the Step 2 (d) of the algorithm for filtering by

(d-S) *Generate $\{(s_{1|n}^{(j)}, \cdots, s_{n-1|n}^{(j)}, s_{n|n}^{(j)})^T, \ j = 1, \ldots, m\}$ by the resampling of $\{(s_{1|n-1}^{(j)}, \cdots, s_{n-1|n-1}^{(j)}, p_n^{(j)})^T, \ j = 1, \ldots, m\}$.*

In this modification, the particles of the past state $s_{1|n-1}^{(j)}, \ldots s_{n-1|n-1}^{(j)}$, are preserved and $\{(s_{1|n-1}^{(j)}, \ldots, s_{n-1|n-1}^{(j)}, p_n^{(j)})^T, j = 1, \ldots, m\}$ is resampled with the same α weights as the one obtained by step 2 (d).

It is worth noting that this algorithm realizes fixed interval smoothing for nonlinear non-Gaussian state space model. However, in practice, since the number of realizations is finite, the repetition of the resampling (d-S) will gradually decrease the number of different realizations in $\{s_{i|n}^{(1)}, \ldots, s_{i|n}^{(m)}\}$ and eventually loses the accuracy of the distribution.

To mitigate this problem, the step (d-S) needs to be replaced by:

(d-L) *For fixed* L, *generate* $\{(s_{n-L|n}^{(j)}, \cdots, s_{n-1|n}^{(j)}, s_{n|n}^{(j)})^T, j = 1, \ldots, m\}$ *by the resampling of* $\{(s_{n-L|n-1}^{(j)}, \cdots, s_{n-1|n-1}^{(j)}, p_n^{(j)})^T, j = 1, \ldots, m\}$ *with* $f_n^{(j)} = s_{n|n}^{(j)}$.

Interestingly, this is equivalent to applying the L-lag fixed lag smoother (Anderson and Moore 1979). The increase of lag, L, will improve the accuracy of the $p(x_n|Y_{n+L})$ as an approximation to $p(x_n|Y_N)$, while it is very likely to decrease the accuracy of $\{s_{n|N}^{(1)}, \cdots, s_{n|N}^{(m)}\}$ as representatives of $p(x_n|Y_{n+L})$. Since $p(x_n|Y_{n+L})$ usually converges rather quickly to $p(x_n|Y_N)$, it is recommended to take L not so large.

3. Identification of Time Series Model

3.1 Likelihood of the Model and Parameter Estimation

The state space model, (1) and (2), usually contains several unknown parameters such as the variances of the noises and the coefficients of the functions F_n and H_n. The vector consisting of such unknown parameters is hereafter denoted by θ. Then $p(x_n|Y_{n-1}, \theta)$ defines the density of the one-step-ahead predictor. On the other hand, in general, the likelihood of time series model specified by the parameter θ is obtained by

$$L(\theta) = p(y_1, \cdots, y_N|\theta) = \prod_{n=1}^{N} p(y_n|Y_{n-1}, \theta), \tag{10}$$

where $p(y_n|Y_{n-1}, \theta)$ is the conditional density of y_n given Y_{n-1} and is obtained by

$$p(y_n|Y_{n-1}, \theta) = \int p(y_n|x_n, \theta) p(x_n|Y_{n-1}, \theta) dx_n. \tag{11}$$

When we used the Monte Carlo filter, the conditional density in the right hand side of (11) can be approximated by

$$p(y_n|Y_{n-1}, \theta) = \frac{1}{m} \sum_{j=1}^{m} \alpha_n^{(j)}. \tag{12}$$

3.2 Model Selection by Information Criterion AIC

The maximum likelihood estimate of the parameter θ is obtained by maximizing the log-likelihood

$$\ell(\theta) = \log L(\theta) = \sum_{n=1}^{N} \log p(y_n | Y_{n-1}, \theta) \tag{13}$$

If there are several candidate models, we can evaluate the goodness of the models and find the best one by finding the model with smallest value of AIC defined by

$$\text{AIC} = -2\max\ell(\theta) + 2\#(\theta) \tag{14}$$

where $\#(\theta)$ denotes the dimension of the parameter θ.

3.3 Self-Adjusting State Space Model

If the non-Gaussian filter is implemented by the Monte Carlo filter, the sampling error due to the approximation (13) sometimes renders the maximum likelihood method impractical. In this case, instead of estimating the parameter θ by the maximum likelihood method, we consider a Bayesian estimation by augmenting the state vector as

$$z_n = \left[\begin{array}{c} x_n \\ \theta \end{array} \right]. \tag{15}$$

The state space model for this augmented state vector z_n is given by

$$\begin{aligned} z_n &= F^*(z_{n-1}, v_n) \\ y_n &= H^*(z_n, w_n) \end{aligned} \tag{16}$$

where the nonlinear functions $F^*(z, v)$ and $H^*(z, w)$ are defined by $F^*(z, v) = [F(x, v), \theta]^T$, $H^*(z, w) = H(x, w)$.

Assume that we obtain the posterior distribution $p(z_n | Y_N)$ given the entire observations $Y_N = \{y_1, \cdots, y_N\}$. Since the original state vector x_n and the parameter vector θ are included in the augmented state vector z_n, it immediately yields the marginal posterior densities of the parameter and of the original state.

This method of Bayesian simultaneous estimation of the parameter of the state space model can be easily extended to a time-varying parameter situation where the parameter $\theta = \theta_n$ evolves with time n.

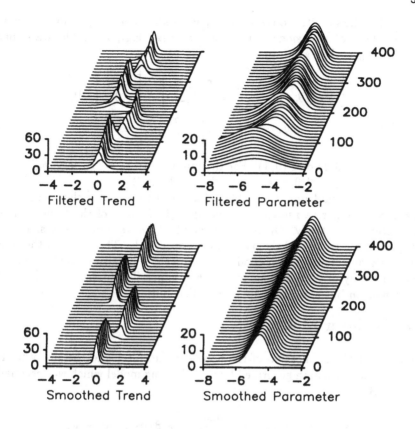

Figure 1: Marginal posterior densities obtained by self-organizing state space model (Kitagawa JASA(1998))

4. Examples

4.1 Trend Estimation

For the estimation of the trend of a nonstationary time series, we consider a simple first order trend model

$$x_n = x_{n-1} + v_n, \quad y_n = x_n + w_n, \tag{17}$$

where w_n is a Gaussian white noise with mean 0 and variance 1. For automatic detection of sudden changes of the trend, we consider the following two distributions for the system noise v_n; $N(0, \tau^2)$ and $C(0, \tau^2)$. Here $C(0, \tau^2)$ denotes the Cauchy distribution with density function $\tau\pi^{-1}(v^2 + \tau^2)^{-1}$. The only unknown model parameter is τ^2. The maximum likelihood estimate of the parameter τ^2 is 1.40×10^{-2} for the Gaussian model, and 3.53×10^{-5} for the Cauchy model.

For the simultaneous estimation of the state and the parameter $\theta = \tau^2$, we used the two dimensional state vector $z_n = [x_n, \log_{10} \theta_n]^T$. The state space model for z_n is then given by

$$\begin{bmatrix} x_n \\ \log_{10} \theta_n \end{bmatrix} = \begin{bmatrix} x_{n-1} \\ \log_{10} \theta_{n-1} \end{bmatrix} + \begin{bmatrix} \sqrt{\theta_n} \\ 0 \end{bmatrix} v_n,$$

$$y_n = [1, 0] \begin{bmatrix} x_n \\ \log_{10} \theta_n \end{bmatrix} + w_n, \tag{18}$$

where $v_n \sim N(0, 1)$ or $C(0, 1)$. For the initial state $z_0 = (x_0, \log_{10} \theta_n)^T$, it is assumed that $x_0 \sim N(0, 4)$ and $\log_{10} \theta_0 \sim U([-4, 0])$.

By this modeling, the sudden changes of the trend is clearly detected. In Figure 1, upper two plots show the filtered densities of the trend and the parameter, respectively. On the other hand, the lower plots show smoothed densities of the trend and the parameter, respectively. The jumps of the trend are clearly detected. This method can be further generalize to a more general Pearson family of distributions by which it is possible to determine not only the dispersion but also the shape of the distribution.

4.2 Nonlinear Model

We consider the application of the non-Gaussian filter and the smoother for the analysis of the data generated by the following model which was originally used by [3],

$$x_n = \frac{1}{2} x_{n-1} + \frac{25 x_{n-1}}{1 + x_{n-1}^2} + 8 \cos(1.2n) + v_n$$

$$y_n = \frac{x_n^2}{20} + w_n, \tag{19}$$

with $v_n \sim N(0, 0.1)$ and $w_n \sim N(0, 1)$. The problem is to estimate the unobserved true signal x_n (Figure 2 A) only from the sequence of observations $\{y_n\}$ (Figure 2 B) assuming that the model (19) is known. It should be emphasized here that since the value of the state x_n is squared in the observation equation in (19), it is quite difficult to identify whether the state x_n is positive or negative. The non-Gaussian or Monte Carlo filter/smoother can be applied to this problem and we can get a reasonable estimate of the signal (Figure 2 C).

4.3 Time Series with Trend and Stochastic Volatility

In financial engineering, estimation and analysis of volatility of the series are important problems. Our nonlinear state space model can be applied to this problem, since the time series with stochastic volatility can be expressed by a nonlinear observation model,

$$y_n = \sigma_n w_n, \quad w_n \sim N(0, 1). \tag{20}$$

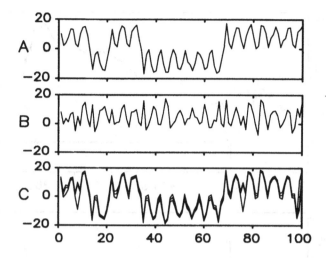

Figure 2: Nonlinear smoothing (Kitagawa JCGS(1996))

Therefore, by assuming a model for the change of the stochastic volatility, for example,

$$\log \sigma_n = \log \sigma_{n-1} + v_n, \quad w_n \sim N(0, \tau^2), \tag{21}$$

we obtain a state space model for the time series. In this case, since the state is defined by $x_n = \log \sigma_n$, by estimating the state vector, the estimation of the volatility is automatically achieved.

Stock price data or exchange rate data have significant characteristics that the volatility increases after a significant change of trend. For modeling such series, we can generalize the above model to

$$y_n = t_n + \sigma_n w_n, \tag{22}$$

where t_n is a trend, σ_n is the volatility and w_n is a standard Gaussian white noise sequence. For the simultaneous estimation of trend and variance, we introduce the following smoothness prior models

$$
\begin{aligned}
t_n &= 2t_{n-1} - t_{n-2} + \varepsilon_n \\
\log \sigma_n &= 2\log \sigma_{n-1} - \log \sigma_{n-2} + \delta_n,
\end{aligned} \tag{23}
$$

where ε_n and δ_n are white noise sequences specified below. Log-transformation is used to assure the positivity of the volatility σ_n. The models (22) and (23), can be expressed in a state space model (1) and (2) with the 4-dimensional state vector $x_n = (t_n, t_{n-1}, \log \sigma_n, \log \sigma_{n-1})^T$, 2-dimensional system noise $v_n = (\varepsilon_n, \delta_n)^T$ and a nonlinear function $H(x_n, w_n) = H((t_n, t_{n-1}, \log \sigma_n, \log \sigma_{n-1})^T, w_n) = t_n + \sigma_n w_n$.

380

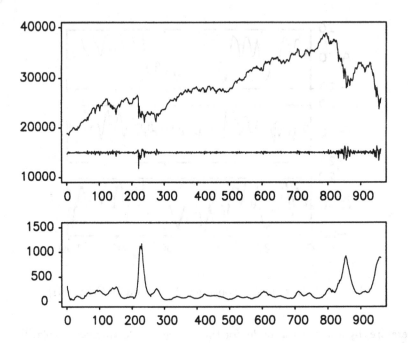

Figure 3: Estimation of the volatility of stock price data (Kitagawa JCGS(1996))

The upper plot of Figure 3 shows the original Nikkei 225 stock price data and the estimated noise sequence by this model. The lower plot shows the estimated volatility, σ_n. Increase of volatility after the black Monday around at $n = 200$ and the crush of bubble around at $n = 800$ is clearly detected.

5. Conclusion

General state space model is useful platform for Bayesian time series modeling. Recursive filtering and smoothing formula corresponding to the Kalman filter and smoother for linear Gaussian model can be obtained for the general state space model, and can be realized by the implementations based on numerical integration or Monte Carlo approximation.

Selection among the possible candidate models can be done by the information criterion AIC. When we need to use the Monte Carlo filter, we may have difficulty in parameter estimation, since the computed log-likelihood contains a sampling error. For such a situation, we also developed a self-organizing state space model, with which the state and unknown parameters are simultaneously estimated. Numerical examples show the usefulness of the developed method.

Appendix: Derivation of Monte Carlo Filter

In the following two subsections, we briefly show a Monte Carlo based algorithm for recursive evaluation of the predictor and the filter, separately.

3.1 One Step Ahead Prediction

Let $\{f_{n-1}^{(1)}, \ldots, f_{n-1}^{(m)}\}$ and $\{v_n^{(1)}, \cdots, v_n^{(m)}\}$ be independent realizations of $p(x_{n-1}|Y_{n-1})$ and the system noise v_n. Namely, for $j = 1, \ldots, m$

$$f_{n-1}^{(j)} \sim p(x_{n-1}|Y_{n-1}), \quad v_n^{(j)} \sim q(v). \tag{24}$$

Then the predictive distribution $p(x_n|Y_{n-1})$ can be expressed by

$$
\begin{aligned}
p(x_n|Y_{n-1}) &= \int\int p(x_n, x_{n-1}, v_n|Y_{n-1}) dv_n\, dx_{n-1} \\
&= \int\int p(x_n|x_{n-1}, v_n) p(v_n) p(x_{n-1}|Y_{n-1}) dv_n\, dx_{n-1} \\
&= \int\int \delta(x_n - f(x_{n-1}, v_n)) p(v_n) p(x_{n-1}|Y_{n-1}) dv_n\, dx_{n-1},
\end{aligned}
\tag{25}
$$

where $\delta(x)$ denotes the delta function. Therefore, if we define $p_n^{(j)}$ by

$$p_n^{(j)} = f(f_{n-1}^{(j)}, v_n^{(j)}), \tag{26}$$

$\{p_n^{(1)}, \cdots, p_n^{(m)}\}$ can be considered as independent realizations from $p(x_n|Y_{n-1})$.

3.2 Filtering

Given the observation y_n and the particle, $p_n^{(j)}$, compute the importance factor $\alpha_n^{(j)}$ of the particle $p_n^{(j)}$ with respect to the observation y_n by

$$\alpha_n^{(j)} = p(y_n|p_n^{(j)}).$$

The posterior probability of the particle is then obtained by

$$
\begin{aligned}
\Pr(x_n = p_n^{(i)}|Y_n) &= \Pr(x_n = p_n^{(i)}|Y_{n-1}, y_n) \\
&= \frac{p(y_n|p_n^{(i)})\Pr(x_n = p_n^{(i)}|Y_{n-1})}{\sum_{j=1}^m p(y_n|p_n^{(j)})\Pr(x_n = p_n^{(j)}|Y_{n-1})} \\
&= \frac{\alpha_n^{(i)}}{\sum_{j=1}^m \alpha_n^{(j)}}.
\end{aligned}
\tag{27}
$$

This means that the distribution function of $\Pr(x_n = p_n^{(i)}|Y_n)$ can be expressed by a step function

$$\frac{1}{\sum_{j=1}^m \alpha_n^{(j)}} \sum_{i=1}^m \alpha_n^{(i)} I(x, p_n^{(i)}), \tag{28}$$

which has jumps at $p_1, \cdots, p_m$ with step sizes proportional to $\alpha_n^{(1)}, \cdots, \alpha_n^{(m)}$.

For the next step of prediction, it is necessary to represent this distribution function by an empirical distribution of the form

$$\frac{1}{m}\sum_{i=1}^{m} I(x, f_n^{(i)}). \tag{29}$$

This can be done by generating particles $\{f_n^{(1)}, \cdots, f_n^{(m)}\}$ by the resampling of $\{p_n^{(1)}, \cdots, p_n^{(m)}\}$ with probabilities

$$\Pr(f_n^{(i)} = p_n^{(j)}|Y_n) = \frac{\alpha_n^{(j)}}{\alpha_n^{(1)} + \cdots + \alpha_n^{(m)}}. \tag{30}$$

References

[1] Alspach, D. L. and Sorenson, H. W., 1972, "Nonlinear Bayesian Estimation Using Gaussian Sum Approximations", *IEEE Transactions on Automatic Control*", Vol. AC-17, No.4, 439–448.

[2] Anderson, B. D. O. and Moore, J. B., 1979, *Optimal Filtering*. Prentice-Hall, New Jersey.

[3] Andrede Netto, M. L. Gimeno, L. and Mendes, M. J., 1978, "On the optimal and suboptimal nonlinear filtering problem for discrete-time systems", *IEEE Trans. Automat. Control*, Vol. 23, 1062–1067.

[4] Gordon, N. J., Salmond, D. J. and Smith, A. F. M., 1993, "Novel approach to nonlinear/non-Gaussian Bayesian state estimation", *IEE Proceedings-F*, Vol. 140, No. 2, 107–113.

[5] Kitagawa, G., 1987, "Non-Gaussian state-space modeling of nonstationary time series" (with discussion). *Journal of the American Statistical Association*, Vol. 82, No. 400, 1032–1063.

[6] Kitagawa, G., 1991, "A nonlinear smoothing method for time series analysis", *Statistica Sinica*, Vol. 1, No. 2, 371–388.

[7] Kitagawa, G., 1996, "Monte Carlo filter and smoother for non-Gaussian nonlinear state space model", *Journal of Computational and Graphical Statistics*, Vol. 5, 1-25.

[8] Kitagawa, G., 1998, "Self-organizing state space model", *Journal of the American Statistical Association*, Vol. 93, No. 443, 1203–1215.

[9] Kitagawa, G. and Gersch, W., 1996, *Smoothness Priors Analysis of Time Series*, Springer-Verlag, New York.

[10] Ljung, L., 1979, "Asymptotic behavior of the extended Kalman filter as a parameter estimator for linear systems", *IEEE Transactions on Automatic Control*, AC-24, No. 1, 36–50.

[11] Solo, V., 1980, "Some aspect of recursive parameter estimation", *International Journal of Control*, Vol. 32, 395–410.

[12] West, M., Harrison, P. J. and Migon, H. S., 1985, "Dynamic generalized linear models and Bayesian forecasting (with discussion)", *Journal of the American Statistical Association*, 80, 73–97.

Noise Induced Congestion in Coupled Map Optimal Velocity Model of Traffic Flow

S. Tadaki[1], M. Kikuchi[2], Y. Sugiyama[3] and S. Yukawa[4]

[1]Department of Information Science, Saga University, Saga 840-8502, Japan
[2]Department of Physics, Osaka University, Toyonaka 560-8531, Japan
[3]Division of Mathematical Science, City College of Mie, Mie 514-0112, Japan
[4]Department of Applied Physics, University of Tokyo, Bunkyo 113-8656, Japan

Abstract: The optimal velocity model is one of the car following models which describe the behavior of cars by differential equations. In that model, each car controls its speed toward an optimal (safety) value depending on the headway. We construct a new car following type simulation model for traffic flow in a coupled map form based on optimal velocity functions. We can simulate open road systems with the model. The emergence of weakly congested flow induced by noise is investigated. We observe the enhancement of the car density induced by noise. We also discuss the possibility of spontaneous formation of strong traffic jams.

1. Introduction

Understanding properties of the traffic flow and traffic networks is one of the important research projects from the viewpoint of, for example, social interests: increasing importance of transportation under restricted social and natural resources. It is also interesting in the viewpoint of non-equilibrium statistical mechanics. Physical modeling of the transportational phenomena is a recently developing research area [1, 2].

One of the simplest modeling of transportational phenomena is based on cellular automata (CA). The rule-184 elementary CA is the simplest choice[3]. A variety of CA models of the traffic flow has been reviewed by Nagel in Ref. [4].

Cellular automaton models of the traffic flow can be extended to two-dimensional ones as a simple model of city traffic or traffic networks[5]. Self-organization of jam clusters have been observed under open boundaries[6]. Some CA traffic models have been applied to real urban traffic networks in cities [2, 7].

Another approach to transportational phenomena is based on the idea that the behavior of each car is described by differential equations. Each driver controls a car under the stimuli from the preceding car, which can be expressed by the function of the headway distance or the relative velocity of two successive cars. Models of this type are called *car following* models[8].

A different car following model, called the *Optimal Velocity* (OV) model, has been proposed by Bando *et al*[9, 10]. The central idea of the OV model is the introduction of the OV function, which decides the optimal (safety) velocity according to the headway. They have executed computer simulations of the OV model under the periodic boundary condition and found the existence of the phase transition from a freely-moving phase to a jamming phase by changing the car density[9, 10, 11]. The model is very simple and thus is highly suitable for analytical treatments[12, 13]. On the other hand, the sequence of cars can not be changed as long as the model is described in the form of differential equations. Therefore it is not easy to extend it to realistic traffic situations.

We have defined a coupled map traffic flow model based on optimal velocity functions (CMOV model)[14] according to the idea of coupled map (CM) modeling[15, 16]. By discretization, CM-based optimal velocity models are suitable for computer simulations. Moreover, rule-based modeling is easily incorporated in CM-based models.

Our model can be easily applied to open road with an inflow and an outflow. The strong traffic jam, which is observed in the original OV model with the periodic boundary condition, does not appear in open systems without traffic blockades. Besides it, we can observe the weak traffic jam called *noise induced congested flow*. The purpose of this paper is to investigate the noise induced congested traffic flow. We also discuss the possibility of spontaneous emergence of the strong traffic jam.

2. The Model

In the optimal velocity model, the motion of each car is described by the following differential equation for the position x of the car:

$$\frac{d^2x}{dt^2} = \alpha \left[V_{\text{optimal}} \left(\Delta x \right) - \frac{dx}{dt} \right], \tag{1}$$

where Δx is the headway distance to the preceding car and α is the sensitivity (Fig. 1). Indices of x to distinguish cars are omitted in Eq. (1) and hereafter. Each driver controls a car with acceleration or deceleration to tune the velocity to the optimal (safety) value given by the optimal velocity (OV) function $V_{\text{optimal}}(\Delta x)$. In the OV model, the optimal velocity is determined solely by the headway Δx.

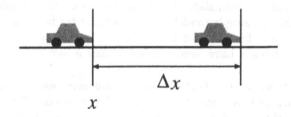

Figure 1: Schematic view of the definition of the headway. The headway is measured from the front of the car to the front of the preceding car.

The form of the OV function can be chosen arbitrarily under some conditions. Here we employ a sigmoidal one after Ref. [9] (Fig. 2):

$$V_{\text{optimal}}(\Delta x) = \frac{v_{\text{max}}}{2}\left[\tanh\left(2\frac{\Delta x - d}{x_{\text{width}}}\right) + c_{\text{bias}}\right]. \tag{2}$$

The parameters, v_{max}, d, x_{width} and c_{bias}, can be determined from observed data[11]. The delay in acceleration is naturally introduced to car motions, because the behavior of cars is described by second-order differential equations, in contrast to the conventional car following models based on delayed differential equations.

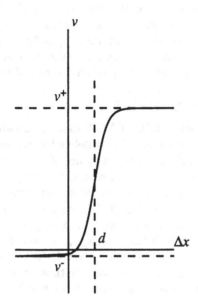

Figure 2: Schematic view of the optimal velocity function. $v^+ = v_{\text{max}}(1+c_{\text{bias}})/2$ and $v^- = v_{\text{max}}(-1 + c_{\text{bias}})/2$ are the maximum and the minimum velocity respectively.

As discussed in the previous section, we reformulate the original OV model into the coupled map form. To this end, we discretize the original OV model. It is stressed that the aim of the discretization of the OV model is not to integrate the differential equations numerically but to make the new model applicable to various realistic traffic situations.

The coupled map traffic flow model based on OV functions (CMOV model) is defined as follows: when the headway distance is sufficiently large, the position x and velocity v of a car at $t + \Delta t$ are given by

$$x(t + \Delta t) = x(t) + v(t)\Delta t, \tag{3}$$
$$v(t + \Delta t) = v(t) + \alpha\left[v_{\text{optimal}}\left(\Delta x(t)\right) - v(t)\right]\Delta t. \tag{4}$$

In the CMOV model, the time step Δt is a fixed parameter. The optimal velocity v_{optimal} is given by the OV function:

$$v_{\text{optimal}}(\Delta x) = V_{\text{optimal}}(\Delta x). \qquad (5)$$

To avoid rear-end collisions and backward motions, cars are forced to stop if the headway distance is less than the minimum value Δx_{min}: for $\Delta x < \Delta x_{\text{min}}$ eqs. (3) and (4) are replaced by

$$\begin{aligned} x(t + \Delta t) &= x(t), \qquad (6) \\ v(t + \Delta t) &= 0. \qquad (7) \end{aligned}$$

One of the most important improvements upon the OV model is that the CMOV model allows us to rearrange the sequence of cars. As a result, the CMOV model can be easily simulated under open boundary conditions, where new cars are created and added to the tail of the sequence of cars and cars are deleted from the sequence when they reach the end of the system.

3. Traffic Flow Simulator

In the computer simulations of the CMOV model, cars are expressed as almost independent data sets and created and deleted dynamically. These types of models are suitable to be implemented by use of object-oriented programming schemes. We implemented our model with C++ programming language.

We simulate the system under open boundary conditions. The schematic view of the system is shown in Fig. 3. The course consists of three parts: the left and right parts are for relaxation, whose lengths are L_1 and L_2, respectively. The middle part with length L_{observe} is used for measurement. The parameters used in the simulations are given in Table. 1. All parameters except Δt, Δx_{min} and L_{observe} are equivalent to those in Ref. [11], where they are determined so as to fit to the data observed in Chuo Expressway in Japan[17].

Figure 3: Schematic view of the system. Cars are injected from the left side with zero speed and ejected away from the right side.

The system has no car at the initial state. From the left side, the upper stream, a car is injected every second if the distance between the left side and the tail of the sequence of cars is larger than Δx_{min}. The initial speed of the injected car is zero. This injection method means that an injected car escapes from a virtual traffic jam with infinite length[18].

We observe mainly the trajectories of the test car in the headway-velocity plane. In the periodic system with the original OV model, the strong traffic jam has been observed as a hysteresis loop in the headway-velocity plane[9, 10]. We have observed the same hysteresis loop in the open system with a traffic blockade (Fig. 4)[14].

Table 1: The values of the parameters.

parameter	value	(unit)
d	25.	m
x_{width}	23.3	m
v_{max}	33.6	m/sec
α	2.	sec^{-1}
c_{bias}	0.913	
Δt	0.1	sec
Δx_{min}	7.02	m
L_{observe}	1	km

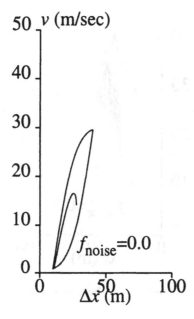

Figure 4: The trajectory of a test car in the headway-velocity plane in the open system with a traffic blockade.

4. Noise Induced Congested Flow

We investigate the effect of noise in the coupled map traffic flow model based on optimal velocity functions in the open system without traffic blockades. One of the possible origins of noise is an error in acceleration processes to the optimal velocity. To treat this effect, we impose a multiplicative random noise to velocity. We replace Eq. (4) by

$$v(t + \Delta t) = [v(t) + \alpha \left(v_{\text{optimal}} - v(t) \right) \Delta t] \times (1 + f_{\text{noise}} \xi), \qquad (8)$$

where $\xi \in [-0.5, 0.5]$ is a uniform random variable. The noise level is given by the parameter f_{noise}. The multiplicative form of the noise is a reasonable choice, because the errors are enhanced in the high speed region and suppressed in the low speed region.

Without noise injected cars simply accelerate to catch up preceding cars. In that case no traffic jam occurs. We observe the increase of the car density with the imposed noise level according to Eq. (8). The trajectory in the headway-velocity plane also fluctuates with noise. If the imposed noise exceeds a threshold, the trajectory shows a loop structure (Fig. 5), which shows the emergence of congested flow.

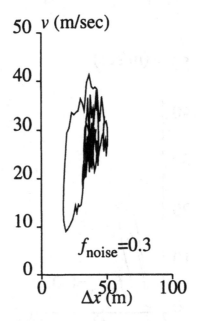

Figure 5: The trajectory of a test car in the headway-velocity plane with the noise $f_{noise} = 0.3$ in velocity.

The car density is enhanced by the noise to approach to a critical value. In the state near the critical density, a noise causes an avalanche of the short headway and induces congested flow. As a consequence, the traffic flow becomes to bear power law spectra. Other features of noise induced congested flow have been observed in Ref [19].

Another origin of noise is included in the observation of the headway distance to the preceding car. To take this effect into account, we impose a multiplicative random noise to the headway distance by replacing Eq. (5) by

$$v_{optimal}(\Delta x) = V_{optimal}\left(\Delta x \left(1 + f_{noise}\xi\right)\right). \tag{9}$$

Also in this case, we observe the similar noise-induced congested flow (see Fig. 6).

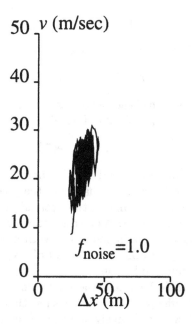

Figure 6: The trajectory of a test car in the headway-velocity plane with the noise $f_{\text{noise}} = 1.0$ in the headway distance.

5. Spontaneous Formation of Strong Traffic Jam

We observed weak traffic jam called noise-induced congested flow. There is another traffic jam called strong traffic jam, in which cars decelerate to almost zero speed. In the simulations based on optimal velocity functions, strong traffic jam has been observed in periodic systems[9, 10, 11] and open systems with traffic blockades (Fig. 4)[14]. In our model, we can not observe the spontaneous emergence of strong traffic jam in open systems without traffic blockades.

One of the reasons for the nonexistence of strong traffic jam is the high value of the sensitivity α. If the sensitivity is low enough, the strong traffic jam will happen. The traffic jam with the low value of α, however, corresponds to traffic accidents by rear-end collisions.

To clarify the difficulty of the low value of α, we discuss an exactly solvable case with a simple optimal velocity function:

$$V_{\text{optimal}} = \begin{cases} 0 & \Delta x < d, \\ v'_{\text{max}} & \Delta x \geq d. \end{cases} \tag{10}$$

We can obtain the exact solution for an equilibrium state[12]. We denote the headway distance in equilibrium jam clusters as Δx_{jam} and that in freely-running regions as Δx_{free}. They obey the following relations:

$$\Delta x_{\text{jam}} + \Delta x_{\text{free}} = 2d, \tag{11}$$

$$-\Delta x_{\text{jam}} + \Delta x_{\text{free}} = v'_{\text{max}}\tau, \tag{12}$$

where τ is a non-trivial solution of the equation:

$$\alpha\tau = 2\left(1 - e^{-\alpha\tau}\right). \tag{13}$$

To avoid rear-end collisions, the headway distance in jam clusters, Δx_{jam}, must be larger than the length Δx_{car} of cars. This gives the lower limit of α as

$$\alpha > \alpha_c = 1.28\ldots\sec^{-1}, \tag{14}$$

if we use the parameters in Table 1 and $\Delta x_{\text{car}} = 5$m. Therefore the sensitivity must be larger than α_c to follow the behavior of the preceding car.

To describe traffic flow phenomena, the optimal velocity model requires the lower limit on the sensitivity α. As a result the spontaneous formation of strong traffic jams hardly occurs in the model with open boundary systems. In this line, strong traffic jams appear as a result of traffic blockade structures: traffic junctions, speed limit and so on.

Even if there is a traffic blockade, the traffic flow with high flow rate is needed to form strong traffic jam. Our results show that noises can induce the traffic flow with high flow rate. Namely noise-induced congested flow is a key concept for understanding the formation of traffic jam.

References

[1] D. E. Wolf, M. Schreckenberg and A. Backem ed., *Traffic and Granular Flow* (World Scientific, Singapore, 1996).

[2] M. Schreckenberg and D. E. Wolf ed., *Traffic and Granular Flow '97* (Springer, Singapore, 1998).

[3] S. Wolfram, Rev. Mod. Phys. **55** pp.601, 1983.

[4] K. Nagel, Phys. Rev. E **53** pp.4655, 1996.

[5] O. Biham, A. A. Middleton and D. Levine; Phys. Rev. A**46** pp.6124, 1992; T. Nagatani, J. Phys. Soc. Jpn. **62** pp.2625, 1993; M. Fukui and Y. Ishibashi, J. Phys. Soc. Jpn. **62** pp.3841, 1993; J. A. Cuesta, F. C. Martínez, J. A. Molera and A. Sánchez, Phys. Rev. E **48** pp.5175, 1993; S. Tadaki and M. Kikuchi, Phys. Rev. E **50** pp.4546, 1994.

[6] S. Tadaki, Phys. Rev. E **54** pp.2409, 1996; J. Phys. Soc. Jpn. **66** pp.514, 1997.

[7] P. M. Simon and K. Nagel, Phys. Rev. E **58** pp.1286, 1998.

[8] The conventional forms of car following type models are found in textbooks on traffic flow. See, for example, W. Leutzbach, *Introduction to the Theory of Traffic Flow* (Springer-Verlag, Berlin, 1988).

[9] M. Bando, K. Hasebe, A. Nakayama, A. Shibata and Y. Sugiyama, Phys. Rev. E **51** pp.1035, 1995.

[10] M. Bando, K. Hasebe, A. Nakayama, A. Shibata and Y. Sugiyama, Jpn. J. Ind. Appl. Math. **11** pp.203, 1994.

[11] M. Bando, K. Hasebe, K. Nakanishi, A. Nakayama, A. Shibata and Y. Sugiyama, J. Phys. I France **5** pp.1389, 1995.

[12] Y. Sugiyama and H. Yamada, Phys. Rev. E **55** pp.7749, 1997; K.Nakanishi, K. Itoh, Y. Igarashi and B. Bando, Phys. Rev. E **55** pp.6519, 1997.

[13] T. Komatsu and S. Sasa, Phys. Rev. E **52** pp.5574, 1995.

[14] S. Tadaki, M. Kikuchi, Y. Sugiyama and S. Yukawa, J. Phys. Soc. Jpn. **67** pp.2270, 1998.

[15] S. Yukawa and M. Kikuchi, J. Phys. Soc. Jpn. **64** pp.35, 1995; **65** pp.916, 1996.

[16] K. Kaneko ed., *Theory and Applications of Coupled Map Lattices* (John Wiley & Sons, Chichester, 1993).

[17] M. Koshi, M. Iwasaki and I. Ohkura, *Some Findings and an Overview on Vehicular Flow Characteristics*, in V. F. Hurdle *et al.* eds. *Proc. 8th Intl. Symp. on Transp. and Traffic Theory* (1983); T. Oba, *An Experimental Study on Car-Following Behavior*, Thesis of Master of Engineering, Univ. of Tokyo (1988).

[18] K. Nagel and M. Paczuski, Phys. Rev. E **51** pp.2909, 1995.

[19] S. Tadaki, M. Kikuchi, Y. Sugiyama and S. Yukawa, J. Phys. Soc. Jpn. **68** No.9, 1999.

Geometrical View on Mean-Field Approximation for Solving Optimization Problems

Toshiyuki Tanaka

Department of Electrical Engineering, Graduate School of Engineering,
Tokyo Metropolitan University,
1-1, Minami-Osawa, Hachioji-shi, Tokyo, 192-0397 Japan
tanaka@eei.metro-u.ac.jp

Abstract: When one wishes to solve optimization problems by simulated annealing, the naive mean-field approximation provides a practical way of doing it. Extensions of the naive approximation by including higher-order terms have been proposed in the prospect of improving accuracy of the approximation. It has been reported, however, that higher-order approximations do not work well, especially in low temperature regions. We present an analytical argument and a geometrical view on this contradictory observation based on information-geometry, and give an intuitive explanation as to why the naive approximation does work well when it is applied to solving optimization problems.

1. Introduction

Simulated annealing [1] has been recognized as a tool for solving optimization problems in a stochastic manner. It includes the so-called Markov-chain-Monte-Carlo (MCMC) procedure to evaluate expectations of relevant quantities with respect to a "thermal equilibrium" distribution, and therefore, it suffers from the same difficulty as the one for the MCMC that in general it requires a huge amount of computation for sampling states with respect to the "thermal equilibrium" distributions.

Mean-field approximation [2] has been proposed as one way to circumvent this difficulty, allowing us to evaluate expectations by a far less amount of computation. It has been widely used in practice since it is often almost the only way to handle large-scale problems. From the theoretical viewpoint, however, it has now been known that the accuracy of the approximation, in its usual, naive form, is not so good [3]. In the theoretical context, the naive mean-field approximation can be regarded as the "lowest-order" approximation. Inclusion of higher-order terms has been proposed with the intention of improving the accuracy [3–7]. As for simulated annealing, however, it is observed that higher-order methods do not give good results compared with the naive method, especially in the low-temperature region [7]. The objective of this paper is to provide a geometrical view to this apparently contradictory observation.

2. Formulation

We consider an optimization problem in the state space $S = \{-1, 1\}^N$. Let $s = (s_i) \in S$ be the state variable. The loss function, or the *energy function*, $E(s)$ is a real-valued function on S to be minimized, and it is assumed to have the following quadratic form:

$$E(s) = -\sum_i h_i s_i - \sum_{\langle ij \rangle} w_{ij} s_i s_j \tag{1}$$

$\langle ij \rangle$ means that the sum should be taken over all distinct pairs of indices. The optimization problem here is to find $s_{\min}$ which gives the minimum of the energy function $E(s)$.

Simulated annealing tries to solve the problem by simulating dynamics of a physical system whose energy function is given by E. When the "inverse temperature" of the system is β, the system takes the state s according to the Boltzmann-Gibbs distribution,

$$p(s) = Z^{-1} \exp(-\beta E(s)), \tag{2}$$

where $Z = \sum_s \exp(-\beta E(s))$ is a normalizing coefficient, called the partition function. When the temperature $1/\beta$ is high, the state distribution is rather uniform over S. On the other hand, if the temperature becomes low enough, the state distribution is well confined to the state with the minimum energy, $s_{\min}$. Thus, what one has to do is to realize the thermal equilibrium at a low enough temperature: In order to do it, one first starts simulating the system at high enough temperature and then gradually lowers the temperature to achieve the thermal equilibrium at a low enough temperature. Once it is achieved, then one can pick up a state according to the thermal equilibrium distribution, so that with sufficiently high probability it gives the minimum energy state $s_{\min}$, which solves the problem.

3. Mean-Field Approximation

In the case where the temperature is low enough, the expectation $\langle s \rangle$ should be close to $s_{\min}$, because the distribution is almost confined to the minimum-energy state. Therefore, if one can evaluate the expectation $\langle s \rangle$, doing it at a low temperature would also give the solution. This is the place where mean-field approximation comes into play in solving the optimization problem: If one can expect that mean-field approximation gives a good estimation of the expectation $\langle s \rangle$, then doing the estimation while gradually decreasing the temperature will eventually give the solution.

In the naive mean-field approximation, one can obtain an estimate $m = (m_i)$ of the expectation $\langle s \rangle$ as a solution of the so-called mean-field equation,

$$\tanh^{-1} m_i - \beta \left[h_i + \sum_{j \neq i} w_{ij} m_j \right] = 0. \tag{3}$$

It should be noted that this equation can be transformed into the following form,

$$m_i = \tanh\left[\beta\left(h_i + \sum_{j\neq i} w_{ij}m_j\right)\right], \tag{4}$$

which is the frequently used, common form of the mean-field equation.

According to a rigorous theory of mean-field approximation [8], however, the naive mean-field approximation will provide a poor approximation especially at low temperature region, as confirmed by numerical experiments [3]. If one wishes to improve the approximation, one has to incorporate into the mean-field equation higher-order terms. Inclusion of the next-lowest-order terms into the mean-field equation yields the following formula [3].

$$\tanh^{-1} m_i - \beta\left[h_i + \sum_{j\neq i} w_{ij}m_j - \beta\sum_{j\neq i}(w_{ij})^2(1 - m_j^2)m_i\right] = 0 \tag{5}$$

We call this approximation the second-order approximation (which has been sometimes *inappropriately* termed TAP approximation), since it includes the terms up to the second order with respect to $w = (w_{ij})$. The naive mean-field approximation can be regarded as the first-order approximation. The true formula, which is expected to give the true expectation m provided that the relevant power series in terms of w does converge, can be expressed as:

$$\tanh^{-1} m_i - \beta\left[h_i + \sum_{j\neq i} w_{ij}m_j - \beta\sum_{j\neq i}(w_{ij})^2(1 - m_j^2)m_i\right] = O(|\beta w|^3) \tag{6}$$

4. Analytical Result

In this section we show an analytical result concerning the estimate obtained from the naive mean-field approximation. Assume that $|\beta w|$ be sufficiently small, and let $m^{(1)}$ denote the estimate obtained from the naive mean-field approximation. Then, from perturbation calculations we have (see Appendix for the derivation)

$$m_i^{(1)} = \left[1 + \beta^2 \sum_{j\neq i}(w_{ij})^2(1 - m_i^2)(1 - m_j^2)\right]m_i + O(|\beta w|^3), \tag{7}$$

which states that *the naive mean-field approximation gives overestimation* when considering up to the order $O(|\beta w|^2)$, since the term $\beta^2 \sum_{j\neq i}(w_{ij})^2(1 - m_i^2)(1 - m_j^2)$ is always nonnegative.

5. Information-Geometrical View

The analytical result presented in the previous section is valid only when $|\beta w|$ is sufficiently small. For the general case where $|\beta w|$ is not sufficiently small, theoretical arguments become almost infeasible, so that we can only provide qualitative descriptions. In this section we present an information-geometrical view about the overestimation.

In the following information-geometrical arguments, we will ignore β without loss of generality, since it can be effectively incorporated into the parameters h and w. Let P be a set of all possible Boltzmann-Gibbs distributions realizable by the system whose energy function is of the form of Eq. (1). Although it seems natural to use the parameters (h, w) in order to specify a Boltzmann-Gibbs distribution $p \in P$, from information-geometrical argument [8], when one discusses mean-field approximation it turns out to be natural to use the *mixed* coordinate system (m, w) to specify a Boltzmann-Gibbs distribution [8]. Using this coordinate system, the problem of estimating m is translated into the problem of evaluating the *projection* of a given distribution $p \in P$ onto the "factorizable" submanifold P_0, which is defined by

$$P_0 = \{p \in P \mid w(p) = 0\}. \tag{8}$$

Figure 1 (a) illustrates the geometry for the simplest, 2-unit case, showing the set of systems whose expectation m satisfies $|m|^2 = 0.5$. This set is expressed as an upright cylinder in the (m_1, m_2, w_{12})-coordinate. Note that the use of the mixed coordinate system makes the difficulty in solving the problem less obvious: Indeed, the difficulty arises from the fact that, even though what we want to know is m, which would become apparent once we can specify the mixed coordinate of the distribution of interest, what we actually know is h and w and evaluation of m from h and w is in general computationally very hard.

Mean-field approximation gives estimates of m from h and w as solutions of the mean-field equations. Figure 1 (b) shows the set of systems for which the first-order, or naive, approximation gives the estimate $m^{(1)}$ with $|m^{(1)}|^2 = 0.5$. The fact, analytically derived in the preceding section, that the first-order approximation gives overestimation, is visualized in the figure as the convexity of the set around $w_{12} = 0$: As w_{12} grows away from 0, the exact value of m becomes smaller, that is, $|m|^2 < 0.5$, giving $|m^{(1)}| > |m|$. Although the analytical argument assumes that $|w|$ is small, it seems that this holds beyond small-$|w|$ range. This property of the first-order approximation is indeed attractive in applying the first-order approximation to solving optimization problems: In optimization problems one wishes to obtain binary results ($s_i = \pm 1$), and the first-order approximation amplifies the expectation m so that it approaches binary vectors.

Figure 1 (c) shows the similar result for the second-order approximation. One can observe that the set shown in this case is neither convex nor concave near $w_{12} = 0$, implying the second-order approximation will give better estimations than the first-order approximation. However, investigation in the larger-w_{12} case reveals the following fact (see Fig. 1 (d)): When w_{12} is large, even if the exact expectations m_1 and m_2 are close to ± 1, the second-order approximation can give considerably smaller values as the estimates of them. This means that in such cases the second-order approximation gives *underestimation*. Since the cases where w_{12} is large correspond to the cases where the temperature is low, this observation implies that the second-order approximation will have poor accuracy in low-temperature stages

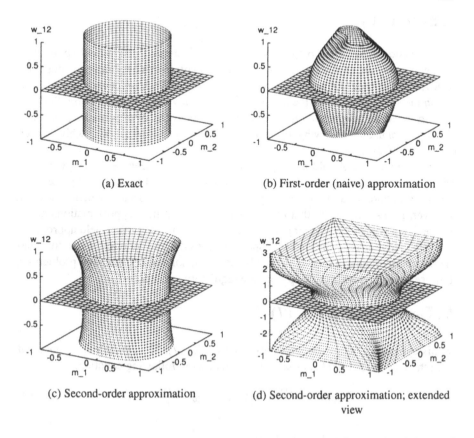

(a) Exact

(b) First-order (naive) approximation

(c) Second-order approximation

(d) Second-order approximation; extended view

Fig. 1. Sets of systems for which mean-field approximation yields estimates m satisfying $|m|^2 = 0.5$, shown in the mixed coordinate system (m_1, m_2, w_{12}), in the 2-unit case. (a) shows the case where the exact estimation is used, (b) and (c) show the cases where the first-order (naive) and second-order approximations are used, respectively. (d) shows the same result as (c) except that the range of w_{12} is extended to $[-3, 3]$. The plane in each figure shows the set P_0.

in simulated annealing, where m_i should become close to ± 1. This is qualitatively in agreement with the result reported by Hofmann and Buhmann [7]. Considering the nature of mean-field approximation that it is essentially a Taylor expansion with respect to w [8], approximations of still higher orders will yield poor accuracy as well in low-temperature cases, because accuracy of the Taylor expansion will be lost as $|w|$ becomes large.

6. Summary

We have addressed the problem of why the effectiveness of mean-field approximation, when they are applied to solving optimization problems by simulated annealing, is different between the naive and higher-order ones. We have analytically showed that in the small-$|\beta w|$ cases the naive approximation yields overestimation. We have also provided an information-geometrical view about the overestimation, and have given an explanation as to why the naive approximation does work well while higher-order ones do not.

In summary, we should not expect that higher-order mean-field approximations might work well when using any of them blindly as a tool for improving the naive mean-field approximation to solve optimization problems. This statement does not, however, necessarily mean that that higher-order mean-field approximations are of no use when we consider solving optimization problem using mean-field approximation: Since higher-order mean-field approximations do give better estimates for intermediate temperature ranges, they may be useful in solving optimization problems if they are combined with some appropriate regulatory systems.

A. Derivation of Eq. (7)

In this appendix, I show the brief derivation of Eq. (7). The estimate $m_i^{(1)}$ obtained from the naive mean-field approximation satisfies the following equation.

$$\tanh^{-1} m_i^{(1)} - \beta\Big[h_i + \sum_{j \neq i} w_{ij} m_j^{(1)}\Big] = 0 \tag{9}$$

Subtracting Eq. (6) from Eq. (9) yields

$$\tanh^{-1} m_i^{(1)} - \tanh^{-1} m_i - \beta\Big[\sum_{j \neq i} w_{ij}(m_j^{(1)} - m_j) + \beta \sum_{j \neq i}(w_{ij})^2(1 - m_j^2)m_i\Big]$$

$$= O(|\beta w|^3). \tag{10}$$

Let $\varepsilon_i \equiv m_i^{(1)} - m_i$. It can be shown that $\varepsilon_i = O(|\beta w|^2)$, and therefore $\sum_{j \neq i} w_{ij}\varepsilon_j = O(|\beta w|^3)$. Taylor expanding the term $\tanh^{-1} m_i^{(1)}$ at m_i gives

$$\frac{\varepsilon_i}{1 - m_i^2} - \beta^2 \sum_{j \neq i}(w_{ij})^2(1 - m_j^2)m_i = O(|\beta w|^3), \tag{11}$$

from which Eq. (7) immediately follows.

References

[1] S. Kirkpatrick, C. D. Gelatt, Jr., and M. P. Vecchi, "Optimization by simulated annealing," *Science*, vol. 220, no. 4598, pp. 671–680, 1983.

[2] C. Peterson and J. R. Anderson, "A mean field theory learning algorithm for neural networks," *Complex Syst.*, vol. 1, pp. 995–1019, 1987.

[3] T. Tanaka, "Mean field theory of Boltzmann machine learning," *Phys. Rev. E*, vol. 58, no. 2, pp. 2302–2310, Aug. 1998.

[4] C. C. Galland, "The limitations of deterministic Boltzmann machine learning," *Network*, vol. 4, no. 3, pp. 355–379, 1993.

[5] H. J. Kappen and F. B. Rodríguez, "Efficient learning in Boltzmann machines using linear response theory," *Neural Computation*, vol. 10, no. 5, pp. 1137–1156, 1998.

[6] H. J. Kappen and F. B. Rodríguez, "Boltzmann machine learning using mean field theory and linear response correction," in M. I. Jordan, M. J. Kearns, and S. A. Solla (Eds.), *Advances in Neural Information Processing Systems* 10, pp. 280–286, The MIT Press, 1998.

[7] T. Hofmann and J. M. Buhmann, "Pairwise data clustering by deterministic annealing," *IEEE Trans. Patt. Anal. & Machine Intell.*, vol. 19, no. 1, pp. 1–14, Jan. 1997; Errata, *ibid.*, vol. 19, no. 2, p. 197, Feb. 1997.

[8] T. Tanaka, "A theory of mean field approximation," M. S. Kearns, S. A. Solla, and D. A. Cohn (Eds.), *Advances in Neural Information Processing Systems* 11, The MIT Press, pp. 351–357, 1999.

Probabilistic Computational Method in Image Restoration based on Statistical-mechanical Technique

Kazuyuki Tanaka,
Department of Computer and Mathematical Sciences,
Graduate School of Information Sciences,
Tohoku University, Sendai 980-8579, Japan
(E-mail address: kazu@statp.is.tohoku.ac.jp)

Abstract

An image restoration can be often formulated as an energy minimization problem. When an energy function is expressed by using the hamiltonian of a classical spin system only with finite range interactions, the probabilistic model, which is described in the form of Gibbs distribution for the energy function, can be regarded as a Markov random field (MRF) model. In the MRF model, we have to determine not only the minimum-energy configuration but also hyperparameters. We have a constrained optimization and a maximum likelihood (ML) estimation as mathematical frameworks to determine the hyperparameters. In this paper, some probabilistic computational methods for the search of minimum-energy configuration and the estimation of hyperparameters are proposed in the standpoint of statistical-mechanics. We summarize the mathematical framework of probabilistic computational method based on the constrained optimization and reformulate the framework of ML estimation as a hyperparameter estimation method at a finite temperature in the standpoint of the constrained optimization. The probabilistic computational algorithms for natural image restorations are constructed from the mean-field approximation, mean-field annealing (MFA), iterative conditional modes (ICM) and cluster zero-temperature process (CZTP).

1. Introduction

Recently, many authors have shown an interest in image restoration using the Markov random field (MRF) model, in which the configuration of a lattice site is dependent only on the configurations of the nearest neighbours [1, 2, 3, 4, 5, 21]. The MRF model, can be regarded as one of classical spin systems with finite range interactions and non-uniform external fields on a finite square lattice. The MRF model can be regarded as a classical spin system in statistical mechanics.

The authors have proposed a new method for systematically constructing the energy function [7], which is based on constrained optimization. In the image restoration, the constraints are introduced as *a priori* information on the original image. By introducing a Lagrange multiplier for each of the constraints, the image restoration is reduced to an energy minimization problem [7, 8, 9]. In Ref. [10], we described a classical spin system, which is applicable to gray-level image restoration. In the standpoint of statistics, Besag [11], and Lakshmanan and Derin [12] introduced the generalized likelihood function, which is the joint probability distribution of original and degraded images, in order to determine the hyperparameters. In their framework, the restored image and the hyperparameters are determined from the coupled equation constructed from the equation for the maximization of the *a posteriori* probability distribution and the generalized likelihood function when the degraded image is given. The hyperparameter estimation can be regarded as an extension of the standard ML estimation and it is called generalized maximum likelihood estimation[1].

In the search for an optimal solution of MRF model, many authors applied the iterated conditional modes (ICM) algorithm [11]. Though the ICM algorithm is simple algorithm and can erase noise when the noise is in an isolated site, it is difficult to avoid the local minimum and then it cannot erase noise when two or more successive sites are affected by the noise. to the search of the minimum-energy configuration. In order to avoid the local minimum, the cluster type Monte Carlo simulation [13], and the cluster type mean field approximation [14] are also applied to the optimization of the MRF model. In the statistical mechanics, we have some important fluctuation effects to avoid the local minimum. One of them is thermal effect. In order to adopt the thermal fluctuation effect as annealing procedure, we introduce a temperature in the form of Gibbs distribution. Geiger and Girosi[15], and Zhang[16, 17] proposed a deterministic algorithm, which is based on the mean-field approximation. On the other hand, we have general methodology of construction of high-level effective-field approximation in statistical mechanics, which is called cluster variation method (CVM) [18]. We gave a simple example of its application to the MRF model [19]. Moreover, in the standpoint of the CVM, we proposed the methodology in the extension of the ICM to the cluster type algorithm, which is called the cluster zero-temperature process (CZTP) and is obtained as the zero-temperature limit in the optimization algorithm of the CVM [20].

In this paper, we clarify the mathematical frameworks for the image restoration by means of the MRF model in the standpoints of both the constrained optimization [10] and the ML estimation [11, 12], and propose some probabilistic computational algorithms by means of the ICM, the CZTP and the MFA. In Sec. 2, we explain the energy function and the probabilistic model constructed by means of the mathematical framework of Ref.[10], and reformulate the generalized ML estimation, which was proposed in Refs.[11] and [12], in the standpoint of Ref.[10]. In Sec. 3, we give some probabilistic computational algorithms for the search of minimum-energy configuration and the

---

[1]In the standard ML approach, when many degraded images are produced from the same original image in a given degradation process with some hyperparameters, the parameters are determined so as to maximize the likelihood function, which is constructed from the degradation process.

hyperparameter estimation. In Secs. 4 and 5, we give some numerical experiments and some concluding remarks, respectively.

2. Markov Random Field Model

In this section, we give the energy function of image restoration for natural images with the aid of the mathematical framework in Ref.[10].

We consider a digital image with q grades on a $M \times N$ finite square lattice

$$\mathbf{I} \equiv \Big\{(i,j)\Big| \, i = 1, 2, \cdots, M, \ j = 1, 2, \cdots, N\Big\},$$

and denote the set of the nearest-neighbour pair of pixels by $\mathbf{B}$. We express the random fields of an original and a degraded images by $\mathbf{X} \equiv \{X_{i,j} | (i,j) \in \mathbf{I}\}$ and $\mathbf{Y} \equiv \{Y_{i,j} | (i,j) \in \mathbf{I}\}$, respectively. The random variables $X_{i,j}$ and $Y_{i,j}$ on a pixel (i,j) takes a value from $\mathbf{Q} \equiv \{0, 1, 2, \cdots, q-1\}$. For the degradation process, we assume that a degraded image $\mathbf{y} \equiv \{y_{i,j} | (i,j) \in \mathbf{I}\}$ is obtained from the original image $\mathbf{x} = \tilde{\mathbf{x}} \equiv \{\tilde{x}_{i,j} | (i,j) \in \mathbf{I}\}$ by changing the state of each pixel to another state by the same probability $\tilde{p}$, independently of the other pixel. The conditional probability for a degraded image $\mathbf{y}$ when the original image is $\tilde{\mathbf{x}}$, $\Pr\{\mathbf{Y} = \mathbf{y} | \mathbf{X} = \tilde{\mathbf{x}}, \tilde{p}\}$, is given by

$$\Pr\Big\{\mathbf{Y} = \mathbf{y} | \mathbf{X} = \tilde{\mathbf{x}}, \tilde{p}\Big\} \equiv \prod_{(i,j) \in \mathbf{I}} \Big(\tilde{p}^{1 - \delta(\tilde{x}_{i,j}, y_{i,j})}(1 - q\tilde{p} + \tilde{p})^{\delta(\tilde{x}_{i,j}, y_{i,j})}\Big)$$

$$= \frac{\exp\Big(-\frac{1}{T_{\tilde{p}}} d(\mathbf{x}, \mathbf{y})\Big)}{\sum_{\mathbf{y}} \exp\Big(-\frac{1}{T_{\tilde{p}}} d(\mathbf{x}, \mathbf{y})\Big)}, \tag{1}$$

where

$$d(\tilde{\mathbf{x}}, \mathbf{y}) \equiv \sum_{(i,j) \in \mathbf{I}} \Big(1 - \delta(\tilde{x}_{i,j}, y_{i,j})\Big), \quad T_{\tilde{p}} \equiv \frac{1}{\ln\Big(\frac{1 - q\tilde{p} + \tilde{p}}{\tilde{p}}\Big)}. \tag{2}$$

Here, $\delta(a, b) = \delta_{ab}$ is the Kronecker's delta. In the present paper, we treat only the case in which $T_{\tilde{p}}$ is positive such that $\tilde{p} < 1/q$.

In the Bayes formula, the *a posteriori* probability distribution for the original image $\mathbf{x} \equiv \{x_{i,j} | (i,j) \in \mathbf{I}\}$ when the degraded image is $\tilde{\mathbf{y}}$, $\Pr\{\mathbf{X} = \mathbf{x} | \mathbf{y} = \tilde{\mathbf{y}}\}$, is expressed as

$$\Pr\Big\{\mathbf{X} = \mathbf{x} | \mathbf{y} = \tilde{\mathbf{y}}\Big\} = \Pr\Big\{\mathbf{X} = \mathbf{x} | \mathbf{Y} = \tilde{\mathbf{y}}, \tilde{p}\Big\}$$

$$= \frac{\Pr\Big\{\mathbf{Y} = \tilde{\mathbf{y}} | \mathbf{X} = \mathbf{x}, \tilde{p}\Big\} \Pr\Big\{\mathbf{X} = \mathbf{x}\Big\}}{\sum_{\mathbf{x}} \Pr\Big\{\mathbf{Y} = \tilde{\mathbf{y}} | \mathbf{X} = \mathbf{x}, \tilde{p}\Big\} \Pr\Big\{\mathbf{X} = \mathbf{x}\Big\}}, \tag{3}$$

where $\Pr\{\mathbf{X} = \mathbf{x}\}$ is the *a priori* probability distribution that the original image is $\mathbf{x}$. We assume that the original image is the one of images that are produced with the large probability in the *a priori* probability distribution $\Pr\{\mathbf{X} = \mathbf{x}\}$. In the maximum *a posteriori* (MAP) estimation, the restored image $\hat{\mathbf{x}} \equiv \{\hat{x}_{i,j} | (i,j) \in \mathbf{I}\}$ is estimated as follows:

$$\hat{\mathbf{x}} = \operatorname*{argmax}_{\mathbf{X}} \Pr\left\{\mathbf{X} = \mathbf{x} \middle| \mathbf{Y} = \tilde{\mathbf{y}}, \tilde{p}\right\}. \tag{4}$$

In Ref.[10], it is assumed that we know the following quantities for the true original image $\tilde{\mathbf{x}}$:

$$\sigma_2(\tilde{\mathbf{x}}) \equiv = \sum_{(i,j) \in \mathbf{I}} \left(2 - \delta(\tilde{x}_{i,j}, \tilde{x}_{i+1,j}) - \delta(\tilde{x}_{i,j}, \tilde{x}_{i,j+1})\right), \tag{5}$$

$$\sigma_{2n}(\tilde{\mathbf{x}}) \equiv \sum_{(i,j) \in \mathbf{I}} \left(\delta(|\tilde{x}_{i,j} - \tilde{x}_{i+1,j}|, n) + \delta(|\tilde{x}_{i,j} - \tilde{x}_{i,j+1}|, n)\right),$$

$$(n = 1, 2, \cdots, k-1). \tag{6}$$

The *a priori* probability distribution, that the original image is $\mathbf{x}$, is given by

$$\Pr\left\{\mathbf{X} = \mathbf{x}\right\} \equiv \frac{\delta\left(\sigma_2(\mathbf{x}), \sigma_2(\tilde{\mathbf{x}})\right) \prod_{n=1}^{k-1} \delta\left(\sigma_{2n}(\mathbf{x}), \sigma_{2n}(\tilde{\mathbf{x}})\right)}{\sum_{\mathbf{x}} \left\{\delta\left(\sigma_2(\mathbf{x}), \sigma_2(\tilde{\mathbf{x}})\right) \prod_{n=1}^{k-1} \delta\left(\sigma_{2n}(\mathbf{x}), \sigma_{2n}(\tilde{\mathbf{x}})\right)\right\}}. \tag{7}$$

The image restoration in natural images is formulated as the following conditional optimization problem:

$$\hat{\mathbf{x}} = \arg \min_{\mathbf{X}} \left\{d(\mathbf{x}, \mathbf{y}) \middle| \sigma_2(\mathbf{x}) = \sigma_2(\tilde{\mathbf{x}}),\right.$$

$$\left. \sigma_{2n}(\mathbf{x}) = \sigma_{2n}(\tilde{\mathbf{x}}) \ (n = 1, 2, \cdots, k-1)\right\} \tag{8}$$

In order to ensure the constrained conditions $\sigma_2(\mathbf{x}) = \sigma_2(\tilde{\mathbf{x}})$ and $\sigma_{2n}(\mathbf{x}) = \sigma_{2n}(\tilde{\mathbf{x}})$ $(n = 1, 2, \cdots, k-1)$, we introduce the Lagrange multipliers J_k and $J_k - J_n$ $(n = 1, 2, \cdots, k-1)$. The conditional optimization (8) can be reduced to the following energy minimization problem:

$$\mathbf{x}(\mathbf{J}) = \operatorname*{argmin}_{\mathbf{X}} H(\mathbf{x}|\tilde{\mathbf{y}}, \mathbf{J}), \tag{9}$$

where

$$H(\mathbf{x}|\tilde{\mathbf{y}}, \mathbf{J}) \equiv \sum_{(i,j) \in \mathbf{I}} \left(-\delta(x_{i,j}, y_{i,j}) + \phi(x_{i,j}, x_{i+1,j}|\mathbf{J}) + \phi(x_{i,j}, x_{i,j+1}|\mathbf{J})\right), \tag{10}$$

$$\phi(m, n|\mathbf{J}) \equiv \sum_{l=1}^{k-1} J_l \delta(|m-n|, l) + J_k \sum_{l=k}^{q-1} \delta(|m-n|, l), \tag{11}$$

$$\mathbf{J} \equiv \left\{ J_n \middle| n = 1, 2, \cdots, k \right\}. \tag{12}$$

Here the notation $\arg \min_x f(x)$ means any minimizer of a function $f(x)$. The optimal parameters $\hat{\mathbf{J}} \equiv \{\hat{J}_n | n = 1, 2, \cdots, k\}$ should be determined so as to satisfy the following constraints:

$$\sigma_2(\mathbf{x}(\hat{\mathbf{J}})) = \sigma_2(\tilde{\mathbf{x}}), \tag{13}$$

$$\sigma_{2n}(\mathbf{x}(\hat{\mathbf{J}})) = \sigma_{2n}(\tilde{\mathbf{x}}) \quad (n = 1, 2, , \cdots, k - 1). \tag{14}$$

The restored image obtained by using Eqs. (9)-(14) is denoted by $\hat{\mathbf{x}} = \mathbf{x}(\hat{\mathbf{J}})$. The energy minimization problem (9) is equivalent to the following probability maximization problem:

$$\mathbf{x}(\mathbf{J}) = \operatorname*{argmax}_{\mathbf{x}} \rho(\mathbf{x}|\mathbf{y}, T, \mathbf{J}), \tag{15}$$

where

$$\rho(\mathbf{x}|\mathbf{y}, T, \mathbf{J}) \equiv \frac{\exp\left(-\frac{1}{T} H(\mathbf{x}|\tilde{\mathbf{y}}, \mathbf{J})\right)}{\sum_{\mathbf{x}} \exp\left(-\frac{1}{T} H(\mathbf{x}|\tilde{\mathbf{y}}, \mathbf{J})\right)} \quad (T > 0), \tag{16}$$

and T is a temperature. The probabilistic model described by the probability distribution (16) with Eq.(10) can be regarded as the MRF model.

As the familiar techniques of the estimation of hyperparameters in the statistics, we have the generalized ML estimation [11, 12]. We reformulate the generalized ML estimation under the assumption that the form of the *a priori* probability distribution, that the original image is $\mathbf{x}$, is given as follows:

$$\begin{aligned} \Pr\left\{\mathbf{X} = \mathbf{x}\right\} &= \Pr\left\{\mathbf{X} = \mathbf{x}\middle|\mathbf{K}\right\} \\ &\equiv \frac{\exp\left(-K_k \sigma_2(\mathbf{x}) - \sum_{n=1}^{k-1} K_n \sigma_{2n}(\mathbf{x})\right)}{\sum_{\mathbf{x}} \exp\left(-K_k \sigma_2(\mathbf{x}) - \sum_{n=1}^{k-1} K_n \sigma_{2n}(\mathbf{x})\right)}, \end{aligned} \tag{17}$$

where

$$\mathbf{K} \equiv \left\{ K_n \middle| n = 1, 2, \cdots, k \right\}. \tag{18}$$

Even though the true original image $\tilde{\mathbf{x}} = \{\tilde{x}_{i,j} | (i,j) \in \mathbf{I}\}$ itself is unknown, we assume that we know and the quantities $d(\tilde{\mathbf{x}}, \tilde{\mathbf{y}})$, $\sigma_2(\tilde{\mathbf{x}})$ and $\sigma_{2n}(\tilde{\mathbf{x}})$ ($n = 1, 2, \cdots, k - 1$) in the true original image $\tilde{\mathbf{x}}$.

By substituting Eqs. (17) and (1) into Eq. (3), the *a posteriori* probability distribution is given by

$$\begin{aligned} \Pr\left\{\mathbf{X} = \mathbf{x}\middle|\mathbf{Y} = \mathbf{y}\right\} &= \Pr\left\{\mathbf{X} = \mathbf{x}\middle|\mathbf{Y} = \mathbf{y}, p, \mathbf{K}\right\} \\ &= \frac{\exp\left(-\frac{1}{T_p} d(\mathbf{x}, \mathbf{y}) - K_k \sigma_2(\mathbf{x}) - \sum_{n=1}^{k-1} K_n \sigma_{2n}(\mathbf{x})\right)}{\sum_{\mathbf{x}} \exp\left(-\frac{1}{T_p} d(\mathbf{x}, \mathbf{y}) - K_k \sigma_2(\mathbf{x}) - \sum_{n=1}^{k-1} K_n \sigma_{2n}(\mathbf{x})\right)}. \end{aligned} \tag{19}$$

For various values of p and $\mathbf{K}$, the optimal image is given by

$$\hat{\mathbf{x}}(p, \mathbf{K}) = \arg\max_{\mathbf{x}} \Pr\left\{\mathbf{X} = \mathbf{x} \middle| \mathbf{Y} = \mathbf{y}, p, \mathbf{K}\right\}. \tag{20}$$

By introducing hyperparameters $\mathbf{J} \equiv \{J_n | n = 1, 2, \cdots, k\}$ instead of T_p and $\mathbf{K}$, respectively,

$$J_n \equiv T_p(K_k - K_n) \ (n = 1, 2, \cdots, k-1), \quad J_k \equiv T_p K_k, \tag{21}$$

we can rewrite the *a posteriori* probability distribution given by Eq. (19) as follows:

$$\Pr\left\{\mathbf{X} = \mathbf{x} \middle| \mathbf{Y} = \mathbf{y}, p, \mathbf{K}\right\} = \frac{\exp\left(-\frac{1}{T_p} H(\mathbf{x}|\tilde{\mathbf{y}}, \mathbf{J})\right)}{\sum_{\mathbf{x}} \exp\left(-\frac{1}{T_p} H(\mathbf{x}|\tilde{\mathbf{y}}, \mathbf{J})\right)}. \tag{22}$$

It is obvious that the optimal image $\hat{\mathbf{x}}(p, \mathbf{K})$ which is determined by Eq. (20) is dependent on only new hyperparameter $\mathbf{J}$. Hence the search of optimal image $\hat{\mathbf{x}}(p, \mathbf{K})$ for a fixed values of p and $\mathbf{K}$ can be reduced to the search of the optimal image $\hat{\mathbf{x}}(\mathbf{J})$ which minimizes the energy function given by Eq. (9) with Eqs. (10) and (11) and it is equivalent to the maximum probability problem given in Eqs. (15) and (16) for any positive value of T. We should remark that the temperature T in Eq. (16) is just a tool for the annealing procedure and plays a different role from T_p.

In the generalized ML estimation the hyperparameters p and $\mathbf{K}$ are determined from the joint probability distribution for the original image $\mathbf{x}$ and degraded image $\mathbf{y}$, which is given by

$$\Pr\left\{\mathbf{X} = \mathbf{x}, \mathbf{Y} = \mathbf{y}\right\} = \Pr\left\{\mathbf{X} = \mathbf{x}, \mathbf{Y} = \mathbf{y} \middle| p, K_1, \cdots, K_k\right\}$$
$$\equiv \Pr\left\{\mathbf{Y} = \mathbf{y} \middle| \mathbf{X} = \mathbf{x}, p\right\} \Pr\left\{\mathbf{X} = \mathbf{x} \middle| K_1, \cdots, K_k\right\}. \tag{23}$$

We can regard the joint probability distribution as a generalized likelihood function of hyperparameters p and $\mathbf{K}$ when the true original and the given degraded images are $\tilde{\mathbf{x}} \equiv \{\tilde{x}_{i,j} | (i, j) \in \mathbf{I}\}$ and $\tilde{\mathbf{y}} \equiv \{\tilde{y}_{i,j} | (i, j) \in \mathbf{I}\}$, respectively. The estimates of the hyperparameters p and $\mathbf{K}$ are denoted by $\hat{p}$ and $\hat{\mathbf{K}}$. In the standpoint of the generalized ML estimation, the estimates of hyperparameters, $\hat{p}$ and $\hat{\mathbf{K}}$ are determined by

$$(\hat{p}, \hat{\mathbf{K}}) = \arg\max_{(p, \mathbf{K})} \Pr\left\{\mathbf{X} = \tilde{\mathbf{x}}, \mathbf{Y} = \tilde{\mathbf{y}} \middle| p, \mathbf{K}\right\}. \tag{24}$$

From the extreme condition of $\Pr\{\mathbf{X} = \tilde{\mathbf{x}}, \mathbf{Y} = \tilde{\mathbf{y}} | p, \mathbf{K}\}$ at $p = \hat{p}$ and $\mathbf{K} = \hat{\mathbf{K}}$, the deterministic equations for the estimates $\hat{p}$ and $\hat{\mathbf{K}}$ can be reduced to the following equations:

$$|\mathbf{I}|(n-1)\hat{p} = d(\tilde{\mathbf{x}}, \tilde{\mathbf{y}}), \tag{25}$$

$$\sum_{\mathbf{x}} \sigma_2(\mathbf{x}) \Pr\left\{ \mathbf{X} = \mathbf{x} \middle| \mathbf{K} = \hat{\mathbf{K}} \right\} = \sigma_2(\tilde{\mathbf{x}}), \tag{26}$$

$$\sum_{\mathbf{x}} \sigma_{2n}(\mathbf{x}) \Pr\left\{ \mathbf{X} = \mathbf{x} \middle| \mathbf{K} = \hat{\mathbf{K}} \right\} = \sigma_{2n}(\tilde{\mathbf{x}}) \quad (n = 1, 2, \cdots, k-1). \tag{27}$$

Because we assume that the values of $d(\tilde{\mathbf{x}}, \tilde{\mathbf{y}})$, $\sigma_2(\tilde{\mathbf{x}})$ and $\sigma_{2n}(\tilde{\mathbf{x}})$ $(n = 1, 2, \cdots, k-1)$ are known, the estimates $\hat{p}$ and $\hat{\mathbf{K}}$ can be determined by Eqs. (25) and (26). It is difficult to calculate the left-hand sides of Eqs. (26) and (27) exactly. The statistical-mechanical techniques for calculating them approximately are given in the next section.

3. Probabilistic Computational Algorithms

In this section, we give three approximate optimization algorithms for the energy minimization problem (9), such that ICM [11], CZTP [20] and MFA [15]. Both the ICM and the CZTP are iterative algorithms at zero temperature and the MFA is one at a finite temperature. The approximate minimum-energy configuration obtained by means of an annealing procedure is closer to the true minimum-energy configuration than by means of the ICM.

First we explain the iterative algorithm at zero temperature. The optimization problem (9) can be reduced to the following iterative equation:

$$x_{i,j}(\mathbf{J}) = \arg \min_{x_{i,j} \in \mathbf{Q}} H\left(x_{i,j} \middle| x_{i',j'} = x_{i',j'}(\mathbf{J}), (i',j') \in \mathbf{I} \backslash (i,j) \right). \tag{28}$$

From the iterative equation, we can construct the ICM algorithm [11], which is the simplest algorithms for the optimization (9). Morita and the present author extend it to the cluster version ICM algorithm, which is constructed from the following iterative equation:

$$\mathbf{x_c}(\mathbf{J}) = \underset{\mathbf{x_c}}{\arg\min} H\left(\mathbf{x_c} \middle| x_{i',j'} = x_{i',j'}(\mathbf{J}), (i',j') \in \mathbf{I} \backslash \mathbf{c} \right), \tag{29}$$

where

$$\mathbf{x_c} \equiv \left\{ x_{i,j} \middle| (i,j) \in \mathbf{c} \right\}, \quad \mathbf{x_c}(\mathbf{J}) \equiv \left\{ x_{i,j}(\mathbf{J}) \middle| (i,j) \in \mathbf{c} \right\}.$$

Morita and the present author called it CZTP [20]. Here $\mathbf{c}$ is a set of pixels, and the square CZTP can be constructed by setting $\mathbf{c} = \{(i,j), (i+1,j), (i+1,j+1), (i,j+1)\}$.

Second, we explain the MFA algorithm which is the most familiar deterministic annealing algorithm for the optimization (9). The annealing algorithm is able to avoid local minima. In order to adopt the MFA algorithm for the search of minimum-energy configuration $\mathbf{x}(\mathbf{J})$, we introduce the one-body marginal probability distributions:

$$\rho_{i,j}(n) \equiv \sum_{\mathbf{x}} \rho(\mathbf{x}|\tilde{\mathbf{y}}, T, \mathbf{J}) \delta(x_{i,j}, n) \quad (n \in \mathbf{Q}, \ (i,j) \in \mathbf{I}). \tag{30}$$

In the mean-field approximation, the probability distribution $\rho(\mathbf{x}|\bar{\mathbf{y}}, T, \mathbf{J})$ is approximately expressed as

$$\rho(\mathbf{x}|\bar{\mathbf{y}}, T, \mathbf{J}) \simeq \prod_{(i,j)\in\mathbf{I}} \rho_{i,j}(x_{i,j}). \tag{31}$$

The Gibbs distribution $\rho(\mathbf{x}|\bar{\mathbf{y}}, T, \mathbf{J})$ satisfy the variational principle for the following free energy minimum:

$$\rho(\mathbf{x}|\bar{\mathbf{y}}, T, \mathbf{J}) = \min_f \mathcal{F}[f], \tag{32}$$

where

$$\mathcal{F}[f] \equiv \sum_{\mathbf{x}} f(\mathbf{x}) \Big(H(\mathbf{x}|\bar{\mathbf{y}}, T, \mathbf{J}) + T\ln(f(\mathbf{x})) \Big), \tag{33}$$

under the normalization condition for $\rho(\mathbf{x}|\bar{\mathbf{y}}, T, \mathbf{J})$. By substituting Eq.(31) into the free energy $\mathcal{F}[\rho]$ and by taking the first variation of the free energy $\mathcal{F}[\{\rho_{i,j}\}]$ with respect to $\rho_{i,j}(n)$ under the normalization condition for the marginal probability distribution, the deterministic mean-field equations for the set of one-body marginal distribution functions $\{\rho_{i,j}(n)|(i,j)\in\mathbf{I}, n\in\mathbf{Q}\}$ are obtained as follows:

$$x_{i,j}(\mathbf{J}) = \operatorname*{argmax}_{n\in\mathbf{Q}} \rho_{i,j}(n), \tag{34}$$

where

$$\rho_{i,j}(n) = \frac{\exp\left(-\frac{1}{T}H_{i,j}(n)\right)}{\sum_{m\in\mathbf{Q}}\exp\left(-\frac{1}{T}H_{i,j}(m)\right)}, \tag{35}$$

$$H_{i,j}(n) = -\delta(n, y_{i,j}) + \sum_{(i',j')\in\mathbf{c}_{i,j}}\sum_{m=1}^{k-1} J_m\Big(\rho_{i',j'}(n+m) + \rho_{i',j'}(n-m)\Big)$$

$$+ J_k \sum_{(i',j')\in\mathbf{c}_{i,j}}\sum_{m=k}^{q-1}\Big(\rho_{i',j'}(n+m) + \rho_{i',j'}(n-m)\Big), \tag{36}$$

$$\mathbf{c}_{i,j} = \Big\{(i+1,j),(i-1,j),(i,j+1),(i,j-1)\Big\}. \tag{37}$$

By solving Eqs.(34)-(37) at a sufficiently small positive value of T by using the annealing procedure, we obtain the approximate optimal solution of Eq.(9).

We give also the deterministic method to calculate the left-hand sides of Eqs. (26) and (27) by using the mean-field approximation in high accuracy.

The mean field equation for the probabilistic model described by $\Pr\{X = x|K = \hat{K}\}$ can be obtained by replacing $H(x)$ by

$$H_{\text{PR}}(x) \equiv \sum_{(i,j)\in\mathbf{I}} \Big(\phi(x_{i,j}, x_{i+1,j}|\mathbf{J}) + \phi(x_{i,j}, x_{i,j+1}|\mathbf{J})\Big). \tag{38}$$

Because the probabilistic model which is given by the *a priori* probability distribution is spatially uniform, the one-body marginal distribution function for the *a priori* probability distribution should be independent of a pixel (i,j) and is defined by

$$\rho^{(1)}(n) \equiv \sum_{\mathbf{x}} \delta(x_{i,j}, n)\Pr\Big\{X = x\Big|K\Big\} \quad (n\in\mathbf{Q}), \tag{39}$$

The deterministic mean-field equations for the one-body marginal distribution function $\rho^{(1)}(n)$ in the probabilistic model with the energy function $H_{\text{PR}}(x)$ are given as

$$\rho^{(1)}(n) = \frac{\exp\Big(-\frac{4}{T}\sum_{m\in\mathbf{Q}}\phi(n,m)\rho^{(1)}(m)\Big)}{\sum_{n\in\mathbf{Q}}\exp\Big(-\frac{4}{T}\sum_{m\in\mathbf{Q}}\phi(n,m)\rho^{(1)}(m)\Big)} \quad (n\in\mathbf{Q}). \tag{40}$$

We can solve the simultaneous nonlinear equations by the numerical iteration method at any value of T for fixed hyperparameters $\mathbf{J}$. The left-hand sides of Eqs. (26) and (27) are approximately expressed as

$$\sum_{\mathbf{x}} \sigma_2(\mathbf{x})\Pr\Big\{X = x\Big|K_l = \frac{J_k - J_l}{T} \ (l = 1, 2, \cdots, k-1), K_k = \frac{J_k}{T}\Big\}$$
$$= |\mathbf{B}| \sum_{m\in\mathbf{Q}}\sum_{m'\in\mathbf{Q}} (1 - \delta(m, m'))\rho^{(1)}(m)\rho^{(1)}(m'), \tag{41}$$

$$\sum_{\mathbf{x}} \sigma_{2n}(\mathbf{x})\Pr\Big\{X = x\Big|K_l = \frac{J_k - J_l}{T} \ (l = 1, 2, \cdots, k-1), K_k = \frac{J_k}{T}\Big\}$$
$$= |\mathbf{B}| \sum_{m\in\mathbf{Q}}\sum_{m'\in\mathbf{Q}} \delta(|m - m'|, n)\rho^{(1)}(m)\rho^{(1)}(m'). \tag{42}$$

By substituting the solution $\rho^{(1)}(n)$ of Eq. (40) to Eqs. (41) and (42), we can calculate the left-hand sides of Eqs. (26) and (27) at $T = \hat{T}_p$.

4. Numerical Experiments

In this section, we give some numerical experiments for the original image $\hat{\mathbf{x}}$ given in Fig.1a. The degradation process is subject to the probability given in Eq.(1). Here, we set $(q-1)\tilde{p} = 0.1, 0.3$ and 0.5 where $q = 8$ and $M = N = 64$.

(a)   (b)

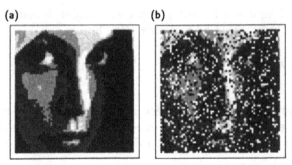

Figure 1: Original and degraded images. (a) Original image $\tilde{\mathbf{x}}$ ($\sigma_2(\tilde{\mathbf{x}}) \simeq 0.2764$, $\sigma_2(\tilde{\mathbf{x}}) \simeq 0.2390$). (b) Degraded image $\tilde{\mathbf{y}}$ ($(q-1)\tilde{p} = 0.3$, $\bar{d}(\tilde{\mathbf{x}}, \mathbf{y}) \simeq 0.2937$).

For $(q-1)\tilde{p} = 0.3$, the degraded images $\tilde{\mathbf{y}}$ obtained from the original image $\tilde{\mathbf{x}}$ in Fig.1a are given in Fig.1b. The quantities

$$\bar{\sigma}_{2n}(\tilde{\mathbf{x}}) \equiv \frac{1}{|\mathbf{B}|}\sigma_{2n}(\tilde{\mathbf{x}}) \quad (n = 0, 1, \cdots, q-1),$$

are shown in Table 1 where $\sigma_{20}(\tilde{\mathbf{x}}) \equiv 1 - \sigma_2(\tilde{\mathbf{x}})$. We see that the quantities $\sigma_{20}(\tilde{\mathbf{x}})$ and $\sigma_{21}(\tilde{\mathbf{x}})$ are especially important in images with 8 grades. In the energy function (10), we set $k = 2$. In Table 2a, we give the values of optimal parameters $\hat{J}_1$ and $\hat{J}_2$, which are estimated by means of the constrained optimization in Eqs.(9), (13) and (14) and of quantities

$$\bar{d}(\hat{\mathbf{x}}, \tilde{\mathbf{x}}) \equiv \frac{1}{|\mathbf{I}|}d(\hat{\mathbf{x}}, \tilde{\mathbf{x}}), \quad \bar{\sigma}_2(\hat{\mathbf{x}}) \equiv \frac{1}{|\mathbf{B}|}\sigma_2(\hat{\mathbf{x}}), \quad \bar{\sigma}_{21}(\hat{\mathbf{x}}) \equiv \frac{1}{|\mathbf{B}|}\sigma_{21}(\hat{\mathbf{x}}),$$

in the restored image $\hat{\mathbf{x}}$, which is obtained for $(q-1)\tilde{p} = 0.1$, 0.3 and 0.5 by using the ICM, the CZTP and the MFA. For $(q-1)\tilde{p} = 0.3$, the obtained restored images $\hat{\mathbf{x}}$ are shown in Fig.2. In Table 2b, we give the values of optimal parameters $\hat{J}_1$ and $\hat{J}_2$, which are estimated by means of the generalized ML estimation in Eqs.(25), (26) and (27) with the mean field approximation, and of quantity $\bar{d}(\hat{\mathbf{x}}, \tilde{\mathbf{x}})$ in the restored image $\hat{\mathbf{x}}$, which is obtained for $(q-1)\tilde{p} = 0.1$, 0.3 and 0.5 by using the ICM, the CZTP and the MFA. For $(q-1)\tilde{p} = 0.3$, the obtained restored images $\hat{\mathbf{x}}$ are shown in Fig.3. In the CZTP, we set

$$\mathbf{c} \equiv \Big\{(i,j), (i+1,j), (i+1,j+1), (i,j+1)\Big\}.$$

5. Concluding Remarks

In this paper, we summarize the situation of image restorations by means of the MRF model in the standpoint of the constrained optimization [10]

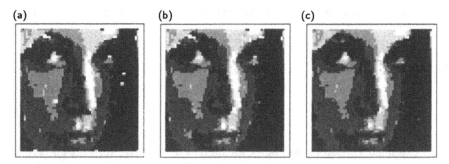

Figure 2: Restored images $\hat{\mathbf{x}}$ obtained for the values of hyperparameters $\hat{J}_1$, $\hat{J}_2$ estimated by using the constrained optimization from the degraded image $\tilde{\mathbf{y}}$ given in Fig. 1b. (a) ICM. (b) CZTP. (c) MFA.

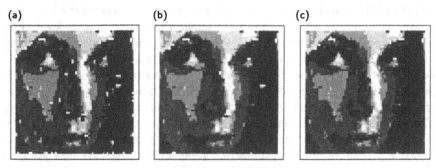

Figure 3: Restored images $\hat{\mathbf{x}}$ obtained for the values of hyperparameters $\hat{J}_1$, $\hat{J}_2$ estimated by using the generalized ML estimation from the degraded image $\tilde{\mathbf{y}}$ given in Fig. 1b. (a) ICM. (b) CZTP. (c) MFA.

and some probabilistic computational algorithms based on the statistical-mechanical techniques for the energy minimization problem formulated for the image restoration. On the other hand, the generalized ML estimation [11, 12] are also reformulated in the standpoint of the constrained optimization.

In the constrained optimization in Ref. [10], both the hyperparameters and the restored image are estimated at the zero-temperature $T = 0$ because the mathematical framework is constructed at the zero-temperature. The temperature is introduced in order to employ the annealing procedure for the search of minimum-energy configuration. In the generalized ML estimation which is reformulated in the present paper, while the restored image are estimated at the zero-temperature $T = 0$, the hyperparameters are determined at a finite temperature because the temperature itself is also hyperparameter. Recently, Nishimori and Wong insisted that the restored image obtained by a maximum posterior marginal (MPM) estimation are closer to the original image than by the MAP estimation (15) by proving an inequality for binary image restorations [5]. This means that the restored image should be searched not at the zero temperature but at a finite temperature. The authors also gave

Table 1: Value of $\bar{\sigma}_{2n}(\tilde{\mathbf{x}})$ for the true original image $\tilde{\mathbf{x}}$ given in Fig. 1a.

n	$\bar{\sigma}_{2n}(\tilde{\mathbf{x}})$	n	$\bar{\sigma}_{2n}(\tilde{\mathbf{x}})$
0	0.7224	4	0.0039
1	0.2351	5	0.0031
2	0.0204	6	0.0033
3	0.0067	7	0.0051

the extended version of inequality for multi-valued image restoration [21].

In some numerical experiments, we show that the algorithms are applicable to the image restoration of natural image. We remark that the ICM algorithm can erase when the noise is in an isolated pixel, but it cannot erase when two or more successive pixels are affected by the noise. On the other hand, the MFA and the CZTP algorithms can deal with this problem. However the MFA need a large memory. If we apply it to the image restoration of gray-level image with 256 grades, we have to treat a $256 \times 256 \times 256$ dimensions for the one-body marginal probability distribution $\rho_{i,j}(n)$. The ICM and CZTP algorithms do not need so many memory because most of the data stored are in integer. The present author are still studying the problem of finding a computer algorithm giving the same results in much shorter time and in much smaller memory. The present author and Horiguchi proposed a new optimization method, in which we replaced the thermal fluctuation by a quantum fluctuation in the iterative procedure of the optimization problem [22]. We are now studying also to extend the quantum iterative algorithm proposed for binary image restoration to more practical algorithms in natural image restorations.

ACKNOWLEDGEMENT

The author is thankful to Professor T. Morita of Department of Computer Science, Nihon University in Japan, and Professor T. Horiguchi of Graduate School of Information Sciences, Tohoku University in Japan, Professor D. M. Titterington of the Department of Statistics, University of Glasgow in UK, for some valuable advises on the manuscript. This work was partly supported by the Grants-In-Aid (No.10780159) for Scientific Research from the Ministry of Education, Science, Sports and Culture of Japan.

References

[1] R. Chellappa and A. Jain (eds), *Markov Random Fields: Theory and Applications*, Academic Press, New York, 1993.

[2] H. Derin, H. Elliott, R. Cristi and D. Geman, "Bayes smoothing algorithms for segmentation of binary images modeled by Markov random fields," *IEEE Trans. Pattern Anal. Mach. Intell.*, vol.6, pp.707-720, 1984.

Table 2: Values of optimal parameters $\hat{J}_1$ and $\hat{J}_2$, which are estimated by means of the constrained optimization and the generalized ML estimation, and values of $\bar{d}(\hat{x},\bar{x})$, $\bar{\sigma}_2(\hat{x})$ and $\bar{\sigma}_{21}(\hat{x})$ in the restored image $\hat{x}$, which is obtained for $(q-1)\tilde{p} = 0.1$, 0.3 and 0.5 by using the ICM, the CZTP and the MFA.

(a) Constrained Optimization.

$(q-1)\tilde{p}$		$\hat{J}_1$	$\hat{J}_2$	$\bar{d}(\hat{x},\bar{x})$	$\bar{\sigma}_2(\hat{x})$	$\bar{\sigma}_{21}(\hat{x})$
	ICM	0.2500	0.5015	0.0427	0.2782	0.2385
0.1	CZTP	0.2500	0.5600	0.0420	0.2750	0.2395
	MFA	0.2541	0.4511	0.0374	0.2772	0.2295
	ICM	0.2533	0.8800	0.1211	0.2769	0.2361
0.3	CZTP	0.2500	0.6200	0.1187	0.2815	0.2356
	MFA	0.2695	0.7631	0.1055	0.2734	0.2394
	ICM	0.4611	1.4611	0.2793	0.2763	0.2416
0.5	CZTP	0.2599	0.8899	0.2327	0.2632	0.2354
	MFA	0.2785	0.7701	0.2210	0.2752	0.2394

(b) Generalized Maximum Likelihood Estimation.

$(q-1)\tilde{p}$		$\hat{J}_1$	$\hat{J}_2$	$\bar{d}(\hat{x},\bar{x})$	$\bar{\sigma}_2(\hat{x})$	$\bar{\sigma}_{21}(\hat{x})$
	ICM	0.1722	0.3799	0.0554	0.3242	0.2515
0.10	CZTP	0.1722	0.3799	0.0491	0.3154	0.2531
	MFA	0.1722	0.3799	0.0479	0.3151	0.2528
	ICM	0.2579	0.5596	0.1345	0.3099	0.2286
0.30	CZTP	0.2579	0.5596	0.1179	0.2756	0.2238
	MFA	0.2579	0.5596	0.1128	0.2900	0.2405
	ICM	0.3712	0.8089	0.2615	0.3047	0.2142
0.50	CZTP	0.3712	0.8089	0.2227	0.2289	0.1891
	MFA	0.3712	0.8089	0.2183	0.2275	0.1899

[3] F. C. Jeng and J. W. Woods, "Compound Gauss-Markov random fields for image estimation," *IEEE Trans. Signal Processing*, vol.39, pp.683-697, 1991.

[4] J. Marroquin, S. Mitter and T. Poggio, "Probabilistic solution of ill-posed problems in computer vision," *Journal of the American Statistical Association*, vol.82, pp.76-89, 1987.

[5] H. Nishimori and K. Y. M. Wong, "Statistical mechanics of image restoration and error-correcting codes," *Physical Review E*, vol.60, pp.132-144, 1999.

[6] K. Tanaka, "Statistical-mechanical method in image restoration," *BUTSURI*, vol.54, pp.25-33, 1999 (in Japanese).

[7] T. Morita and K. Tanaka, "Determination of parameters in an image recovery of statistical-mechanical means," *Physica A*, vol.223, p.244-262, 1996.

[8] K. Tanaka, M. Ichioka and T. Morita, "Statistical-mechanical iterative method in image restoration," *Trans. Inst. Electron. Inf. Commun. Eng. A*, vol.J80-A, pp.280-286, 1997 (in Japanese).

[9] K. Tanaka, M. Ichioka and T. Morita, "Statistical-mechanical algorithm in MRF model based on variational principle," *Proceedings of 13th International Conference on Pat-*

tern Recognition, vol.II, Track B: Pattern Recognition and Signal Analysis, Vienna, pp.381-388, 1996 (IEEE Computer Society Press).

[10] K. Tanaka and T. Morita, "Statistical-mechanical iterative algorithm for image restoration of a gray level image," *Trans. Inst. Electron. Inf. Commun. Eng. A*, vol.J80-A, pp.1033-1037, 1997 (in Japanese).

[11] J. Besag, "On the statistical analysis of dirty picture (with discussion)," *J. Royal Statistical Society Ser. B*, vol.48, pp.259-302, 1986.

[12] S. Lakshmanan and H. Derin, "Simultaneous parameter estimation and segmentation of Gibbs random fields using simulated annealing," *IEEE Trans. Pattern Anal. and Mach. Intel.*, vol.11, pp.799-813, 1989.

[13] W. Qian and D. M. Titterington, "Stochastic relaxations and EM algorithms for Markov random fields," *J. Statist. Comput. Simul.*, vol.40, pp.55-69, 1991.

[14] C.-h. Wu and P.C.Doerschk, "Cluster expansions for the deterministic computation of Bayesian estimators based on Markov random fields," *IEEE Trans. Pattern Anal. Mach. Intel.*, vol.17, pp.275, 1995.

[15] D. Geiger and F. Girosi, "Parallel and deterministic algorithms from MRF's: surface reconstruction," *IEEE Trans. Pattern Anal. Mach. Intell.*, vol.13, pp.401-412, 1991.

[16] J. Zhang, "The mean field theory in EM procedures for Markov random fields," *IEEE Trans. Signal Processing*, vol.40, pp.2570-2583, 1992.

[17] J. Zhang, "The application of the Gibbs-Bogoliubov-Feynman inequality in mean field calculations for Markov random fields," *IEEE Trans. Image Processing*, vol.5, pp.1208-1214, 1996.

[18] T. Morita, "General structure of the distribution functions for the Heisenberg model and the Ising model," *J. Math. Phys.*, vol.13, pp.115-123, 1972.

[19] K. Tanaka and T. Morita, "Cluster variation method and image restoration problem," *Physics Letters A*, vol.203, pp.122-128, 1995.

[20] T. Morita and K. Tanaka, "Cluster ZTP in the recovery of an image," *Pattern Recognition Letters*, vol.18, pp.1479-1493, 1997.

[21] K. Tanaka, "Maximum Posterior Marginal Estimate in Statistical Method for Image Restoration," *Trans. Inst. Electron. Inf. Commun. Eng. A*, vol.J82-A, no.10, 1999 (in Japanese), to appear.

[22] K. Tanaka and T. Horiguchi, "Quantum statistical-mechanical iterative method in image restoration," *Trans. Inst. Electron. Inf. Commun. Eng. A*, vol.J80-A, pp.2117-2126, 1997 (in Japanese).

Bayesian Neural Networks: Case Studies in Industrial Applications

Aki Vehtari[1] and Jouko Lampinen[1]

[1]Laboratory of Computational Engineering, Helsinki University of Technology,
P.O.Box 9400, FIN-02015 HUT, FINLAND

Abstract: We demonstrate the advantages of using Bayesian neural networks in industrial applications. The Bayesian approach provides consistent way to do inference by combining the evidence from data to prior knowledge from the problem. A practical problem with neural networks is to select the correct complexity for the model, i.e., the right number of hidden units or correct regularization parameters. The Bayesian approach offers efficient tools for avoiding overfitting even with very complex models, and facilitates estimation of the confidence intervals of the results. In this contribution we review the Bayesian methods for neural networks and present comparison results from case studies in prediction of the quality properties of concrete (regression problem), electrical impedance tomography (inverse problem) and forest scene analysis (classification problem). The Bayesian neural networks provided consistently better results than other methods.

1 Introduction

In classification and non-linear function approximation neural networks have become very popular in recent years. With neural networks the main difficulty is in controlling the complexity of the model. Another problem of standard neural network models is the lack of tools for analyzing the results (confidence intervals, like 10 % and 90 % quantiles, etc.).

The Bayesian approach provides consistent way to do inference by combining the evidence from data to prior knowledge from the problem. Bayesian methods use probability to quantify uncertainty in inferences and the result of Bayesian learning is a probability distribution expressing our beliefs regarding how likely the different predictions are. Predictions are made by integrating over the posterior distribution. The main advantages of using Bayesian methods are:

- Automatic complexity control: Values of regularization coefficients can be selected using only the training data, without the need to use separate training and validation data.
- Possibility to use prior information and hierarchical models for the hyperparameters.
- Predictive distributions for outputs.

In this contribution we demonstrate the advantages of Bayesian neural networks [1] in three case problems. First we briefly review Bayesian methods for MLP neural networks in section 2. In sections 3, 4 and 5 we present results using Bayesian neural network in prediction of the quality properties of concrete (regression problem), electrical impedance tomography (inverse problem) and forest scene analysis (classification problem).

2 Bayesian Learning for MLP

Bayesian methods can be used for many types of neural networks, but we concentrate here to one hidden layer MLP network with hyperbolic tangent (tanh) activation function. See [2] for thorough introduction to MLP. Basic MLP network model with m hidden units and k outputs is

$$f_k(\mathbf{x}, \mathbf{w}) = w_{k0} + \sum_{j=1}^{m} w_{kj} \tanh \left(w_{j0} + \sum_{i=1}^{d} w_{ji} x_i \right), \tag{1}$$

where $\mathbf{x}$ is a d-dimensional input vector, $\mathbf{w}$ denotes the weights, and indices i and j correspond to hidden and output units, respectively.

Next we review main ideas of Bayesian learning briefly. See, e.g., [3] for good introduction to Bayesian methods.

2.1 Bayesian Learning

Consider a regression or classification problem involving the prediction of a noisy vector $\mathbf{y}$ of target variables given the value of a vector $\mathbf{x}$ of input variables.

The process of Bayesian learning is started by defining a model, $\mathcal{M}$, and prior distribution $p(\theta)$ for the model parameters θ. Prior distribution expresses our initial beliefs about parameter values, before any data has observed. After observing new data $D = \{(\mathbf{x}^{(1)}, \mathbf{y}^{(1)}), \ldots, (\mathbf{x}^{(n)}, \mathbf{y}^{(n)})\}$, prior distribution is updated to the posterior distribution using Bayes' rule

$$p(\theta|D) = \frac{p(D|\theta)p(\theta)}{p(D)} \propto L(\theta|D)p(\theta), \tag{2}$$

where the likelihood function $L(\theta|D)$ gives the probability of the observed data as function of the unknown model parameters.

To predict the new output $\mathbf{y}^{(n+1)}$ for new input $\mathbf{x}^{(n+1)}$, predictive distribution is obtained by integrating the predictions of the model with respect to the posterior distribution of the model parameters

$$p(\mathbf{y}^{(n+1)}|\mathbf{x}^{(n+1)}, D) = \int p(\mathbf{y}^{(n+1)}|\mathbf{x}^{(n+1)}, \theta)p(\theta|D)d\theta. \tag{3}$$

2.2 Models

Statistical model is defined with the likelihood function. If we assume that the n data points $(\mathbf{x}^{(i)}, \mathbf{y}^{(i)})$ are exchangeable we get

$$L(\theta|D) = \prod_{i=1}^{n} p(\mathbf{y}^{(i)}|\mathbf{x}^{(i)}, \theta). \tag{4}$$

The term $p(\mathbf{y}^{(i)}|\mathbf{x}^{(i)}, \theta)$ in (4) depends on our problem. In regression problems, it is generally assumed that the distribution of target data can be described by a deterministic function of inputs, corrupted by additive Gaussian noise of a constant variance. Probability density for a target y_j is then

$$p(y_j|\mathbf{x}, \mathbf{w}, \sigma) = \frac{1}{\sqrt{2\pi}\sigma_j} \exp\left(-\frac{1}{2\sigma_j^2}[y_j - f_j(\mathbf{x}, \mathbf{w})]^2\right), \tag{5}$$

where σ_j^2 is the noise variance for the target. See [1] for per-case normal noise variance model and [4] for full covariance model assuming correlating residuals. For a two class classification (logistic regression) model, the probability that a binary-valued target, y_j, has the value 1 is

$$p(y_j = 1|\mathbf{x}, \mathbf{w}) = [1 + \exp(-f_j(\mathbf{x}, \mathbf{w}))]^{-1} \tag{6}$$

and for many class classification (softmax) model, the probability that a class target, y, has value j is

$$p(y = j|\mathbf{x}, \mathbf{w}) = \frac{\exp(f_j(\mathbf{x}, \mathbf{w}))}{\sum_k \exp(f_k(\mathbf{x}, \mathbf{w}))}. \tag{7}$$

In (5), (6) and (7) function $f_j(\mathbf{x}, \mathbf{w})$ is in this case an MLP network. Traditionally in many methods one of the problems has been to find a good topology for the MLP. In Bayesian approach we could use infinite number of hidden units. We do not need to restrict the size of the MLP based on the size of the training set, but in practice, we will have to use finite number of hidden units due to computational limits [1].

2.3 Priors

Next, we have to define the prior information about our model parameters, before any data has been seen. Usual prior is that the model has some unknown complexity but the model is not constant or extremely flexible. To express this prior belief we set hierarchical model specification.

For small weights the network mapping (1) is almost linear and has low effective complexity, since the central region of sigmoidal activation function can be approximated by linear transformation. The complexity of the MLP network can be controlled by controlling the size of the weights $\mathbf{w}$. This can be achieved by using e.g. Gaussian prior distribution for weights **w** given hyperparameter α

$$p(\mathbf{w}|\alpha) = (2\pi)^{-m/2} \alpha^{m/2} \exp(-\alpha \sum_{i=1}^{m} w_i^2/2). \tag{8}$$

This prior states that smaller weights are more probable, but how much more is determined by the value of hyperparameter α. Since we do not know the correct value for hyperparameter α, we set a vague hyperprior $p(\alpha)$ expressing our belief that complexity controlled by α is unknown but the model is not constant or extremely flexible. A convenient form for this hyperprior is vague Gamma distribution with mean μ and shape parameter a

$$p(\alpha) \sim \mathrm{Gamma}(\mu, a) \propto \alpha^{a/2-1} \exp(-\alpha a/2\mu). \tag{9}$$

In order to have prior for weights which is invariant under the linear transformations of data, separate priors (each having its own hyperparameters α_i) for different weight groups in each layer of a MLP are used [1].

In MLP networks, the weights from less important inputs are typically smaller than weights from more important inputs[1]. Prior belief that some inputs are likely to be more relevant than others can be implemented by using different priors for weight groups from each input, and hierarchical hyperpriors for these priors. The posteriors for hyperparameters should then adjust according to relevance of the inputs. This prior is called Automatic Relevance Determination (ARD) [5, 1, 6].

For regression models we need prior for noise variance σ in (5), which is often specified in terms of corresponding precision, $\tau = \sigma^{-2}$. As for α, our prior information is usually quite vague, stating that noise variance σ is not zero or extremely large. This prior can be expressed with vague Gamma-distribution with mean μ and shape parameter a

$$p(\tau) \sim \mathrm{Gamma}(\mu, a) \propto \tau^{a/2-1} \exp(-\tau a/2\mu). \tag{10}$$

2.4 Prediction

After defining the model and prior information, we combine the evidence from the data to get the posterior distribution for the parameters

$$p(\mathbf{w}, \alpha, \tau|D) \propto L(\mathbf{w}, \alpha, \tau|D) p(\mathbf{w}, \alpha, \tau). \tag{11}$$

Predictive distribution for new data is then obtained by integrating over

---

[1]Note that in the non-linear network the effect of an input may be small even if the weights from it are large and vice verse, but in general the size of the weights roughly reflects the relevance of the input.

this posterior distribution

$$p(\mathbf{y}^{(n+1)}|\mathbf{x}^{(n+1)}, D) = \int p(\mathbf{y}^{(n+1)}|\mathbf{x}^{(n+1)}, \mathbf{w}, \alpha, \tau) p(\mathbf{w}, \alpha, \tau|D) \, d\mathbf{w}\alpha\tau. \quad (12)$$

We can also evaluate expectations of various functions with respect to the posterior distribution for parameters. For example in regression we may evaluate the expectation for a component of $\mathbf{y}^{(n+1)}$

$$\hat{\mathbf{y}}_k^{(n+1)} = \int f_k(\mathbf{x}^{(n+1)}, \mathbf{w}) p(\mathbf{w}, \alpha, \tau|D) \, d\mathbf{w}\alpha\tau, \quad (13)$$

which corresponds to the best guess with squared error loss.

The posterior distribution for the parameters $p(\mathbf{w}, \alpha, \tau|D)$ is typically very complex, with many modes. Evaluating the integral of (13) is therefore a difficult task. The integral can be approximated with parametric approximation as in [7] or with numerical approximation as described in next section.

2.5 Markov Chain Monte Carlo method

Neal has introduced implementation of Bayesian learning for neural networks in which the difficult integration of (13) is performed using Markov Chain Monte Carlo (MCMC) methods [1]. In [8] there is a good introduction to basic MCMC methods and many applications in statistical data analysis.

MCMC methods make no assumptions about the form of the posterior distribution. They may in some circumstances require a very long time to converge to the desired distribution.

The integral of (13) is the expectation of function $f_k(\mathbf{x}^{(n+1)}, \mathbf{w})$ with respect to the posterior distribution of the parameters. This and other expectations can be approximated by Monte Carlo method, using a sample of values $\mathbf{w}^{(t)}$ drawn from the posterior distribution of parameters

$$\hat{\mathbf{y}}_k^{(n+1)} \approx \frac{1}{N} \sum_{t=1}^{N} f_k(\mathbf{x}^{(n+1)}, \mathbf{w}^{(t)}). \quad (14)$$

Note that samples from the posterior distribution are drawn during the "learning phase" and predictions for new data can be calculated quickly using the same samples and (14).

In the MCMC, samples are generated using a Markov chain that has the desired posterior distribution as its stationary distribution. Difficult part is to create Markov chain which converges rapidly and in which states visited after convergence are not highly dependent.

When the amount of data increases, the evidence from data causes the probability mass to concentrate to the smaller area and we need less samples from the posterior distribution. Also less samples are needed to evaluate the mean of the predictive distribution than the tail-quantiles like, 10% and 90%

quantiles. So depending on the problem 10–200 samples may be enough for practical purposes (given that samples are not too highly dependent).

In our examples of Bayesian learning for neural networks with MCMC we have used Flexible Bayesian Modeling (FBM) software[2], which implements the methods described in [1].

3 Case I: Concrete Quality Estimation

The goal of the project was to develop a model for predicting the quality properties of concrete. The quality variables contained, e.g., compression strengths and densities for 1, 28 and 91 days after casting, bleeding (water extraction) and spread and slump that measure softness of the fresh concrete. These quality measurements depend on the properties of the stone material (natural or crushed, size and shape distributions of the grains, mineralogical composition), additives, and the amount of cement and water. In the study we had 7 target variables and 19 explanatory variables.

Collecting the samples for statistical modeling is rather expensive in this application, as each sample requires preparation of the sand mixture, casting the test pieces and waiting for 91 days for the final tests. Thus available samples must be used as efficiently as possible, which makes Bayesian techniques tempting alternative, as they allow fine balance of prior assumptions and evidence from samples. In the study we had 149 samples designed to cover the practical range of the variables, collected by a concrete manufacturer company.

MLP networks containing 6 hidden units were used. Different MLP models tested were:

MLP ESC : Early stopping committee of 20 MLP networks, with different division of data to training and stopping sets for each member. The networks were initialized to near zero weights to guarantee that the mapping is smooth in the beginning. (see [2, 6, 9] for more about MLP ESC).

Bayes MLP : Bayesian neural network with FBM-software, using vague priors and MCMC-run specifications similar as used in [1, 6]. 20 networks from the posterior distribution of network parameters were used.

Bayes MLP +ARD: Similar Bayesian neural network to the previous, but using also the ARD prior.

Error estimates for predicting the slump are collected in Table 1. Note that use of ARD prior gives much better results.

4 Case II: Electrical Impedance Tomography

In this section we report results on using Bayesian neural networks for solving the ill-posed inverse problem in electrical impedance tomography, EIT. The full report of the proposed approach is presented in [10].

---

[2]<URL:http://www.cs.toronto.edu/~radford/fbm.software.html>

Table 1: Ten fold cross-validation error estimates for predicting the slump of concrete.

Method	Root mean square error
MLP ESC	37
Bayes MLP	34
Bayes MLP +ARD	27

Figure 1: Example of image reconstructions with MLP ESC (upper row) and the Bayesian MLP (lower row)

The aim in EIT is to recover the internal structure of an object from surface measurements. Number of electrodes are attached to the surface of the object and current patterns are injected from through the electrodes and the resulting potentials are measured. The inverse problem in EIT, estimating the conductivity distribution from the surface potentials, is known to be severely ill-posed, thus some regularization methods must be used to obtain feasible results [11].

In [10] we proposed a novel feedforward solution for the reconstruction problem. The approach is based on transformation of both input and output data by principal component projection and application of the neural network in this lower dimensional eigenspace.

The reconstruction was based on 20 principal components of the 128 dimensional potential signal and 60 eigenimages with resolution 41 × 41 pixels. The training data consisted of 500 simulated bubble formations with one to ten overlapping circular bubbles in each image. To compute the reconstructions MLP networks containing 30 hidden units were used. Models tested were *MLP ESC* and *Bayes MLP* (see section 3).

Fig. 1 shows examples of the image reconstruction results. Table 2 shows the quality of the image reconstructions with models, measured by error in the void fraction and percentage of erroneous pixels in the segmentation, over the test set.

An important goal in the studied process tomography application was to estimate the void fraction, which is the proportion of gas and liquid in the image. With the proposed approach such goal variables can be estimated directly without explicit reconstruction of the image. The last column in

Table 2: Errors in reconstructing the bubble shape and estimating the void fraction from the reconstructed images. See text for explanation of the models.

Method	Classifica-tion error %	Relative error in VF %	Relative error in direct VF %
MLP ESC	6.7	8.7	3.8
Bayes MLP	5.9	8.1	3.4

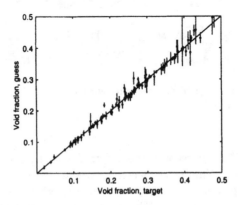

Figure 2: Scatterplot of the void fraction estimate with 10% and 90% quantiles.

Table 2 shows the relative absolute error in estimating the void fraction directly from the projections of the potential signals.

With Bayesian methods we can easily calculate confidence intervals for outputs. Fig. 2 shows the scatter plot of the void fraction versus the estimate by the Bayesian neural network. The 10% and 90% quantiles are computed directly from the posterior distribution of the model output.

See [10] for results for effect of additive Gaussian noise to the performance of the method.

5 Case III: Forest Scene Analysis

In this section we report results of using Bayesian neural networks for classification of forest scenes, to accurately recognize and locate the trees from any background.

Forest scene classification task is demanding due to the texture richness of the trees, occlusions of the forest scene objects and diverse lighting conditions

Table 3: CV error estimates for forest scene classification. See text for explanation of the different models.

	Error%
KNN LOOCV	20
CART	30
MLP ESC	13
Bayes MLP	12
Bayes MLP +ARD	11

under operation. This makes it difficult to determine which are optimal image features for the classification. A natural way to proceed is to extract many different types of potentially suitable features.

In [12] we extracted total of 84 statistical and Gabor features over different sized windows at each spectral channel. Due to great number of features used, many classifier methods would suffer from the curse of dimensionality, but Bayesian neural networks manage well in high dimensional problems.

The image data for teaching and testing of the classifiers were collected by using an ordinary digital camera in varying weather conditions. Ideal weather conditions were not searched, as the aim was to test the viability and the robustness of the methods. Total of 48 images were taken. The labeling of the image data was done by hand via identifying many types of tree and background image blocks with different textures and lighting conditions. In this study only pines were considered.

In addition to 20 hidden unit MLP models *MLP ESC*, *Bayes MLP* and *Bayes MLP +ARD* (see section 3) the models tested were:

KNN LOOCV : K-nearest-neighbor. K is chosen by leave-one-out cross-validation on the training set.

CART : Classification And Regression Tree [13].

Eight folded cross-validation (CV) error estimates are collected in Table 3. Fig. 3 shows example image classified with different methods.

6 Summary discussion

Reviewed case problems illustrate the advantages of using Bayesian neural networks in industrial applications. The Bayesian approach contains automatic complexity control and it gives the predictive distributions for outputs, which can be used to estimate reliability of the predictions.

Acknowledgments

This study was partly funded by TEKES Grant 40888/97 (Project *PROMISE, Applications of Probabilistic Modeling and Search*).

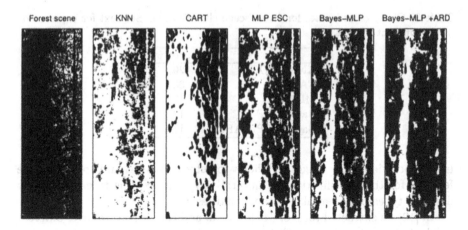

Figure 3: Examples of classified forest scene. See text for explanation of the different models.

References

[1] Neal, R. M., 1996. *Bayesian Learning for Neural Networks*. Springer-Verlag.

[2] Bishop, C. M., 1995. *Neural Networks for Pattern Recognition*. Oxford University Press.

[3] Gelman, A., Carlin, J. B., Stern, H. S. *et al.*, 1995. *Bayesian Data Analysis*. Chapman & Hall.

[4] Vehtari, A. & Lampinen, J., 1999. Bayesian neural networks with correlating residuals. In *Proceedings of the IJCNN'99*.

[5] MacKay, D. J. C., 1994. Bayesian non-linear modelling for the prediction competition. In *ASHRAE Transactions, V.100, Pt.2*, pp. 1053–1062.

[6] Neal, R. M., 1998. Assessing relevance determination methods using DELVE. In C. M. Bishop, ed., *Neural Networks and Machine Learning*. Springer-Verlag.

[7] MacKay, D. J. C., 1992. A practical Bayesian framework for backpropagation networks. *Neural Computation*, 4(3):pp. 448–472.

[8] Gilks, W., Richardson, S. & Spiegelhalter, D., eds., 1996. *Markov Chain Monte Carlo in Practice*. Chapman & Hall.

[9] Vehtari, A. & Lampinen, J., 1999. Bayesian neural networks for industrial applications. In *Proceedings of SMCIA'99*, pp. 63–68.

[10] Lampinen, J., Vehtari, A. & Leinonen, K., 1999. Using Bayesian neural network to solve the inverse problem in electrical impedance tomography. In *Proceedings of SCIA'99*, pp. 87–94.

[11] Vauhkonen, M., Kaipio, J. P., Somersalo, E. *et al.*, 1997. Electrical impedance tomography with basis constraints. *Inverse Problems*, 13(2):pp. 523–530.

[12] Vehtari, A., Heikkonen, J., Lampinen, J. *et al.*, 1998. Using Bayesian neural networks to classify forest scenes. In vol. 3522 of *Proceedings of SPIE*, pp. 66–73.

[13] Breiman, L., Friedman, J., Olshen, R. *et al.*, 1984. *Classification and regression trees*. Chapman and Hall.

Chapter 5: Hybrid Methods, Chaos, and Immune Networks

Papers:

Multivariable Predictive Control Based on Neural Network Model and Simplex-Evolutionary Hybrid Optimization
L. S. Coelho and A. A. R. Coelho

Permeability Prediction in Petroleum Reservoir using a Hybrid System
Y. Huang, P. M. Wong, and T. D. Gedeon

A Performance Comparison of Chaotic Simulated Annealing Models for Solving the N-Queen Problem
T. Kwok and K. A. Smith

Study on the Idiotypic Network Model for the Feature Extraction of Patterns
T. Shimooka, Y. Kikuchi, and K. Shimizu

Multivariable Predictive Control Based on Neural Network Model and Simplex-Evolutionary Hybrid Optimization

Leandro dos Santos Coelho and Antonio Augusto Rodrigues Coelho

Federal University of Santa Catarina, Department of Automation and Systems
P.O. Box 476, ZIP CODE 88040.900, Florianópolis, SC, BRAZIL
E-mail: {lscoelho, aarc}@lcmi.ufsc.br

Abstract: Generalized predictive controller (*GPC*) has robust performance when faced to control non-minimum phase and open-loop unstable processes. *GPC* has been originally developed with linear predictor models which leads to a formulation that can be solved analytically. If a nonlinear model is used then nonlinear optimization and nonlinear modeling techniques are necessary. This paper presents a new method of multivariable *GPC* design for nonlinear systems. The proposed *GPC* design incorporates the neural network model based on radial basis function neural network and optimization methodology of cost function by Lamarckian hybrid method using evolutionary programming and Nelder-Mead's simplex algorithm. Steps of the control law implementation and intelligent procedure are discussed. Simulation results on a nonlinear *MIMO* (multi-input multi-output) system are presented to show the effectiveness of the neuro-evolutionary *GPC* method. The new *GPC* design is applicable for improve the approximation of nonlinear systems with unknown interactor matrix using neural network model and an efficient control law optimization task by simplex-evolutionary hybrid methodology.

1. Introduction

In recent years, as one of the modern computer optimization control techniques, Model-Based Predictive Control (*MBPC*) has achieve a significant level of acceptability and great development in control theory and practical process control applications. *MBPC* algorithms are versatile and robust in applications where processes are difficult to control, with features such as open-loop unstable, having nonminimum phase response, and with dead-time. Several *MBPC* methods have evolved — Dynamic Matrix Control, Extended Predictive Self Adaptive Control, Model Algorithm Control, Extended Horizon Adaptive Control, *GPC*, and others — but techniques have a common underlying framework. The main differences between methods are the ways by which future predictions are calculated, namely, the type of model used and how the cost function is defined. These control techniques have been widely used in refining, petrochemicals, chemicals, gas, pulp and paper areas and other industrial processes around the world [1],[2]. Most practical multivariable processes are inherently nonlinear. However, a majority of controller design approaches are based on linear models. Although linear-model-based-control methods are well established, they do not give satisfactory control performance when the controlled process is highly nonlinear. Some nonlinear model-based design techniques have been developed with complete knowledge of the controlled process. Unfortunately, the nonlinearities of the controlled process are usually complex. Issues such as stability of multivariable systems under certain feedback control action, decoupling, multivariable time-delay compensation, optimal closed-loop response and robustness of *MIMO* systems are intimately related to the structure of the process transfer matrix [3]. It is difficult for the designer to derive the interactor matrix of the process that is a generalization of the time delay presents in monovariable plant cases.

Recently, there has been considerable interest in the application of computational intelligence paradigms [4], such as genetic algorithms [5], fuzzy systems [6] and artificial neural networks [7] to nonlinear predictive control applications. This paper deals with the *GPC* design based on radial basis function neural network (*RBF-NN*) model with optimization by hybrid method using evolutionary programming and Nelder-Mead's simplex algorithm. As an example model, a simulated *MIMO* nonlinear process has been modeled and controlled with the proposed *GPC*. The paper is organized as follows. Section 2 shows the concept of *GPC* and its *MIMO* formulation. In section 3, identification and prediction model concepts by *RBF-NN* are presented. Hybrid evolutionary optimization for *GPC* is described in section 4. The performance of neuro-evolutionary predictive controller is shown in section 5 with a *MIMO* nonlinear process. Conclusion and future works are presented in section 6.

2. Generalized Predictive Control

The *GPC* algorithm minimizes a cost function over a finite prediction horizon. Consider the following quadratic cost function

$$J(N_1, N_2, N_u, k) = \varepsilon \left\{ \sum_{i=k+N_1}^{N_2} \left\| \hat{y}(k+i) - y_r(k+i) \right\|_{\Gamma_e}^2 + \sum_{i=1}^{N_u} \left\| \Delta u(k+i-1) \right\|_{\Gamma_u}^2 \right\} \quad (1)$$

where N_1 and N_2 are initial and output prediction horizons, N_u is the control horizon, $y(k)$ is the process output, $y_r(k)$ is the setpoint, $\hat{y}(k)$ is the predicted process output, Γ_e and Γ_u are weighting matrices of output and control, Δu is the incremental control derived as

$$\Delta u(k) = u(k) - u(k-1) \quad (2)$$

GPC utilizes the system model to predict the future behavior of the process into the controller design procedure. This *GPC* design and implementation usually comprises:

- a model, which describes the process, often a discrete linear system model obtained experimentally;

- a prediction equation, defined by an objective function usually quadratic, to assess future process output errors and controls. This is run forward for a fixed number of time steps to predict the process behavior;
- a known future reference trajectory;
- optional constraints on the system and control variables;
- an optimization procedure.

Most nonlinear control techniques are based on models, and thus their success depends substantially on the quality of the used model. This is especially true in the case of model-based predictive control, where a good dynamic model of the process is essential. The quality of the process model affects the accuracy of a prediction. With a linear plant there are techniques available to make modeling easier [1], but when the process is nonlinear this task can become difficult. For nonlinear process, the ability of the *GPC* to make accurate predictions can be enhanced if an artificial neural network (*ANN*) is used to learn the dynamic of the process instead of standard identification methodologies. Another question is how to select an adequate optimization method. Evolutionary algorithms (*EAs*) are optimization methods and their ability to find the optimum of functions where classical methods have difficulties (e.g. non derivative functions), is one of the most important characteristics of this methodology [5]. The conventional iterative optimization methods are very sensitive to the initialization of the algorithm and usually lead to unacceptable solutions due to convergence to local optimal. Next, the prediction model based on *RBF-NN* and simplex-evolutionary optimization method of cost function are presented.

2.1 Prediction Model by Radial Basis Function Neural Network

ANNs are relevant techniques to be applied to nonlinear identification and control system applications because of their universal mapping characteristic, learning ability, parallel processing and fault tolerance [8]. An *ANN* architecture for implementing nonlinear multivariable input-output mapping is the *RBF-NN*. The behavior of a discrete dynamic system can be described by a *RBF-NN* as:

$$y(k+1) = f(u(k), u(k-1),..., u(k-n), y(k), y(k-1),..., y(k-m)) \qquad (3)$$

where $f(\cdot)$ is the function that mapped the input and output vectors and can be a linear or a nonlinear function. *RBF-NN* is a network of symmetric basis functions which can be e.g. Gaussian, inverse multiquadratic, thin-plane splines functions and others. *RBF-NN* implement the mapping $f: \Re^n \rightarrow \Re^m$ such that

$$f_k(\vec{x}) = w_{k0} + \sum_{j=1}^{M} w_{kj} \Phi\left(\frac{\|\vec{x} - \vec{\mu}\|}{\sigma_j}\right) = y_k(\vec{x}), \quad 1 \le k \le m \qquad (4)$$

where $\vec{x}$ is an input vector in $\Re^n$, $y_k(\vec{x})$ is the kth network output, w_{kj} denotes the weight connecting hidden node j and output node k, $\vec{\mu}_j$ denotes the jth function center, and $\bar{\sigma}_j$ denotes the jth function width. The utilized basis function is the Gaussian function, i.e.,

$$\Phi(r) = e^{-r^2/2} \tag{5}$$

In this *RBF-NN* the centers are adjusted using the k-means clustering algorithm, the widths through *p-nearest neighbor* technique [9] and weights adjusted by pseudo-inverse technique.

2.2 Optimization by Simplex-Evolutionary Lamarckian Method

EAs are computer-based problem-solving systems based on principles of evolution theory. The interest in *EAs* is increasing very fast, their robust and powerful adaptive search mechanisms. *EAs* have been used in many problems, dealing with multidimensional and multimodal search. There are a variety of evolutionary models that have been proposed and studied, such as genetic algorithms, evolution strategies, evolutionary programming (*EP*), genetic programming, which are referred as *EAs*. They share a common conceptual base of simulating the evolution of individual structures via selection and reproduction procedure. The basic idea is to maintain a population of candidate solutions which evolve under selective pressure that favors better solutions. *EP* is a stochastic optimization strategy similar to genetic algorithms, which placed emphasis the behavioral link between an individual and its offspring, rather than seeking to emulate specific genetic operators as observed in nature. *EP* is different from conventional optimization methods. It is a useful method of optimization because does not need to differentiate cost function and constraints. The *EP* operates as follows: The initial population (real representation) is selected at random and is scored with respect to a given cost function. Offspring are created from these parents through random mutations, i.e., each component is usually perturbed by a Gaussian random variable with mean zero and an adaptable standard deviation term. It uses probability transition rules to select generations. Selection is based on a probabilistic tournament where each individual competes with other individuals in a combined population of the old generation and the mutated old generation. The competition results are valued using a probabilistic rule. The winners of same number as the individuals in the old generation constitute the next generation. The fast *EP* which use Cauchy mutations [10] is employed in this work. The idea of Cauchy mutation was originally inspired by fast simulated annealing [11]. The relationship between the classical *EP* using Gaussian mutation and the fast *EP* using mutation Cauchy is analogous to that between classical simulated annealing and fast simulated annealing. *EP* using Gaussian mutation is conducted by Box-Muller algorithm [12] for generation of random values with normal distribution, where

$$p(y)dy = \frac{1}{\sqrt{2\pi}} e^{-y^2/2} dy \tag{6}$$

The mutation operator of *EP* with Cauchy distribution presents a one-dimensional Cauchy density function centered at the origin and defined by

$$f_t = \frac{1}{\pi} \frac{t}{t^2 + x^2}, \qquad -\infty < x < \infty \tag{7}$$

where $t > 0$, is the scale parameter. The corresponding distribution function is

$$F_t(x) = \frac{1}{2} + \frac{1}{\pi} arctan\left(\frac{x}{t}\right) \tag{8}$$

The variance of the Cauchy distribution is infinite. Many studies have indicated the benefits of variance are increased in Monte-Carlo algorithms [13]. The hybrid method, proposed in this paper is composed by *EP* and a simplex method. This approach combines local and evolutionary searches that characterize a form of Lamarckian evolution. Hybrid algorithms can combine global search using *EAs* and local search using individual learning algorithms using individual learning algorithms. Hybrid evolutionary algorithms can exploit learning either actively via Lamarckian inheritance or passively via the Baldwin effect [14]. In the 19th century, Darwin's theory was challenged by Jean Baptiste Lamarck, who proposed that environmental changes throughout an organism's life cause structural changes that are transmitted to offspring. This theory lets organisms pass along the knowledge and experience that they acquire in their lifetime [15],[16]. This is analogous to Lamarckian inheritance in evolutionary theory, whereby characters acquired during a parent's lifetime are passed on their offspring [14].

Conventional *EP* produce the offspring of the next generation through the evaluation function to determine its fitness. This procedure can benefit from the advantages of Lamarckian theory. By letting some of the organism's "experiences" be passed along to future organisms. Following a Lamarckian approach, first would try inject some "smarts" into the offspring organism before returning it be evaluated. A traditional hill-climbing routine could use the offspring organism as a starting point and perform quick, localized optimization. The hill-climbing procedure is realized by Nelder-Mead's simplex approach. Nelder-Mead's simplex method (*NMSM*) [17] is local search method that uses the evaluation of the current data set to determine the promising direction search. This method is a direct search procedure based on heuristic ideas and its strengths are that it requires no derivatives to be computed. The basic procedure of the simplex method is to start with an initial basic feasible solution, i.e., at the vertex of the convex polyhedron. A simplex is the structure formed by ($n+1$) points, not in same plane, but in a n-dimensional space. The essence of the algorithm is as follows: the function is evaluated at each point (vertex) of the simplex and the vertex having the highest function value is replaced by a new point with a lower function value. This is done in such a way that the simplex adapts itself to the local landscape, and contracts on the final minimum. There are four main operations which are made on the simplex: reflection, expansion, reduction and contraction. Details of the *NMSM* can be find in [12]. The hybrid method of *EP* with *NMSM* is useful when addressing heavily optimization problems both in terms of computational efficiency and solution accuracy in mathematical form of the nonlinear quadratic objective function to be minimized for *GPC*. The *EP* with *NMSM* procedure is listed as follows (Figure 1):

Step 1: *Initialization*: Distribute the solutions (individuals - control signals) in accordance with uniform distribution ranging over [-10, 10].

Step 2: *Fitness of parent individuals*: Evaluate each parent solution with respect the cost function. The future control input sequence, Δu, is obtained by minimising of equation (1).

Step 3: *Offspring creation*: Generate one offspring for each parent by adding a Cauchy random number.

Step 4: *Competition*: Each individual in the combined population has to compete with some other individuals to have a chance to be transcribed to the next generation. The $2N_p$ individuals with best fitness values are selected to form a survivor set according to a stochastic decision rule.

Step 5: *Fitness of offspring individuals*: Calculate the fitness for each offspring;

Step 6: *Selection*: The individuals in the survivor set are new parents.

Step 7: *Lamarckian operator application*: Copy (overwrite) best N_p individuals (parameter design) from old population into bottom portion of new pop. The N_t individuals are optimized *NMSM*.

Step 8: *Stopping rule*: The procedures of generating new trials and selecting with best function are realized (steps 2-8) until the desired number of generations is achieved.

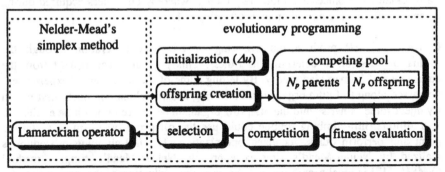

Figure 1. Schematic diagram of the Simplex-Evolutionary Lamarckian Method

2.3 Complete Procedure of Neuro--Simplex-Evolutionary *GPC*

The neuro-evolutionary-simplex *GPC* (*NES-GPC*) design for multivariable processes has the following steps (see Figure 2):

Step 1: Obtain the process measurements: input and output data.

Step 2: Determination of the *RBF-NN* topology and configuration of input nodes with past values of process inputs and outputs.

Step 3: Identification of multivariable process by *RBF-NN* training.

Step 4: Configuration of *GPC* parameters, i. e., N_1, N_2, Γ_e and Γ_u.

Step 5: Determination of setpoint signals.

Step 6: Start with the previous control input vector, and to predict the performance of the process using the *RBF-NN* model. It is realized for the output horizon, N_2, of the *RBF-NN* model and hence, determine the cost function, *J*.

Step 7: Repeat step 6 using simplex-evolutionary optimization algorithm until the desired minimization is achieved.

Step 8: Apply the control input, *Δu*, that minimizes the cost function of the process.

Step 9: Repeat the steps 5 to 8 for each time, *k*.

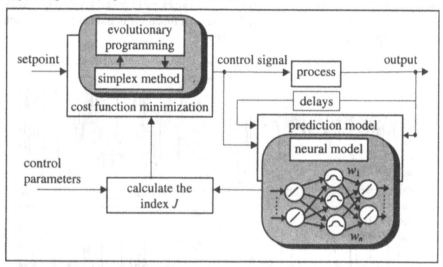

Figure 2. Block scheme of the *NES-GPC* configuration.

3. Experiments and Results

As an example, it is considered an unknown *MIMO* nonlinear system described by the following discrete dynamic equations [18]:

$$x_1(k+1) = 0.9x_1(k)\sin[x_2(k)] + \left[2 + 1.5\frac{x_1(k)u_1(k)}{1+x_1^2(k)u_1^2(k)}\right]u_1(k) +$$

$$\left[x_1(k) + \frac{2x_1(k)}{1+x_1^2}\right]u_2(k) \qquad (9)$$

$$x_2(k+1) = x_3(k)\left[1 + \sin[4x_3(k)] + \frac{x_3(k)}{1+x_3^2(k)}\right] \qquad (10)$$

$$x_3(k+1) = [3 + \sin[2x_1(k)]]u_2(k) \qquad (11)$$

$$y_1(k) = x_1(k); \qquad\qquad y_2(k) = x_2(k) \qquad\qquad\qquad (12)$$

An input signal for process excitation is required to adequately excite the nonlinear process dynamics over the whole region of interest. The data set for estimation (samples from 1 to 1000) and validation (samples from 1001 to 1500) phases of identification via *RBF-NN* are generated through of composition of a random input with uniform distribution and step response signals, where $u_1(k)$, $u_2(k) \in [-0.8, 0.8]$. The results for series-parallel multivariable identification by *RBF-NN* are shown in Figure 3. *RBF-NN* is configured with 10 inputs $\{y_1(k), y_2(k), y_1(k-1), y_2(k-1), y_1(k-2), y_2(k-2), u_1(k), u_2(k), u_1(k-1), u_2(k-1)\}$, 15 Gaussian functions in hidden layer for each *RBF-NN* input, and 2 outputs $\{y_1(k+1), y_2(k+1)\}$.

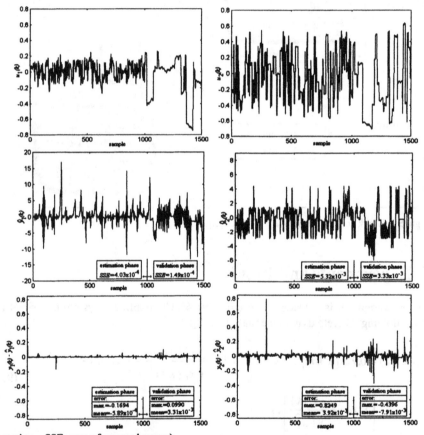

(notation - *SSE*: sum of squared errors)
Figure 3. Result of identification via *RBF-NN*.

Experiments for setpoint changes and load disturbances of *NES-GPC* are shown in Figure 4. The employed *NES-GPC* has the following parameters configuration:

- *GPC*: $N_1=1$, $N_2=2$, $N_u=1$, $\Gamma_e = 1$, $\Gamma_u = 0.001$;
- *EP*: $N_p=10$, number of generations = 20 (stopping criterion); *NMSM*: $N_s=5$;

- setpoint changes: samples from 001 to 100: $y_{r1}(k) = 0.5$; $y_{r2}(k) = 1.0$;
  samples from 101 to 200: $y_{r1}(k) = 1.0$; $y_{r2}(k) = 0.5$.
- load disturbances:
  sample no. 40: $y_1(k) = y_1(k) + 0.5$;   sample no. 70: $y_2(k) = y_2(k) + 0.5$;
  sample no. 140: $y_1(k) = y_1(k) + 1.0$;   sample no. 170: $y_1(k) = y_1(k) - 1.0$.

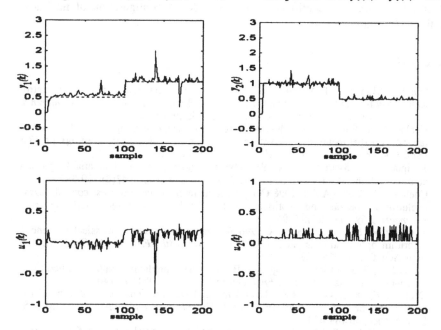

Figure 4. Results of *NES-GPC* applied to *MIMO* system in setpoint changes and load disturbances.

4. Conclusion

The proposed *NES-GPC* technique is a flexible methodology for designing the controllers. The *NES-GPC* has been successfully applied to the *MIMO* control in tests for setpoint changes and load disturbances. There are two essential aspects for performance's *NES-GPC*. Firstly, the nonlinear identification obtained by *RBF-NN* without the necessity of calculating the unknown interactor matrix. *RBF-NN* presents also approximation accuracy of a complex *MIMO* nonlinear mapping with cross-coupling between variables. Another aspect is the robustness to select of the control input values by simplex-evolutionary Lamarckian optimization.

The *NES-GPC* presents limitations because the *EA* needs to assess all solutions at each sampling time. However, this technique can be applied to *MIMO* processes with relatively slow dynamics, such as distillation columns, and that necessity of long control horizons. Parallel computer platforms are also expected to reduce the computing time and heuristic creation of the initial population of the candidate controllers (from previous off-line experiments) results in a significant reduction in computational time. Further works needs to be done in area of optimization combining *EAs* with self-adaptation mechanisms and hill-climbing methodologies and on-line parameters identification by *RBF-NN* for configuration of nonlinear adaptive *MBPC* applications.

References

1. Qin S J, Badgwell T A 1997 An overview of industrial model predictive control technology In: Kantor C J, Garcia C E, Carnahan B (eds) *5th International Conference on Chemical Process Control*, AIChE and CACHE, pp 232-256
2. Coelho L S, Simas H, Coelho, A A R 1999 Inexpensive apparatus for control laboratory experiments using advanced control methodologies *Proceedings of 38th IEEE Conference on Decision and Control*, Phoenix, Arizona
3. Tsiligiannis C, Svoronos S A 1988 Dynamic interactors in multivariable process control — I. the general time delay case *Chemical Eng Science* 43(2):339-347
4. Coelho L S, Coelho A A R 1998 Computational intelligence in process control: fuzzy, evolutionary, neural, and hybrid approaches *Int J of Knowledge-Based Intelligent Engineering Systems* 2(2):80-94
5. Martínez M, Senent J, Blasco X 1996 A comparative study of classical vs genetic algorithm optimization applied in GPC controller *13th IFAC World Congress*, San Francisco, CA, pp 327-332
6. Sousa J M, Babuska R, Verbruggen H B 1997 Fuzzy predictive control applied to an air-conditioning system *Control Engineering Practice* 5(10):1395-1406
7. Declercq F, De Keyser R 1996 Neural model based predictive control a simulated bioreactor *CESA, Symposium on Control, Optimization and Supervision*, IMACS, Lille, France, pp 489-494
8. Furuta T, Shibata T 1992 Theory and applications of neural networks for industrial control systems *IEEE Trans on Industrial Electronics* 39(6):472-489
9. Moody J, Darken C J 1989 Fast learning in networks of locally-tuned processing units *Neural Computation* 1(2):281-294
10. Yao X, Liu Y 1996 Fast evolutionary programming In: Fogel L J, Angeline P J, Bäck T (eds) *Proceedings of the 5th Annual Conference on Evolutionary Programming*, San Diego, CA, MIT Press, pp 451-460
11. Szu H H, Hartley R L 1987 Nonconvex optimization by fast simulated annealing *Proceedings of the IEEE* 75:1538-1540
12. Press W H, Teukolsky S A, Vetterling W T et al 1994 *Numerical Recipes in C: The Art of Scientific Computing*, Cambridge Press, 2nd ed
13. Ingber L 1989 Very fast simulated re-annealing, *Math Comp Modelling* 12:967-973
14. Anderson R W, Fogel D B, Schütz M 1997 Other operators In: Bäck T, Fogel D B, Michalewicz Z (eds) *Handbook of Evolutionary Computation* IOP Publishing, Bristol, Philadelphia and Oxford University Press, Oxford, pp C3:4:1-C3:4:15
15. Kennedy S A 1993 Five ways to a smarter genetic algorithm *AI Expert*, December, 35-38
16. Whitley D, Gordon S, Mathias K 1994 Lamarckian evolution, the Baldwin effect and function optimization In: Davidor Y, Schwefel H P, Manner R (eds) *Parallel Problem Solving for Nature*, Springer-Verlag, Berlin, pp 6-15
17. Nelder J A, Mead R 1965 A simplex method for function minimisation *Computer J* 7:308-313
18. Narendra K S, Mukhopadhyay S 1993 Adaptive control of nonlinear multivariable systems using neural networks *Proceedings of the 32nd IEEE Conference on Decision and Control*, San Antonio, TX, pp 3066-3071

Permeability Prediction in Petroleum Reservoir using a Hybrid System

Y. Huang[1#], P.M. Wong[2] and T.D. Gedeon[3#]

[1]TechComm Simulation Pty Ltd, 1/53 Balfour Street
Chippendale NSW 2008, Australia (yuantuh@techsim.com.au)
[2]School of Petroleum Engineering, University of New South Wales
Sydney NSW 2052, Australia (pm.wong@unsw.edu.au)
[3]School of Information Technology, Murdoch University
Murdoch WA 6150, Australia (t.gedeon@murdoch.edu.au)

Abstract: This paper introduces and demonstrates a hybrid soft computing system for predicting reservoir permeability of sedimentary rocks in drilled wells in the petroleum exploration and development industry. The method employs Takagi-Sugeno's fuzzy reasoning, and its fuzzy rules and membership functions are automatically derived by neural networks and floating-point encoding genetic algorithms. The method is trained with known data and tested with unseen data. The results show that the hybrid system has a good generalisation capability and is effective for industrial applications.

1. Introduction

1.1 Petroleum Reservoir

A petroleum reservoir is a volume of porous sedimentary rock which has been filled with hydrocarbons, such as oil and gas. Reservoir permeability (a measure of fluid conductance in porous media) is an important parameter for characterising complex geological formation, reserves estimation and production forecasting. This property is commonly obtained from drilled wells. Electronic equipment is used to *log* the well in such a way that multi-type digital measurements or *well logs* are obtained as a function of drilled depth.

Permeability can be estimated by correlating well logs with the laboratory measured permeability data obtained from rock samples or *cores*. Retrieving cores for laboratory testing is tedious and expensive. It is only practiced at selected depths/intervals. Therefore, permeability prediction at the *un-cored* wells relies strongly on functional transformation of well logs developed at the *cored* wells.

[#] Paper prepared at the University of New South Wales, Australia.

1.2 A Regression Problem

The problem is traditionally solved by developing simple transfer functions, which attempt to match the target permeability values by manually adjusting constants and exponents of the functions. In many cases, these equations perform unsatisfactorily. Multivariate statistical techniques have offered a new insight and provided a potential solution by multiple regression. The regression approach directly takes the target values to minimise the prediction error. The transfer functions, however, oversimplify the natural complexity of the reservoir data. The advent of soft computing techniques (e.g. neural networks, fuzzy logic and genetic algorithms) offers powerful tools for further improving permeability predictions.

The regression approach requires a set of training patterns with known inputs (well logs) and targets (permeability). These patterns can be obtained from cored wells. The permeability required at the un-cored wells can be estimated by using the same types of well logs as inputs to the transfer function developed. If multiple cored wells are available, we may develop a transfer function at each cored well. An appropriate weighting scheme can be used to weigh the importance of each of the transfer functions and obtain the final (averaged) prediction at the un-cored well.

Figure 1 illustrates the problem by using three wells, namely W1, W2 and W3, drilled in the same petroleum reservoir. W1 and W2 are cored and W3 is assumed to be un-cored. We define the input-output relation as follows:

$$y_3 = \frac{\sum_{i=1}^{2} \mu_i^{\beta_i} \cdot f(\mathbf{x}_3, \theta_i)}{\sum_{i=1}^{2} \mu_i^{\beta_i}} \tag{1}$$

where $\mathbf{x}_3 = (x_1, ..., x_m)_3$ denotes the input log data (m types) at W3, θ_i denotes a set of parameters of the transfer function $f(.)$ developed at cored well i, $\mu = (\mu_1, \mu_2)$ is a weighting vector, $\beta = (\beta_1, \beta_2)$ is a real number in ($-\infty, +\infty$) and y_3 denotes the output permeability.

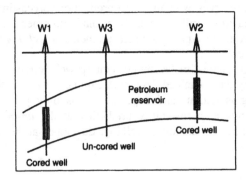

Figure 1. Permeability prediction problem.

Given the training patterns at W1 and W2, $\{x_i, t_i\}, i = 1, 2$ where t is the target value, we may apply regression to optimise θ by minimising the following objective function, O:

$$O(\theta_i) = \sum_{j=1}^{n_i} (f(x_i, \theta_i)_j - t_{i,j})^2 \qquad (2)$$

where n_i is the number of training patterns at cored well i.

In this paper, we propose to use a hybrid soft computing system to optimise θ, μ, β. The system combines neural networks, fuzzy reasoning and genetic algorithms for an industrial application.

1.3 Previous Works

There are many solutions to the above problem, ranging from multiple regression, neural networks, and neural-fuzzy techniques. Both multiple regression [1][2] and backpropagation neural networks [3][4] minimise Equation (2) using a combined data set (W1 and W2) and generate y_3 without considering μ, β. The major problem of the previous techniques is the use of a combined data set, which essentially averages the location-specific input-output relations.

We have avoided the averaging problem by training a separate network for each cored well [5][6]. The final prediction was obtained by using the Takagi-Sugeno's fuzzy reasoning [7] as shown in Equation (1). In [5][6], we used the separation distances between W1 and W3, and W2 and W3 as the fuzzy membership function values $\mu = (\mu_1, \mu_2)$ and set $\beta = (-1, -1)$. Hence, Equation (1) becomes the conventional inverse-distance estimator, which assigns a higher weighting to the estimate from the cored well closer to the un-cored well. We showed that this approach avoids the averaging problem. However, it ignores the internal similarity of the data sets among the drilled wells. It may be ineffective when applied to complex reservoirs.

In the past years, we have also looked at the optimisation of the fuzzy membership function values and the connection weights of the multi-layer perceptrons. We have used a neural-driven fuzzy reasoning method [8] to optimise the membership function values by genetic algorithms (GAs) with dramatically improved results [9]. In [10], we applied binary encoding GAs [11] to optimise the connecting weights. Our results were improved compared to weight training by the backpropagation algorithm (BP), but were computationally more expensive. One potential solution is to use the connection weights trained by BP to initialise the chromosomes in GAs. This requires floating-point encoding GAs which are fast and accurate [12], but are not popular for industrial applications to date.

1.4 Objective

The objective of this paper is to develop a hybrid soft computing system for permeability prediction which combines our previous work on neural-fuzzy estimators and floating-point encoding genetic algorithms. The major contribution of this work is the incorporation of Takagi-Sugeno's fuzzy reasoning to account for the internal similarity of data sets using $\mu = (\mu_1, \mu_2)$.

Section 2 presents the hybrid system with a detailed description of the floating-point encoding GAs. Section 3 applies the hybrid system to predict permeability in an oil and gas well located in the North West Shelf, offshore western Australia. Data from three cored wells are available for this study. The first two data sets are used for training/validation, while the third set is treated as unseen data and is used to test the performance of the proposed method (see Figure 1). We will also compare the performance of the proposed method with the backpropagation neural networks alone.

2. Hybrid System

2.1 Neural Networks

A standard three layer (input, hidden, and output) neural network is used to generate $f(.)$ in each cored well. The number of input neurons equals to the number of well logs used, and one output neuron is used to represent permeability. The number of hidden neurons is obtained by trial and error.

In order to avoid over-fitting, we apply early-stopping by using a validation set, i.e. to terminate training when the minimum error on the validation set is reached. We use the data from one cored well as the training set and to develop $f(.)$. The data from the other cored well is used as the validation set. Swapping the use of the data sets give two generalised transfer functions. The total root mean square errors (training plus the validation errors) are used as $\beta = (\beta_1, \beta_2)$. Note that the smaller the β value, the better the estimator, and the higher the weighting μ^β as $\mu = (0,1)$.

2.2 Fuzzy Reasoning

To generate μ, a similar multi-layer perceptron is used. The inputs are identical, but we use two output neurons. The following targets (s_1, s_2) are used:

$$(s_1, s_2) = \begin{cases} (1,0) & \text{if} \quad x \in W1 \\ (0,1) & \text{if} \quad x \in W2 \end{cases} \tag{3}$$

This training scheme estimates the degree of membership of each input pattern "belonging" to each of the cored wells. The value of the membership function is defined as the output of the trained neural network, i.e., $\mu_i = \hat{s}_i$, where $\hat{s}_i$ denotes the output from the trained network.

The above weighting scheme examines the similarity of the inputs of the training data sets and the inputs of the unseen data set. It is superior to the one proposed in [5][6] which used only the inverse-distances.

2.3 Genetic Algorithms

Genetic algorithms (GAs) mimic processes observed in nature evolution, and are stochastic global search methods. Individuals in a population are called *chromosomes*. A genetic representation for a potential solution to a problem is encoded as a chromosome. A good initial population of potential solutions can result in fast convergence with higher accuracy for real world problems. The steps of a typical GA are listed as follows:

a. Initialise a population of chromosomes within the range of potential solutions.
b. Evaluate each chromosome in the population based on a *fitness* function.
c. Select the parent chromosomes to reproduce (according to the fitness values).
d. Apply genetic operators to the parent chromosomes to produce children so as to generate a new population.
e. Evaluate the chromosomes in the new population.
f. Stop and return the best chromosomes as the final solution if a termination condition is satisfied; otherwise, go to step c.

Generally speaking, the parent chromosome selection for both binary and floating-point encoding is similar. However, the genetic operators used for floating-point encoding are different from these for binary encoding. In our problem we will use the connection weights trained by BP as one of the chromosomes in the initial population, and hence, floating-point encoding is required. The following steps outline the general structure of floating-point encoding GAs:

2.3.1 Population Initialisation
Let $\theta = (w_1, w_2,..., w_n)$ denotes a parameter vector of connection weights trained by BP, where n is the number of total connection weights in the network. We define a real number parameter "swing" as Δw. The use of this parameter avoids the need to define the upper and lower bounds for the parameters to be optimised. Typically the value of Δw for our problem is $[\Delta w_{min} = 0.5, \Delta w_{max} = 10]$. Assuming the size of a population is m and we keep θ as one of the chromosomes in the initialised population, the remaining $m-1$ chromosomes are randomly generated, and the range of the ith parameter in a chromosome is $[w_i - \Delta w, w_i + \Delta w], i = 1,...n$.

2.3.2 Performance Evaluation
A fitness function is used to evaluate the performance of each of the chromosomes. A typical fitness function is defined as follows:

$$F(\theta_j) = \frac{10}{1 + E(\theta_j)} \qquad (4)$$

and,

$$E(\boldsymbol{\theta}_j) = \sum_{k=1}^{N} (f(\mathbf{x}_k, \boldsymbol{\theta}_j) - t_k)^2 \tag{5}$$

where $F(\boldsymbol{\theta}_j)$ with $\boldsymbol{\theta}_j = (w_{j1}, w_{j2}, ..., w_{jn})$ is the fitness value for the jth $(j = 1, ..., m)$ chromosome, $E(\boldsymbol{\theta}_j)$ is the sum of squared errors of the target data t_k $(k = 1, ..., N)$ and the predictions $f(.)$ and N is the total number of training patterns. The higher the $F(\boldsymbol{\theta}_j)$ value, or the lower the $E(\boldsymbol{\theta}_j)$, the better the solution.

2.3.3 Reproduction
Selecting parents for reproduction is a very important aspect of GAs. There are many selection methods. A parent solution can be selected more than once. The most popular selection method is the Goldberg's roulette wheel parent selection [13]. The roulette wheel has slots sized according to the fitness of each chromosome. The purpose is to give more reproductive chances to those population members who are the most fit. After the roulette wheel parent selection, we copy the best chromosome twice to replace two of the worst.

In order to accelerate convergence, before crossover and mutation we dynamically update the swing and the range of the parameters for the chromosomes as follows:

$$\Delta w = \Delta w_{\min} + \frac{(M - k) \cdot (\Delta w_{\max} - \Delta w_{\min})}{M} \tag{6}$$

where M is the desired number of iterations and $k (\leq M)$ is the current iteration counter. Let $\boldsymbol{\theta} = (w_1, w_2, ..., w_n)$ be the best chromosome, then the new range of the ith parameter in a chromosome for a new population is $[w_i - \Delta w, w_i + \Delta w], i = 1, ...n$.

2.3.4 Crossover
Crossover produces offspring by exchanging genetic information between the selected parent solutions. The selection criteria are based on a user-defined probability for crossover, P_c (generally between 0.5 to 0.9). This probability defines the number of candidates for crossover. For example, if P_c is 0.7, it means that 70% of the parent chromosomes in the population will be selected randomly and mated in pairs. In this paper, we use two-point arithmetical crossover.

Let the ith chromosome $(w_{i1}, w_{i2}, ..., w_{in})$ and the jth chromosome $(w_{j1}, w_{j2}, ..., w_{jn})$ be selected for crossover between the pth and the qth parameters $(1 \leq p \leq q \leq n)$. The offspring becomes $(w_{i1}, ..., w_{ip-1}, w_{ip}^j, ..., w_{iq}^j, w_{iq+1} ..., w_{in})$ and $(w_{j1}, ..., w_{jp-1}, w_{jp}^i, ..., w_{jq}^i, w_{jq+1} ..., w_{jn})$ where $w_{ik}^j = \alpha \cdot w_{ik} + (1 - \alpha) \cdot w_{jk}$ and

$w^i_{jk} = \alpha \cdot w_{jk} + (1-\alpha) \cdot w_{ik}$ for $p \leq k \leq q$ and α is a uniform random number in $[0,1]$.

2.3.5 Mutation

The reproduction and crossover operation would only exploit the known regions in the solution space, which could lead to premature convergence for the fitness function with the consequence of missing the global optimum by exploiting some local optimum. Mutation is a genetic process to avoid such a problem. This process allows the introduction of new characteristics to the offspring, which are unrelated to the parent solutions. It first requires a user-defined probability for mutation P_m (generally between 0.01 to 0.2). Here, we use non-uniform arithmetical mutation. Let the jth parameter in the ith chromosome w_{ij} is selected for mutation. Thus, the new parameter is $w^*_{ij} = \alpha \cdot w_{ij} + (1-\alpha) \cdot v$, where v is a uniform random number in $[w_i - \Delta w, w_i + \Delta w]$. More details about genetic operators for floating-point encoding can be found in [14].

2.4 Neural-Driven Fuzzy Reasoning

After generating the functions $f(.)$ with optimised θ, we can extract two fuzzy rules from the two cored wells, W1 and W2:

$$Rule\ 1: \quad \text{If } x_3 \in W1 \quad \text{then } y_1 = f(x_3, \theta_1)$$
$$Rule\ 2: \quad \text{If } x_3 \in W2 \quad \text{then } y_2 = f(x_3, \theta_2)$$

Similar optimisation routines can be run to obtain the membership function values $\mu = (\mu_1, \mu_2)$ in Equation (3). Equation (1) can then be applied to obtain the final estimate.

3. Field Example

In this case study, data from three wells, W1, W2 and W3, located in the North West Shelf, offshore western Australia, were used. The well logs available for the analyses were: gamma ray (GR), deep resistivity (LLD), sonic travel time (DT), bulk density (RHOB) and neutron porosity (NPHI). The 11 rock classifications were converted to 11 values spread evenly within the interval $[0,1]$. Permeability measurements were available at selected well depths. There are a total of six inputs and one output. All the well log data were normalised in the range of $[0,1]$. All the permeability values were normalized in the range of $[0.1,0.9]$. The number of data pairs in each well was 152, 156, and 140 points, respectively. We aim to predict the permeability values at W3 using the transfer functions developed at W1 and W2.

In this study, a three-layer neural network with five hidden neurons was found to be the best structure. Due to the presence of the bias weights (only in the input layer), $f(.)$ had a total of 40 connection weights. For the network of $\mu = (\mu_1, \mu_2)$, the use of two output neurons resulted in 45 connection weights. The GA configuration is shown in Table 1.

Table 1: GA configuration.

Maximum iteration	5,000
Population size, m	20
Probability for crossover, P_c	0.8
Probability for mutation, P_m	0.1

Table 2 shows the training and validation errors for $f(.)$, optimised by the floating-point encoding GAs using BP trained weights as the initial population. The results using BP alone are also tabulated in the same table. From these results, the errors from BP followed by GAs were smaller (10% and 7% lower respectively) and hence the rules were more reliable for prediction. The corresponding $\beta = (\beta_1, \beta_2)$ was also calculated. Note that the similarity of the β values showed that the two transfer functions had similar prediction reliability.

Table 2. Root mean square errors from BP followed by GA.
The bracketed values are results from BP alone.

	Training error	Validation error	β value
W1 for training	0.065	0.075	0.140
W2 for validation	(0.072)	(0.084)	(0.156)
W2 for training	0.073	0.076	0.149
W1 for validation	(0.075)	(0.086)	(0.161)

Figure 2 shows the trained hyper-surface membership function values $\mu = (\mu_1, \mu_2)$ obtained by GAs. These values were calculated by using x_3 (140 points in total), the input patterns of W3 in the fuzzy reasoning step. The plots show the degree of membership of W3 patterns "belonging" to W1 and W2 respectively. We can see that the majority of the W3 patterns are similar to the W2 patterns.

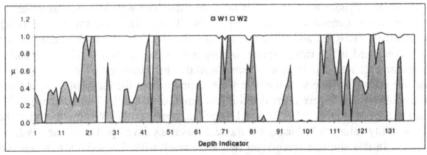

Figure 2. Membership of the W3 patterns "belonging" to W1 and W2.

4. Results

The results from the hybrid system are tabulated in Table 3. The results from using the separate rules trained by BP alone are also shown in the same table. Clearly, the hybrid system gave the smallest total sum of error squares (TSS) and the highest correlation coefficient (r^2). When comparing the predictions from those of the separate rules, the TSS from the hybrid system was 26% and 20% smaller than rule 1 and rule 2 respectively. Similarly, the r^2 was 11% and 16% higher respectively. Figure 3 shows the predictions at W3 along with the actual permeability data. The predictions matched very well with the actual values.

Table 3. Comparison of TSS and r^2 of 140 data points at W3
using the hybrid system and the separate rules developed by BP alone.

	TSS	r^2
Hybrid system	0.939	0.73
Rule 1 from BP alone	1.273	0.66
Rule 2 from BP alone	1.180	0.63

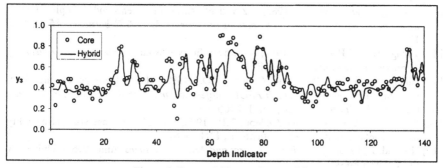

Figure 3. Permeability profiles at W3.

The performance of the hybrid system was good compared to our previous approach. It has not only the intrinsic advantages of the neural-driven fuzzy reasoning, but also incorporates the floating-point encoding GAs to further optimise the neural networks trained by BP. Because the chromosomes of the initial population in GAs come from the BP-trained weights, the final results are never worse than those obtained by BP alone. Therefore, the combination of the GAs with BP can provide fast and accurate results.

5. Conclusions

In this paper, we present the use of a hybrid soft computing system to predict permeability in a petroleum reservoir in offshore western Australia. The system uses the neural-driven fuzzy reasoning combining with the floating-point encoding genetic algorithms. It is a robust and flexible estimator for industrial applications. In the field example, the results showed that the hybrid system provided the smallest error and the highest correlation coefficient on the unseen data compared to the conventional method using backpropagation neural networks alone.

References

[1] Bloch, S., 1991, Empirical prediction of porosity and permeability in sandstones. *AAPG Bulletin*, **75**, 1145-1160.

[2] Jian et al., 1994, A genetic approach to the prediction of petrophysical properties. *Journal of Petroleum Geology*, **17**, 71-88.

[3] Wong, P.M. Taggart, I.J. and Gedeon, T.D., 1995, Use of neural network methods to predict porosity and permeability of a petroleum reservoir. *AI Applications*, **9**(2), 27-38.

[4] Mohaghegh et al., 1996, Petroleum reservoir characterization with the aim of artificial neural networks. *Journal of Petroleum Science and Engineering*, **16**, 263-274.

[5] Huang, Y., Wong, P.M. and Gedeon, T.D., 1997, Spatial interpolation in log analysis using neural-fuzzy technique. *59th EAGE Conference & Technical Exhibition*, Geneva, Extended Abstracts, vol. 1, P174.

[6] Gedeon et al., 1997, Two dimensional neural-fuzzy interpolation for spatial data. *Proceedings of GIS AM/FM ASIA '97 & Geoinformatics '97*, Taipei, Taiwan, vol. 1, 159-166.

[7] Takagi, T. and Sugeno, M., 1985, Fuzzy identification of systems and its applications to modeling and control. *IEEE Transactions on Systems, Man & Cybernetics*, **15**(1), 116-132.

[8] Takagi, H. and Hayashi, I., 1991, NN-driven fuzzy reasoning. *International Journal of Approximate Reasoning*, **5**, 91-212.

[9] Huang Y., Wong, P.M. and Gedeon T.D., 1998, Neural-fuzzy-genetic-algorithm interpolator in log analysis. *60th EAGE Conference and Technical Exhibition*, Leipzig, Germany, Extended Abstracts, vol. 1, P106.

[10] Huang, Y., Wong, P.M. and Gedeon, T.D., 1998, Prediction of reservoir permeability using genetic algorithms. *AI Applications*, **12**(1-3), 67-75.

[11] Holland, J., 1975, *Adaptation in Natural and Artificial Systems*. Ann Harbor: University of Michican Press.

[12] Wang, X. and Elbuluk, M., 1996, Neural network control of induction machines using genetic algorithm training. *Proceedings of the 31st IEEE IAS Annual Meeting*, 3, 1733-1740.

[13] Goldberg, D.E., 1989, *Genetic Algorithms in Search, Optimization, and Machines Learning*. Addison-Wesley, Reading, MA.

[14] Michalewicz, Z., 1994, *Genetic Algorithms + Data Structures = Evolution Programs*. Springer-Verlag, Berlin.

A Performance Comparison of Chaotic Simulated Annealing Models for Solving the N-queen Problem

Terence Kwok[1] and Kate A. Smith[2]

School of Business Systems, Faculty of Information Technology, Monash University, Clayton, Victoria 3168, Australia
[1]terence.kwok@infotech.monash.edu.au
[2]kate.smith@infotech.monash.edu.au

Abstract: Chaotic neural network models employing two chaotic simulated annealing (CSA) schemes, one with decaying self-couplings and the other with a decaying time-step, are used to solve the N-queen problem. Their optimisation performances are compared in terms of feasibility, efficiency, robustness and scalability in a two-parameter domain chosen for each model. Computational results show that the decaying self-coupling approach offers better feasibility, robustness and scalability, with efficiency being comparable for the two models. Correlation between feasibility and efficiency illustrates some chaotic search characteristics common to both models.

1. Introduction

Artificial neural networks have been used to solve combinatorial optimisation problems since the Hopfield network was used to solve the Travelling Salesman Problem (TSP) [1]. Because of its steepest descent approach to optimisation, the Hopfield network often arrives at local minimum solutions and requires tedious trial and error tuning of penalty parameters. For complex problems, the network becomes impractical to use. With the advances in chaos theory and its increasing relevance in diverse fields, its role in neural networks is an important area of research. A chaotic neuron model and the chaotic neural network (CNN) were proposed by Aihara et al [2] and have been used to solve the TSP in various investigations to study its dynamics and properties [3,4,5]. Computer simulations have shown that chaotic search for the global minimum is most effective at the edge of chaos, the transition region between the ordered phase and disordered phase [3,6]. A theoretical chaotic search mechanism based on crisis-induced intermittency was proposed by Tokuda et al [7]. A different approach to incorporate chaos into neural networks is to add chaotic noise [8,9,10], which helps escape from local minima. However, more investigation is needed into the role of chaos in this method.

   Chaotic simulated annealing (CSA) was proposed by Chen & Aihara as a means to utilise chaotic search with convergent properties [11,12], which is achieved by adding a decaying self-coupling term to the Hopfield network. The method was used to solve the TSP, maintenance scheduling problem [11], and the N-queen problem [13], with a far superior optimisation ability than the traditional Hopfield network. A different approach to CSA was proposed by Wang & Smith [14,15], where chaos is induced internally by varying the time-step on an Euler discretisation of the continuous Hopfield network, and the method was implemented to solve the N-queen problem [16].

This paper compares the two CSA approaches, namely the decaying self-coupling CSA and decreasing time-step CSA. The investigation involves performance issues such as feasibility and efficiency in a chosen parameter space. Robustness is looked at in terms of flexibility in parameter choice for optimal performance and its scalability as problem size increases.

2. Two Approaches to Chaotic Simulated Annealing

Section 2.1 is an introduction to Chen & Aihara's CSA with decaying self-coupling and Section 2.2 is for Wang & Smith's CSA with decaying time-step. These two methods are chosen for comparison because they are both modified on the original Hopfield network but with different means of incorporating CSA. An implementation to solve the N-queen problem is given in Section 2.3.

2.1 CSA with Decaying Self-coupling

The formulation of Chen & Aihara's CSA [11] is as follows,

$$x_i(t) = \frac{1}{1 + e^{-y_i(t)/\varepsilon}} \tag{1}$$

$$y_i(t+1) = ky_i(t) + \alpha\left(\sum_j^N w_{ij}x_j(t) + I_i\right) - z(t)(x_i(t) - I_0) \tag{2}$$

$$z(t+1) = (1 - \beta)z(t) \tag{3}$$

y_i = internal state of neuron i; x_i = output state,

α = positive scaling input parameter,

β = damping factor of $z(t)$, $0 \le \beta \le 1$,

k = damping factor of nerve membrane, $0 \le k \le 1$,

E = energy function to be derived in later section,

w = connection weight matrix, which satisfies

$$\sum_{j=1}^N w_{ij}x_j + I_i = -\frac{\partial E}{\partial x_i}$$

$z(t)$ = self-feedback connection weight (refractory strength) ≥ 0,

I_i = input threshold of neuron i,

I_0 = positive parameter.

$z(t)$ is the self-feedback parameter that determines the size of the self-coupling term that induces chaos in the network [11]. The CSA scheme requires $z(t)$ to be decreasing in time. By updating (1) to (3), the neural network can achieve chaotic dynamics while gradually approaching the convergent Hopfield network because of the decaying self-coupling term. A smaller β means a slower "annealing" schedule, and vice versa. In general, values of $k, \alpha, \beta, \varepsilon$ and $z(0)$ should be selected for good optimisation ability with asymptotic convergence. In this paper, α and $z(0)$ are chosen for performance and scalability analysis of the network because they affect the size of the steepest descent term and the self-feedback term respectively.

The existence of transient chaos and the convergence of this transiently chaotic neural network (TCNN) have been proved by Chen & Aihara [12]. The stability condition for an asynchronous updating of (2) is:

1. $\frac{1}{3} \ge k \ge 0$, $4(1-k)\varepsilon > -\min\{\alpha w_{ii} - z\}$, or

2. $1 \ge k \ge \frac{1}{3}$, $8k\varepsilon > -\min\{\alpha w_{ii} - z\}$, or

3. $k > 1$, $8\varepsilon > -\min\{\alpha w_{ii} - z\}$. $\hspace{2cm}$ (4)

where w is a symmetric weight matrix.

2.2 CSA with Decaying Time-step

The formulation of this CSA scheme is based on the Euler approximation of the continuous Hopfield network [14]:

$$y_i(t + \Delta t) = \left(1 - \frac{\Delta t}{\tau}\right) y(t) + \Delta t \left(\sum_j w_{ij} x_j(t) + I_i\right) \hspace{1cm} (5)$$

Here, Δt is a time-varying quantity that starts with a chosen $\Delta t(0)$ and then gradually decreased:

$$\Delta t(t + 1) = (1 - \beta)\Delta t(t) \hspace{1cm} (6)$$

where $0 < \beta < 1$. With a sufficiently large initial time-step $\Delta t(0)$, the network undergoes a reverse bifurcation process [15], which provides a transiently chaotic period. The network approaches the continuous Hopfield network as $\Delta t \to 0$ which has guaranteed convergence and a tendency to minimise the energy function E. A continuous activation function as in (1) is used in this paper.

For the network to gain stability Δt needs to be sufficiently small, with the following criteria [14]:

$$\Delta t \leq \tau, \; \Delta t < \frac{2\tau}{1-\mu_{max} w_{ii,}} \; \text{ and } \; w_{ii} \leq 0 \tag{7}$$

where w is a symmetric matrix, $\tau > 0$, μ_{max} is the maximum slope of the activation function, and an asynchronous updating for the network states is to be used. In general, parameters to be adjusted are τ, $\Delta t(0)$ and ε. Advantages of this method over Chen & Aihara's model are: it requires fewer parameters that need to be adjusted, and with more compact stability regions (c.f. (7), (4)) that lie closer to the bifurcation point [14].

Despite apparent differences in the means of incorporating chaos and the associated parameter and stability requirements, the two models are similar to each other in the sense that their formulations are related to each other [14], and that both induce transient chaos as the main component for optimisation.

2.3 Solving the N-queen Problem

The N-queen problem is an example of a constraint satisfaction problem which is frequently encountered in engineering and business applications. The aim is to place N queens onto an N by N chessboard without attacking each other, which is enforced by having 1) one queen in each row; 2) one queen in each column; 3) one queen in each diagonal (more than two), and 4) exactly N queens on the chessboard. With an $N \times N$ output matrix x representing the chessboard, a neuronal output of 1 represents a queen placed in that position, and 0 means no queen is placed.

Research in solving the N-queen problem with artificial neural networks has involved the use of Hopfield network [17], Cauchy machines [18], Gaussian machines [19], and more recently with CSA approaches for 10-queen problem [13,16]. This paper differs from previous research in the aspect of emphasizing scalability issues, as well as feasibility, efficiency and effectiveness, in comparing the performance of CSA approaches across several problem sizes.

Since both methods of CSA in this paper can be considered as modifications to the original Hopfield network, they share the same formulation of the weight matrix w and energy E. In order to satisfy the N-queen constraints, a cost function is constructed such that its value increases if a constraint is violated:

$$f = \frac{A}{2}\sum_{i,j}\sum_{l \neq j} x_{ij}x_{il} + \frac{B}{2}\sum_{i,j}\sum_{k \neq i} x_{ij}x_{kj} + \frac{C}{2}\left(\sum_{i,j} x_{ij} - N\right)^2$$

$$+ \frac{D}{2}\left(\sum_{i,j}\sum_{\substack{p \neq 0 \\ 1 \leq i-p \leq N \\ 1 \leq j-p \leq N}} x_{ij}x_{i-p,j-p} + \sum_{i,j}\sum_{\substack{p \neq 0 \\ 1 \leq i-p \leq N \\ 1 \leq j+p \leq N}} x_{ij}x_{i-p,j+p}\right) \tag{8}$$

The first and second terms of (8) corresponds to the constraints of having only one queen in each row and each column respectively; the third term corresponds to having exactly N queens; and the last term corresponds to the diagonal constraint. A, B, C and D are positive parameters. Expanding (8), neglecting the term $CN^2/2$ and compare (8) to the energy function of the Hopfield network [16]:

$$E = -\frac{1}{2} \sum_{i,j,k,l} x_{ij} x_{kl} w_{ijkl} - \sum_{i,j} I_{ij} x_{ij} \tag{9}$$

gives

$$E = \frac{A}{2} \sum_{i,j,k,l} \delta_{ik} (1 - \delta_{jl}) x_{ij} x_{kl} + \frac{B}{2} \sum_{i,j,k,l} \delta_{jl} (1 - \delta_{ik}) x_{ij} x_{kl}$$
$$+ \frac{C}{2} \left(\sum_{i,j,k,l} x_{ij} x_{kl} - 2N \sum_{i,j} x_{ij} \right) \tag{10}$$
$$+ \frac{D}{2} \left(\sum_{i,j,k,l} x_{ij} x_{kl} \left(\delta_{i+j,k+l} + \delta_{i-j,k-l} \right) \right) (1 - \delta_{ik})$$

and

$$w_{ijkl} = -A \delta_{ik} (1 - \delta_{jl}) - B \delta_{jl} (1 - \delta_{ik}) - C$$
$$- D(1 - \delta_{ik}) \left(\delta_{i+j,k+l} + \delta_{i-j,k-l} \right) \tag{11}$$

which gives the coefficients of the quadratic terms in (9). For the remaining term,

$$I_{ij} = CN \tag{12}$$

3. Results

To compare the optimisation performance of the two CNNs, namely, Chen & Aihara's CSA with decaying self-coupling and Wang & Smith's CSA with decaying time-step, the N-queen problem is solved by computer simulation with N = 5, 10 and 15. For each CNN, feasibility and efficiency are recorded on a parameter space of two chosen parameters of the CNN. Feasibility is the proportion of globally optimal solutions obtained in 100 simulations, each with different randomised initial neuron states centered around the unit hyper-cube. For example, a feasibility of 0.8 means 80 solutions out of the 100 runs are globally optimal. Efficiency is measured by the average number of iterations required for the network to converge to a stable state (within a tolerance of 5×10^{-5}). Only asynchronous cyclic updating of neuron states is used in this study, and A = B = C = D = 1.

The focus of the simulations is on the variation of feasibility and efficiency in the parameter space for each network, and the effect of changing problem size on these measures.

3.1 CSA with Decaying Self-coupling

Chen & Aihara's CSA mentioned in Section 2.1 is used, with $k = 1 - \alpha$ [4]. Same Parameter values as Chen & Aihara [11] are used: $\beta = 0.001$, $I_0 = 0.65$ and $\varepsilon = 0.004$. Results of feasibility and efficiency are plotted on a 2-D parameter space of α and $z(0)$. Fig. 1(a), (b) & (c) shows the feasibility with $N = 5$, 10 & 15 respectively. Fig. 1(d) shows the average number of iterations for $N = 10$ as an example of measuring efficiency.

3.2 CSA with Decaying Time-step

Wang & Smith's CSA described in Section 2.2 with an exponential decay rule as (6) is used. For all simulations, $\varepsilon = 0.01$, $\beta = 0.001$ unless specified otherwise. Results of feasibility and efficiency are plotted on a parameter space of τ and $\Delta t(0)$. Fig. 2(a), (b) & (c) shows the feasibility with $N = 5$, 10 & 15 respectively. Fig. 2(d) shows the average number of iterations for $N = 10$ and Fig. 3 shows the feasibility with $N = 15$ and $\beta = 0.01$, i.e. a faster annealing schedule compared to Fig. 2(c). The energy function E as described in (9) is plotted for the following pairs of parameters: $(\Delta t(0), \tau) = (0.4, 0.25)$, $(0.4, 0.6)$ as Fig. 4 (a) & (b) respectively. The first point $(0.4, 0.25)$ is chosen from the high feasibility region in the $N = 10$ parameter space, and the other point is from a low feasibility region (Fig. 2 (b)).

4. Discussions

From the simulation results shown in Fig. 1, it can be observed that Chen & Aihara's CSA offers robustness in two aspects: high feasibility (80 ~ 90%) is obtained for most values of α and $z(0)$ for a given problem size; and the size and shape of the high feasibility region persists as problem size increases. These two aspects are important for using the neural network to solve practical problems. Note that this high feasibility is achieved without any adjustment to the penalty parameters A, B, C and D, an advantage the Hopfield network cannot offer. From the comparison between Fig. 1(b) and (c), it can be seen that the network offers a slightly higher overall feasibility as the problem size grows from 10 to 15, which means the optimisation ability of the network does not necessarily decrease with problem size.

In terms of efficiency, the network requires fewer iterations to converge when the parameters are in the high feasibility region than in the low feasibility region (Fig. 1(b), (d)), which is interesting as one may expect that it takes longer to arrive at a global minimum than a local minimum. This correlation between feasibility and efficiency maintains for the two other problem sizes as well. This may be due to the particular properties of a chaotic search.

A common feature in Fig. 1(a), (b) and (c) is the low feasibility region in the left side of the parameter space. As α and $z(0)$ represents the steepest descent and nonlinear feedback dynamics respectively, that region suggests interesting behaviours as transition from high to low feasibility takes place as a result of the competition of the two dynamics. It should be noted that $z(0) = 0$ means that no chaotic behaviour is present, and the network is a steepest descent Hopfield network.

For the method of CSA with decaying time-step, the feasibility diagrams show a distinct wedge-shaped region of high feasibility (Fig. 2(a)-(c)). Above that is a low feasibility region and below that is a region which gives divergent properties to the network with weak problem solving ability if any. This correlates with the convergence criteria described in Section 2.2. Although the central wedge-shaped high feasibility region has the same extent as the problem size increases, feasibility gradually drops, especially towards the end with higher values of τ and $\Delta t(0)$. An overall drop in feasibility also occurs to the low feasibility region in the top-left half of the parameter space. When compared to CSA with decaying self-coupling, this method is less robust in the sense that given a problem size, the choice of parameters giving high feasibility is more limited. Scalability is also slightly weaker as the overall feasibility decreases more rapidly with increasing problem size.

In terms of efficiency (Fig. 2(d)), the number of iterations required in the high feasibility region is comparable to that of the decaying self-coupling method. Another difference is the lack of the clear correlation between feasibility and efficiency seen previously.

To see the effect of a faster annealing schedule on feasibility, a larger value of β is used (Fig.3). The wedge-shaped region becomes less distinct and the overall feasibility is much lower. This shows that a sufficiently slow annealing schedule is required to utilise the searching property of this method. The same effect has been observed for the decaying self-coupling method in other research [11,13].

The particular wedge-shape of the high feasibility region is most likely due to stability conditions of the network (7). The top left region corresponds to $\Delta t(0) < \tau$ which is one of those conditions. Fig. 4(a), (b) illustrates the energy evolution for two typical points, one lies within the wedge-shape region and the other one above that. Inside the region (Fig. 4(a)), there is strong transient fluctuations of the energy, which is far weaker outside the region (Fig. 4(b)). This difference in dynamics and the resulting difference in feasibility in the two regions supports the idea that chaotic transient behaviours are largely responsible for the problem solving ability of chaotic neural networks [3,6].

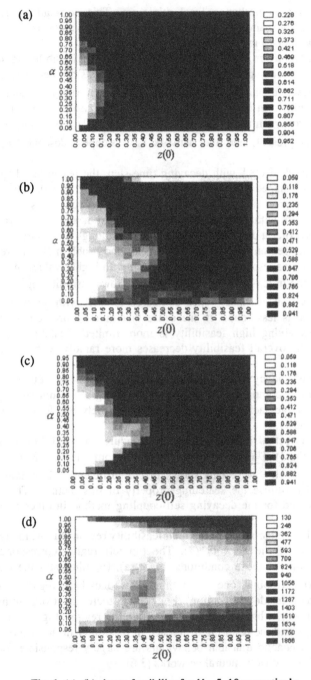

Fig. 1. (a), (b) shows feasibility for $N = 5$, 10 respectively.
(c) feasibility for $N = 15$. (d) average number of iterations for $N = 10$.

455

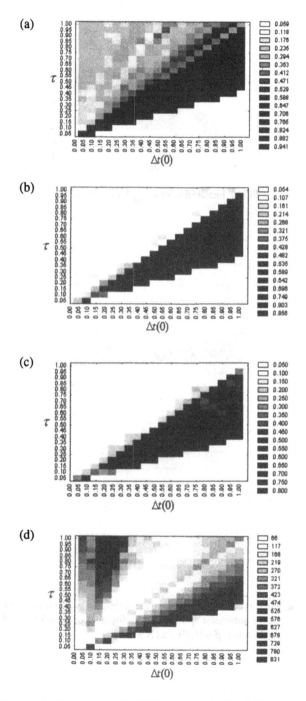

Fig. 2. (a), (b), (c) shows feasibility for $N = 5$, 10 and 15 respectively.
(d) average number of iterations for $N = 10$.

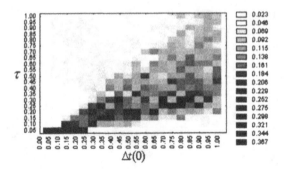

Fig. 3. Feasibility for $N = 15$ with $\beta = 0.01$.

(a) $\tau = 0.25$

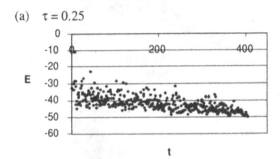

(b) $\tau = 0.6$

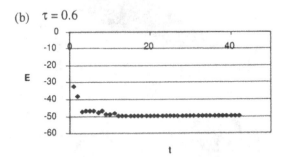

Fig. 4. Time evolution of the energy function for two chosen points in the parameter space (t denotes number of iterations), with $\Delta t(0) = 0.4$, $\beta = 0.001$ and $N = 10$.

5. Conclusions

Two chaotic neural networks, one using CSA with decaying self-coupling and the other one CSA with decaying time-step, are compared on feasibility, efficiency robustness and scalability on a two-parameter space. The N-queen problem with various values of N is solved using the two approaches. The decaying self-coupling approach is found to offer better robustness and feasibility, both for a given problem size and when the problem size increases. Efficiency is comparable for the two methods when high feasibility is achieved. Correlation between feasibility and efficiency for the decaying self-coupling method, together with the optimisation role of the selective transient dynamics of the decaying time-step method, reveals interesting properties of chaotic neural networks for optimisation.

Future work should look into a possible relationship between the high feasibility region and stability properties of both methods, and the transition of network dynamics across the boundary separating high and low feasibility.

References

[1]   Hopfield, J. J., Tank, D. W., (1985). "Neural" Computation of Decisions in Optimization Problems. *Biol. Cybern.*, Vol. 52, pp. 141-152.

[2]   Aihara, K., Takabe, T., Toyoda, M., (1990). Chaotic Neural Networks. *Physics Letters A*, Vol. 144, No. 6, 7, Mar., pp. 333-340.

[3]   Hasegawa, M., Ikeguchi, T., Matozaki, T., (1995). Solving Combinatorial Optimization Problems by Nonlinear Neural Dynamics. *Proc. IEEE Int. Conf. Neural Networks*, Vol. 6, pp. 3140-3145.

[4]   Nozawa, H., (1994). Solution of the Optimization Problem Using the Neural Network Model as a Globally Coupled Map. *Towards the Harnessing of Chaos*, pp. 99-114.

[5]   Yamada, T., Aihara, K., Kotani, M., (1993). Chaotic Neural Networks and The Travelling Salesman Problem. *Proc. 1993 Int. Joint Conf. Neural Networks*, Vol. 2, pp. 1549-1552.

[6]   Ishii, S., Sato, M. A., (1997). Chaotic Potts Spin Model for Combinatorial Optimization Problems. *Neural Networks*, Vol. 10, No. 5, pp. 941-963.

[7]   Tokuda, I., Nagashima, T., Aihara, K., (1997). Global bifurcation Structure of Chaotic Neural Networks and its Application to Traveling Salesman Problems. *Neural Networks*, Vol. 10, No. 9, pp. 1673-1690.

[8]   Asai H., Onodera, K., Kamio, T., Ninomiya, H., (1995). A Study of Hopfield Neural Networks with External Noises. *Proc. IEEE Int. Conf. Neural Networks*, Vol. 4, pp. 1584-1589.

[9]   Hasegawa, M., Ikeguchi, T., Matozaki, T., Aihara, K., (1997). An Analysis on Additive Effects of Nonlinear Dynamics for Combinatorial Optimization. *IEICE Trans. Fundamentals*, Vol. E80-A, Iss. 1, pp. 206-213.

[10] Hayakawa, Y., Marumoto, A., Sawada, Y., (1995). Effects of the Chaotic Noise on the Performance of a Neural Network Model for Optimization Problems. *Physical Review E*, Vol. 51, No. 4, Apr., pp. 2693-2696.

[11] Chen, L., Aihara, K., (1995). Chaotic Simulated Annealing by a Neural Network Model with Transient Chaos. *Neural Networks*, Vol. 8, No. 6, pp. 915-930.

[12] Chen, L., Aihara, K., (1997). Chaos and asymptotical stability in discrete-time neural networks. *Physica D*, Vol. 104, pp. 286-325.

[13] Kwok, T., Smith, K. and Wang, L., (1998). Solving Combinatorial Optimization Problems by Chaotic Neural Networks. *Intelligent Engineering Systems Through Artificial Neural Networks: Neural Networks, Fuzzy Logic, Evolutionary Programming, Data Mining, and Rough Sets*, vol. 8. C. Dagli et al. (eds.) , pp. 317-322.

[14] Wang, L., Smith, K., (1998). On Chaotic Simulated Annealing. *IEEE Trans. Neural Networks*, Vol. 9, No. 4, pp. 716-718.

[15] Wang, L., Smith, K., (1998). Chaos in the Discretized Analog Hopfield Neural Network and Potential Applications to Optimization. *Proc. Int. Conf. Neural Networks*, pp. 1679-1684.

[16] Kwok, T., Smith, K., Wang, L., (1998). Incorporating Chaos into the Hopfield Neural Network for Combinatorial Optimization. *1998 World Multiconference on Systemics, Cybernetics and Informatics*, Vol. 1, pp. 659-665.

[17] Tagliarini, G. A., Page, E. W., (1987). Solving Constraint Satisfaction Problems with Neural Networks. *Proc. IEEE Int. Conf. Neural Networks*, III-741 - III-747.

[18] Takefuji, Y., Szu, H., (1989). Design of Parallel Distributed Cauchy Machines. *Proc. IJCNN.*, I-529 – I-532.

[19] Akiyama, Y., Yamashita, A., Kajiura, M., Aiso, H., (1989). Combinatorial Optimization with Gaussian Machines. *IJCNN*, Vol. 1, pp. I-533 -- I-540.

Study on the Idiotypic Network Model for Feature Extraction of Patterns

Toshiyuki Shimooka[1], Yasufumi Kikuchi[2], Kouichi Shimizu[3].

[1] Graduate School of Eng., Hokkaido Univ., Kita 13, Nishi 8, Kita-ku, Sapporo, Japan, e-mail address: shimo@bme.eng.hokudai.ac.jp
[2] Graduate School of Eng., Hokkaido Univ., Kita 13, Nishi 8, Kita-ku, Sapporo, Japan.
[3] Graduate School of Eng., Hokkaido Univ., Kita 13, Nishi 8, Kita-ku, Sapporo, Japan, e-mail address: shimizu @bme.eng.hokudai.ac.jp

Abstract: The dynamics of an idiotypic network was studied for the purpose of the feature extraction of patterns. When observed patterns are considered antigens, the immune network is the pattern-matching system using antibodies as partial templates. Antigens are represented by steady solutions as internal images via the interaction of antibodies. Farmer's mathematical model and modified model with antibody-antibody complex were adopted, and the dynamics of simple cyclic networks were simulated. Orbits in the phase plots or steady states could represent differences in the network structures. The periodic orbits of the dynamics of the network will be useful for feature extraction of patterns.

1. Introduction

The immune system is a highly evolved biological system and is capable of learning and memory. Many kinds of idiotypic network based on Jerne's theorem[1] have been studied by mathematical methods. In recent years, applications of artificial idiotypic networks have been proposed for such fields as sensor diagnosis, control of mobile robots, etc.[2].

We have attempted to use an idiotypic network model for pattern recognition. The function of the immune system is to identify and eliminate foreign molecules. The antibody response system has many similar points to the pattern recognition method using a set of partial templates. Pattern recognition is a reasonable application of the antibody responses system and is suitable for the study of the potential performance of an artificial immune system in the engineering field. Observed patterns are regarded as foreign antigens. An antigen is usually much bigger than an antibody and has many kinds of epitopes on its surface. Antibodies bind specifically to the epitopes on the basis of the structure of them. When an antibody is regarded as a partial template, binding of the paratope and epitope is exactly complementary template matching (Fig.1).

In Jerne's idiotypic network theorem, an antibody is recognized as an antigen by another antibody, which is called an anti-idiotypic antibody, and a network of antibodies is constructed. When the epitope of the original antigen and the idiotope of anti-idiotypic antigen have the same structure, the anti-idiotypic antibodies are considered as internal images of the parts of the antigen. Invasion of an antigen causes the interaction of

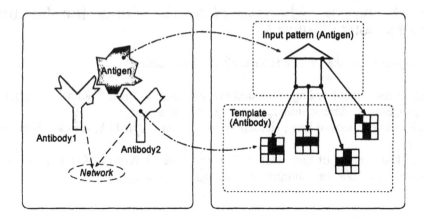

Fig.1 Antibodies and partial templates

many antibodies in the idiotypic network. The state of network, that is, the concentrations of antibodies, is dynamically modified and moves to a new steady state. Therefore, the steady state of the network represents the internal image of the whole antibody.

We focused on network dynamics and the concept of internal images. We investigated the usefulness of the network dynamics, that is, changes in the concentration of antibodies, for extracting pattern features and for classifying them. In this paper, our trial plan for feature extraction is presented.

2. Models and Methods

The outline of the procedures of our present plan is as follows.

1. The observed pattern is matched by partial templates, which have given affinities. The initial values of the concentration of each template (antibody) are obtained.

2. The dynamics of the idiotypic network model is calculated with the initial values.

3. After sufficient calculation, the obtained solution is regarded as the feature of the observed pattern.

4. The obtained solution is compared with the solutions of standard patterns that are obtained in advance by the same procedures.

In this paper, we deal with the network dynamics that are related to steps 2 and 3. Many idiotypic network models are nonlinear simultaneous differential equations. On the assumption that the network is stable like a natural immune system, some classes of equilibrium points will appear in the phase space; point attractor, center (eddy), stable node, stable spiral, etc. It is considered that the property and the state itself represent the internal image of the observed pattern. A different observed pattern is expected to cause different orbits and have different steady solutions. Our recognition problem is whether the original pattern can be distinguished or not when the steady solution is given. We must therefore study the specificity of the solution of the idiotypic network model in detail.

2.1 Idiotypic Network Model

The mathematical model by Farmer[3] was used as a trial model. This model is one of the simplest antibody response systems. It represents only interaction of antigens, antibodies and anti-idiotypic antibodies. The existence of cells such as lymphocytes and that of cytokines is disregarded. The response is not the same as that of natural body but is relatively easy to analyze.   The model is given by

$$\frac{d}{dt}Abi = c\left[\sum_{j=1}^{N} mjiAbiAbj - k_1 \sum_{j=1}^{N} mijAbiAbj + \sum_{j=1}^{n} mijAbiAgj\right] - k_2 Abi \qquad (1)$$

where Abi is the concentration of the antibody i; mij is the matching specificity or affinity of antibody j to antibody i; Agj is the original antigen bound by antibody i; N is the number of classes of antibody, which is equal to the number of classes of partial template; and c, k_1, and k_2 are constants to adjust the rate of response. The final term models the tendency of cells to die at a rate determined by k_2. Since we are not interested in the rate of response, c and k1 are fixed at 1. k_2 is fixed at 0. If k_2 is not 0, all concentrations of antibodies decrease to 0 in a steady state in our simulation.

2.2 Structure and Affinity of an Antibody

Representation of antibodies (templates) and design of the affinity between antibodies are significant points for the response of the network. In this paper, it is assumed that the shape of the template can be represented as a binary string, which is widely used in many studies. It is supposed that each template has a different class of strings and that affinity between templates is automatically determined on the basis of matching of strings. However, when one simply uses the whole string of the template as a paratope of the antibody, the relation of mij$_j$=mji holds because the paratope of an antibody become the epitope for the anti-idiotypic antibody. M(=[mij]) becomes a symmetric matrix. This is a very specific case. For example, in the case of k_1=1, which is a natural assumption because of no superiority between antibodies, the first and second terms of the right side of (1) are eliminated and the model can not work as an immune network. Farmer et al. assigned two binary strings to each antibody, one representing the paratope and the other an epitope or idiotope[3]. In this method, the paratope and the idiotope are independent. Therefore, the designer can arbitrarily define the affinity between two antibodies. This means that the designer has to determine the appropriate affinities for a good response of the network.

The modeling of affinity or specificity of an antibody is also an important factor in the design of the network structure[4]. When the specificity is strong, an antibody can bind to only one or a few epitopes. The generated network is expected to be a relative simple structure. When the specificity is weak, a class of antibodies can bind to many classes of epitopes or idiotopes. The interaction of antibodies makes a highly connected network. One epitope will affect almost all antibodies via the network. It is expected that this type of network made complicated responses. We used Farmer's structure for the structure of the antibody: That is, an antibody consists of one paratope and one epitope. They have

462

the same size of strings; an anti-idiotypic antibody is bound to one antibody. As for the affinity of an antibody, strong specificity is adopted, because that we attempted to study the response of the simple structure networks in this paper. However, it is not clear at the present time whether strong or weak specificity is better for feature extraction of patterns. There are many points to consider regarding the structure and the specificity of an antibody.

3. Results of Numerical Simulation and Discussion

As mentioned above, it is assumed that the partial pattern matching is completed and that the initial concentration of each antibody is given. The class of the bound antibody determines the structure of the network but it is not explicit. The difference in initial values causes a different orbit of solution, and we expected to use the difference in orbits in a steady state or periodic oscillation to extract the feature of the pattern. One of the assumptions of the network model is that the system is not unstable. We started the study on simple cyclic networks. The networks have feedback loop structures that may be able to repress the extreme increase of the concentration of antibodies. Therefore, the network is expected to be not unstable in many cases. The dynamics of the network was calculated by the Lunge-Kutta method. Antigens are eliminated in a relatively short time. Therefore, the network can be regarded as an autonomous system after the antigen is completely eliminated. In the

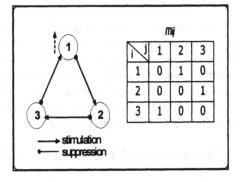

Fig.2 Three-antibodies cyclic network model.

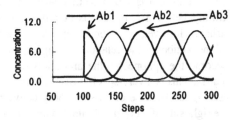

Fig.3 Dynamics of three-antibodies model.

following calculations, it was assumed that the antigens were already replaced by the corresponding antibody and the term of the antigen was ignored.

Fig.2 shows the structure of three-antibodies cyclic network model and affinity matrix M. The value of mij is restricted to 1 or 0, and there is no intermediate value. This means the strongest specificity. An example of the response, i.e., the change in the concentration of antibodies, of the network is shown in Fig.3. The initial concentrations of antibodies were 1.0(arbitrary unit). After 100 steps of calculation, Ab1 was abruptly increased to 10.0. The responses showed oscillation. Each concentration of the antibody has the same waveform, but the phase of them is different.

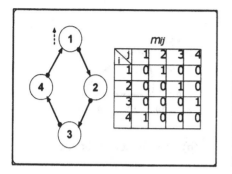

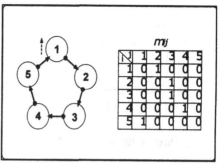

Fig.4 Four-antibodies cyclic network model

Fig.5 Five-antibodies cyclic network model

The four-antibodies cyclic network model and the five-antibodies cyclic network model with the affinity matrix M are shown in Fig.4 and Fig.5, respectively. The conditions of the antibodies for calculation were similar to those mentioned above; all initial values of antibodies are 1.0, Ab1 (100)=10.0.

An example of the response of the four-antibodies model is shown in Fig.6(a). There is a difference in the amplitude between that of Ab1 and Ab2. The waveform of Ab1 and Ab3 or that of Ab2 and Ab4 was almost the same. Fig.6(b) shows an example of the response of Ab1 in the five-antibodies model. Two waves with different amplitude appeared by turns and the amplitude has a relatively long periodic change. Ab1 of the five-antibodies model had a quite different response compared with that of the three- or four-antibodies model.

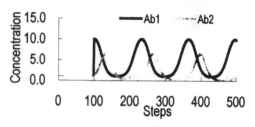

(a) Four-antibodies model.

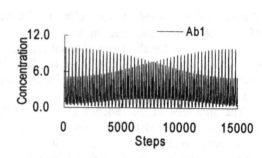

(b) Five-antibodies model.

Fig.6 Transitions of the concentrations of antibodies

Phase plots of Ab1 and Ab2 of three-, four- and five-antibodies models are shown in Fig.7(a), (b), and (c), respectively. As mentioned above, we try to use these orbits in phase plane as the features of the network structures, which are the internal mages of the observed pattern. The periodic parts of the solutions are shown in these figures. The closed curves of the four-antibodies and five-antibodies models are simple closed curves. However, the shapes of the curves were clearly different and easy to be distinguished.

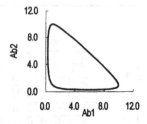

(a) Three-antibodies model.

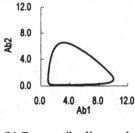

(b) Four-antibodies model.

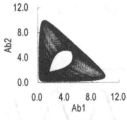

(c) Five-antibodies model.

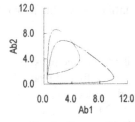

(d) Short term plot in five-antibodies model

Fig.7 Phase plots of dynamics of cyclic network models.

The plot of the five-antibodies model has a quite different property compared with that of the three- or four-antibodies model. It is also a periodic solution, but the period is long. However, in the short-term observation, the orbit looks rather simple. The short-term orbit is shown in Fig.7(d). If we can find this simple orbit from long periodic orbits, it will be useful to deal with the orbit as the feature of the pattern.

The results so far are summarized as follows.

1) The difference in the structure of the network is represented by the geometric property in the phase plot of the solution.

2) The orbits obtained from only two antibodies have to be monitored to distinguish the structures of the network, that is, the observed pattern.

The state of the dynamics is described by the concentration of antibodies. Therefore, the dimension of state space is equal to the number of the class of partial templates in equation(1). However, the dimension of the phase plot which is used in this paper is only second order. It is thought that the phase plot of dynamics of a network is very useful for recognizing the difference in the structure of the network.

The five-antibodies cyclic network, even a simple loop structure, shows a rather complicated response. The following example is the response of a two-loops structure. The structure and affinity matrix M are shown in Fig.8. The model has a loop containing three antibodies and a loop containing four antibodies, and they are connected via Ab3. Again, all initial concentrations of antibodies were 1.0, and Ab1(100) was 10.0. The responses of Ab1 and Ab2 are shown in Fig.9(a). The orbit in the phase plot (Ab1-Ab2)

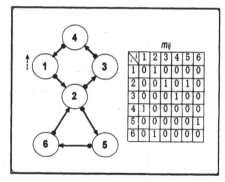

Fig.8 Two-loops model.

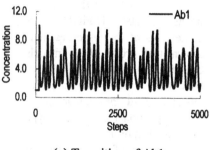

(a) Transition of Ab1

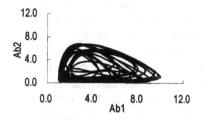

(b) Phase plot

Fig.9 Dynamics in Two-loops model.

is shown in Fig.9(b). The solution did not diverge, but the periodicity of the solution did not appear, even though the network has only two small loops in that structure. In this case, the shape of orbit is clearly different compared with the other examples. However, it seems to be difficult to use this orbit in order to classify the structures. The orbit showed a chaos-like dynamics. We are not sure whether the solution is exactly non-periodic, but for our purposes, it is sufficiently non-periodic. In the case of a non-periodic solution, our plan, i.e., using steady orbits as feature of patterns will not work well.

To improve time responses, we tested introduction of the dissociation between antibody-antibody complexes. In the natural body, it is well known the existence of reversible binding reactions between an antigen and antibodies. So there will be same reaction between an antibody and an anti-antibody. This reaction makes the function as a buffer, therefore the antibody response is expected to become more stable. Equation (2) was added and equation(1) was modified as equation(3).

$$\frac{d}{dt} Acij = mijAbiAbj - rAcij \tag{2}$$

$$\frac{d}{dt} Abi = \sum_{j=1}^{N} mjiAbiAbj - k_1 \sum_{j=1}^{N} mijAbiAbj + r \sum_{j=1}^{N} (Acij + Acji) - k_2 Abj \tag{3}$$

where Acij is a concentration of antibody i -antibody j complex, and r is a dissociation constant. r was 0.01 and k_1 was 3.0 in this study. As to k_1, because of the equilibrium condition of the system, the parameter has to be the value.

Examples of responses of four- and five-antibodies model and two-loops model are shown in Fig.10(a), (b) and (c), respectively. Oscillations were apparently attenuated. Concentrations of antibodies moved to steady states very rapidly; several ten steps after the input of Ab1. Ab1 of steady points were about 1. Fig.11 shows phase plots of the responses. It is shown that orbits of all models converged on steady states. Obits of three-,four- and five-antibodies model showed the similar shape. Two-loops model had another type of shape of an orbit.

Steady points of the response of each model are shown in Fig.12. Though potions on a phase plot were rather similar, each model had different a steady point. Introduction of antibody-antibody complex can reduce the steps of calculations and make the response have steady state. This method is expected to be useful for feature extraction.

The results in this paper indicate that the idiotypic network has sufficient potential performance to extract the structure of the patterns. The qualitative property of the solution of a non-linear differential

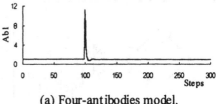

(a) Four-antibodies model.

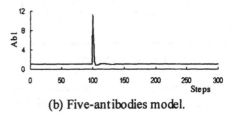

(b) Five-antibodies model.

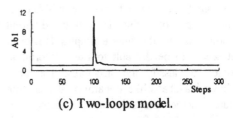

(c) Two-loops model.

Fig.10 Transitions of Ab1 of antibody-antibody complex model.

equation generally depends on the equilibrium point and the eigen values of the Jacobian matrix of the right side of equation(1). Based on the affinity matrix and the initial state, many classes of equilibrium points will appear. We analyzed three-antibodies and four-antibodies cyclic network models, and the results show that the eigen values of the models are purely imaginary numbers. The models, therefore, have centers with every initial state and have closed orbits as shown in the phase plots. In this case, the solutions have no inset in the phase plot, and a different initial state makes a different orbit. If the solutions of standard patterns have several equilibrium points and those have insets, such as point attractors, the observed pattern would be classified on the basis of the insets. In order to design the property of the equilibrium point, the differential equations or the affinity matrix has to be modified to change the eigen values. Segel reviewed the idiotypic network model that has multiple attractors [5]. Several models must have multiple attractors. These models may be useful for our purposes.

Of course, there are many problems to be considered. One problem is as follows. One of the assumptions of Farmer's model is that the system is well mixed. Therefore, all antibodies exist in the same space or at one point and have no distinction. This means that the results depend only on the number of matching antibodies, and the connections or relative positions of them are ignored. When the observed patterns have the same

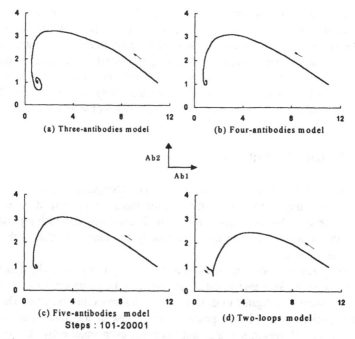

Fig.11 Phase plots of responses of antibody-antibody complex models.

parts with the same concentration, they are all recognized as the same pattern. To deal with this problem, we are planning to investigate the following systems. One is the adoption of the diffusion of antibodies, that is, the system is regarded as a distributed system. In this system, several domains are set to recognize one observed pattern, and a sub-network system is constructed in each domain. It is assumed that distance exists between the domains and antibodies diffuse from one domain to others. The whole system is not mixed, but the subsystem is well mixed in each domain. It is expected that the diffused antibodies

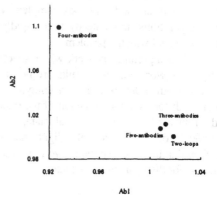

Fig.12 Steady points of antibody-antibody complex models.

contain information on the location of the parts of the pattern. When the domains are small enough and the diffusion is appropriately restricted, the existence of a simple structure of networks, which is adopted in this paper, is probable. Therefore, the results of this study will be useful in each domain. Another system is a hierarchical structure of antibodies. The assumption in this system is that the complex of an antibody and an anti-idiotypic antibody can become a new idiotope. The antibody corresponding to the complex may not exist in the natural body. However, it will be useful for our purposes, because the existence of the antibody is essentially the information of the structure. The existence of the anti-complex antibody will be designed when the standard pattern is handled. A system like that of the thymus, which educates and selects useful lymphocytes, will be needed. It is possible that the idiotypic network can deal the structural information via the dynamics of the concentration of antibodies.

4. Concluding Remarks

The dynamics of an idiotypic network, i.e., changes in the concentration of antibodies, was studied for the application of the feature extraction of patterns. The periodical or steady solution is considered as the internal image, which represents the structure of the network or initial values of antibody concentrations, that is, the feature of the observed pattern.

Farmer's mathematical model was adopted as the idiotypic network. The dynamics of the small cyclic networks were simulated, and orbits in the phase space of the antibody concentrations were investigated to distinguish the difference in structures. The results suggest that periodic orbits in the phase plane are useful for discrimination of the network structures. However, it is also indicated that even small networks are able to have a non-periodic solution.

Introduction of antibody-antibody complex, which works as a buffer, made long or non-periodic solutions move steady state rapidly. This model is expected to be a useful measure to deal with the problem.

This study is a preliminary work with respect to the formation of internal images by the idiotypic network, and the results are not sufficient to discuss the feasibility for real application. In order to design an idiotypic network, many significant points must be considered; for example, selection of the mathematical model of the network, structure and affinity of antibodies, initial state, etc. The property of the solution will be completely different with only slight changes in these design factors. However, we consider that the fact, that the network can have a variety of responses, suggest the high potential for distinguishing the patterns.

References

[1] N. K. Jerne, The immune system, Sci. Am. 229:52-60, 1973

[2] Y. Yoshida, et al., Immunity-based system and its application, Korona Publishing, 1998 (in Japanese)

[3] J.D. Farmer and N. H. Packed, The immune system, adaptation, and machine learning, Physica 22D:187-204, 1986

[4] A.S. Perelson, Immune network theory, Immunological Reviews 110:5-36, 1989

[5] L.A.Segel, Multiple attractors in immunology: theory and experiment, Biophysical Chemistry, 72: 223-230, 1998

Chapter 6: Rough Sets

Papers:

Rough Set Based Uncertainty Management for Spatial Databases and Geographical Information Systems

T. Beaubouef, F. Petry, and J. Breckenridge

Soft Computing for Evolutionary Information Systems – Potentials of Rough Sets

A. B. Patki, G. V. Raghunathan, S. Ghosh, S. Sivasubramanian, and A. Khurshid

Towards Rough Set Based Concept Modeler

A. B. Patki, G. V. Raghunathan, S. Ghosh, and S. Sivasubramanian

Rough Set Based Uncertainty Management for Spatial Databases and Geographical Information Systems

Theresa Beaubouef[1], Frederick Petry[2], John Breckenridge[3]

[1]Computer Science Dept., Southeastern La. University, Hammond, LA 70402 USA
 Email: tbeaubouef@selu.edu
[2]Dept. of Elect. Eng. and Computer Science, Tulane University, New Orleans, LA 70118
[3]Naval Research Laboratory, Stennis Space Center, MS 39529 USA

Abstract: Uncertainty management is necessary for real world applications. This especially holds true for database systems. Spatial data and geographic information systems in particular require some means for managing uncertainty. Rough set theory has been shown to be an effective tool for data mining and for uncertainty management in databases. This paper addresses the particular needs for management of uncertainty in spatial data and GIS and discusses ways in which rough sets can be used to enhance these systems.

1. Introduction

A spatial database is a collection of data concerning objects located in some reference space, usually, but not necessarily, referring to positions on and about the earth. Data for various attributes are stored for all items of interest, as is in the case for "ordinary" databases such as those found in businesses. Spatial data, however, is interrelated within the context of some physical "space", so additional information must be stored in the database to position data items in the space and to capture spatial relationships among the various data items.

Geographical Information Systems (GIS) are complete systems typically including a spatial database component and various display and data analysis functions provided in some type of user environment. These systems are used in a variety of applications including remote sensing, forestry, land-use mapping, urban planning, census taking, cartography, and many others.

A database is simply a collection of data stored in some schema, which attempts to model some enterprise in the real world. The real world abounds in uncertainty, and any attempt to model aspects of the world should include some mechanism for incorporating uncertainty. There may be uncertainty in the understanding of the enterprise or in the quality or meaning of the data [1]. There may be uncertainty in the model, which leads to uncertainty in entities or the attributes describing them. And at a higher level, there may be uncertainty about the level of uncertainty prevalent in the various aspects of the database.

In relational databases it has been demonstrated that uncertainty may be managed via rough set techniques by incorporating rough sets into the underlying data model [2] and through rough querying of crisp data [3]. In this paper we point out those areas peculiar to spatial databases and GIS which are in need of uncertainty management and suggest ways in which rough sets techniques may be used to alleviate the problems to result in a better overall system. Most of these unresolved problems are mentioned in one form or another in [4], which also provides an excellent background for spatial information systems. Implementation of the rough set techniques may be accomplished via methods established in [2,3], details of which are beyond the scope of this paper.

2. Background: Rough Sets

Rough set theory, introduced by Pawlak [5] and discussed in greater detail in [6,7,8], is a technique for dealing with uncertainty and for identifying cause-effect relationships in databases as a form of database learning [9]. It has also been used for improved information retrieval [10] and for uncertainty management in relational databases [2,3].

Rough sets involve the following:

> U is the *universe*, which cannot be empty,
> R is the *indiscernibility relation*, or equivalence relation,
> $A = (U,R)$, an ordered pair, is called an *approximation space*,
> $[x]_R$ denotes the equivalence class of R containing x, for any element x of U,
> *elementary sets* in A - the equivalence classes of R,
> *definable set* in A - any finite union of elementary sets in A.

Therefore, for any given approximation space defined on some universe U and having an equivalence relation R imposed upon it, U is partitioned into equivalence classes called elementary sets which may be used to define other sets in A. Given that $X \subseteq U$, X can be defined in terms of the definable sets in A by the following:

> *lower approximation of X in A* is the set $\underline{R}X = \{x \in U \mid [x]_R \subseteq X\}$
>
> *upper approximation of X in A* is the set $\overline{R}X = \{x \in U \mid [x]_R \cap X \neq \emptyset\}$.

Another way to describe the set approximations is as follows. Given the upper and lower approximations $\overline{R}X$ and $\underline{R}X$, of X a subset of U, the R-positive region of X is $POS_R(X) = \underline{R}X$, the R-negative region of X is $NEG_R(X) = U - \overline{R}X$, and the boundary or R-borderline region of X is $BN_R(X) = \overline{R}X - \underline{R}X$. X is called R-definable if and only if $\underline{R}X = \overline{R}X$. Otherwise, $\underline{R}X \neq \overline{R}X$ and X is rough with respect to R. In Figure 1 the universe U is partitioned into equivalence classes denoted by the squares. Those elements in the lower approximation of X, $POS_R(X)$, are denoted with the letter P and elements in the R-negative region by the letter N. All other classes belong to the boundary region of the upper approximation.

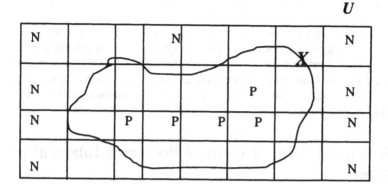

Figure 1. Example of a rough set X.

Consider the following example:

Let U = {tower, stream, creek, river, forest, woodland, pasture, meadow}.

Let the equivalence relation R be defined as follows:

R* = {[tower], [stream, creek, river], [forest, woodland], [pasture, meadow]}.

Given some set X = {medium, small, little, tiny, big, huge}, we would like to define it in terms of its lower and upper approximations:

$\underline{R}X$ = {tower, stream, creek, river}, and

$\overline{R}X$ = {tower, stream, creek, river, forest, woodland, pasture, meadow}.

A *rough set in A* is the group of subsets of U with the same upper and lower approximations. In the example given, the rough set is

{{tower, stream, creek, river, forest, pasture}
{tower, stream, creek, river, forest, meadow}
{tower, stream, creek, river, woodland, pasture}
{tower, stream, creek, river, woodland, meadow}}.

The major rough set concepts of interest are the use of an indiscernibility relation to partition domains into equivalence classes and the concept of lower and upper approximation regions to allow the distinction between certain and possible, or partial, inclusion in a rough set.

The indiscernibility relation allows us to group items based on some definition of 'equivalence' as it relates to the application domain. We may use this partitioning to increase or decrease the granularity of a domain, to group items together that are considered indiscernible for a given purpose, or to "bin" ordered domains into range groups.

In order to allow *possible* results, in addition to the obvious, certain results encountered in querying an ordinary spatial database system, we may employ the use of the boundary region information in addition to that of the lower approximation region. The results in the lower approximation region are certain. These correspond to exact matches. The boundary region of the upper approximation contains those results that are possible, but not certain.

3. Rough Sets for Management of Problems Inherent in Spatial Data

We now identify and discuss particular problems inherent in spatial data and suggest rough set approaches to managing uncertainty for each. We have classified these problems into three general categories: problems related to data quality, problems related to precision and granularity, and problems associated with topology. There will obviously be some overlap of the categories mentioned in the motivating examples, but each subsection below will focus on issues relevant to the particular category being discussed. It should also be noted that the types of problems discussed are not all encompassing for spatial information systems, but only representative of some of commonly encountered problems.

3.1 Data problems

Many of the problems associated with data are prevalent in all types of databases systems. Spatial databases and GIS contain descriptive as well as positional data. The various forms of uncertainty may occur in both types of data, so many of the issues discussed below apply to ordinary databases as well. See [2, 3] for in-depth discussion of implementation of rough set uncertainty in (non-spatial) databases. These same techniques may be employed for spatial contexts.

In spatial information systems data is collected and input into the system in a variety of ways. Data should be verified for correctness before being accessible to users of the system. However, it may be the case that the only data available in answer to a particular request is as yet unverified. Consider the case where a survey ship is collecting data from various sensors for pre-defined attributes and compiling this data into a larger system with verified data. Now suppose there is some type of emergency situation that requires knowledge of information about some data area, part of which has just been collected, an area that has not already been included as verified data in the database. We have several options available. We may report that there is no data available for the prescribed area, which is not the best solution since there *is* some data about the area; it just hasn't yet been verified. We may offer the user the opportunity to refine the query to include only the area in which verified data is available, a better solution, but still possibly not very helpful. Better yet, we may use the approximation region concept from rough sets theory to allow results to contain both *certain*, or verified, data and *possible*, or unverified data. In all cases the user is informed about the certainty or uncertainty of results.

A similar problem occurs in systems that model phenomena that vary or change over time such as ocean currents and tides. In such a system we would like data to be current, or up-to-date. If the only data available, however, is "old" data which may or may not be valid any longer, we may still be able to gain some insight or knowledge by viewing results obtained from this old data, keeping in mind that it is uncertain in terms of the present situation. A person knowledgeable in the domain area could perhaps use the information to aid in making predictions about the present.

Similar data could be entering the system from various sources. For example, there may be multiple sensors surveying some area. Or we may have historical ground truth data for land use in some area of study and data resulting from classification of satellite imagery of the same area. It is likely that these data will not be consistent for the entire spatial extent due to changes in land use over time, human or sensor error, or errors resulting from the classification scheme. In such situations where there is conflicting data from multiple sources, we cannot be certain about the validity of either. Both are possible. Therefore, the user should be presented with both the *certain* results obtained from nonconflicting sources along with all *possible* results.

In the previous examples uncertainty in data may be managed by the approximation region concept of the rough set theory. Certain results belong to the lower approximation regions and possible results belong to the boundary region of the upper approximation. Recall, however, that these regions are inherently based on the concept of the indiscernibly relation. Some types of data quality problems can be managed through the representation of indiscernibility. These include, among others, imprecision in measurement as may be obtained directly through sensors or recorded by humans and imprecise or inconsistent wording of descriptive data entered by humans. For example, we may use instruments that measure earth data at a precision which is greater than that required by the model, we may have built-in error tolerances, or we may have decided to group or bin the data into certain categories such as {[high reflectivity], [moderate reflectivity], [low reflectivity]} or {[northerly], [easterly], [southerly], [westerly]}. In all of these, we define our indiscernibility relation to partition our domain of data values into classes that are appropriate for the application.

When humans are making judgments for data entry, they often differ in the words selected to describe a particular entity. For example, one individual may refer to a flowing waterway as a "river" and another may refer to it as a "creek". In some applications the distinction between the two may be necessary, but for others where it is only necessary to know that the water flow is at a particular location the terms "river", "creek", and "stream" are considered indiscernible and therefore should be treated as equivalent by the database system. It is also possible for data wording to differ between the person or persons entering the data and those trying to access the data. For example, an area may have been viewed as a meadow by a person out in the field surveying the area. Later, if a user were trying to access spatial extents, which may have been used for "pasture", it would probably be desirable to return those designated "meadow" as well. If these values were identified in the spatial database indiscernibility relation as being equivalent for the application at hand, then it would not matter which term was used in the data entry or the query since any one of the terms would represent them all.

3.2 Scale, resolution, granularity problems

Often spatial data is associated with a particular grid. The positions are set up in a regular matrix-like structure and data is affiliated with point locations on the grid. This is the case for raster data and for other types of non-vector type data such as topography or sea surface temperature data. There is a tradeoff between the resolution or the scale of the grid and the amount of system resources necessary to store and process the data. Higher resolutions provide more information, but at a cost of memory space and execution time.

If we approach the data from a rough set point of view, we can see that there is indiscernibility inherent in the process of gridding or rasterizing data. A data item at a particular grid point in essence may represent data near the point as well. This is due to the fact that often point data must be mapped to the grid using techniques such as nearest-neighbor, averaging, or statistics. We may set up our rough set indiscernibility relation so that the entire spatial area is partitioned into equivalence classes where each point on the grid belongs to an equivalence class. If we change the resolution of the grid, we are in fact, changing the granularity of the partitioning, resulting in fewer, but larger classes.

There are many example of the utilization of this type of indiscernibility in spatial data. In viewing an image on a display monitor, for example, a user may zoom in or out to view an area at different resolutions. In statistical or other data analysis procedures, it may be desirable to incorporate only some of the data taken from regularly spaced intervals in an effort to improve the speed and decrease the memory requirements of the analysis procedure. In both cases, the user is aware of the loss of resolution. He or she is in effect changing the granularity of the partitioning of the underlying data, data that at full resolution, still contains inherent uncertainty. The actual data remains unchanged.

The approximation regions of rough sets come into play when information concerning sizes, lengths, and other areal properties of spatial data features are calculated or displayed.

Consider an areal feature such as a lake. One can reasonably conclude that any grid point identified as "lake" that is surrounded on all sides by grid points also identified as "lake" is in fact a point represented by the feature "lake". However, consider points identified as lake that are adjacent to points identified as land. Is it not possible that these points represent land area as well as lake area but were identified as lake in the classification process? Likewise, consider those points identified as "land" but adjacent to "lake" points. Is it not also possible that some of these points may represent areas that contain part of the lake? For some applications it would be useful to represent this type of boundary or uncertain region. This uncertainty maps naturally to the use of the approximation regions of the rough set theory, where the lower approximation region represents the certain data and the boundary region of the upper approximation represents the uncertain data.

In the preceding example, if we consider changing the granularity of the partitioning, we see that increasing the grid resolution (forcing a finer granulation) will result in a smaller boundary region and that decreasing the resolution (a coarser granulation) will result in a larger boundary region. As the partitioning becomes finer and finer, eventually a point is reached where the boundary region is non-existent. In this case, the upper and lower approximation regions are the same and there is no uncertainty in the spatial data.

Next consider an application which models a curvilinear feature such as a coastline. When data for the coastline is gridded, information associated with position of points between the gridded points is lost. These points can later be approximated by one of several techniques including that of connecting known points by straight line segments or by lines derived from cubic splines or other functions. As in the lake example, the granularity of the partitioning determines to some extent the uncertainty of the data and as the partitioning becomes finer and finer the uncertainty approaches zero. The line approximation techniques also play a part in the uncertainty.

What really matters is the amount of uncertainty acceptable for a particular application. A display of a locator map might allow for a good bit of uncertainty. However, suppose it is desirable to compute the length of some coastline for the purpose of determining the number of plants to purchase for an erosion-control project. In this case a better estimate of length is necessary. If the exact length of the coastline cannot be determined (by actually measuring it by hand) approximations for the range of minimum possible to maximum possible lengths can be computed based on the indiscernibility level in use. None of the predictions will give the exact length of the coastline, but having the predictions and knowing that all of the predictions are uncertain is a form of information, which an analyst can use as an aid in decision making.

3.3 Topological structure problems

A rough set approach is also quite useful for managing the types of uncertainty related to topology. These include concepts such as nearness, contiguity, connection, orientation, inclusion, and overlap of spatial entities.

In GIS or spatial databases, it is often the case that we need information concerning the relative distances of objects. Is object A *adjacent to* object B? Or, is object A *near* object B? The first question appears to be fairly straightforward. The system must simply check all the edges of both objects to see if any parts of them are coincident. This gives the *certain* results. However, often in GIS data is input either automatically via scanners or digitized by humans, and in both cases it is easy for error in position of data objects to occur. Therefore, we may also want to have the system check to see if object B is very near object A, to derive the *possible* result. If so, the user could be informed that "it is not certain, but it is possible, that A is adjacent to B." Assume we want to know whether a cliff is next to the sea. If the system returns the results that it is possible, but not certain, that the cliff is adjacent to the sea, we may be led to investigate the influence of the tides in the area to determine whether low beaches alongside the cliffs are exposed at low tide.

Now we must address the concept of nearness. Nearness is a relative term that cannot be represented by any exact spatial query mechanism for the database. However, the approximation regions of rough sets can be used to determine nearness as follows. For a particular application, the designer/user of the system decides what distance constitutes nearness based on the units of measurement in use. These definitions may vary for different applications and scales. "Is the tower near the canal?" and "Is the factory near a major city?" are examples of how the meaning of nearness can vary. Objects that are adjacent and objects that are spaced less than some prescribed distance from each other are "near". The lower approximation region of rough sets represents this "nearness". We expand the distance to include object distances that are considered "relatively near" for the boundary region of the upper approximation, which represents possible results.

The concepts of connection and overlap can be managed by rough sets in a similar manner to the above. Connection is similar to adjacency, but related to line type objects instead of area objects. Overlap can be defined in a manner similar to that of nearness with the user deciding how much overlap is required for the lower approximation. Coincidence of a single point may constitute *possible* overlap, as can very close proximity of two objects, if there is a high degree of positional error involved in the data.

Inclusion is related to overlap in the following way. If an object A is completely surrounded by some object B, perhaps we can conclude certainly that A is included in B, lacking additional information about the objects. If the objects overlap, then it may be possible that one object includes the other. Approximation regions can be defined to reflect these concepts as well.

Orientation is another key area requiring uncertainty management in spatial data systems. It is often the case that it is desirable to determine the orientation of one object with respect to another. For example, we may want to know which towns are south of the airport, or whether or not the sugar cane field is east of the industrial waste site. In such cases, we are dealing with areas and general directions, not with particular point locations and exact compass headings.

Rough set techniques can be utilized to manage the uncertainty in direction, defining conditions for lower and upper approximation regions for each of the directions north, south, east, and west. Let us consider the direction "east" and define approximation regions to denote "east of." The other directions can be defined in an analogous manner.

We want to determine whether some object A is east of object B. If the entire extent of an area (object A) is between northeast and southeast of the reference (object B), then we determine that A is east of B. This is denoted by the lower approximation region, which gives the certain results. The upper approximation may define "east of" in a broader sense as follows: If any part of object A is at any direction ranging from true north through east to true south of object B, then A is east of B. This additional *possible* information could prove very useful in spatial queries involving direction.

In topological queries the approximation regions of the rough sets theory can be used to improve recall by returning possible, in addition to certain, results. The user, having knowledge of the certainty of these results, is exposed to greater information than could be extracted in a spatial data system without rough set uncertainty management.

4. Conclusion

Spatial and geographical information systems will continue to play an ever-increasing role in applications based on spatial data. Improvements in technology, the widespread use of global positioning systems, and interest in virtual reality applications will result in spatial information systems being applied to new areas of research and everyday life. Uncertainty management will be necessary for any of these applications.

In this paper we identified several aspects of spatial databases in need of uncertainty management and demonstrated the use of rough set techniques for management of this uncertainty. Examples from the real world were used as motivation for each of the areas needing improvement.

Rough sets are an approach that fits in well with spatial data and its inherent problems. We have shown how the rough sets concepts of indiscernibility and approximation regions can be used to manage uncertainty in spatial data and aid in spatial data querying. We are also investigating the use of fuzzy set approaches and combinations of rough and fuzzy sets theory for possible application to these and related problems.

References

[1] Strong, D., Lee, Y., and Wang, R., 1997, "10 Potholes in the Road to Information Quality," *IEEE Computer*, vol. 30, no. 8, pp. 38-46.

[2] Beaubouef, T., Petry, F., and Buckles, B., "Extension of the Relational Database and its Algebra with Rough Set Techniques," *Computational Intelligence*, Vol. 11, No. 2, May 1995, pp. 233-245.

[3] Beaubouef, T. and Petry, F., "Rough Querying of Crisp Data in Relational Databases," *Third International Workshop on Rough Sets and Soft Computing (RSSC'94)*, San Jose, California, November 1994.

[4] Laurini, R. and Thompson, D., 1992, *Fundamentals of Spatial Information Systems*, Academic Press, London.

[5] Pawlak, Z., 1984, "Rough Sets," *International Journal of Man-Machine Studies*, vol. 21, pp. 127-134.

[6] Grzymala-Busse, J., 1991, *Managing Uncertainty in Expert Systems*, Kluwer Academic Publishers, Boston.

[7] Pawlak, Z., 1991, *Rough Sets: Theoretical Aspects of Reasoning About Data*, Kluwer Academic Publishers, Norwell, MA.

[8] Komorowski, J., Pawlak, Z., Polkowski, L., et al, 1999, "Rough Sets: A Tutorial," in *Rough Fuzzy Hybridization: A New Trend in Decision-Making* (ed. S. K. Pal and A. Skowron), Springer-Verlag, Singapore, pp. 3-98.

[9] Slowinski, R., 1992, "A Generalization of the Indiscernibility Relation for Rough Sets Analysis of Quantitative Information," *First International Workshop on Rough Sets: State of the Art and Perspectives*, Poland.

[10] Srinivasan, P., 1991, "The importance of rough approximations for information retrieval," *International Journal of Man-Machine Studies*, 34, pp. 657-671.

Soft Computing for Evolutionary Information Systems - Potentials of Rough Sets

A.B. Patki[1], G.V. Raghunathan[1], Soumik Ghosh[2], S. Sivasubramanian[1], Azar Khurshid[3]

[1]Department of Electronics, 6, C G O Complex, New Delhi-110003, INDIA.
[2]Software Engineer, Schlumberger: Tests and Transactions, Besancon, FRANCE.
[3]School of Computing, University of Plymouth, Plymouth PL48AA, U.K.

Abstract: The paper discusses the issues involved in design of Integrated Intelligent Information Systems (IIIS) wherein information is viewed as a function of concepts and which has **open ended query** handling features. Need for evolutionary software - tools and a new programming language which could support soft computing with special reference to Rough sets as potential candidate has been examined. A shift from large scale computing to large scope computing in the context of Multimedia and Multilingual databases is highlighted.

1. Introduction

In the past decade, the growth in the use and popularity of the PC has been unprecedented. Also, concurrently there has been a tremendous growth in the use of computers for entertainment, information services and in corporate computing. All these have led to people expecting more from computers. However the varying needs of the professionals and the general public have not been met to their full expectation, mainly due to absence of sectoral information in consolidated form. With the advancement in technology it has been possible to create and maintain information in text, audio, video, graphics and image form. Despite all these, the users requirement has not been fully satisfied because of lack of support facilities in respect of information systems with analysis, modeling, simulation, forecasting, automation, decision support etc., for handling concepts and knowledge based on concepts. In any society driven by information, there are certain basic issues which need to be tackled. a) Generation of information, determining it's content, etc. b) Identifying specific users of various information. c) Service provider(s) to provide general/specific information to general/specific users through a network or delivery medium d) Availability of Connectivity Systems/Devices to launch the information into the network through suitable switches and to help the users to access information from the network with in-built provisions for accounting, management, etc. [1]. To provide information for the masses by the turn of the century and to have dynamic user responsive systems, it would be necessary to shift from database management systems to Integrated Intelligent Information Systems (IIIS) which would pave the way for Multimedia-Multilingual Information Services Network Operating System (MISNOS) [2]. The IIIS should be capable of learning and analyzing the user query requirement and generating the required information.

In contrast to the existing database management systems including OLAP/OLTP, the starting point of such Integrated Intelligent Information systems would be a database management system which would contain Multimedia and Multilingual data. Multimedia and Multilingual databases as required for such information systems do not exist today. It is envisaged that such databases and information systems would be the first step towards developing systems which would provide intelligent information, guidance and help in the virtual reality mode [2] *While the following features would be desirable but not exhaustive in such an IIIS, these would become the essential prerequisites and the barest minimum features:

- Informative, interactive and responsive.
- Prior training should not be required for the user for system operation.
- System handling should not demand the use of a user's manual.
- System guided data and information insertion by the user.
- System generated information modules for different type of queries.
- Networking.

In the following sections of the paper, these issues are discussed and the need for intelligent modules, intelligent tools, etc., which would use soft computing with specific reference to rough sets are brought out.

2. Past Efforts

In the past, efforts have been made to develop sectoral based Intelligent Information System - Software Industry Intelligent Information System (SIIIS) which is capable of providing dynamic industry user responses [3]. A particular system for software industry was designed which contained a Database Management system, a Knowledge Base, a Processed Data and Knowledge Base Integrator; and a Information Processor which took raw information as input. These efforts also looked into the issues of data overload, One time Referral situations, raw data, processed data, Raw Information and Intelligent Information. These efforts have highlighted the problems of data overload primarily due to two main causes:

1. Data is imprecise, incomplete, vague, etc.
2. Inadequacy to represent data on non-trivial cognitive terms.

On the basis of limited work carried out for SIIIS [3] for meeting the requirements of MISNOS [2], a tentative list of software tools under the following categories were also identified for taking up development work: a) Analysis and design (for structured, object oriented and client/server methodologies) tools; b) Documentation tools; c) Project Planning tools; d) Benchmarking tools; e) Quality assurance and improvement tools; f) Standardizing tools; etc.

Even though the SIIIS paper [3] discussed the two causes for data overload problems, the paper had not addressed data overload due to "concept definition", information as a function of concepts; and the knowledge derived out of concept and information in a seamless integrated manner. It had not addressed issues pertaining to:

A) Open ended query-processing capabilities.

B) Modelling the unstructured and unmodeled data, which will be flowing in as Multimedia and Non-Multimedia objects.

C) Deriving concept definitions from unstructured queries.

D) Mapping the concepts on the data.

The present paper discusses an effort to bring in an Integrated Intelligent Information System (IIS), which would have the Intelligent Information System shown in the Figure 1. [3], as one of it's modules. In addition it would have the two modules - Query processor and Intelligent modelling Unit.

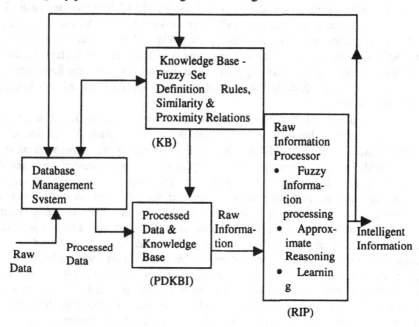

Fig. 1 Intelligent Information System (IIS)

3. Multimedia and Multilingual Databases

As a DBMS forms the entry point for an IIS [3], it is necessary to study their structure and form for IIIS. Other than classifying databases as Relational and Non-Relational, we can also classify them as Multimedia and Non-Multimedia databases. Multimedia databases as required for IIIS do not exist today. A Multimedia database would consist of data in the form of Text, pictures - motion and still, computer generated images and drawings, audio - music and speech, and a combination of these. These Multimedia objects would be analogous to the data fields in such databases and we would have to do operations ON these objects in contrast to the existing system where the operations are done WITH the data objects in totality. e.g. like in ACCESS, a full picture can be stored in a data field as an object. As it exists today, this could only be retrieved. There is no provision to perform operations on these objects. The idea of textual databases along with limited audio-video data objects can be viewed as method of expression. In order to meet the requirements of information for the masses as brought out in earlier sections, the Multimedia should enhance its scope to cover sensory stimuli as method of expression. The efforts in these direction have been restricted to audio, video and textual sensory stimuli and in order to give them true power of expression a concept mapping needs to be viewed as a core, instead of looking at these objects merely as collection of symbols, bitmaps or audio signal. Thus existing approaches to Multimedia database could be viewed as pseudo-Multimedia database. This situation is similar to the case of availability of Visual Basic, Visual C++, etc., in the case of pseudo visual programming, since Visual Basic/Visual C++ have only provided superficial input-output capabilities. In the context of textual databases, the natural language processing has taken recourse to intermediate language representation. We illustrate the concept mapping approach for natural language processing problem.

Consider the following example: A businessman in Japan makes the following query in Japanese - " What are the strategic plans for growth of the software industry in France, Britain, Russia and Germany? What prospects do these plans hold for my company? What softwares do they consider are probable candidates for Development for future? ". The query is conceptual, requires conceptual analysis and the concept is context sensitive. So, as a first step the contextual concept from the query has to be derived. Experts study various companies and prepare strategic reports on these. The strategic analysis section of this report is likely to contain the required material in textual and/or in a combination of textual and graphical form. This entire report itself could be existing as an attribute value (as an object) in a relation in the database itself. These reports are existing in Britain in English, in France in French, in Russia in Russian and in Germany in German. This could be viewed as relations in multilingual domains. Say a relation in Britain (R_B) would be defined over a set of attributes: $R_B = <A_1, A_2, ...A_n>$ values/objects of which are derived from set of domains { $D_1, D_2, ...D_n$} where $D_i \neq D_j$.

The textual material could be one of the objects in the relation existing in tuples. These are in the form of text, which is unmodeled .The text, itself (the strategic report) could be viewed as the approximation space on which the concepts have to be defined as objects. Information extraction would involve extracting the concept from the query and assessing the data/knowledge requirement and matching with the dynamically extracted concept from the textual data at various locations. The information derived from text at each location (by using the IIS) has to be unioned and modeled into the final information and then given to the Businessman in Japanese.

The first step in making a database consists of modelling the real world into a form representable on a computer. For making Relational databases, the EAR model is followed but for making the proposed Multimedia and Multilingual Databases, no modelling method exists and the data at the input stage is unmodelled and unstructured. Going by the very nature of the data to be modelled, it is seen that some new methodologies have to be defined and some new software tools have to be made for modelling and entering such data. Another problem with such databases is that they would take up a lot of physical storage space. A Multimedia Database which would be logically small and would have very little information content would take up a lot of space in storage terms to become a physically large database and operations ON such databases would be a cumbersome or maybe an impossible process with the existing software and hardware technologies. e.g. Consider a bitmap which contains just the colour blue in it. It takes up some storage space but has no information content in it.

4. Large Databases and intelligent queries

Large Databases found in the services sector, large industrial houses, libraries, etc., face problems of data overload [3] . Even though large volumes of data are available, many queries remain unanswered due to the static nature of data [3]; queries which could have otherwise been answered by approximate reasoning employing soft computing techniques. Present day RDBMS systems like ORACLE, INGRESS etc., are very-large Database management systems and not information systems or information management systems. Databases made using these platforms have the following problems:

- Static in nature [3]
- Provide a static query system based on the system (situation) analysis.
- Essentially deal with the Quantity of data rather than the Quality of data.
- Any real world situation where data is imprecise, incomplete and vague cannot be modeled into a useful database, as such a database would have tuples with blank or vague entries. This leaves untapped information, which can only be tapped approximately with human expertise.
- They cannot cater to conceptual queries.

These database systems have a very rigid query structure. Queries are designed using SQL and are based on the attributes in the database table. Once a query structure with its report has been designed, it cannot be changed dynamically, i.e. there is no provision for *open ended queries* or *queries on concepts*. e.g. queries on concepts could be in the following form :

a) " Out of all the works of Picasso, which are the ones that art critics think are outstanding?"

b) A set of resume of various students as input data is circulated to the experts. These experts grade them and assess the suitability of the candidate on the basis of the student resume. The same process to be handled as a query to the Information system which contains the student's resume, would need a mapping of the overall grading as average, good, very good, etc. as concepts on the resume as approximation space. e.g. the user asks - " Who are the candidates who are good and are suitable for entry into course X ?" . This query needs to be answered in context of the course for which the candidates are being considered and to what extent the student's all round overall performance and his areas of interest (presented by the candidate in a one page write-up) maps onto the requirement of the course.

In an information system it is necessary to have a flexible query system which would be able to cater to queries which cannot be directly answered from the database table. However, a restriction has to be put that the query should be within the scope of the data in the table (The concept of scope is discussed later in the paper). It is this property of an information system wherein it can answer some query using the data in a data table, and not reproduce the data as an answer, which makes a system 'Intelligent'. In other words the system is able to 'approximately reason' out some conclusion from the data. In an *open ended query* we are looking at deriving concepts from a query in an infinite universe, bringing those concepts into a finite universe and the mapping them onto the data relevant to the query.

5. Integrated Intelligent Information System

The proposed Integrated Intelligent Information System shown in Figure.2 would contain three different modules. a) A Query processor; b) An Intelligent modeller; and c) The Intelligent Information System. Of these, the Intelligent Information System has been discussed in paper [3]. Information as derived from the IIIS would be a function of concepts to be mapped as objects onto the data. The information derived forms a picture would be the concept or meaning contained in the picture. Thus the information extracted form any data would be the concept derivation from the data and conceptual analysis based on the derived concepts. In effect this means that data modeling and the data need has to be driven by the query. At present the data modeling is independent of the query and is not query driven (as in OLTP/OLAP).

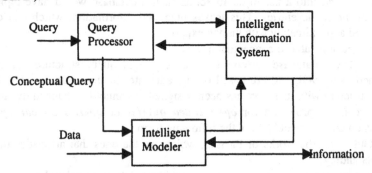

Figure 2. Integrated Intelligent Information System.

The query processor would be required for dealing with ' *open ended queries* '. We assume that an *open ened query* can consist of just about anything. It would be the job of the query processor to

- derive the concept of the query ; and
- decide whether the query is a conceptual one or one which can be answered directly from the DBMS or addressed to any of the constituent units of IIS .

If the query is based on the concept, the processor has to decide whether the concept is vaguely or crisply definable. If the concept is crisply definable it can be answered by the PDKBI module in the IIS. If the concept is vaguely definable, then it has to be answered by the raw information processor. For concept derivation form the query, the query processor has to be attached to the KB and RIP modules in the IIS. Figure 3 gives the flow chart for the query processor.

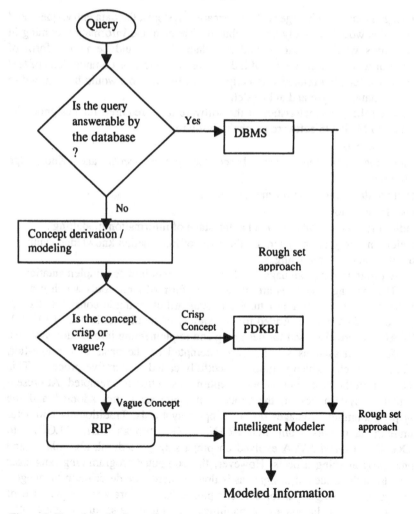

Figure 3. Flowchart for Query Processor

The Intelligent Modeller (IM) is required to model the data and address data overload problems, which are dealt with at this stage itself. The concept of the query has to be mapped onto the data and the concept derived form the data has to be modelled for making or modifying a database. It is the job of the IM to check whether the data associated with a query is in the modelled or unmodelled form, or is it in a modelled form but in need of some modifications. The IM has to also model the results of a query in a format convenient for the user.

6. Software Tools for Integrated Intelligent Information Systems

It is envisaged that in such large scale information systems, the data entry, query and retrieval nodes would naturally be distributed. The data and information coming in at these points would be unmodelled raw data. This would be in the form of paragraphs of text (as is the format of industry reports, etc.), as combinations of text and graphics, as context related video clips etc. Software tools would be required to look at these data, analyse and aid modeling .

The salient characteristics of the software tools required for designing the IIIS leading to MISNOS software are their ability to,

♦ Model Concepts.
♦ Decide the data requirement based on the concepts and Knowledge requirements.
♦ Model the data based on information needs.
♦ Model Information.
♦ Handle Information/Data overload at the stage of information modelling .
♦ Develop an information system on the basis of information/knowledge synthesis and not on iterative modeling.

A programming language level support is required for implementation of the IIIS. The existing languages are more in the form of large tools which provide support for a variety of Data structures and standard algorithmic constructs like *if-then-else*, case, *do-loop*, etc. Added to this, variety of smaller tools like MATLAB, MATHCAD etc. are also used for designing and implementing information systems. So far information systems solutions are attempted on the basis of a rough-fair, rough-fair, approach , which is more respectfully called as iterative process . This may not result in the best solution but an optimum solution is attempted. At present in information system design, the process of synthesis is not adopted and the existing programming languages do not support synthesis. Traditionally computer programming languages from FORTRAN, LISP, through PL1, ALGOL, to PROLOG, C, C++ and JAVA evolved encompassing procedural, algorithmic and functional programming aspects. However, the computer programming languages did not address the issues of designs, as is done in hardware description languages. Thus programming languages for soft computing have to draw the experience of hardware description languages, capabilities of rough sets, fuzzy logic, etc.

Hardware description language like VHDL has the capability to support structural modeling, behavioral modeling, mixed modeling as well as synthesis of the VLSI design. This is made possible as VHDL supports standard building blocks like AND, OR, NOR, XOR, etc., required for VLSI synthesis [4]. This feature is possible if basic building blocks are predefined. In the context of the information system and information processing, it would be difficult to predefine basic building blocks, as the building blocks would have to be dynamically defined and synthesized. This explains why computer programming languages in the past could not support synthesis capabilities. However, this could be made possible because of the inherent capability of Rough sets to extract elementary sets. Let the approximation space be defined over R as A = (U/R) where U/R are the elementary sets or partitions in A, denoted by $E_1, E_2, ... E_n$. An object could be synthesized as a combination of these elementary sets dynamically [7],[8],[9]. Hence for soft computing, such features need to be supported at the language level itself.

Authors observed that for implementing IIIS leading to MISNOS, the off-the-shelf tools available are not able to cater to the above requirements since these are capable of analyzing the existing data for it's structures. Though the tools claim to model the real world, they do not satisfy the requirements as brought out in [2],[3],[5]. This has given the authors an idea to look at soft computing techniques with an emphasis on rough sets as a potential candidate capable of modeling the real world for IIIS and MISNOS. As sufficient research work has not been carried out worldwide with respect to soft computing in general and particularly in rough sets, it is not possible as well as not desirable to freeze the specs of tools for developing MISNOS. In addition, potential of rough set techniques for modeling of data present in unstructured form, approximation techniques for information retrieval would definitely be necessary. Whether rough sets in its present mathematical form would be able to achieve this or some modification is required is not known.

7. Use of Rough Sets as an Evolutionary Tool

As brought out earlier, it is not possible as well as not desirable to freeze the specs of tools for developing IIIS and MISNOS. The set of Tools required would include:
I. Concept Modeller.
II. Data requirement extractor.
III. Overload Analyzer and optimizer.
IV. Information Modeller.
V. Data/Information Synthesizer.
VI. Concept refiner and synthesizer.
Hence these works are being taken up on an evolutionary basis. The foregoing brings out rough sets as a potential candidate methodology for developing software tools for building MISNOS. These tools are expected to be more powerful than the existing CASE tools as they are expected to handle problems of data overload and Real world Modeling.

In the design of platforms for making intelligent information systems, rough sets comes in as a very useful mathematical tool for extraction of information from vague, imprecise and inadequate data. The advantage of using rough sets is that ,

1. They do not need any prior information about data.
2. They can bring out the minimum attribute dependencies of data
3. Can give definite and possible results from the relationships of attributes.

For modeling the real world into Multimedia and Multilingual data types, which are inherently vague, and inadequately represented, rough-sets is a possible evolutionary mathematical tool for developing and modeling such databases.

In the case of very large Databases however, at a cursory look one feels that rough sets may not serve as an approximation tool for the databases *per se*. A very large database consists of millions of entries. Information extraction form a database using rough sets consists of,

1. Making partitions in the database using the possible attributes for answering a query.
2. Finding the lower and upper approximations.
3. Making reducts to find the minimal attribute dependencies.

The various operations of a relational database, particularly join, select, divide, etc., add to the data processing overload problems. Thus one may feel that if the operations of information extraction using rough sets were to be carried out on large logical databases in totality, then it would only enhance the problem of data overload. However these problems are essentially due to lack of programming language constructs for implementing database softwares. The programming languages used for developing DBMS packages, have constructs like if-then-else, iterative loops etc. which are used to implement relational algebra and are not capable of handling soft-computing features or rough set operations. In this context, the need for primitive data-type for soft computing for fuzzy operations have already been reported [6]. Similar exercises have to be carried out for rough sets datatypes and operations, including lower approximation, upper approximation, reducts etc. If soft computing has to flourish, research work has to be carried out at programming language level itself.

Almost all large and very large databases are distributed systems, logically forming one single database but physically present at different places. Information could however be derived from these databases by putting a rough-preprocessor at the physical location of these distributed systems. The partitions made on these databases would reduce their size, and the approximation space of the total logical database would be

$A = \quad U \quad \{a_i\}$ where a_i is the approximation space (partition)of one physical database.

$\quad i = 1$ to n

For example if the set of attributes in a data table are $\{a1,a2 ,a3 ,a4,a5\}$ with attribute values v0,v1,v2. The minimum possible attribute dependencies are say

$a3(v1) \rightarrow a5(v0)$;

$a4(v0) \rightarrow a5(v0)$;

$a4(v2) + a2(v2) \rightarrow a5(v1)$;

$a4(v1) + a3(v1) \rightarrow a5(v1)$;

$a3(v2) \rightarrow a5(v2)$;

based on a partition of a5 as the query was related to it but could not be answered by the data table. If similar dependencies are formed from other tables, the whole set could be made into a attribute-decision table with a5 as the decision attribute (Figure 4.). The process of reasoning, which is bi-directional, could then be done, depending on the query. The forward direction of reasoning is inferencing which is done by making partitions on the condition attributes. The reverse process of reasoning i.e. synthesis is done by making partitions on the decision attributes. The attribute dependency table generated becomes input to the Raw Information Processor module as raw information.

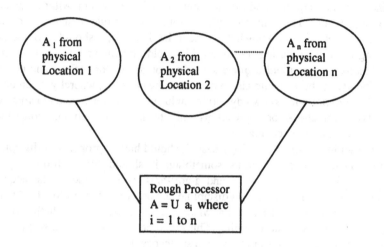

Figure 4. Making Approximations in large databases.

Once a larger database is reduced to a smaller one in such a manner, it would be possible to make reducts and derive attribute dependencies. Also it would be possible to introduce such systems in machines with smaller RAM's. As mentioned earlier, in order to answer *'open ended queries'*, it has to be determined what is the scope of the data in the database. This is absolutely essential if we are looking for such approximate analysis. Thus there has to be a definite shift from large-scale databases to *large scope databases*. How the scope may possibly be determined using rough sets is illustrated:

Say, there are three independently existing databases; One belonging to a library management system, second belonging to a publisher and the third belonging to a paper Export-Import (Exim) house. From the library database point of view, the publisher's database *definitely* lies in it's scope and the Exim house's *possibly* lies it's scope. This analysis is possible from dependency relationships existing in an attached Knowledge base.

Once the scope of a database is approximately determined, we would then have truly interactive distributed systems, wherein one particular database would derive information from another database, leading to cooperative computing on a limited basis which have attribute dependencies to answer *real time event driven queries*. This will also make significant contribution in making the Internet popular.

8. Role of Knowledge Base

For such *interactive distributed systems* involving Intelligence and *cooperative computing*, existence of a knowledge base is a must. The knowledge base for such large systems would be initially built by expert training. Periodic interaction with the trainer would help to modify the contents of the system without actually changing its structure. This training could be done with the rule inducing algorithms based on rough sets. For the knowledge base to be dynamic, it should have methods of acquiring knowledge during run time. This could be achieved by inducing rules from the results of ' *open ended queries* ', using rough sets for definite and possible rules. A Knowledge base should also have reasoning capacity, whereby it can derive conclusions from previous knowledge (rules induced by inference and existing facts & concepts), it can check for implications, and check for the truth and consistency of an inference by cross checking.

For an intelligent Knowledge base, it should have a property of disciplined inquisition. That is once it learns something, it should try to follow up the knowledge, looking for rules, facts and algorithms in the scope of the acquired knowledge. This has to be done with discipline and authorization protocols. Another desirable characteristic in such a knowledge base is *intuition,* which involves deriving conclusions without reasoning. This could be done with rough sets but with yet to be defined modifications in the rough set theory [5].

9. Issues on experimental implementation

In order to familiarize with the rough sets capability and its suitability to develop required tools for modeling the IIIS leading to MISNOS, a pilot study was undertaken. In order to provide continuity with earlier work done with SIIIS [3], PROLOG was taken up for implementing Rough Set tools. Parallely efforts have been initiated to examine the suitability of an object oriented language like Visual C++ for implementing the Rough Set tools. This would become essential as many of the data items such as Video, photographs, paintings, animations, computer generated images, audio, text etc. would be handled as objects in totality with the features of Inheritance, Polymorphism, function and operator overloading of Visual C++. It was possible to induce rules out of concepts, thereby the concept could be defined as an object either roughly, definitely or possibly. If the data requirement is from the definite rule, the data overload is the barest minimum. If data requirement is from the possible rules, the data overload increases. If the data requirement is from all the rules, then data overload is maximum. This directly helped in modeling the data requirement for a concept and potential of its application for modeling the data is under investigation [7].

10. Conclusion

The above study has brought out that Rough Set is a potential candidate for developing the required software tools for implementing IIS and MISNOS. It has also emerged that there is a need to examine the suitability of rough relations by implementing it in software form. The issue of development of suitable programming language(s) for soft-computing environment needs to be addressed urgently.

Acknowledgments

Authors wish to acknowledge useful discussions with R&D professionals of Software Industry. The authors also acknowledge fruitful discussions with Mr. R. Bandyopadhyay , Director , DOE & Mr. W. R. Deshpande , Director , DOE and wish to thank them for their valuable comments . The authors would like to thank Dr. A.K. Chakravarty , Advisor and Dr. U.P. Phadke , Senior Director , Department of Electronics , Government of India , for encouraging interdisciplinary activities . Special thanks are due to Mr. A.Q. Ansari , Reader , Department of Electrical Engineering , Faculty of Engineering & Technology , Jamia Millia Islamia University, for permitting Mr.Soumik Ghosh to carry out investigations in Department of Electronics as Student Trainee .

References

[1] S.M. Prasad , A.K. Chakravarty. Rights-of-Way for Fibre Optic Cables. *Electronics Information & Planning* , Vol. 23 , No. 10 , July 1996, pp 571-584 .

[2] R. Bandyopadhyay. Multimedia Multilingual Information Services Network Operating System (MISNOS) Software. *Electronics Information & Planning* , Vol. 23 , No. 4, July 1996,pp 205-230 .

[3] S. Sivasubramanian , A.B. Patki (1996) .Software Industry Intelligent Information System *Electronics Information & Planning* , Vol. 23 , No.9 , pp 513-518 .

[4] G.V. Raghunathan , A.B. Patki , U.P. Phadke , N. Gopalaswami . Modeling Considerations for the performance of Public Sector Enterprises - Fuzzy Theoretical Approach. *WSC1 - proceedings of the first online workshop on soft computing,* 19-30 August 1996, Nagoya University, Nagoya, Japan. pp 168-173.

[5] T.Y.Lin . Inferences in Finite Fuzzy Universe - A View From Rough Set Theory . *WSC1 - proceedings of the first online workshop on soft computing,* 19-30 August 1996, Nagoya University, Nagoya, Japan. pp 134-138.

[6] A.B. Patki , G.V. Raghunathan , N. Narayanan . On Data Types for Object Oriented Methodology for Fuzzy Software Development . *WSC1 - proceedings of the first online workshop on soft computing,* 19-30 August 1996, Nagoya University, Nagoya, Japan. pp - 163-167.

[7] Soumik Ghosh Exploration of rough set potential - PROLOG Implementation . *Student training report,* MDD/2/97.

[8] Z.Pawlak (1982). Rough sets. *International Journal of Computer & Information Sciences.* Vol. 11 , No. 5 , pp 342-356 .

[9] Z.Pawlak (1984). Rough Classification. *International Journal of Man-Machine Studies.* pp 469-483.

Towards Rough Set Based Concept Modeler

A.B. Patki[1], G.V.Raghunathan[1], Soumik Ghosh[2], S. Sivasubramanian[1]

[1]Department of Electronics, 6, C G O Complex, New Delhi – 110003, INDIA.
[2]Software Engineer, Schlumberger, Besancon, FRANCE.

Abstract: The paper reports a rough set approach for modeling concepts encountered in evolutionary information systems for corporate computing. Features like extracting concepts from textual queries and Multimedia are illustrated in the context of open-ended queries. Concept transformation issues with its engineering applications are discussed. Authors suggest concept modeler as a methodology for large-scale production of training and educational material for Web based distance learning programmes incorporating guaranteed minimum accuracy measure for concept introduction and absorption.

1. Introduction

In the daily operations routine, corporate computing relies heavily on generation and collection of large volumes of data and this raw data needs to be processed. As sometimes defined in the literature, the methods deployed for obtaining useful information implicit in data is called data mining. By accessing the data mining functions and algorithms, smart data analysis is carried out using toolkit packages like IBM's Intelligent Miner. While such an approach extends the analytical capabilities by implementing data-driven approach for distillation of information, the absence of query based data modeling, does not provide the complete solution. It is felt that more advanced solutions for handling concepts is a pre-requisite for largescope computing as against supercomputing.

Evolutionary information systems, therefore, call for more demanding support from the operating systems. Even the next version of Windows 95 appears to be more like an extended version of desktop operating system [1]. Operating systems likely to be commercially available in next two to three years are merely considering integration of web browsers as a part of operating system support. Authors feel that Windows 98 with HTML powered shell including Java environment may not address the real issues involved in large volume intranet applications for corporate computing world. The approach of shifting from RDBMS to Integrated Intelligent Information Systems (IIIS) is of utmost importance when we consider the impact of Intranets in the corporate computing world. When we consider the growing population of World Wide Web (WWW) users and analyse their satisfaction level with the performance of various Query Processing Engines, it becomes obvious that even for simple query based approach to surfing on the network, there is a strong need to introduce more capable query processing inference engines. The problem is of utmost significance for Intranet applications using electronic commerce, web shopping, corporate computing and other similar transborder computer network applications. For corporate organisations using intranets incorporating IIIS, the requirement of learning and analysing the user query requirement and generating the required information need not be re-emphasised.

The information derived from the IIIS would be a function of concepts to be mapped as objects onto the data. In such a system, data modeling has to be driven by the query, instead of the query being based on the data available. Thus the concepts in such "open ended queries" have to be modeled and the information derived also has to be modeled as a function of concepts. Various tools that have been envisaged for the development of such an IIIS make use of concept processing. For concept modeling, these tools call for support for dynamic concept analysis and synthesis properties. Data and query input to the IIIS would be in the form of unmodeled raw data whose concept modeling has to be done. Applications of rough set techniques have been thought of as a possible tool for developing the concept modeler [2].

2. Concept Modeler

The functional framework of the concept modeler has been thought of so as to operate it in an integrated as well as in a stand-alone environment. Hence the concept modeler could be used as an add-on programmable hardware so also as a software module. In an integrated environment, as in the working of the IIIS, the concept modeler would be part of the Intelligent Modeler as well as the query processor [Fig 1.], interacting with all the three modules. In the query processor, the concept modeler would be in the software module, incorporating rough set algorithms among other methods to model the concept in a query. The input for the software concept modeler would be unmodeled raw data, in the form of paragraphs of text, or combinations of text, graphics, video, animation, audio, etc., and unmodeled, unstructured queries. Using approximation techniques as described by rough sets, the nearest possible concept with threshold accuracy levels would be derived and given to the output by the modeler. The threshold could be fixed or programmable depending upon the user preferences. For giving the output accuracies, the modeler has to interact with the knowledge-base, which may be local or distributed. In the IIIS, the system would have to learn from the user query as well the query results. As the system progresses, its granularity increases and the concept gets refined, and we get concepts having improved accuracies. In the software form, the concept modeler could give as an output, the accuracy of the concept derived from the query and if acceptable to the user gives the concept modeled result with its accuracy. In the case of large databases or Multimedia databases, concept modelling or a part of the processes involved may have to be dealt at the hardware stage. In a large database, the knowledge base, may be

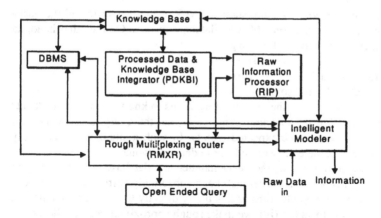

Figure 1. Block Diagram Showing various modules of the IIIS.

distributed and for concept synthesis some preliminary multiplexing may have to be done before the software stage comes into play. In case of multimedia data or multimedia rule-base, using Rough Sets approach [3] it may be necessary to do partitioning and generation of Lower Approximation (LA) & Upper Approximation (UA) at the hardware level itself. Thus in systems like the IIIS the concept modeler could be present at the hardware and software levels together in an integrated environment.

The concept modeler existing as a hardware add-on card, when plugged in with an existing multimedia system could help in concept analysis and synthesis of the information in the multimedia objects with the help of a software knowledge base. Software modules of the concept modeler along with the supporting knowledge base could be integrated into the present day databases to build up the required knowledge base for the evolutionary IIIS or could be used to answer conceptual queries in a limited fashion. In addition to the rough set algorithms for concept derivation, the modeler would have to have hardware modules in the form of speech recognizers and vision capabilities for concept synthesis in the multimedia mode.

3. Open Ended Queries

As mentioned, the IIIS is a methodology, which is driven by the user query rather than the data in the database. Thus we are dealing with open ended queries which could have almost anything as its contents. i.e., the query can be in the form of paragraphs of text in the natural language form, or it could be in the form of speech, or visual actions etc. As a simple example,

The query in a structured form may be posed as:

"Name(list) the black cats which eat white fish" - (i)

This query may be answered by using standard NLP techniques to model the query to get an answer from the relation table of a database on cats. On the other hand, an unstructured query may take any form, e.g.

"A cat eats"   -   (ii)
"A black cat eats"   -   (iii)
"A cat eats fish" and so on. -   (iv)

In all these sentences, the user possibly wants to know about the black cats which eat white fish, but the meaning is not clear, i.e. the query is imprecise and vague in the perspective of the expected results. In such a situation we can only hope to give an approximate answer and our endeavour would be to do so without causing data overload. Let us say the query handler in a database takes its query in the format as presented in the statement (i), i.e. an absolutely correct query has been typed in. So the concept modeler throws this query directly to the query handler. In case the query is of the form (ii)-(iv), we make rough approximations as follows:

Say the query is of the form: "a cat eats", which has three words in the form:

Article   Noun   Verb

The elementary sets are formed according to the grammatical rules of the language. e.g. a simple sentence like,

"A black cat eats small fish"

has two parts, a noun phrase as : "A black cat" & a verb phrase as : "eats small fish". So this has six elementary sets as:

E1={Article}; E2={Adjective}; E3={Noun}; E4={Verb}; E5={Adjective}; E6={Noun}

Now, though E2 & E5 both have Adjective as their element, they are not the same as one belongs to a noun phrase while the other belongs to the verb phrase. So the query has the form:

E1=Article(At)   E3=Noun(N)   E4=Verb(V)

Thus, for the combination At:N:V ; LA = {E1, E3, E4}
                                                        UA = {E1, E3, E4}
for the combination At:V:N :   LA = {E4,E6}
                                                        UA = {E1, E4, E6}

The LA for these processes are determined based on the Elementary sets and grammar rules from the knowledge base. The UA is also then dynamically generated. The query is sent to the module whose semantics fit into the dynamically generated UA and into which the dynamically generated LA fits and the boundary (LA-UA) is minimum. Obviously, this is the most accurate concept that can be derived with the highest accuracy. In our case the UA itself may well fit into any one of the semantics. This is due to lack of granularity in the query. As the query is more granularized, the concept becomes more refined. So in this case we try to choose the module for which the difference between the semantics and the UA is minimum.

When we go from one level to another with increasing granularity of the query, at every level we get more number of combinations of elementary sets of which the LA is decided by the grammar rules and the UA being decided at the entry points of the modules by the module semantics and the English language grammar rules for formation of the elementary sets. This increase in granularity increases or decreases the accuracy of the derived concept from the query but this change in accuracy from one level to another is not continuous, but rather in discrete quanta. To go back to our example, let us say that the query is acceptable to a module in the following format:

" A black cat eats small fish "

This then is the UA for the query. We illustrate how accuracy is computed at each stage while processing the query. Three cases arising in concept accuracy evaluation process are presented below.

Case I:

Consider at the first stage, LA = {cat} & the concept accuracy is say $x\%$

At the second stage, LA = {cat, eats} & the concept accuracy is $y\%$ $(y>x)$. It is assumed that to improve the accuracy, y has necessarily to be greater than x. This can be guaranteed if the starting element was one of the approved tokens as per the English grammar. Such tokens are typically noun or pronoun giving the idea of "subject" (in a typical subject | predicate format).

Case II:

Consider at the first stage, LA = {A} & the concept accuracy is say $x\%$

At the second stage, LA = {A, eats} & the concept accuracy is $y\%$. This is a typical case of $y<x$; since, the subsequent state did not improve the concept accuracy. Thus it is not the length of LA that contributes to the accuracy but the proper sequencing of the tokens. In this case, the starting token is not from the approved tokens, hence, $y < x$. The second stage has failed to pass on the newly acquired token to any of the routing lines branching out of the Multiplexer.

Case III:

Consider at the first stage, LA = {A} & the concept accuracy is say $x\%$

At the second stage, LA = {A, small} & the concept accuracy is $y\%$ $(y>x)$.

At the next stage, LA = {A, small, black} & the concept accuracy is $z\%$ $(z = y)$. The cases of $z = y$ appear when "qualifier" type of tokens are encountered.

It is assumed that the words are processed with proper algorithm so that the English grammar rules are strictly adhered to at each stage in each of the cases mentioned above. These are the cases where the query is not entered by a human operator but is generated by one processor and passed on to another processor for further handling.

It is seen typically x, y, z are neither irrational nor imaginary numbers. Obviously this characteristic lends to representing the accuracy in discrete quantum i.e. "quantum granularity". But as the granularity has increased in a quantum (number of words) the accuracy also changes in a quantum. This leads to an idea that there is a bare minimum level of granularity in an Information system, which can be thought of as the basic unit of information in an information system beyond which no useful or meaningful granularity is possible. So it becomes necessary to define a unit for information system for concept derivation and refinement.

This method of analysis also brings us to the concept of rough equality, both laterally (i.e. at the level of same granularity) and vertically (i.e. between two level of granularity). Let us consider our query again. Say at the first stage, the query is only "cat".

Hence, we have N{E3} ; LA=UA={E3}

at the next level the query becomes "A cat" or "cat a"

i.e.   At:N ; LA={E1, E3}
            UA={E1, E3}

or   N:At ; LA={E3}
            UA={E1, E3}

But, we see that in going from "cat" to "A cat"/ "cat a" there is little change of meaning i.e. they are roughly equal.

At the same level of granularity, say the query is "black cat" or "cat black" for which we have:

Ad:N ; LA={E2, E3} UA={E2, E3} &

N:Ad ; LA={E3}   UA={E2, E3}

But, again we see that the meanings of the two sentences are nearly equal and we have lateral rough equality. The same methodology is to be applied while multiplexing when selecting the module for processing the query.

      The routing of the query by the concept handler is done as follows:

* If the query is in the exact format of a query handler in a DBMS then it is directly routed to the DBMS.

* In case the format does not match that of the DBMS as above, then the concept for the query has to be determined.

          - a crisp concept goes into the PDKBI.

          - a vague concept goes into the RIP.

      As an example of a "roughly equal process", consider opening a document in Wordstar. We have to undergo the following steps:

(a)Type ws <return>

(b)Type D

(c)Type the file name, say rough.set <return>

      Alternatively, the same result of opening the file could be obtained by typing the following command:

(a)ws rough.set

      Thus three steps are brought down to only one step. The two routes to the same solution are thus roughly equal. This example shows the problem of query overload.

      To further the concept of "open ended queries" in the Multimedia domain, let us consider a situation of a visual quiz, in which there is a picture covered by a square, the square being cut into four sectors [Fig 2]. Removing one sector partially reveals the picture giving a clue to its contents. Removing one particular sector may give more clue about the picture than others (leading to greater accuracy) or the clue supplied by all of them individually may be "roughly equal". Based on this clue the computer (or the user, as the case may be) tries to guess about the contents of the picture with a certain amount of accuracy (from 0 to 100%). The guess would be based on yet-to-determined elementary sets in the multimedia domain; and LA, UA derived thereof. On removal of two sectors, more information is available and it is possible to take a more accurate guess than in the previous stage. Thus with the increase in granularity of information from one stage to another, it is possible to answer a query with a greater amount of accuracy. This is brought out in figure 2.

In case we divide the square into more than four sectors and then remove the sectors one by one, we get a more granular system and at each stage we get different levels of accuracy, which may be greater or less than the previous stage, but the change in accuracy is always in a discrete quantum than being continuous. The effect of changeover from square to circle as a basic projection unit will not have much impact on LA & UA. Thus the primary shape of the projection unit as a regular polygon is not a pre-requisite though it is preferred in practice when the shape is generated / selected by user for edutainment.

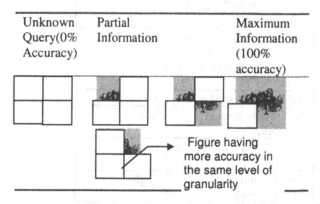

Figure 2. Depiction of granularity and accuracy

4.Concept Transformation

The concept transformation is important in assessing the performance of IIIS, Robot based assembly line work in manufacturing processes, and web based education. We look at the problem of concept transformation. Say children's story like "Alice in wonderland" is to be told to children on the web. This story can be told in media forms of text, speech, video or a combination of these. Let us consider a portion of this story for analysis

" Alice was small girl who loved to play. One day she went with her sister to the lakeside where her sister started reading a book with no pictures. Alice got bored and lay down on the grass. Suddenly she saw a white rabbit wearing a coat and fully dressed, hurrying towards the bushes. As Alice watched in amazement, the rabbit took out a watch from it's waistcoat and exclaimed "I'm Late!!". Alice decided to follow the rabbit"

This is the story, given in text format, which in the Multimedia version would contain some pictures to supplement the story along with background music and narration. The author (i.e. the Web designer) by use of these three techniques (text, sound, and picture) has tried to convey some concept to the viewer. The problem of concept transformation comes into picture when we want to know what the viewer has understood (which may also be the computer - for robotic applications). The level of concept transformation will also depend on the viewer's ability to absorb the concept; thus if the story is told to children in the age group of three to seven, and to another set of children in the age group of seven to eleven the accuracy of concept transformation required will be different. The accuracy measurements could be carried out by adopting suitable techniques available in the field of educational psychology. An illustration is shown in table below:

No.	Media	Concept	Accuracy
1.	Text	When a white rabbit with pink eyes ran past Alice, she saw something very remarkable	low
2.	Speech	Narrative(soft Music Throwing the ambience +voice of rabbit+voice of Alice)	Low
3.	Visual	Rabbit+astonished Alice	Medium
4.	Text/ Speech	Text+Speech(more relatable)	Low
5.	Text/ Visual	Text+Rabbit+ astonished Alice	High
6.	Visual/ Speech	Rabbit+astonished Alice+Speech	High

A Multimedia object has three parts:
1. Properties: which would contain information/data about the object.
2. Methods: which tells how to handle properties
3. Events: which tells when the object responds to changes.
   In our case the object would have three properties
(a) Sound, with arguments of amplitude, frequency, noise, etc.
(b) Picture with arguments of image resolution (say 640 by 480), colours (using RGB model with 24-bit colors), brightness, etc.
(c) Text characterized by font type (say Times New Roman), font size (say 12 point), font color etc.

The picture shown in Figure. 3 is supplementing the paragraph of text on "Alice in Wonderland". The picture is divided in to various object numbered 1,2,3,4,5,6 as shown. To determine which is the best method for concept transformation for the various age groups and for machine learning (needed for IIIS and AI in Robotics problems) we determine an experimental tutorial based on the following algorithms:

** Algorithm for making the training module for concept transformation which is typical algorithm in an object-oriented programming language supporting Multimedia, like JAVA.

** An algorithm to separate out the various objects in the picture / text and put them in a jumbled form as in a jigsaw puzzle for the trainee to reconstruct.

** An algorithm to match the original picture with the reconstructed one to measure the accuracy of the transformation of concept.

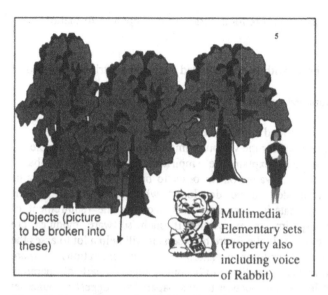

Figure 3. Elementary sets for Multimedia Objects

The steps involved in running the tutorial are as follows:

A. Text

1. The text is shown to the user for a given period of time, say five minutes (in case of the concept modeler in the query processor, the time for learning from an open ended query and the answer to the query would be in milliseconds as the exposure time would be very small). The user reads the text in this time and understands whatever he/she can.

2. The text is wiped off the screen and is replaced by keywords from the text, to aid in reconstruction of the text.

3. The user types in whatever he has understood from the text, with the help of the keywords.

4. The screen clears and the accuracy score is displayed.

The analysis of text in the paragraph form would be done with the help of the rough set technique shown above along with standard NLP algorithms and grammar rules.

B. Picture

1. Only the picture is shown without any text or sound attributes. The picture is composed of its various objects in their proper places. The objects in the picture are the most basic components in the picture or the atomic components of the picture, which would be the unit of the graphical information system in lines with the unit of textual information system mentioned above.

2. The screen is again cleared after a time interval and the jumbled up objects are given to the user for reconstruction.

3. The user drags and drops to reconstruct whatever he understood and retained, of the image.

4. The screen clears and shows a score for accuracy of concept transformation.

C. Sound
1. The picture is again shown with sound included (the sound attribute attached to the various objects).
2. The picture is again jumbled up and given for sound identification.
3. The user relates the sounds with various objects to which they belong.
4. The accuracy score for the sound concept transformation is recorded.
Similar exercises would be done for the various combinations shown in the table.

The above paragraphs explain the empirical approach for web-based education and training material evaluation. In order to bring uniformity of results systematic methods are needed to be developed which can permit a-priori estimation of the training material accuracy; thus avoiding the requirement of conducting the actual experiments everytime. If such software modules are introduced as built-in provisions in the future browsers, it will help a lot to asses the effectiveness of web based cyber training material. These considerations are more important since the web browsers are likely to become integrated parts of operating systems as brought out in the introduction to this paper. We suggest a rough set based, object oriented methodology for developing such software modules [6].

5. Conclusions

This paper brings out the methodology for applications of rough sets in hardware and software in an integrated fashion for variety of applications. In contrast to the present mathematical approaches as indicated in public domain literature, the present research work focuses on the technology and engineering applications of rough sets. In this context, the 'Concept Modeler' as an implimentable tool both in hardware and software is described. The work related with interfacing and integrating as a retrofit solution to the existing SQL is planned in the near future. It is believed that with the availability of platform independent compilers like Java with features for 16-bit primary data unit, such efforts will be helpful for developing more potential Multilingual Multimedia Intelligent Information Systems.

Acknowledgements

Authors wish to acknowledge the fruitful discussions with Mr. R. Bandhyopadyay, Director, DOE and Mr. W.R. Deshpande, Director, DOE and wish to thank them for their valuable comments. The Authors would also like to thank Dr. A.K. Chakravarty, Advisor and Dr. U.P. Phadke, Senior Director DOE, for encouraging interdisciplinary activities.

References

[1]Michael Caton, Memphis fills gaps, gains IE Web browser: Reviews, *PCWeek Asia*, 11(15):8 August 8-21, 1997.

[2]A.B. Patki, G.V. Raghunathan, Soumik Ghosh, S. Sivasubramanian, Azar Khurshid, Soft computing for Evolutionary Information Systems - Potentials of rough sets, *WSC4 – 4ᵗʰ On-line World Conference on Soft Computing in Industrial Applications (WSC4) Hosted on Internet, 21- 30, Sept., 1999 Muroran, Helsinki, Nagoya, Cranfield, Bath.*

[3]Z.Pawlak, Rough Sets, *International Journal of Computer & Information Sciences*, II(5):342-356, 1982.

[4]S. Sivasubramanian, A.B. Patki, Software Industry Intelligent Information System, *Electronics Information & Plannig Journal*, 23(9): 513-518, 1996.

[5] Won Kim, Frederick H. Lochovsky, *Object Oriented Concepts, Databases and appications*, ACM Press, New York, Addison Wesley Publishing Company, 1989.

[6] Soumik Ghosh, Software considerations of Rough Set based Concept modeler, *Technical report*, May 1997.

Chapter 7: Image Processing

Papers:

Still Images Compression Using Fractal Approximation, Wavelet Transform and Vector Quantization
K. A. Saadi, Z. Brahimi, and N. Baraka

N-dimensional Frameworks for the Application of Soft Computing to Image Processing
M. Koeppen, L. Lohmann, and P. Soille

Computational Autopoiesis for Texture Analysis
J. Ruiz-del-Solar

Novel Approach in Watermarking of Digital Image
C. C. Wah

A Fuzzy Region-Growing Algorithm for Segmentation of Natural Images
J. Maeda, S. Novianto, S. Saga, and Y. Suzuki

Still Images Compression Using Fractal Approximation, Wavelet Transform and Vector Quantization

Karima Ait saadi[1], Zahia Brahimi[2], and Noria Baraka[3]

Centre de Développement des Technologies Avancées, 128 Chemin Mohamed Gacem B.P-245 El Madania-Alger ALGERIE, [1]ait_saadi@yahoo.com, [2]zbrahimi@yahoo.com, [3]baraka2@yahoo.com

Abstact: A new hybrid approach for still image compression based on Vector Quantization (VQ), Fractal Approximation and Wavelet Transforms is proposed. It aims to improve in terms of bit rate and Peak-to-Peak Signal-to-Noise (PSNR) the hybrid approach for image compression based on VQ and fractal approximation developed in the literature. Also it tends to limit the blocking effects which usually appears at low bit rates by partitioning the Wavelet Transforms domain instead of the spatial domain. By this approach, the constraint of contraction mapping is not required, fractal approximation that uses the self-similarity of grey patterns works on the approximated image.

1. Introduction

A fundamental goal of data compression is to reduce the bit rate for transmission or storage while maintaining an acceptable fidelity or image quality [3]. Many compression methods have been developed. One usually distinguishes between different coding schemes as fractal image coding. Generally, interest bit rates are reported for still monochrome images without a great loss of visual quality. Higher ratios can be obtained with hybrid coders incorporating different techniques with respect to local image properties. Our paper deals with this area of research, in which, we introduce a new hybrid approach based on fractal image coding in association with VQ and wavelet transform. The aim of this work is:

- To speed up the search time for the domain block by eliminating the contraction mapping for the fractal approximation,
- To reduce the complexity coding phase,
- To improve in terms of PSNR and bit rate the hybrid approach for image compression based on VQ and Fractal approximation which employs the DCT for approximation of an input image [1] [2].

The balance of the paper is organized as follows: In Section II, we show the procedure of the proposed method using a domain pool that is constructed by wavelet transform, VQ then decimation. Section III, present experimental results of the proposed method and the performance comparison in terms of bit rate and PSNR with other hybrid approach based on fractal approximation and VQ which employ the DCT for approximating an input image [2].

2. Proposed Fractal Image Coding

The proposed fractal coding method is based on the concept of self-similarity of an image. For generating the domain pool, instead of using the grey patterns of an image with the contraction mapping like in the conventional fractal coding algorithms, we first approximate an image by transform vector quantization applied on the wavelet coefficients (WTVQ) [4]. Then decimate the approximated image by a factor of four in order to use its grey patterns. By this technique, it is more flexible than the conventional algorithms that used an input image with contraction mappings. In addition, it is a noniterative method because it doesn't satisfy the contractivity constraint and it can take a full advantage by using a grey patterns of a domain pool. This approach adopts the idea of VQ to approximate an input image, so the learning procedure is required.

The codec scheme is illustrated in Figure 1.

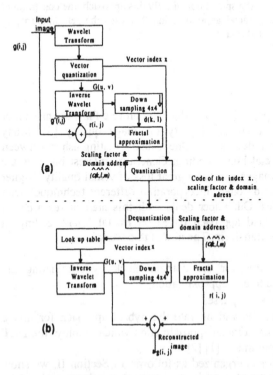

Figure 1. Block diagram of the proposed method Figure 1. Block diagram of the proposed method (WTVQ), (a) encoder, (b) Decoder.

The encoder shown in Figure 1(a) consists in first of all, to approximate the input image by using the multiscale pyramidal decomposition followed by the VQ. The second step, consists of encoding the residual image by the fractal approximation. Figure 1(b) shows a block diagram of the decoder. This one is fairly simple. The reconstructed image is the addition of the approximated image WTVQ and reconstructed one by fractal approximation.

The first one is generated by applying the entropy decoding on the vectors index x by consulting the codebooks and then inverse wavelet transform followed by a decimation by a factor of four, The second one is obtained by applying the entropy decoding on the fractal parameters.

2.1 Generation of the domain pool

In this method the approximated image WTVQ is generated by applying a VQ on the wavelet coefficients resulting from the wavelet transform of the trained image. This instead of generating the approximated image by using the VQ of the DCT coefficient like in [1] [2].

Why using wavelet transform than DCT transform?

It is observed in [3] that the DCT based algorithm can introduce block distortion that is unacceptable in high picture quality applications. On the other hand, use of VQ in high quality image coding requires very large codebooks, which typically lead to an unaffordable increase of complexity. In addition to the decorrelation of the image data, 2-D wavelet transforms have another important property. Despite the low correlation among themselves, bands of same orientation look like scaled versions of each other. That is, their edges are approximately in the same corresponding positions; hence their nonsignificant coefficients are approximately in the same corresponding locations. By this technique, it's more flexible than the conventional fractal coding algorithm that used an input image with contraction mapping. In addition, it's a no iterative method because it doesn't satisfy the contractivity constraint and it can take full advantage by using grey patterns of a domain pool. Figure 2 represents one stage in a multiscale pyramidal decomposition of an input image. The filters h and g are one-dimensional filters. The reconstructed scheme of the image is presented in Figure 3. The wavelet coefficients are then quantized using VQ, where the vector index x is transmitted to the receiver by which we can generate a domain pool at a receiver.

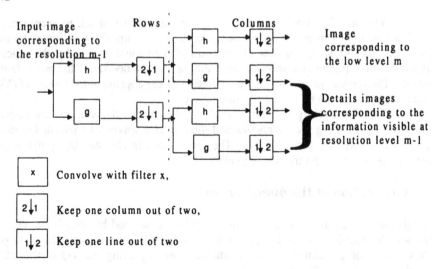

Figure 2. One stage in a multiscale image decomposition.

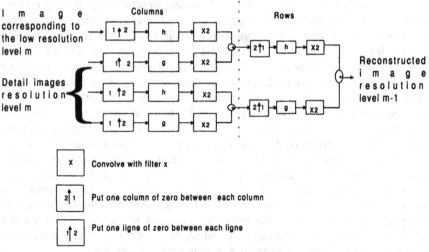

Figure 3. One stage in a multiscale image reconstruction.

The Figure 4 demonstrates the use of wavelet transform together with VQ. Approximating the sequence to be encoded by a vector in the codebooks performs the encoding. Several techniques and methods are known for creating these codebooks. One of them is the LBG algorithm [5].

The codebooks are created and optimized using classification based training set comprised of vectors belonging to different images; it converges iteratively to locally optimal codebooks.

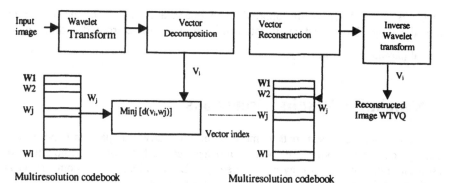

Figure 4. Block diagram of the approximated image.

2.2 Fractal Approximation

The residual image r(i, j), which is the difference image between an input image g(i, j) and an approximated one g'(i, j) is coded by fractal approximation. The last one corresponds to the range image. The block domain is formed by the approximated and then the decimated image d(k, l). The domain pool is formed respectively by r(i, j) and d(k, l). Because fractal coding uses an input image as a codebook without a learning procedure, the shuffle function employed in Jacquin's method is used to generate various codevectors. However, to reduce the search time, we use only six most frequent shuffle functions rather than eight shuffle functions used in Jacquin's algorithm [6]. The distortion function h_n used for determining the fractal mapping is defined by:

$$h_n = \sum_{i=0}^{L-1}\sum_{j=0}^{L-1} \left\| r_n(i,j) - \alpha_n G_n[\Theta_n\{d_n(k_n,l_n)\}] \right\|^2 \quad (1)$$

Where L denotes the block size of range block r_n and domain block d_n for a w_n. G_n represents one of six shuffle functions used. Θ_n is an orthogonalization operator used to remove the predictable component in $r_n(i, j)$ to $d_n(.,.)$. Thus, in order to make the grey pattern of the domain block similar to that of the range block. The orthogonalization process is done by Gram-Schmidt approach [7]. It consists to multiply the range block $d_n(., .)$ by an orthogonalization operator Θ_n defined by:

$$\Theta_n = I_n - b_1.b_1^T \quad (2)$$

Where I is the identity matrix of size $L^2 \times L^2$ and b_1 the basis vector of size L^2. The position (k_n, l_n) of a domain block, the scaling factor α_n, and the index of a shuffle function m_n are chosen so that they minimize the distortion function h_n. The set of four selected fractal parameters $(\hat{\alpha}, \hat{k}, \hat{l}, \hat{m})$ are coded with Huffman coding scheme and transmitted to the receiver. As we mentioned below, the reconstructed image is the addition of the approximated image and the reconstructed one by fractal coding. The second one is obtained by applying the entropy decoding on the fractal parameters $(\hat{\alpha}, \hat{k}, \hat{l}, \hat{m})$.

3. Simulation Results Discussions

In this section, we present both numerical and qualitative comparison between our coding scheme and other previously published results [2]. For the simulation we used the 256x256x8 bits grey scale version of Lena and City. The comparison is made using the two images taken outside the training set.

The numerical evaluation of the proposed coder's performance is archieved by computing the bit rate and the PSNR between the original image and the coded one. The PSNR is defined as:

$$PSNR = 10 \log_{10} 255^2 \Big/ MSE \qquad (3)$$

Where

$$MSE = \frac{1}{M^2} \sum_{i=0}^{M-1} \sum_{j=0}^{M-1} g\left[g(i, j) - \hat{g}(i, j) \right]^2 \qquad (4)$$

Where MxM represents the size of an input image g(i, j), and ĝ(i, j) denotes the original image and the reconstructed one by the proposed method.

The approximated image WTVQ is generated following the below scheme (Figure 4). We used the two-dimensional wavelet transform defined by Meger and Lemarié [8] together with its implementation as described by Mallat [9]. For the test images the bit assignement is presented in Figure 5. Resolution 1 (diagonal orientation is discarded). Resolution 1 (horizontal and vertical orientations) and resolution 2 (diagonal orientation) are coded using 256 vector codebooks (codeword size 4x4) resulting in 0.05 b/pixel rate, while resolution 2 (horizontal and vertical orientation) is coded at a 2 b/pixel rate using 256 vector codebooks (codeword size 2 by 2). Finally, the lowest resolution is coded at 8 b/pixel.

The wavelet decomposition of an input image enables the generation of a codebook containing two-dimensional vectors for each resolution level and preferential direction (horizontal, vertical and diagonal). Each of these subcodebooks Figure 6 is generated using the LBG algorithm. The training set is comprised of vectors belonging to eleven 256x256 images (Baboon, Boat, Desk, Home, Clown, Girl, Flower, Hotel, Lisa, Tiger, Zelda) corresponding to the resolution and orientation under consideration.

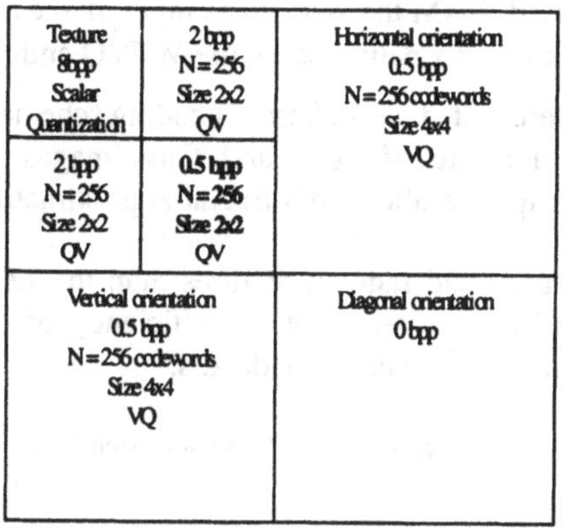

Texture
8bpp
Scalar
Quantization
2 bpp
N=256
Size 2x2
QV
Horizontal orientation
0.5 bpp
N=256 codewords
Size 4x4
VQ
2 bpp
N=256
Size 2x2
QV
0.5 bpp
N=256
Size 2x2
QV
Vertical orientation
0.5 bpp
N=256 codewords
Size 4x4
VQ
Diagonal orientation
0 bpp

Figure 5. Subimages bit rate allocation.

Resolution
M codebook
Horizontal orientation sub-codebook
Vertical orientation sub-codebook
Diagonal orientation sub-codebook
M -1
codebook
Horizontal orientation sub-codebook
Vertical orientation sub-codebook
Diagonal orientation sub-codebook
Resolution
one
codebook
Horizontal orientation sub-codebook
Vertical orientation sub-codebook
Diagonal orientation sub-codebook

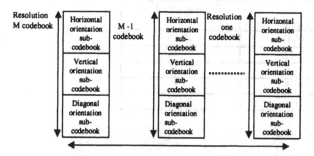

Figure 6. Multiresolution codebook.

The initial codebook is generated by splitting the centroid (center of gravity) of this training set [10]. A multiresolution codebook is obtained by assembling all these resulting subcodebooks. Each subcodebook has a low distortion level and contains few words, which clearly facilitates the search for the best coding vector. The approximated image computational load is reduced because only the appropriate subcodebook is checked for each input vector. In addition, the quality of the approximated image WTVQ is better than using the approach that combines DCT and VQ. To reduce the block effect in generating a domain pool, we first average the approximated image WTVQ with 4x4 nonoverlapping blocks then down-sample the averaged image by a factor of four.

For the fractal approximation, the function C_n used in Lepsoy et al's method is applied and six frequent shuffle functions are used. In addition, to reduce the search time, we limited a domain search block in the nearest 16x16 region, center corresponds to that of the range block. The number of blocks domain Q in which, the image domain of size nxn is search is given by:

$$Q = \alpha\left(\frac{MxM/4 - nxn}{\Delta d} + 1\right)^2 \qquad (5)$$

Where α is the shuffle function used, MxM the size of the input image and Δd the step of displacement. The codebook index x delivered by the WTVQ and the quantized fractal parameters $(\hat{\alpha}, \hat{k}, \hat{l}, \hat{m})$ were coded with Huffman coding scheme.

Table I shows the PSNR and bit rate of Lena and City images by approximated WTVQ and for the same images resulted from fractal approximation combined with the DCT and VQ [2].

The City image contains more edges and random textures than the Lena image; thus, the PSNR of the former is lower in terms of the efficiency of the WTVQ. For this the City image requires more bits to preserve details.

Table I: The PSNR and Bit rate of the reconstructed image by WTVQ, the approach based on DCT, VQ, and fractal approximation.

Lena		
	DCT, VQ, and fractal approximation	Proposed method
PSNR (db)	23.41	22.68
Bit rate (bpp)	0.42	0.144
City		
PSNR (db)	21.5	22.81
Bit rate (bpp)	0.61	0.125

Original Lena image   Decomposed Lena Image   Reconstructed Lena Image

Original City image   Decomposed City Image   Reconstructed City Image

Figure 7. Original images, Decomposed images, Reconstructed images by the proposed approach

Figure 7 shows the original images, the decomposed images by wavelet transform and the reconstructed images by the proposed method. We note that, no blocking effects as in DCT-based coders are perceptible in the proposed scheme. As it can be seen, the proposed scheme can achieved a higher compression ratio while maintaining a good reconstruction quality (both objective and subjective).

4. Conclusion

This paper presents an hybrid method coding for still images based on fractal approximation combined with anothers method coding based on block wavelet transform using classified vectors quantization. To obtain pool in fractal coding, we approximate an input image by WTVQ and then decimate. We note that in the proposed algorithm, a smaller search region is employed, and the information of grey pattern of an input image is effectively used without the constraint of contraction mapping. In addition, the algorithm is noniterative, resulting in fast decoding at a receiver. Simulation results shows that the proposed method using Wavelet transform, VQ and fractal approximation limits the blocking effects which usually appears at low bits rates by using the DCT for approximation of an input image. Also it performs better than conventional fractal, and hybrid fractal coding published [2] in terms of the bit rate and PSNR.

References

[1] Kim I. K., Pack R. H., April 1996, *Still image coding on vector Quantization and Fractal Approximation*. IEEE Trans. Image Processing, vol. 5, no. 4, pp. 587 - 597.

[2] Ait saadi K., Brahimi Z. Baraka N., August 31 – September 4, 1998, *Hybrid Approach for Still Image Compression based on Fractal Approximation and Vector Quantification*. Proceeding of the 24th Annual Conference of the IEEE Industrial Electronics Society, IECON'98, Aachen – Germany, vol. 3, pp. 1487-1493.

[3] Antonini M., Barlaud M., Mathieu P. et al. 1992, *Image coding using wavelet transform*. IEEE Trans. Image Processing, vol. 1, no. 2, pp. 205-220.

[4] Gersho A., Gray R. M., 1992, *Vector Quantization and Signal Compression*. Boston, MA: Kluwer.

[5] Linde Y., Buzo A., Gray R. M., January 1980, *An algorithm for vector quantizer design*. IEEE trans. commun, vol. COM-20, n° 1, pp 84-95.

[6] Jacquin A. E., Oct. 1993, *Fractal Image Coding: A review*. Proc. IEEE, vol. 81, pp. 1451-1465.

[7] G. Vines and M. H. Hayes, April 1993, *Adaptive IFS image coding with proximity maps*. in Proc. Int. Conf. Acoust., Speech, Signal Processing'93, vol. V, Minneapolis, MN, pp. 349-352.

[8] P. G. Lemarié, 1988, *Une nouvelle base d'ondelettes de $L^2(\Re)$*. J. Math. Pures et Appl., vol. 67, pp. 227-238.

[9] Mallat S., July 1998, *A theory for multiresolution signal decomposition: The wavelet representation*. IEEE Trans. Pattern Anal. Math. Intel., vol. 11.

[10] R. M. Gray, April 1986, *Vector quantization*. IEEEASSP Mag, pp. 4-29.

N-dimensional Frameworks for the Application of Soft Computing to Image Processing

Mario Köppen[1], Lutz Lohmann[1] and Pierre Soille[2]

[1]Department Pattern Recognition
Fraunhofer IPK Berlin, Pascalstr. 8-9, 10587 Berlin
email: mario.koeppen@ipk.fhg.de
[2]Ecole des Mines d'Alès-EERIE
Parc scientifique George Besse, F-30000 Nîmes, France
email: soille@eerie.fr

Abstract. In this paper, a new class of image processing algorithms, the n-dimensional frameworks, is introduced. Its purpose is to allow image processing algorithms to be better adaptable by soft computing techniques in general. These frameworks perform n image processing chains in parallel and fuse their single results by an appropriate procedure. They can also be considered as biologically-inspired ones. If the fusion operation includes a mapping, the framework as a whole becomes capable of representing a sufficient number of image processing operations, as it is needed for some image processing optimization problems with very large searchspaces. The fusion mapping can be derived from the intermediate result images itself, it can be given by fuzzy rules, or it can be an unmapped, but reliable algorithm. Examples for these frameworks are given by recently proposed image proessing frameworks, which have been proven to give robust and reliable results in real-world applications.

1 Introduction

The automated registration of image processing operations is an important task for the improvement of the robustness, reliability and versatility of technical vision systems. Nearly every approach in this field is based on the classical image processing chain (IPC), which consists of a linear sequence of single image processing operations steps, which are performed in a sequence. Versatility of an IPC for a practical application is achieved by its parametrization. Thus, automated configuration of an IPC can be considered as an optimization problem, whose task is to find the most suitable set of parameter values for the application of the IPC. From this description, the most important drawback of the IPC approach becomes obvious: the IPC as a whole can not perform better than its worst configured part.

Another problem with the IPC approach is not that obvious: each free parameter of an IPC gives an additional dimension of the search space of the corresponding optimization problem. Thus, the search space seems to become quite large. However, a simple quantitative investigation of the case gives, that the search space, if given in terms of the represented image (and not the IPC free parameters) is apparently greater, since it is given by a mapping. The number of possible mappings of n variables with m possible values onto k values is $k^{(m^n)}$. In fact, the set of possible image processing operations, which can be represented by an IPC is of vanishing cardinality with respect to the set of *all* possible image processing operations, even if their scope is restricted to the most simple ones.

When soft computing techniques like genetic algorithms, genetic programming or neural networks are to be applied to the design of image processing operations, the question, how image processing operations can be adequately represented for this purpose becomes very important.

The evolutionarily designed primary visual system of higher mammals gives an important hint on this issue. As was recently modeled by the Boundary Contour System / Feature Contour System [4], there are two independent pathways in the cortical processing of (at least) static images, which are perceived by the retina and cortical cells.

The point of interest in the context mentioned afore is the use of several pathways instead of a single one (i.e. an IPC). Within the final fusion of the processing results of each pathway, a mapping can be included, which dramatically increases the number of representable operations. If the mapping can be specified "in its own terms," the task of image processing registration can be solved much more effectively. This is given in more detail in section 2.

Then, this paper considers some possible twodimensional frameworks based on this general idea, i.e. frameworks, which make use of just two of such pathways, in section 3: a 2D-Lookup framework for texture filter generation [5], a fuzzy edge detector [8] and the texture segregation / region growing framework for texture segmentation [6]. Due to space limitations, the frameworks itself can not be recovered in full detail. More information can be found in the reference.

The paper concludes with a short summary and the reference.

2 Searchspace dimensions

In this section, a rough estimation should be made about the dimensionalities of the search problem involved in an optimization approach to IPC configuration. The basic assumption is, that the result of the IPC should resemble a given goal image as good as possible, or that some properties are fulfilled by the result image. From this, the configuration of an IPC comes out to be an optimization problem.

Consider figure 1, where a simple unit IPC is given. It is assumed, that the computations are restricted within a 3×3 neighborhood at each pixel. All grayvalues at image pixel locations be values between 0 and 255. The result of the operation should be a binary one, i.e. the computations give a value of either 0 or 1. Then, such a unit IPC can be considered as a mapping $f : \{0, \ldots, 255\}^9 \to \{0, 1\}$ from nine grayvalues onto the set of binary values

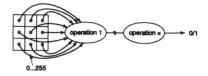

Fig. 1. The unit IPC, which maps nine grayvalues onto two.

$\{0,1\}$. If there is a quality function q for each mapping f, a mapping is searched, for which this quality measure becomes optimal. However, since for the mapping f there are 256^9 function values to specify, each of which can either be 0 or 1, there are $2^{(256^9)}$ possible mappings for the unit IPC. This is the number of elements of the searchspace, too. No optimization procedure can cover this searchspace to a non-vanishing degree!

As an example, consider the class of convolution operations. If a weighted mask is given as the following set

$$M_w = \begin{array}{|c|c|c|} \hline w_{(-1,-1)} & w_{(0,-1)} & w_{(1,-1)} \\ \hline w_{(-1,0)} & w_{(0,0)} & w_{(1,0)} \\ \hline w_{(-1,1)} & w_{(0,1)} & w_{(1,1)} \\ \hline \end{array}$$

with $w_{ij} \in \{0,\ldots,255\}$, then convolving the image I at position (x,y) can be written as

$$R(x,y) = \sum_{(i,j)\in M_w} w(i,j)I(x+i,y+j).$$

By thresholding the result with a value ϑ out of a set of m possible values, a binary image is obtained. For this unit IPC, there are $256^9 \dot{m}$ possible choices for the parameters. Compared with the number of searchspace elements (2^{256^9}), this is as good as nothings. For the simple unit IPC, we would need at least $256^9/8$ parameters to cover the searchspace.

A serious problem arises from this consideration. If adaptive techniques like soft computing methods should be applied to such an image processing problem, the searchspace is much too big to be covered by the search method.

The number of represented operations could be dramatically increased, if a mapping would be involved into the IPC operations. This leads to the definition of n-dimensional frameworks. An n-dimensional framework is a decomposition of the processing flow into n parallel parts op_1 to op_n and a final fusion procedure (see figure 2). Each single operation is applied onto the original image, and then, the n result images are fused by an appropriate algorithm (examples for fusion algorithms will be given below). If the fusion is specified by a mapping of k values out of a set of m values each onto the set $\{0,1\}$, the framework, as seen from the "outside," serves as a unit IPC of the kind given above. There are $2^{(m^k)}$ mappings specified. If m is set to 256 and k to 9, we exactly meet the requirements of the unit IPC! But, for representing

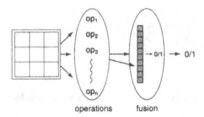

Fig. 2. An n-dimensional framework, which decomposes an IPC into n operations performed in parallel and a final fusion of the operation results. The fusion may include a mapping, which can be specified by other means.

$2^{(256^9)}$ mappings in this manner we would need $2^{(256^9)}$ mappings, and the approach seems to be ill-posed.

It is not so, since the mapping performed by the complete framework and the mapping performed in the fusion unit are qualitatively different. For example, the fusion mapping can be derived from the operation images directly. Or, it can be based on fuzzy rules, if the operation images admit *interpretation modes* (e.g. for one of them to be an edge image). Also, it can be an unmapped algorithm, but which has proven to be a good one in real-world applications. In either case, an adaptive technique simply has to search for good choices for the operation parameters only (or good choices for the operations itself). By such an "intrinsic parallelism," with each choice of the operation parameters, *all* possible settings of the fusion mapping are chosen as well. This is equivalent to the intrinsic parallelism of genetic algorithms, which test all schemata, to which a certain bitstring belongs, at each occasion of testing the bitstring itself. From this, n-dimensional frameworks follow the same searchspace concept as genetic algorithms[1].

This is the key idea of n-dimensional frameworks: they allow for its adaptation as a whole by adapting the parameters of n operations (each of which could be an IPC itself), thereby sampling the searchspace to a non-vanishing degree.

In the following section, three approaches will be given for twodimensional frameworks.

---

[1] For mathematical correctness it had to be shown, that there are not any two different configurations of a framework which lead to the same result image. This resembles the problem of verifying the schemata theorem for genetic progamming, which is still under discussion: no good mathematical tools are available for doing such proofs.

3 Examples for twodimensional frameworks

3.1 The LUCIFER2 framework

The most simple approach is to take two IPCs and merge their outcome (two output images) by means of a two-dimensional lookup procedure. The 2D-Lookup algorithm is an algorithm of mathematical morphology [10], [11]. It was primarily intended for the segmentation of color images. However, the algorithm can be generalized to use two grayvalue images as well.

For starting off the 2D-Lookup algorithm, the two operation images 1 and 2, which are of equal size, need to be provided. These images could be the result images of two image processing algorithms performed in parallel. The 2D-Lookup algorithm goes over all common positions of the two operation images. For each position, the two pixel values at this position in operation images 1 and 2 are used as indices for looking-up the 2D-Lookup matrix. The matrix element, which is found there, is used as pixel value for this position of the result image. If the matrix is bi-valued, the result image is a binary image, as it is required for the unit IPC result image.

Be I_1 and I_2 two grayvalue images, defined by their image functions g_1 and g_2 over their common domain $P \subseteq \mathcal{N} \times \mathcal{N}$. The 2D-Lookup matrix is also given as an image function l, but its domain is not the set of all image positions but the set of tupels of possible grayvalue pairs $[0, \ldots, g_{max}) \times [0, \ldots, g_{max})$. Then, the result image function is given by:

$$r : P \rightarrow [0, \ldots, g_{max})$$
$$r(x, y) = l(g_1(x, y), g_2(x, y)). \tag{1}$$

A twodimensional framework, based on this simple algorithm, was presented in [5]. The operations were designed by means of genetic programming. The framework (see fig. 3) is composed of (user- supplied) original image, filter generator, operation images 1 and 2, result image, (user-supplied) goal image, 2D-Lookup matrix, comparing unit and filter generation signal.

The framework can be thought of as being composed of three (overlapping) layers.

1. The *instruction layer*, which consists of the user-supplied parts of the framework: original image and goal image.
2. The *algorithm layer*, which performs the actual 2D-Lookup, once all of its components (original image, operation 1, operation 2 and 2D-Lookup matrix) are given.
3. The *adaptation layer*, which contains all adaptable components of the framework (operation 1, operation 2, 2D-Lookup matrix) and additional components for the adaptation (filter generator, comparison unit).

For the instruction layer, the user interface has been designed as simple as possible. The user instructs LUCIFER2 by manually drawing a (binary)

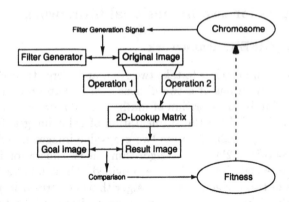

Fig. 3. The Framework for 2D-Lookup based texture filter generation.

goal image from the original image. In this image, texture background is set to White and texture foreground (e.g. the texture fault, handwriting on a textured bankcheck background) to Black. From this, the whole approach is data-driven. No special texture model has to be known by the user. There are no further requirements for the goal image.

The algorithm layer performs the 2D-Lookup algorithm. The algorithm decomposes the filter operation into a set of partial steps, each of which might be adapted to meet the user's instruction.

Adaptation is considered as an optimization problem, and genetic programming [7] is used for performing this adaptation. The fitness function is computed with the degree of resemblance between result image of an individual-specified 2D-Lookup and the goal image.

The 2D-Lookup matrix is the fusion unit of this special 2D algorithm. It represents $2^{(256^2)}$ mappings, what is not the full searchspace of the unit IPC, but a reasonable part of it. It is possible to derive an optimal 2D-Lookup matrix, if operation images 1 and 2 are given. See [5] for details. Figure 4 shows an example for an result achieved by LUCIFER2 for the detetion of a texture fault.

The LUCIFER2 approach is completely data-driven by the original image and the goal image. It has to be noted, that this is accompanied by an undersampling of the searchspace, since there are only as much (or fewer) mapping examples given, as much there are pixels within the original image. This framework has been succesfully applied to problems like texture fault detection and bankcheck background texture removal [3].

3.2   A fuzzy edge detection framework

A biological-inspired fuzzy reasoning system for robust edge detection that "optimizes" the evaluation of the correct results of an edge processing operation by including a fuzzy inference was presented in [9]. Based on the

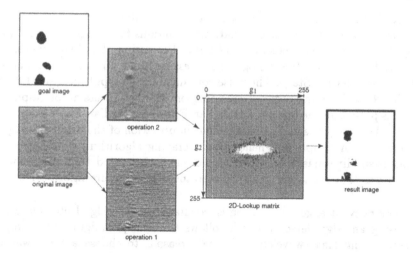

Fig. 4. Adapted 2D-Lookup framework applied to texture fault detection problem.

physiological fact, that edges on dark backgrounds subjectively appear to be of higher contrast than edges on a bright background, the result of a standard edge operator (e.g. the Sobel operator) was fuzzified as well as a measure for background intensity. Then, according to fuzzy rules, the pixel values of the result image were computed by a fuzzy inference of the two fuzzy antecedents edge intensity and background intensity.

In [8], a more general approach was derived from this key idea. The robust edge detection in images of real-world scenes is a very difficult and challenging task, mainly because such kind of images contain incomplete data such as partially hidden areas, and ambiguous data such as distorted or blurred edges, produced by various effects like variable lighting conditions. Figure 5 shows a framework for edge detection in real-world scenes.

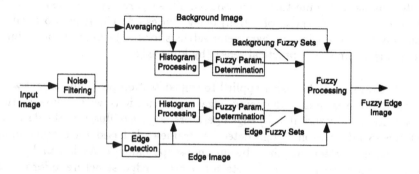

Fig. 5. The fuzzy fusion based framework for edge detection (from [8]).

As it can be seen in this block diagram, the data and control flow are separated. The two-dimensional data flow contains four stages: Noise Filtering, Averaging, Edge Detection and Fuzzy Processing. The two-dimensional control flow contains the Histogram Processing Stage and the Fuzzy Parameter Determination Stage. The noise contained in the input images is filtered in the Noise Filtering Stage. Depending on whether Gaussian or Impulsive noise is present in the images, a Gaussian-Filter or a Median-Filter is applied. In the Averaging Stage, the background information of the original image is determined. A simple and hence fast averaging algorithm, which uses two 5-pixels-size line-masks, one vertical and one horizontal, determines homogenous areas that are larger than the relevant edge structure information.

The relevant edge information is calculated in the Edge Detection Stage by using an edge detector which follows Canny's paradigm and is implemented using Haar wavelets. The main reasons to choose a Haar wavelet implementation are the fast processing and the possibility to develop a multiscale version of the detector in the near future. The proposed filter, called Haar filter (HFilter), has a vertical and a horizontal component of 5 pixel width given by the mask $[1, 2, 0, -2, -1]$. The modulus of both perpendicular components are used to calculate the local maximum of the transform, or equivalently, to find the edge locations. Some optimizations were performed in order to speed up the filter computation (e.g. by using the more simple mask $[1, 1, 0, -1, -1]$).

Based on the previously described data processing stages, the Fuzzy Processing Stage (fusion unit) calculates the final fuzzy edge image by using an Edge Image and a Background Image. This Fuzzy Processing Stage has as inputs data-dependent fuzzy sets, whose membership functions are dynamically determined by using information obtained from the cumulative histograms of the Edge and the Background Images. The fuzzy rules are given by empirical knowledge. It must be pointed out, that in the proposed architecture the fuzzy set parameters are automatically determined and that the end-user must set only very few application-dependent parameters (the size of some filter masks and some threshold values). These parameters depends on the kind of images (on the application) being processed. The framework can be adapted by optimizing the parameter settings for the fusion unit procedures, and it is optimal in a practical sense (by knowledge).

The framework has been applied to edge detection problems in real-world scenes. It gives more convincing edge representations within images than conventional ("stand-alone") edge operators. Figure 6 illustrate the data flow of the two-dimensional edge detection framework based on casting image with an ore defect, acquired by an endoscope camera. As is can be seen, the proposed architecture detects all relevant edge structure information, although the original image is very inhomogeneously illuminated.

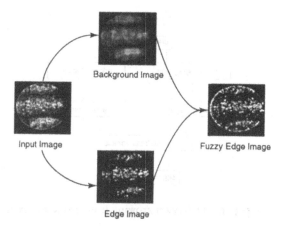

Input Image

Background Image

Edge Image

Fuzzy Edge Image

Fig. 6. The proposed framework applied to a casting image with an ore defect.

3.3 Fusion by watershed transformation

The watershed transformation [1] stems from mathematical morphology [10], [11], [14]. For a detailed description and implementation issues, refer to [2], [12] and [13].

For watershed transformation, an edge image is considered as a topographic surface, the greyvalue of a pixel standing for its elevation. Now, let a drop of water fall onto such a topographic surface. According to the law of gravitation, it will flow down along the steepest slope path until it reaches a given minimum. The whole set of points of the surface whose steepest slope paths reach a given minimum constitutes the catchment basin associated with this minimum. The watersheds are the zones dividing adjacent catchment basins.

In the presented Texture Segregation / Region Growing (TS/RG) framework [6] (see figure 7), the watershed algorithm is supplied with two input images, a label image, which reflects initial catchment bassins without watersheds, and an edge image, the intensities of which control the flooding. The label image can be generated by a segregation procedure (e.g. pixelwise classification), the edge image with either edge operator. Then, the label image is preprocessed in such a way that image minima correspond to pre-assigned texture segments. By a special operation, refered to as emergent opening of all segments, all ambiguously labeled image regions are cleared. In the given implementation, the cleared regions are filled-in by the watershed transformation. The decision about the final labeling of cleared pixel regions is done by the watershed transformation by looking for segments which grow into these regions. But their growing is under control of the intensities of the edge image.

The TS/RG framework clearly belongs to the class of twodimensional frameworks, too. The fusion of the two operation results is achieved by the

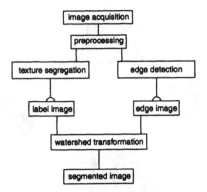

Fig. 7. The texture segregation/region growing framework for texture segmentation (from [6]).

watershed transformation. The two operation images have an intuitive interpretation: one is a segregation image and one is an edge image. The watershed transformation then is able to account for the following possible deficiencies of these images: too much edges, to few edges, virtual edges (as subjective edges between two textures), scattered and isolated wrong-labeled pixels. It has been proven to be a robust and reliable image segmentation technique.

The TS/RG framework was applied to the segmentation of real-world socket scenes taken during sewage pipe inspection, and for textile image segmentation as well.

4   Summary

This paper has introduced a new class of image processing algorithms, referred to as n-dimensional frameworks. Instead of an image operation chain, n operation chains are performed in parallel, and their n result images are fused to the final result image. Some recently proposed algorithms, which proved robustness and reliability in real-world applications, came out to belong to this new class (with $n = 2$). By a theoretical discussion of the possibilities of adapting image processing operations by means of soft computing techniques, it was shown, that n-dimensional frameworks are reasonable approaches to cover the large searchspaces of some image processing optimization problems in a manner similar to the intrinsic parallelism of genetic algorithms. This would be impossible for parameter adaptation of image processing chains alone.

For the future, the given concept of n-dimensional frameworks might help to supply new instances of them, to understand its functionality better, and thus to disclose the field of image processing operation adaptation to soft computing techniques.

References

1. S. Beucher. *Watersheds of functions and picture segmentation.* IEEE Int. Conf. on Acoustics, Speech and Signal Processing, Paris, France, pp. 1928–1931, 1982.

2. S. Beucher, F. Meyer. *The morphological approach to segmentation: the watershed transformation.* In: E. Dougherty (ed.): Mathematical morphology in image processing. Marcel Dekker, chapter 12, pp. 433–481, 1993.

3. K. Franke, M. Köppen. *Towards an universal approach to background removal in images of bankchecks.* Proc. 6. IWFHR, Taejon, Korea, pp. 55–66, 1998.

4. S. Grossberg. *A solution of the figure-ground problem for biological vision.* Neural Networks 6(1993), pp. 463–483, 1993.

5. M. Köppen, M. Teunis, B. Nickolay. *A framework for the evolutionary generation of 2D-Lookup based texture filters.* Proc. IIZUKA'98, Fukuoka, Japan, pp. 965–970, 1998.

6. M. Köppen, J. Ruiz-del-Solar, P. Soille. *Texture segmentation by biologically-inspired use of neural networks and mathematical morphology.* Proc. NC'98, Vienna, Austria, pp. 267–272, 1998.

7. J. Koza. *Genetic programming — On the programming of computers by means of natural selection.* MIT Press, Cambridge, MA, 1992.

8. L. Lohmann, Ch. Nowack, J. Ruiz-del-Solar. *A robust Arcitecture for the Automatic Edge Detection in real-world scenes.* Proc. SOCO'99, Genova, Italy, 1999, pp. 152-159.

9. N. R. Pal, S. Mukhopadhyay. *A psychovisual fuzzy reasoning edge detector.* Proc. IIZUKA'96, Fukuoka, Japan, pp. 201–204, 1996.

10. J. Serra. *Image Analysis and Mathematical Morphology.* Academic Press, New York, 1982.

11. J. Serra. *Image Analysis and Mathematical Morphology: Theoretical Advances.* Academic Press, New York, 1988.

12. P. Soille, L. Vincent. *Determining watersheds in digital pictures via flooding simulations.* In: M. Kunt (ed.): Visual Communications and Image Processing'90. Vol. SPIE-1360, pp. 240–250, 1990.

13. P. Soille. *Morphological partitioning of multispectral images.* Journal of Electronic Imaging 5(3), pp. 252–265, 1996.

14. P. Soille. *Morphological Image Analysis: Principles and Applications.* Springer Verlag, Berlin a.o., 1999.

Computational Autopoiesis for Texture Analysis

Javier Ruiz-del-Solar

Dept. of Electrical Eng., Universidad de Chile

Casilla 412-3, 6513027 Santiago, Chile.

Email: jruizd@cec.uchile.cl

Abstract. The theory of *Autopoiesis* attempts to give an integrated characterization of the nature of the living systems by capturing the key idea that living systems are systems that self-maintain their organization. This theory makes a complementary definition of the concepts of organization and structure of a system. The organization of a system defines its identity as a unity, while the structure determines only an instance of the system organization. In other words, the organization of a system defines its invariant characteristics. In this article the concept of autopoiesis is explored as a tool to analyze the internal organization of a very special kind of systems, which are the images. More specifically, a computational model of autopoiesis is applied for texture identification through the use of an autopoietic-agent.

1. Introduction

The theory of *Autopoiesis*, developed by the Chilean biologists Humberto Maturana and Francisco Varela, attempts to give an integrated characterization of the nature of a living system, which is framed purely with respect to the system in and of itself. The term autopoiesis was coined some twenty-five years ago by combining the Greek **auto** (self-) and **poiesis** (creation; production). The concept of autopoiesis is defined as [6]:

> '*An autopoietic system is organized (defined as a unity) as a network of processes of production (transformation and destruction) of components that produce the components that:*
>
> *1. through their interactions and transformations continuously regenerate and realize the network of processes (relations) that produced them; and*
>
> *2. constitute it (the machine) as a concrete unity in the space in which they [the components] exist by specifying the topological domain of its realization as such a network.*'

In other words an autopoietic system produces its own components in addition to conserving its organization. A network of local transformations produces elements, which maintain a boundary. This boundary captures the domain, in which the local transformations take place. In this context, *life* can be defined as autopoietic organization realized in a physical space. The autopoietic theory describes what the living systems are and not what they do. Instead of investigating the behavior of systems exhibiting autonomy and the concrete implementation of this autonomy (i.e. the system structure), the study addresses the reason why such behavior is exhibited (i.e. the abstract system organization). A complete material concerning the autopoietic theory (tutorials, study plan, bibliography, Internet links, etc.) can be found in the Internet site *Autopoiesis and Enaction: The Observer Web* [8].

The autopoietic theory has been applied in diverse fields such as biology, sociology, psychology, epistemology, software engineering, artificial intelligence and artificial life. In this context, this article tries to explore the use of autopoietic concepts in the field of Image Processing.

At first, the question arises whether images by themselves preserve some kind of autopoietic organization. Because of images generally are considered as static representations of real-world objects, but autopoiesis is constituted by a network of dynamic transformations, the image must be processed by suitable operators in order to reveal possible organizational principles. Thereby, the original image appears to be like a "frozen" state of its intrinsic dynamical processes. Two approaches are possible: the first approach assumes no relation between these dynamics and the real-world objects pictured in the image, in contrary to the second approach, which refers to the kind of features of real-world objects from which the dynamics are driven. In other words, these approaches differ in their interpretation of the domain of the processes (see figure 1). The first approach assumes that the domain of an image is represented only by its gray-value distribution. In order to identify autopoietic organization inside an image's pixel distribution, the *steady state Xor-operation* is identified as the only valid approach for an autopoietic processing of images. This first approach has been explored in [4]. The second approach, presented here, makes use of a second space, the A-space, as autopoietic processing domain. This allows the formulation of adaptable recognition tasks.

Based on this second approach, the concept of autopoiesis as a tool for the analysis of textures is explored in section 2. As a concrete example, a *Texture Retrieval System* based on the use of an autopoietic-agent is presented in section 3. Finally, in section 4 a summary of this work is given.

2. Autopoiesis and Texture Analysis

2.1. Textures and Autopoiesis

Texture perception plays an important role in human vision. It is used to detect and distinguish objects, to infer surface orientation and perspective, and to determine shape in 3D scenes. Even though texture is an intuitive concept, there is no universally accepted definition for it. Despite this fact we can say that textures are homogeneous visual patterns that we perceive in natural or synthetic images. They are made of local micropatterns, repeated somehow, producing the sensation of uniformity [5]. It is important to point out, that textures can not be characterized only by their structure because the same texture, viewed under different conditions, is perceived as having different structures.

In the framework of the theory of autopoiesis, Maturana and Varela make a complementary definition of the concepts of organization and structure of a system. The organization of a system defines its identity as a unity, while the structure determines only an instance of the system organization. In other words, the organization of a system defines its invariant characteristics. Figure 2 shows an example of these two complementary concepts. The concept of autopoiesis captures the key idea that living systems are systems that self maintain their organization (see introduction). In the context of texture analysis, the systems to be analyzed are the textures. As it was established, the concept of organization must be used to characterize a system and in our case to characterize a texture. For this reason, in this section the concept of autopoiesis is explored as a tool for texture identification, which corresponds to an important task in the field of texture analysis. The analogy between the process of autopoietic organization in a chemical medium (i.e. life) and the process of texture identification is used.

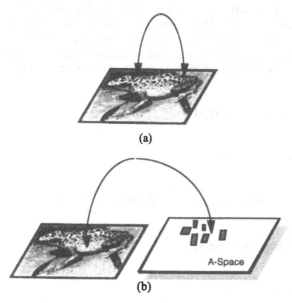

Figure 1. Internal organization of images. (a) The organization of an image is determined only by its internal gray-value distribution. (b) A second space, the A-Space (A=Autopoietic), determines the image organization.

Figure 2: Organization vs. Structure. The figure in the left-image has the basic organization of a face. The figure in the right-image has the same organization (face) but a

different, more complex structure (Source: "Vertumnus", Portrait of Emperor Rudolph II by Giuseppe Arcimboldo, 16th Century).

Before to apply the concept of autopoiesis as a tool for texture identification, a computational model of autopoiesis must be defined. Varela *et al*. developed the first model that was capable of supporting autopoietic organization [7]. Recently, McMullin developed the *SCL* model [1-2], which corresponds to an improvement of the model proposed by Varela. The SCL model from McMullin is used in this work.

2.2. The SCL Model

SCL involves three different chemical elements (or particles): *Substrate (S)*, *Catalyst (K)* and *Link (L)*. These particles move in random walks in a discrete, two-dimensional space. In this space, each position is occupied by a single particle, or is empty. Empty positions are managed by introducing a fourth class of particles: a *Hole (H)*. SCL supports six distinct reactions among particles [2]:

1. Production:
   $$K+S+S \longrightarrow K+L+H$$

2. Disintegration:
   $$L \longrightarrow S+S$$

3. Bonding:
   Adjacent L particles bond into indefinitely long chains.

4. Bond decay:
   Individual bonds can decay, breaking a chain.

5. Absorption:
   $$L+S \longrightarrow L^*$$

6. Emission:
   $$L^* \longrightarrow L+S$$

The autopoietic organization is produced, when a chain of L elements forms a boundary, which defines a concrete unity in the space. Of course, this boundary must be continuously regenerated (see introduction). The L elements are produced only in the presence of a catalyst (*Production* reaction). For this reason, we can say that in this model, an autopoietic organization is produced only in the presence of a catalyst.

2.3. The modified SCL Model

The original SCL model was modified to allow the identification of textures, by introducing the idea of a texture-dependent catalyst. That means, a catalyst that is tuned with a particular texture and that produced an autopoietic organization only in this texture. To implement this idea an autopoietic image $A(i,j)$ is defined for each texture image $T(i,j)$. Each pixel of $A(i,j)$ has a corresponding position in $T(i,j)$ and is represented by 2 bits (enough for representing four particles). A T-Space is associated with the texture image $T(i,j)$ and an A-Space is associated with the autopoietic image $A(i,j)$ (see figure 3). The reactions defined by the SCL model, that is the possible autopoietic organization, take place in the A-Space, but by taking into account information from the T-Space (textures).

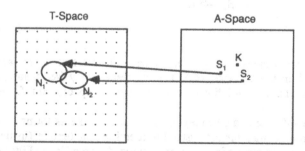

Figure 3. The A-Space, where the autopoietic organization is created, and the T-Space, where convolution between the texture and the Gabor-Filter is performed, are shown.

Gabor-Filters are able to characterize textures by decomposing them into different orientations and frequencies (scales) (see figure 4). In the proposed model, a Gabor-Filter is associated with the catalyst, to allow it (the catalyst) to be tuned with a particular texture. The Gabor-Filter interacts directly with the textures in the T-Space (convolution operation) and the result of this interaction is used to modulate the reactions in the A-Space.

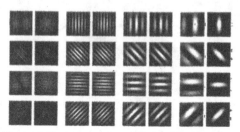

Figure 4. Some examples of Gabor-Filter masks.

From all the reactions defined by the SCL model only the *Production* reaction was modified, because it is the only one where the catalyst operates and the L particles are created. The new *Production* reaction is defined by:

Production:

$$K+S_1+S_2 \text{--------> } K+L+H$$

$$C_1 = N_1 * G_k$$
$$C_2 = N_2 * G_k$$

if $(C_1 > TH$ and $C_2 > TH)$ {
      if $(C_1 > C_2)$ {

$$S_1 \dashrightarrow L$$
$$S_2 \dashrightarrow H$$
$$\}$$
$$\quad \text{else } \{$$
$$S_1 \dashrightarrow H$$
$$S_2 \dashrightarrow L$$
$$\}$$
$$\}$$

where G_k is the Gabor-Filter associated with the catalyst K; N_1 and N_2 are the neighborhood, in the T-Space, of S_1 and S_2, respectively (see figure 3); C_1 and C_2 are the results of the convolution (performed in the T-Space); and TH is a threshold value.

If in the A-Space of a given texture a chain of elements forms a boundary after an interaction time, then the catalyst K has identified the texture (in its T-Space) as corresponding to the class of textures characterized by the Gabor-Filter G_K. Figure 5 shows an example of autopoietic organization that is automatically created by using the modified SCL model.

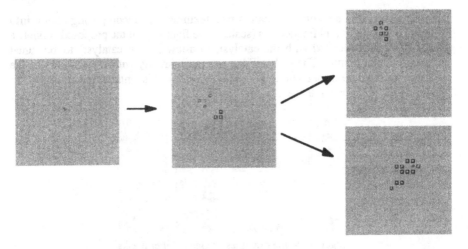

Figure 5. Example of autopoietic organization produced by the modified SCL model. Initialization (left image): The A-Space is initialized with one texture-dependent catalyst (K) and many substrate elements (S). Intermediate State (center image): After some iterations the K-element has created link elements (L), some of them form a chain (these elements are represented as small squares). Final State (right image): Two possible final configurations, where autopoietic-organization was produced, are shown. The autopoietic-organization is created when a group of L-elements forms a chain that encircles the K-element, forming a dynamic boundary. That means that this boundary is continuously broken and regenerated as for example the boundary of a cell.

To illustrate the idea of texture identification by using a computational model of autopoiesis, the proposed model is used in a system for retrieval of textures in image databases.

3. An Autopoietic-Agent for Texture Retrieval

With the new advances in communication and multimedia computing technologies, accessing mass amounts of digital visual information (image databases) is becoming a reality. In this context, textures, due to their esthetical properties, play today an important role in the consumer-oriented design, marketing, selling and exchange of products and/or product information. For this reason, systems that allow the search and retrieval of textures in image databases, the so-called *Texture Retrieval Systems*, are of increasing interest.

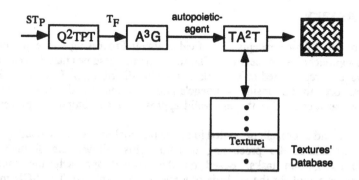

Figure 6. Proposed Texture Retrieval System. Q^2TPT: Qualitative to Quantitative Textural Properties Transformation; A^3G: Automatic Autopoietic-Agent Generator; TA^2T: Textural Autopoietic-Agent Tester; ST_P: Subjective Textural Properties; T_F: Textural Features.

The retrieval of information in a traditional database has always a clear target in mind, this means, it looks for a specific element or record in the database. This is not the case of the retrieval of information in an image database. For this reason, and because the retrieval of textured images must be judged by a human user, it is natural to use human concepts of texture description, texture perception and similarity in the design of these systems. In this context and as part of a main research effort[1], the proposed computational model of autopoiesis is used in a system for retrieval of textures in image databases. The main characteristic of this system is that it accepts human-like descriptions of textures (i.e. using some qualitative properties as business, softness, brightness, roughness, flatness, etc.) as inputs. The block diagram of the proposed system is shown in figure 6.

The texture retrieval process works as follows: A human user makes a query of a texture by using subjective textural properties (ST_P). The Q^2TPT (Qualitative to Quantitative Textural Properties Transformation) module interprets the query and translates it into a quantitative texture description that we call Textural Features (T_F) [3].

The search process is based on the use of an autopoietic-agent (the texture-dependent catalyst described in section 2.3), which is generated in the A^3G

[1] Fondecyt Project 1990595: "Bases de Datos de Texturas: Análisis, Búsqueda y Síntesis de Texturas utilizando Redes Neuronales y Lógica Difusa" (Texture Databases: Analysis, Retrieval and Synthesis of Textures using Neural Networks and Fuzzy Logik).

(Automatic Autopoietic-Agent Generator) module by using the T_F. The autopoietic-agent is tuned with this quantitative texture description, which means it can interact (produce autopoietic organization) only with the texture that corresponds to this description. The autopoietic-agent is sent to every texture in the database (as a processing thread) and is allowed to interact with the substrate particles of the A-Space of the textures. After an interaction time, the texture where an autopoietic organization was produced (in its A-Space), is identified and finally retrieved. This is performed in the TA^2T (Textural Autopoietic-Agent Tester) module.

4. Summary

The use of autopoietic concepts in the field of Image Processing was explored. Two different approaches were identified. The first approach assumes that the organization of an image is represented only by its gray-value distribution. In order to identify autopoietic organization inside an image's pixel distribution, the *steady state Xor-operation* was identified as the only valid approach for an autopoietic processing of images.

The second approach, presented in this paper, makes use of a second space, the *A-space*, as autopoietic processing domain. This allows the formulation of adaptable recognition tasks. Based on this second approach, the concept of autopoiesis as a tool for the analysis of textures was explored. The SCL model, a computational model of autopoiesis, was modified to allow the identification of textures, by introducing the idea of a texture-dependent catalyst. As a demonstrating example, a *Texture Retrieval System* based on the use of an *autopoietic-agent*, the texture-dependent catalyst, was described. Further research must be performed to apply this computational model of autopoiesis in the solution of other kind of image processing problems.

Acknowledgements

This research was supported by FONDECYT (Chile) under Project Number 1990595.

References

[1] B. McMullin, *Computational Autopoiesis: The original algorithm*, Working Paper 97-01-001, Santa Fe Institute, Santa Fe, NM 87501, USA, 1997. http://www.santafe.edu/sfi/publications/Working-Papers/97-01-001/

[2] B. McMullin, *SCL: An artificial chemistry in Swarm*, Working Paper 97-01-002, Santa Fe Institute, Santa Fe, NM 87501, USA, 1997. http://www.santafe.edu/sfi/publications/Working-Papers/97-01-002/

[3] M. Köppen, and J. Ruiz-del-Solar, "Fuzzy-based Texture Retrieval", *Proc. IEEE Int. Conf. on Fuzzy Systems - FUZZ-IEEE 97*, pp. 471-475, July 1-5, Barcelona, Spain, 1997.

[4] M. Köppen, and J. Ruiz-del-Solar, "Autopoiesis and Image Processing I: Detection of Image Structures by using Auto-Projective Operators," Computational Intelligence for Modelling, Control & Automation, M. Mohammadian (Ed.), pp. 66-71, IOS Press, 1999.

[5] J. Ruiz-del-Solar, "TEXSOM: Texture Segmentation using Self-Organizing Maps," *Neurocomputing* (21) 1-3 pp. 7-18, 1998.

[6] F.J. Varela, *Principles of Biological Autonomy*, New York: Elsevier (North Holland), 1979.

[7] F.J. Varela, H.R. Maturana, and R. Uribe, "Autopoiesis: The organization of living systems, its characterization and a model," *BioSystems* 5: 187-196, 1974.

[8] R. Whitaker, *Autopoiesis and Enaction: The Observer Web.* http://www.informatik.umu.se/~rwhit/AT.html

NOVEL APPROACH IN WATERMARKING OF DIGITAL IMAGE

Chan Choong Wah

School of Electrical & Electronic Engineering
Nanyang Technological University,
Nanyang Avenue, Singapore 639798

Abstract: Watermarking techniques, also referred to as digital signature, sign images by introducing changes that are imperceptible to the human eye but easily recoverable by a computer program. In this paper, we present an approach to digital image watermark based on multiresolution wavelet decomposition. The watermarking technique consists of fusing the level-1 watermark coefficients with the host level-4 image coefficients. Three simple fusion algorithms are examined. We discuss the extraction of the watermark and access the retrieved watermark as compared to the original watermark. Simulation results demonstrate the ease of insertion of watermark without image degradation.

1. Introduction

In today's digital world, there is a great wealth of information, which can be accessed in various forms: text, images, audio and video. There is a growing demand for effective techniques to protect the author from having his work stolen, copied, unauthorized duplication and modification. A digital watermark could be used to provide proof of "authorship" of the original piece of work.

As the term 'watermark' suggests, digital watermarking is a kind of technique to add some digital structures to the data to mark the ownership. It is required that this approach must leave the original data *perceptually* unchanged after the watermarking, and in order to discourage the illegal piracy, the removal of the watermark should be relatively difficult unless the exact watermarking method is known.

Generally, the watermark technique is distinguished in three particular algorithms: watermark production algorithm, watermark embedding algorithm and watermark detection/extraction algorithm. The watermark should be characteristic of an author, but a "pirate" should not be able to detect the watermark by comparing several signals belonging to the same author.

The process of digital watermarking involves the modification of the original data to embed a watermark containing key information such as authentication or copyright codes. The embedding methods must leave the original data perceptually unchanged, yet should impose modifications which can be detected by using an appropriate extraction algorithm. Common types of signals to watermark are images, music clips and digital video. In this paper, we concentrate on the application of digital watermarking to still image.

In the next section, a brief background of wavelet transform is described. An overview of Image Watermarking is described in Section 3. We describe the watermark embedding method in Section 4. Section 5 discusses the detection/extraction of digital watermark. The experimental results and discussion are presented in Section 6. Finally, Section 7 concludes the paper and remarks on future work.

2. Wavelet Transform

The wavelet transform is really a family of transforms that satisfy specific conditions. From our prospective, the *wavelet transform* can be described as a transform that has basic functions that are shifted and expanded versions of themselves. Because of this, the wavelet transform contains not just frequency information but spatial information as well. One of the most common models for a wavelet transform uses the Fourier transform, highpass and lowpass filters. To satisfy the conditions for a wavelet transform, the filters must be *perfect reconstruction* filters, which means that any distortion introduced by the forward transform will be cancelled in the inverse transform.

Assuming that the input is a digital image, the wavelet transform breaks the image down into four subsampled, or decimated images. They are subsampled by keeping every other pixel. The results consist of one image that has been highpasss filtered in both the horizontal and vertical directions, one that has been highpass filtered in the vertical and lowpass filtered in the horizontal, one that has been lowpassed in the vertical and highpassed in the horizontal, and one that has been lowpass filtered in both directions.

2.1 2D Discrete-Wavelet Transform

Figure 1 shows the 2D DWT block of the encoder. The 2-D DWT block consists of four levels of decomposition as illustrated in Figure 1(a). Clearly, the specific decomposition used here results in 13 subbands. Each level of decomposition, represented by the operation A in Figure 1(a), is described further in terms of simple operation in Figure 1(b). Specifically, A consists of low-pass and high-pass filtering (L and H) in the row direction and subsampling by a factor of two, followed by the same procedure on each of the resulting outputs in the column direction, resulting in four subbands. The H and L filters are finite-impulse-response (FIR) digital filters. The specific input-output relationship for one level DWT decomposition of a 1-D sequence X(n) can be represented as

$$X_l(n) = \sum_k l_l(2n-k)X(k) \tag{1}$$

$$X_h(n) = \sum_k h_1(2n-k)X(k) \qquad (2)$$

in which $X_l(n)$ and $X_h(n)$ represent, respectively, the output of the low-pass and high-pass filters.

The resulting 2D subbands after the 2D DWT operation are labelled *Subband1* through *Subband13*. Due to the specific form of the decomposition, for a host image of size 256 x 256, *Subband1* through *Subband4* are of 16 x16 pixels size, *Subband5* through *Subband7* are 32 x 32 pixels size, *Subband8* through *Subband10* are 64 x 64 pixels size, and *Subband11* through *Subband13* are of 128 x 128 pixel size.

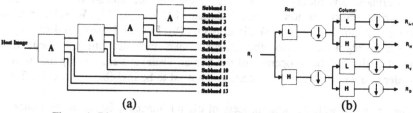

(a)   (b)

Figure 1: Block diagrams of multiresolution image decomposition.

2.2 2D Inverse Discrete-Wavelet Transform

To reconstruct a replica of the image, the subbands are then fed into the 2-D IDWT block. Figure 2 shows the details of the 2-D IDWT operation. The 2-D IDWT block consists of four levels of reconstruction as illustrated in Figure 2(a). Each level of reconstruction, represented by the operation B in Figure 2, is described in terms of simpler operations in Figure 2(b). Specifically, B consists of up-sampling by a factor of two and low-pass and high-pass filtering in the column direction followed by the same procedure on the outputs of this process in the row direction, integrating four subbands into one wider band. The filters used for reconstruction are FIR digital filters. The specific input-output relationship for the reconstruction of the sequence X(n) is represented by

$$X(n) = \sum_k l_2(2k-n)X_l(k) + h_2(2n-k)X_h(k) \qquad (3)$$

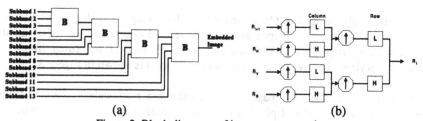

(a)   (b)

Figure 2: Block diagrams of image reconstruction.

3. Overview of Digital Image Watermark

Various authors have proposed varieties of watermarking techniques. The proposed algorithms can be classed in two main classes on the basis of the utilization of the original image during the detection phase. Algorithms proposed in [11, 20, 21, 22, 23, 28] do not require the original image whereas in those presented in [2, 17, 25, 26] the original image is input in the detection algorithm along with the watermarked image. Method proposed in [29] require the original and watermarked images for the extraction of the watermark. Detectors of the second type have the advantage to detect the watermarks in images that have been extensively modified in various ways. However detectors of this kind cannot be combined with web-crawling and automatic watermark searching in a digital library.

Watermark embedding can be done either in the spatial domain or in an appropriate transform domain (DCT domain [2, 23, 24, 25], Wavelet transform domain [17, 26, 29], Fourier Mellin domain [28], Fourier Transform domain [27]). In certain algorithms also, the changes take into account the local image characteristics and the properties of the human visual system (perceptual masking) in order to obtain watermarks that are guaranteed to be invisible [17, 22, 23, 24].

The proposed watermarking process of digital images is based on the concept of multiresolution wavelet fusion [16, 17]. The watermarking technique hides information in digital images so as to discourage unauthorized copying of the origin media. The "host" image and the watermark are decomposed into the discrete wavelet domain. The resulting image pyramids are then fused together in J resolution level. An inverse wavelet reconstruction is performed on the fused image.

Using the wavelet transform approach, the watermark is decomposed to 1-level. Likewise, the host image is decomposed into a level where its coarsest image array has the same size as the decomposed watermark. To embed the watermark into the host image, three algorithms are examined to fuse the decomposed watermark into the decomposed host image. This will form a new decomposed image component. In order to generate the watermarked image, a 2D discrete wavelet reconstruction is performed to the new decomposed image component.

4. Watermark Embedding Method

Watermark embedding is implemented via the following three steps: (i) transform the host and watermark images into the wavelet domain; (ii) fusion of details coefficients of images; and (iii) wavelet reconstruction of the fused image. A general overview of the watermark embedding process is shown in Figure 3.

We assume that the host image and watermark are grey scale images. The host image is an image of $M \times M$ pixels and the watermark is an image of $N \times N$ pixels.

For our simulations, the size of the watermark in relation to the host image is "smaller" than the host image. We use $h(m,n)$ to denote the host image and $w(m,n)$ the watermark. The watermark embedding technique is comprised of the following three steps.

Step 1 - The host image and the watermark are two-dimensional image array. The size of the watermark is chosen such that it is small in relation to the host image. For robustness, the watermark has characteristics, which are "noise-like" and random. We use 'lena' (size of 256 x 256) as the host image while the watermark is of size 16 x 16. The host image and the watermark is shown in Figure 3 of the watermarking process.

The proposed embedding method employs a multiresolution wavelet decomposition of the host image and the watermark as shown in Figure 3. A 1-level 2D wavelet transform decomposition is performed on the watermark. This generates one approximation coefficient and its 3 detail coefficients, corresponding to horizontal, vertical and diagonal details. Similarly, a 4-level 2D wavelet transform decomposition is performed on the host image and a series of 3 detail coefficients and approximation coefficient are produced. At level-4 of the wavelet decomposition, the approximation coefficient and the 3 detail coefficients of the hose image have an array size of 16 x16..

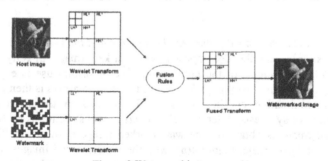

Figure 3 Watermarking process

Step 2 - The fusion of the decomposed watermark with level-4 of the host image is performed at this stage. Three simple fusion algorithms are examined. They are (1) taking sum of the coefficients, (2) taking the difference between them and (3) taking the average between them. The resulted coefficients are used in the reconstruction of the watermarked image at stage 3. The results of the three algorithms are discussed in later session.

Step 3 - The watermarked image is reconstructed at this stage after the merging of the decomposed watermark and the level-4 decomposed host image. A level-4 2D inverse wavelet transformed is performed on the resultant coefficient to reconstruct the watermarked image.

5. Watermark Extraction

The theory behind the watermark extraction is similar to the embedding algorithm. We use both the watermarked image and the host image. After decomposing of the watermarked image and the host image to level-4, we obtain two sets of level-4 approximation coefficients. The difference between these two coefficients is the decomposed watermark. The extracted watermark can be obtained by taking the inverse wavelet transform of the decomposed watermark. Section 7.1 describes the extraction process of the watermark.

6. Simulation Results and Discussion

As described in the watermark embedding process, 2D wavelet transform decomposition is performed on the host image and the watermark. At level-4 decomposition, an approximation and 3 detail coefficients, of size 16x16, of the host image are obtained. At level-1 of the watermark decomposition, the approximation, horizontal, vertical and diagonal coefficients of the watermark is arranged to form a 16x16 array. The newly arranged watermark, wmk', and the level-4 approximate coefficient are of the same size. A simple fusion process of the coefficients is performed. Three fusion algorithms are examined and are described as follows:

i) Taking the sum of the decomposed host image and the watermark
This is to simply add the decomposed watermarked image directly to level-4 approximation coefficients of the host image. The algorithm use is expressed as: $a_{4m(i,j)} = a_{4o(i,j)} + (alpha * wmk_{1(i,j)})$. The sum of these two arrays is then regarded as the new approximation coefficient, $a_{4m(i,j)}$. It is used in the inverse wavelet reconstruction, or synthesis. After a level-4 inverse reconstruction is performed, the watermarked image is obtained. The watermarked image is shown in Figure 4(b). Comparing the host image, Figure 4(a), with the watermarked images, we can see that they are perceptually identical, which means the embedding effect is satisfactory.

ii) Taking the difference of the decomposed host image and watermark
Similarly after the decomposition, instead of adding the watermark to the host image, we subtract the watermark from the host image. The algorithm use is expressed as: $a_{4m(i,j)} = a_{4o(i,j)} - (alpha * wmk_{1(i,j)})$. An inverse wavelet decomposition procedure is then performed. The result of the watermarked image shown in Figure 4(c) is similar to the original host image.

iii) Taking the average values of the decomposed host image and watermark
It is obvious that the only difference between this method and above two methods is to use the average value of the decomposed host image and the watermark as the new approximation coefficients. The algorithm used is expressed as: $a_{4m(i,j)} = (a_{4o(i,j)}$

+ (*alpha* * wmk$_{1(i, j)}$))/2. The result obtained is very poor as shown in Figure 4(d). This is due to the new approximation coefficient has be scaled down. It shows an obvious distortion in the watermarked image.

In all the three equations, the controlling parameter, *alpha*, sets the intensity of the watermark in the fusion process. *Alpha* takes a positive value and is set to 1 for the watermarked images shown in Figure 4.

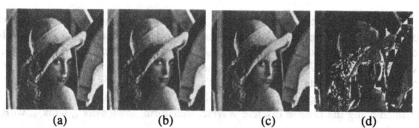

| (a) | (b) | (c) | (d) |

Figure 4: (a) Host image, (b) wmk (add), (c) wmk (subtract), and (d) wmk (average).

Table 1 shows the Correlation and PSNR of the watermarked images as compared with the host image for the three algorithms. The PSNR for the Add and Subtraction algorithms are high as compared to that of the Average algorithm. With a high coefficient value of the host image and a low coefficient value of the watermark, the average of this set of coefficient affect the overall result, hence resulting in a relatively low PSNR.

Table 1: Correlation and PSNR of watermarked image

Algorithm	Correlation	PSNR
Add	0.9987	+ 39.13 dB
Subtract	0.9987	+ 39.13 dB
Average	0.9465	+ 13.68 dB

Figure 5 shows the curve of the PSNR vs. *alpha*. As *alpha* increases from 0.1 to 1, the PSNR decreases for the Add and Subtract algorithms. Since the approximation coefficient of the watermark is low, any change of *alpha* in the Average algorithm does not affect the overall PSNR. The curve remains almost a straight line.

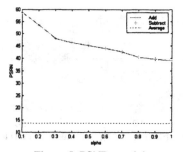

Figure 5: PSNR vs. alpha

6.1 Extraction of watermark

Two watermarks are used to test and validate the watermarking and extraction process. Watermark I consists of a binary level, which has a random pattern. Watermark II is an 8-bit gray scale image with 256 levels. Watermark I can be replaced by an ASCII pattern, which can carriers coded information. A Company logo or any digital image of 256 gray levels can be used as the Watermark II.

The original image is decomposed at level-4, which generate the approximate coefficient, a_{4o}. Likewise, the watermarked image is decomposed at level-4, generating the approximate coefficient, a_{4w}. A watermark, wmk' of size 16 x 16, is extracted using the algorithm $(a_{4w} - a_{4o})$ / *alpha*. The extracted watermark, wmk', holds the approximate coefficient, wmk'$_a$, and three details coefficients, wmk'$_h$, wmk'$_v$, wmk'$_d$. Using the coefficients of wmk', the origianl watermark, wmk, is obtained using the inverse discrete wavelet transform, idwt(wmk'$_a$, wmk'$_h$, wmk'$_v$, wmk'$_d$).

The original and the extracted watermarks are shown in Figure 6. Figure 6(b) and 6(c) shows a similarity to the original watermark. If the extracted watermark of Figure 6(b) and 6(c) are threshold and image processing performed on them to form the binary image, the results are the same as the original watermark. The correlation factor will be near to one.

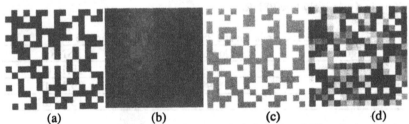

| (a) | (b) | (c) | (d) |

Figure 6: (a) org. wmk (b) wmk. (add) (c) wmk. (subtract) and (d) wmk. (average).

A gray scale intensity watermark, a small section of the wheel of a car, is used in the simulation of the three algorithms. Figure 7 shows the original gray scale watermark and the extracted watermarks.

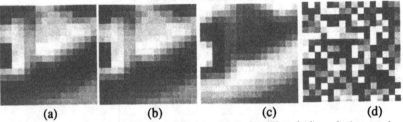

| (a) | (b) | (c) | (d) |

Figure 7: (a) org. wmk, (b) wmk. (add), (c) wmk (subtract) and (d) wmk. (average)

Table 2 shows the correlation and PSNR of the extracted watermark for Watermark I and Watermark II.

Table 2: Correlation and PSNR of Watermarks

Algorithm	Watermark I		Watermark II	
	Correlation	PSNR	Correlation	PSNR
Add	0.9997	+45.15 dB	0.9974	+35.61 dB
Subtract	0.9997	+45.16 dB	0.9974	+35.61 dB
Average	0.9470	+13.66 dB	0.9460	+13.74 dB

As seen from the results of the experiments, the original watermark can be easily extracted from the watermarked image. Hence, the results show that the embedding and extracting algorithms can hide information.

7. Conclusions and Future Development

In this paper we describe a digital image watermarking process based on multiresolution wavelet decomposition. Three methods of fusion algorithms are examined. The results shown that the watermark images are perceptually unchanged, except when using the average algorithm. The extracted watermark shows a close similarity to the original watermark. The methods proposed can be extended to hide critical information in the digital image. Future works involves the formulation of new fusion algorithms, which takes into account of human visual system (HVS). We will also investigate the significant of compression, use of wavelet package and linear filtering to improve the watermarking effects.

References

(1) Nasir Memon and Ping Wah Wong, "Protecting Digital Media Contents," Communication of ACM, July 1998, Vol. 41, No 7., pp. 35-43.

(2) I. J. Cox, J. Kilian, T. Leighton and T. Shamoon, "Secure Spread Spectrum Watermarking for Multimedia," IEEE Trans. on Image Processing, 6, 12, pp. 1673-1687, 1997.

(3) I. J. Cox and J-P Linnartz, "Public watermarks and resistance to tampering," Proc. IEEE Int. Conf. on Image Processing, 1997.

(4) I. J. Cox and J-P Linnartz, "Some general methods for tampering with watermarks," IEEE Journal Selected Areas of Communications (JSAC).

(5) I. J. Cox and M. L. Miller, "A review of watermarking and the importance of perceptual modeling," Proc. SPIE Conf. on Human Vision and Electronic Imaging II, Vol 3016, 92-99, February 1987.

(6) I. J. Cox, J. Kilian, T. Leighton and T. Shamoon, "Secure Spread Spectrum Watermarking for Images, Audio and Video," Proc. of 1996 Int'l. Conf. on Image Processing (ICIP'96), Vol. III, pp. 243-246, 1996.

(7) I. J. Cox, J. Kilian, T. Leighton and T. Shamoon, "A Secure, Robust Watermark for Multimedia," Workshop on Information Hiding, Newton Institute, Univ. of Cambridge, May 1996.

(8) Jean-Paul M. G. Linnartz, A. A. C. Kalker, G. F. G. Depovere and R. A. Beuker, "A Reliability model for detection of electronic watermarks in digital images".

(9) Jean-Paul M. G. Linnartz and Marten van Dijk, "Analysis of the Sensitivity Attack against Electronic Watermarks in Images".

(10) Ching-Yung Lin and Shih-Fu Chan, "A Water-Based Robust Image Authentication Method Using Wavelets".

(11) I. Pitas, "A method for signature casting on digital images," Proc. IEEE Int. Conf. on Image Processing, Vol. 3, pp. 215–218, 1996.

(12) R. B. Wolfgang and E. J. Delp, "A watermark for digital images," Proc. IEEE Int Conf. on Image Processing, Vol. 3, p 219–222, 1996.

(13) G. Vayatzi, N. Nikolaidis and I. Pitas, "Digital Watermarking: An Overview".

(14) M. D. Swanson, B. Zhu, and A. H. Tewfik, "Transparent robust image marking," Proc. IEEE Int. Conf. on Image Processing, Vol 3, pp. 211–214, 1996.

(15) D. Kundur and D. Hatzinakos, "Towards a Telltale Watermarking Technique for Tamper-Proofing," Proc. IEEE Int. Conf. on Image Processing, Chicago, Illinois, October 1998.

(16) D. Kundur and D. Hatzinakos, "Digital Watermarking using Multiresolution Wavelet Decomposition," Proc. IEEE Int. Conf. on Acoustics, Speech and Signal Processing, Seattle, Washington, Vol. 5, pp. 2969-2972, May 1998.

(17) D. Kundur and D. Hatzinakos, "A Robust Digital Image Watermarking Scheme using Wavelet-Based Fusion," Proc. IEEE Int. Conf. on Image Processing, Santa Barbara, California, Vol. 1, pp. 544-547, October 1997.

(18) Houngjyh Mike Wang, "A fast wavelet based watermark embedding and blind detecting algorithm,"@ http://biron.usc.edu/~houngjyh/

(19) Mucahit K. Uner, Liane C. Ramac and Pramod K. Varshney, "Concealed Weapon Detection: An image fusion approach," SPIE Vol. 2942, pp. 123-132.

(20) G.Voyatzis and I.Pitas, "Applications of Toral automorphisms in image watermarking", Proceedings of ICIP'96, vol II, pp. 237-240, 1996.

(21) G. Voyatzis and I. Pitas, "Embedding Robust Logo Watermarks in Digital Images", Proceedings of DSP'97, vol 1, pp. 213-216, 1997.

(22) N. Nikolaidis and I.Pitas "Robust image watermarking in the spatial domain", Signal Proc. special issue on Copyright Protection and Access control, 1998.

(23) A. Piva, M. Barni, F. Bartolini, V. Cappellini, "DCT-based watermark recovering without resorting to the uncorrupted original image", Proceedings of ICIP'97, Santa Barbara, CA, USA, October 26-29, 1997, Vol I, pp. 520-523.

(24) Mitchell D. Swanson, Bin Zhu , and Ahmed H. Tewfik , "Transparent Robust Image Watermarking," Proc. of the 1996 IEEE Int. Conf. on Image Processing, Vol. III, pp. 211-214, 1996.

(25) J. O'Ruanaidh, W. Dowling, F. Boland, "Watermarking digital images for copyright protection", IEE Proceedings on Vision, Image and Signal Processing, 143(4), pp. 250-256, August 1996.

(26) X.-G. Xia , C. G. Boncelet, and G. R. Arce, "A Multiresolution Watermark for Digital Images" Proceedings of ICIP'97, Santa Barbara, CA, USA, October 26-29, 1997, Vol I, pp. 548-551.

(27) J. O'Ruanaidh, W. Dowling, F. Boland, "Phase watermarking of digital images", Proc. 1996 IEEE Int. Conference on Image Processing (ICIP 96), vol III, pp 239-242.

(28) J. O'Ruanaidh, T. Pun, "Rotation, Scale and Translation Invariant Digital Image Watermarking", Proceedings of ICIP'97, Santa Barbara, CA, USA, October 26-29, 1997, Vol I, pp. 536-539.

(29) Chan Choong Wah, "Watermarking Technique for Protection of Digital Images and Extend to Information Hiding," 9th MINDEF-NTU Joint R&D Seminar, NTU, Singapore, 18 January 1999, pp. 130 - 135.

A Fuzzy Region-Growing Algorithm for Segmentation of Natural Images

Junji MAEDA, Sonny NOVIANTO, Sato SAGA and Yukinori SUZUKI

Muroran Institute of Technology, 27–1 Mizumoto–cho, Muroran 050–8585 JAPAN,
junji@csse.muroran–it.ac.jp

Abstract: We present a new method that integrates intensity features and a local fractal-dimension feature into a region growing algorithm for the segmentation of natural images. A fuzzy rule is used to integrate different types of features into a segmentation algorithm. In the proposed algorithm, intensity features are used to produce an accurate segmentation, while the fractal-dimension feature is used to yield a rough segmentation in a natural image. The effective combination of the different features provides the segmented results similar to the ones by a human visual system. Experimental results demonstrate the capabilities of the proposed method to execute the segmentation of natural images using the fuzzy region-growing algorithm.

1. Introduction

The purpose of this paper is to segment natural images with different precision. For a natural image containing houses and trees, we execute an accurate segmentation for a part of the houses and a rough segmentation for a part of the trees. We would like to regard the part of trees including many branches and leaves as the same region as much as possible, while keeping high-precise segmentation at the part of the houses.

It is known that the fractal dimension (FD) of the image is a powerful measure for natural images, since it has been shown that the FD has a strong correlation with human judgement of surface roughness [1]. Although several results of segmentation based on the FD have been reported, the FD alone does not perform a good segmentation because of the low resolution of the FD in natural images [2], [3].

Proposed in this paper is a new segmentation algorithm that integrates intensity features and a FD feature into a fuzzy region-growing algorithm in segmenting natural images. In the proposed method, the intensity features are used to produce an accurate segmentation at the part of non-texture regions such as the houses, while the FD feature is used to yield a rough segmentation at the part of texture regions such as the trees. The low resolution of the FD becomes advantageous in performing a rough segmentation. In this paper we use a blanket method to estimate the local FD [4], [5], [6]. Furthermore, we have estimated an optimum number of the blanket suitable for the local estimation of the FD. We have used a fuzzy set theory in order to integrate different types of features into a region growing algorithm [7], [8].

We present some results of computer simulations that demonstrate the capabilities of the proposed segmentation algorithm in segmenting natural images.

2. Estimation of Local Fractal Dimension

In the blanket method, an upper and lower blanket are grown from the image surface [4]. If ε is the number of the blanket, and u_ε and b_ε are the upper and lower blanket surfaces at position (i, j), then the surface area of the blanket is calculated as follows:

$$A(\varepsilon) = \frac{\sum\sum (u_\varepsilon(i,j) - b_\varepsilon(i,j))}{2\varepsilon} \tag{1}$$

On the other hand, the area of a fractal surface behaves like [9]

$$A(\varepsilon) = F\varepsilon^{2-D} \tag{2}$$

where F is a constant and D is the FD of the image. Therefore, the FD can be estimated from the slope of the straight line if $A(\varepsilon)$ versus ε is plotted on a log-log scale. However, the actual plot is not a straight line but a nonlinear curve especially for the small local area or window. Therefore, the value of the estimated FD will change according to the maximum number of the blanket to be used in the estimation. Since it is desirable to use the local area as small as possible for the purpose of image segmentation, it is necessary to decide the optimum number of the blanket for a small window.

We have evaluated the behavior of the local FD when we change the number of the blanket in calculating the FD for several sizes of window. We have used the window of the following sizes: 3×3, 5×5, 7×7 and 9×9. For a certain size of a window, we estimate each local FD for a fixed number of the blanket by calculating the average of 200 samples taken from a texture image. We have evaluated the sum of the difference (*SOD*) between the global FD (*GFD*) and the local FD (*LFD*) for the several sizes of the window as a function of ϵ:

$$SOD(\epsilon) = \sum_{i=1}^{4} |GFD(\epsilon) - LFD_i(\epsilon)| \tag{3}$$

where i corresponds to the four sizes of the window and global FD means the use of 256×256 window. Figure 1 shows the examples of the estimated local and global FDs for a certain texture image that demonstrate the variation of the estimated value of FD for four sizes of local windows when we change the maximum number of the blanket (ϵ). Figure 2 represents the minimum values of the sum of the difference (*SOD*) between the global FD and the local FD when we change the maximum number of the blanket for 40 kinds of texture images from Brodatz album [10]. This figure shows that the number of the blanket between 30 and 58 demonstrates a minimum variation from the global FD in estimating the local FD. Thus we use 44 as the optimum number of the blanket to calculate the local FD in our algorithm.

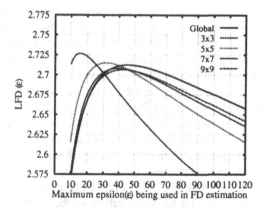

Fig. 1: The estimated local and global fractal dimensions versus the maximum number of the blanket (ϵ) for various sizes of local windows.

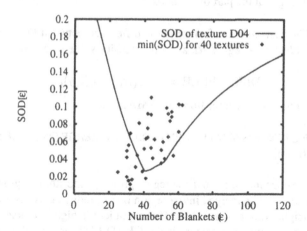

Fig. 2: The minimum values of the sum of the difference between the global and local fractal dimensions versus the maximum number of the blanket for 40 textures.

3. Fuzzy Region-Growing Algorithm

The segmentation procedure in the present investigation is the fuzzy region-growing algorithm that is based on fuzzy rules. Our final objective is to split an original image I into a number of homogeneous but disjoint regions R_j:

$$I = \bigcup_{j=1}^{n} R_j, \quad R_j \cap R_k = \emptyset \quad j \neq k. \tag{4}$$

The region growing is essentially a grouping procedure that groups pixels or subregions into larger regions so that each larger region holds the homogeneity criterion. Starting from a single pixel, a segmented region is created by merging the neighboring pixels or the adjacent regions around a current pixel. The operations are repeatedly performed until there is no pixel that does not belong to a certain region.

Since our strategy in segmenting natural images is an effective combination of an accurate segmentation by the intensity features and a rough segmentation by the FD feature, it is inevitable to employ a technique of information integration. We adopt fuzzy rules to integrate the different features. We use the following criteria where each fuzzy rule has a corresponding membership function. The intensity features are the intensity difference and the intensity gradient.

In the proposed fuzzy rules, we establish a stronger merging rule for the fuzzy set from the FD feature than for the fuzzy sets from the intensity features in order to achieve a rough segmentation at the part of the trees. Since the local FD provides broad edges around the true strong edges, however, we employ the boundary edge [5], [11] as the intensity gradient to protect the unnecessary growth of regions around the true edges at the part of the houses.

[Rule 1] The first intensity feature is the difference between the average intensity value $g_{ave}(R_k)$ of a region R_k and the intensity value of a pixel $g(i,j)$ under investigation:

$$\text{DIFFERENCE} = |g_{ave}(R_k) - g(i,j)| \tag{5}$$

The corresponding fuzzy rule for fuzzy set SMALL is

R1: IF DIFFERENCE IS SMALL THEN PROBABLY MERGE (PM) ELSE PROBABLY NOT_MERGE (PNM).

[Rule 2] The edge information in the region growing algorithm plays an important role. A new pixel may be merged into a region if the gradient between the pixel and the adjacent neighboring region is low. If the gradient is high, the pixel will not be merged. The second intensity feature is the GRADIENT, or the value of boundary edge between the pixel and its adjacent region. We employ the boundary Sobel operator [5] to calculate the gradient and to achieve an accurate segmentation at the part of the houses. The fuzzy rule for fuzzy set LOW becomes

R2: IF GRADIENT IS LOW THEN PROBABLY MERGE (PM) ELSE PROBABLY NOT_MERGE (PNM).

[Rule 3] We incorporate the FD feature that is similar to DIFFERENCE in Rule 1. The difference here is taken between the average local FD value $D_{ave}(R_k)$ of a region R_k and the local FD value $D(i,j)$ of a pixel under investigation:

$$\text{DIMENSION} = |D_{ave}(R_k) - D(i,j)| \tag{6}$$

The corresponding fuzzy rule for fuzzy set SMALL2 is the following one that is a stronger merging rule than Rule1 and Rule2, because the role of the FD feature should be emphasized in the proposed algorithm.

R3: IF DIMENSION IS SMALL THEN MERGE (M) ELSE NOT_MERGE (NM).

[Rule 4] The smaller regions, especially regions that consist of one or two pixels, have to be avoided in the region growing algorithm, since it is preferable to remain few large regions instead of many small regions. Thus a fourth rule is the fuzzy set TINY that has the following simple rule:

R4: IF SIZE IS TINY THEN MERGE (M).

Figure 3 shows the four membership functions corresponding to each fuzzy rule. After the fuzzification by the above four rules, min-max inference takes place using the fuzzy sets shown in Fig. 4. Then the conventional centroid defuzzification method is applied. A pixel is really merged when the homogeneity criterion is satisfied to an extent of 50 % after defuzzification.

The final procedure is the merging of two regions that is not a fuzzy rule but a crisp rule after the grouping procedure by the fuzzy inference. Two regions R_j and R_k are recursively merged if

$$|g_{ave}(R_j) - g_{ave}(R_k)| \leq T \tag{7}$$

is satisfied, where T is a predetermined threshold.

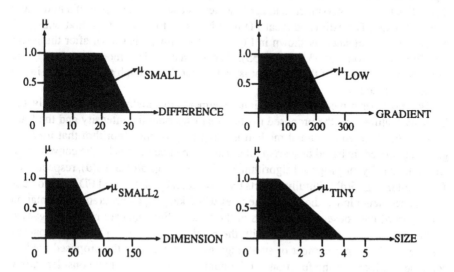

Fig. 3: The membership functions for four fuzzy rules.

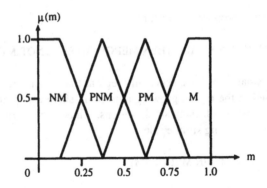

Fig. 4: The fuzzy sets used for inference.

4. Experimental Results and Discussion

To assess the performance of the proposed segmentation method, we have executed the simulated experiments using natural images. We have decided the values of the parameters in the segmentation algorithm empirically, and segmented results are represented by the boundaries of segmented regions. Figure 5(a) is an original natural image that has 400×400 pixels and 256 gray levels and contains a part of a house and a part of trees. The estimated local FD map by using the blanket method with 3×3 window and 44 blankets is shown in Fig. 5(b) (this image is shown after the linear transformation from 2.0~3.0 to 0~255). The estimated FD demonstrates nearly the same value at the part of the trees that is well suited for a rough segmentation in the proposed method.

We have performed the conventional segmentation method that uses only the intensity features and the proposed method that uses both the intensity and the fractal features. The conventional method is a region growing algorithm that uses the grouping procedure based on a crisp rule. The segmented images by the conventional algorithm and by the proposed algorithm are shown in Fig. 5(c) and 5(d), respectively. The result in Fig. 5(d) faithfully reflects the low accuracy of the local FD and provides a rough segmentation at the part of the trees while keeping an accurate segmentation at the part of the house. The portions of the trees in Fig. 5(d) are roughly regarded as the same region in comparison with the result in Fig. 5(c) in which the part of the tree yields a large number of small regions. The result by the proposed method coincides with one of the functions of the human visual system that considers a few trees including lots of branches and leaves as one region. The numbers of segmented region of the resultant images in Fig. 5(c) and 5(d) are 4478 and 101, respectively. The substantial reduction in the number of regions, together with the appearance of the segmented images, clearly indicates the effectiveness of the proposed algorithm in segmenting natural images.

The results of the second experiment are shown in Fig. 6. Figure 6(a) is the second natural image and 6(b) is the estimated local FD. The segmented results by the conventional algorithm and by the proposed algorithm are shown in Fig. 6(c) (no. of regions: 2586) and 6(d) (no. of regions: 79), respectively. The result in Fig 6(d) also demonstrates a rough segmentation at the part of the trees and an accurate segmentation at the part of the house.

5. Conclusion

In this paper we have proposed a method for the segmentation of natural images that integrates the intensity features and the local FD feature into the fuzzy region-growing algorithm. We have estimated the optimum number of the blanket in calculating the local FD by the blanket method. We have investigated the fuzzy-rule based algorithm for integrating different features in the segmentation procedure. Experimental results demonstrate the capabilities of the proposed method to execute the segmentation of natural images with different precision, that is, a rough segmentation at texture regions and an accurate segmentation at non-texture regions simultaneously.

References

[1] A. P. Pentland, Fractal-based description of natural scenes, *IEEE Trans. Pattern Anal. Mach. Intell.*, **PAMI-6**(6), 661-674, (1984).

[2] J. M. Keller, S. Chen and R. Crownover, Texture description and segmentation through fractal geometry, *Comput. Vision, Graphics Image Process.*, **45**, 150-166, (1989).

[3] J. M. Keller and Y. B. Seo, Local fractal geometric features for image segmentation, *Int. J. Imaging Sys. Tech.*, **2**, 267-284 (1990).

[4] S. Peleg, J. Naor, R. Hartly and D. Avnir, Multiple resolution texture analysis and classification, *IEEE Trans. Pattern Anal. Mach. Intell.*, **PAMI-6**(4), 518-523 (1984).

[5] J. Maeda, V. V. Anh, T. Ishizaka and Y. Suzuki, Integration of local fractal dimension and boundary edge in segmenting natural images, *Proc. IEEE Int. Conf. on Image Processing*, vol. 1, pp. 845-848 (1996).

[6] V. V. Ahn, J. Maeda, T. Ishizaka, Y. Suzuki and Q. Tieng, Two-dimensional fractal segmentation of natural images, *Proc. Int. Conf. on Image Analysis and Processing*, vol. 1, pp. 287-294 (1997).

[7] S. G. Kong and B. Kosko, Image coding with fuzzy image segmentation, *Proc. IEEE Int. Conf. on Fuzzy Systems*, pp. 212-220 (1992).

[8] A. Steudel and M. Glesner, Image coding with fuzzy region-growing segmentation, *Proc. IEEE Int. Conf. on Image Processing*, vol. 2, pp. 955-958 (1996).

[9] B. B. Mandelbrot, *The Fractal Geometry of Nature*, Freeman, San Francisco (1982).

[10] P. Brodatz, *Texture: A Photographic Album of Artists and Designers*, Dover, New York (1966).

[11] J. Maeda, T. Iizawa, T. Ishizaka, C. Ishikawa and Y. Suzuki, Segmentation of natural images using anisotropic diffusion and linking of boundary edges, *Pattern Recognition*, **31**(12), 1993-1999 (1998).

(a) (b)

(c) (d)

Fig. 5: Experimental results of segmentation for a natural image: (a) original image; (b) local FD map by using the blanket method; (c) segmented image by the conventional region growing algorithm; (d) segmented image by the proposed algorithm.

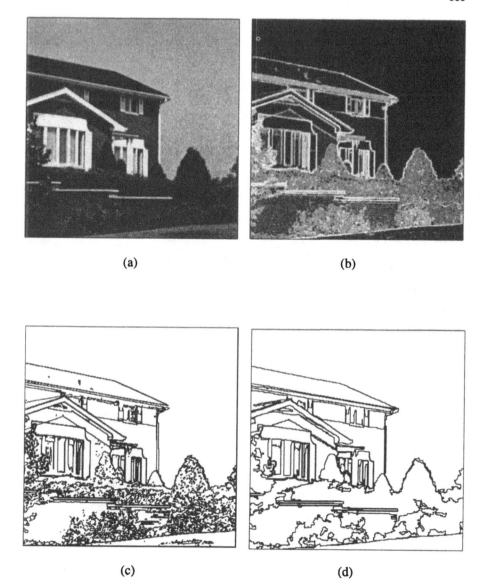

(a) (b)

(c) (d)

Fig. 6: Experimental results of segmentation for a natural image: (a) original image; (b) local FD map by using the blanket method; (c) segmented image by the conventional region growing algorithm; (d) segmented image by the proposed algorithm.

Chapter 8: Human Interfaces

Papers:

Design Issue of Electric Agent for Realizing Biological and Social Coordination with Human and Environment

T. Sawaragi and T. Shiose

An Emergent Approach for System Designs

H. Tamaki and S. Kitamura

Training of Fuzzy Rules in the Freehand Curve Identifier FSCI

S. Saga, S. Mori, and T. Yamaguchi

Intelligent Real-Time Control of Moulding Mixtures Composition in Foundries

J. Voracek

Design Issues of Electric Agent for Realizing Biological and Social Coordination with Human and Environment

Tetsuo Sawaragi and Takayuki Shiose

Dept. of Precision Eng., Graduate School of Eng.,
Kyoto University
Yoshida Honmachi, Sakyo, Kyoto 606-8501, Japan.
{sawaragi, shiose}@prec.kyoto-u.ac.jp

Abstract: In this paper, we discuss about currently ongoing shifts of design principles of human-machine systems for realizing biologically-inspired symbiotic coordination. Especially, we investigate into an architecture of an *interface agent*. Since an agent has to coexist with a human user so that it can evolve by itself as a human user's proficient level improves and can stimulate a human user's creativeness coordinately by changing its role dynamically as a human's associate, some novel architectures from the conventional stand-alone intelligent systems are required. We discuss about this new styles of coordination among human and automated agents from perspectives of social dimensions of the human-machine cooperation, and our ongoing researches on the design of interface agent is presented.

1. Introduction

The introduction of complex and powerful automation to a variety of high-tempo high-risk domains has led to unexpected difficulties which are the result of an increased need for, but lack of support of, human-machine communication and coordination. That is, new computerized and automated devices create new burdens and complexities for the individuals and teams of practitioners responsible for operating and managing high-consequence systems. Wherein, cognitive behaviors and strong affective elements come into play, and unanticipated interactions between the automated system, the human operator and other system in the workplace begin to emerge, causing serious deficiencies that can become apparent after system is delivered and is put to work.

So far many researchers have pointed out such brittleness of automation systems, and to tackle with these automation-induced problems, a concept of *human-centered automation design* has been launched. The ability to accept responsibility and to find innovative ways to fulfill these responsibilities is a unique human characteristic. Herein the design philosophy is "people are in charge" or "human-in-the-loop". The key question is then how to design an artifact system in which a human and an automated system can coexist and collaborate with one another.

In this paper, we discuss about currently ongoing shifts of design principles of human-machine systems for realizing biologically-inspired symbiotic coordination. Especially, we investigate into an architecture of an *interface agent*, which is a semi-intelligent computer programs that can learn by continuously "looking over the shoulder" of the user as he/she is performing actions against some complex artifacts and is expected to be capable of providing the users with adaptive aiding as well as of alternating the activities instead of a human [4], [5]. In this sense, an agent has to coexist with a human user so that it can evolve by itself as a human user's proficient level improves and can stimulate a human user's creativeness coordinately by changing its role dynamically as a human's associate, rather than to replace the human user with itself.

As for an design principle of such an agent, here we put an emphasis on an *ecological* view that intelligent behaviors are produced through dynamical interactions with the surroundings, rather than based on an individual agent's reflection and introspection isolated from the practice. Researchers advocating those put an emphasis on the role of the environment, the context, the social and cultural setting, and the situations in which actors find themselves. Wherein, defining goals, claiming what constitute the facts, and following plans and policies all occur within *nested* activities of an actor's proactively and autonomously forming relations with the surroundings.

In the following of this paper, we list a number of properties that are essential for the attainment of the above goal of the biological coordination between human and an electric agent. Especially, we focus into the following issues that are characteristics of the biological systems:

- management of complexity

- multiresolution and multigranularity

- self-referential architecture

- agent's continuous production of relations via a plasticity of structure

2. Managing Behavioral Complexity by Recursive Symbolization of Perceptual States

As for a design architecture of an intelligent artifact such as an autonomous agent, a *behavior-based* approach is promising. One representative method for this approach is a *reinforcement learning*, but there exist the following disadvantages concerning with this as a method for an agent's learning in a real world environment. That is, the states must be defined rigorously without ambiguity in terms of the physical relations to the environment rather than in terms of the agent's internally grasping world states and/or of its task-relevant distinctions. This sometimes causes an agent to have too many distinctions that necessitate an agent's enormous memory and abundant learning experiences. Moreover, an agent is usually equipped only with a primitive sensing capabilities. Such a perceptual limitation may cause an agent's sensible state space contain too few distinctions, that may hide hidden crucial features of the environment from an agent. Thus, an agent with a limited perception suffers from a many-to-many mapping between agent perception and states of the world (i.e., a *perceptual aliasing* problem).

In this work [6], by adopting an inductive learning method of conceptual formation for adaptively organizing a state space, we propose a new algorithm for an agent to construct its task-relevant state space efficiently through matching state instances that an agent encounters with the similar prior experiences and generalizing them. That is, the agent's percepts are augmented with its history information, that is, short-term memory of past perceptions, actions and rewards, so that the agent can distinguish perceptually aliased states and can then reliably choose correct actions from them. To use memory to overcome

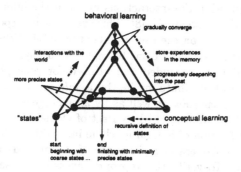

Figure 1: Semiosis joining behavioral and conceptual learning.

the perceptual aliasing problem is well-known and has been attempted by many researchers, but most of the works set an assumption that the memory representation is predefined and fixed. We propose to use a conceptual formation technique called Cobweb [1] for overcoming the aliasing problem. The biggest advantage of using this is that different from a connectionist approaches, a symbolic learning of concept formation enables an agent to learn concepts that has structural properties. This realizes a recursive organization of memory, that functions as a variable-length context avoiding useless memory burden. Fig.1 shows the flow of our learning algorithm for a robot's state grounding. We verified the learning performance under two LPGWs (Locally Perceptual Grid

Worlds) having different complexity, and confirmed that our algorithm can perform tasks with a variety of difficulties in an artificial maze environment. We also presented its robustness even for an agent whose perceptual resources are quite restricted and or bounded.

3. Managing Memory Resource by Evolutional Concept Learning with Attribute-Mining

In most research on concept formation within machine learning and cognitive psychology, the features from which concepts are built are assumed to be provided as elementary vocabulary. We argue that this is an unnecessarily limited paradigm within which to examine concept formation. We believe that any concept formation process relies on the filtered perception and/or interpretation of the world that the observer imposes from a perspective of his own behavioral goals. This problem is closely related with the human skill development, and we have to realize a learning system that is actually *grounded* in its working environment.

We apply such an idea to customization of the human-machine interface by letting it have a learning capability of learning about the concepts that are held by the users by monitoring the interactions between the human users and the status of their operating systems [7], [8]. This system is incorporated with some degree of built-in structure on the user's perceptual level of the observable sensor data, but also with the mechanisms to change and augment the elementary features that were initially provided and utilized for concept formation. We realized this function of attribute-mining by a *salience-based* feature selection for inductive concept formation as well as by an emergent new feature discovery using a genetic algorithm applied to a set of elementary features. That is, we accumulated a collection of human-machine interactions that were observed when a human operator operates a simulator of a generic dynamic production process called Plant. To verify the appropriateness of the selected attributes for that human operator, we carried out an additional experiment by only presenting the information concerning those attributes derived by the the above experiment to the human operator requiring him to operate accordingly. The result was satisfactory and the operator could predict the status of the valve correctly without any fruitless search for added information.

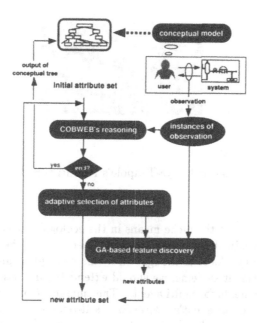

Figure 2: An algorithm for GA-based feature mining.

4. Emergence of Sociality within Multiagent Systems Having a Bi-Referential Architecture

In recent years, practical use of multiagent systems has received much attention. However, the tacit assumption that the whole group of a multiagent may be viewed as a single learning system has made practical use of multiagent system extremely difficult. In the real world, there is the factor of locality, as each agent cannot interact with all others; only with neighboring agents. Thus in multiagent systems, because the global objectives tend to be abstract and individual agents act within a restricted focus, each agent tends to develop a unique perspective on how it is best to achieve their personal goals. Individuals have to learn to become aware of their unique roles within the whole organization; roles which may change according to some temporarily varying context formed out of the individual agents' role fulfilling activities (e.g., an organization has acquired a "social norm"). The above-mentioned capability that each agent can find its own role or niche in its social environment is called "sociality".

In our prior works [9], [?], we defined "sociality" as "the capability to find out one's role or niche in the social environment by subtly shifting one's own

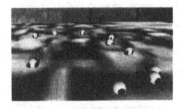

Figure 3: Eye-Tadpole's Primitive Soccer

sense of value". What the niche means in the ecological sense is to find out the agent's own space in the new environment by entrusting the meaning of action to its environment. As it is thought that this concept of niche has important effects for multiagent systems, we should extend the discussion of this concept to social environments for multi agents. The emergence of sociality is due to the dual capabilities of an agent's referencing; self-referential and social-referential abilities. The former is the ability to have conception on what it is doing. Such conceptual constraints enable value judgments about how it uses and evaluates its experiences obtained in the past. The latter is an ability to design inter-actions for itself by considering the prospect of which it is engaged with other agents as well as things in the environment for the future.

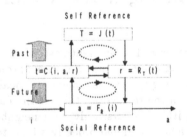

Figure 4: Architecture of the Bi-referential Model

We proposed a learning model of an agent having such dual referencing capabilities as a "bi-referential model" in which each referencing capability is implemented by an evolutional computation method of classifier systems. By relating these two learning components with each other, we realize reciprocal learning. That is, self-referential learning adjusts the evaluation criteria for other social-referential learning, playing the role of switching the latter's learning modes between learning and unlearning. As a result, social-referential learning proceeds under this restriction and varies the agent's interaction with others. The learning performance thus attained is fed back to the self-referential learning to be mediated. Due to this reciprocity, a group of agents can acquire a social norm and individual roles within the group, although the learning occurs solely within the individual agents. We must pay special attention to the relation between the "sociality" that an agent should acquire and the "reciprocity" that an agent should realize.

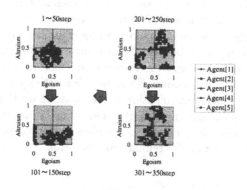

Figure 5: Jostling the Niche on Each Other

5. Managing Computational Complexity through Activating Multigranural Categories

Under such a time-critical situation, an interface agent must be able to flexibly organize the appropriate appearance of the plant status discriminating among what is *now* relevant and what is not for assisting an operator's *situation awareness*. In our developing system [10], a taxonomy of all possible plant anomalies is organized in a hierarchical fashion using a machine learning technique called concept formation as shown in Fig.6. Wherein, the root node represents a class of concepts covering all possible anomaly types, and the leaf nodes represent the individual anomalous cases (i.e., a hierarchy via *is-a* and *subset-of* relations). In terms of this hierarchy, determination of the appearance of the plant concerns with how to determine the appropriate categorization of the plant anomalies out of the hierarchical taxonomy (Fig.6); to find a set of exclusive subtrees whose extensions are regarded as equivalent. The category of the hypothesis on the plausible anomaly types can be defined with a variety of abstraction levels of the hierarchical taxonomy. We call each possible categorization within the taxonomy a *conceptual cover*, that is a categorization of all possible anomalies into mutually exclusive and exhaustive classes.

The selection of the appropriate conceptual cover is important not only to present an affordance-rich information to a human operator but also to consider about an interface agent's own problem solving activity of constructing its decision making model and solving that model. At this time, an agent has to deal with tractability of decision making inference at the expense of decision quality. That is, an agent has to determine the appropriately manageable model by finding the definition of the hypothesis node as an agent's categorization of the environment (i.e., a plant status). Wherein, a resource-constrained agent has to be able to enhance its decisions by expending some resource to consider the tradeoff between the expected benefit of using more detailed models to increase the expected utility of action and the resource costs entailed by computing decisions with a more detailed model. Note that this activity of finding a better conceptual cover for model modification must be done by the interface agent and this is also a time-consuming activity. Therefore, an agent has to take account of the side-effects produced due to the change of conceptual cover as well as the cost performing the model refinement operation.

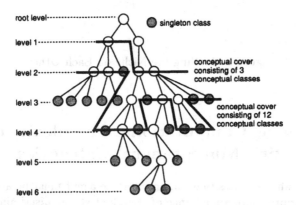

Figure 6: Conceptual cover of anomalous state concepts: a categorization adopted by an interface agent for its decision framing.

In order to rationalize the agent's selecting the appropriate categorization to be presented to the human operator and also to be used for its own problem solving (i.e., diagnosis), we at first consider about a general probabilistic reasoning model (i.e., an *influence diagram* [2]) as an agent's decision model that derives an appropriate action inferring the most plausible plant anomaly type by getting partially observable symptoms from evidential observations obtained

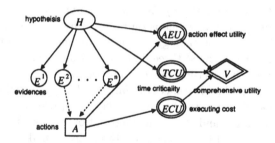

Figure 7: Interface agent's resource-bounded reasoning represented by an influence diagram.

so far. This is illustrated in Fig.7.

Wherein, a chance node H denotes a hypothesis node representing a state of the plant taking the value of anomaly types. Nodes E^i's represent a set of symptoms available to an interface agent, each of which corresponds to the features observable in individual instrumentations of the plant. As knowledge of plant anomalies, an agent has acquired dependency among the classes of anomaly types and their associating evidences as well as the prior probabilities of occurrence of those hypothetical classes through the concept formation process, all of which are stored in a taxonomy hierarchy. Based on this apriori knowledge, getting new observations from the plant, an agent updates its prior belief on probable anomaly types based on a Bayes theorem. Then, these posterior beliefs are used for calculating the expected utilities of the available options defined in the decision node A based on the knowledge that are defined in the diagram defining the interrelationships among nodes of H, A, ACU, TCC, ECU and V. The option having the maximum expected utility would be determined as a recommendation for an interface agent to adopt. A diamond node is a special type of an oval node and is called a value node representing an agent's comprehensive utility. This shows that an agent's utility is determined by the types of anomaly and the adopted action as well as by the delay. This reflects the fact that some anomaly types require an immediate recovering operations as soon as possible and if it is delayed its utility would drastically decrease.

In terms of this decision model, a definition of a domain of a hypothesis node corresponds to a conceptual cover showing an appearance of the plant to an interface agent. This definition reflects the granularities of an agent's recognizing anomaly classifications, so we call this style of decision making as a recognition-primed decision making after Klein's *naturalistic decision making* model [3]. The qualities of each possible conceptual cover must be evaluated in terms of the effectivity brought about by the adopted actions.

We finally present an example using well known database in a machine learning field (i.e., UC Irvine database). We use a database of animals consisting of 100 animal instances, each of which is represented by a set of 17 attributes with their binary values. As an analogy to the plant anomalies and other practical diagnosis, we newly defined three different actions (i.e., *approach*, *stop*, and *escape*) available to the decision maker (and/or an actor) that are commonly extertible by the individual instances of animals. That is, attributes correspond to the evidences available to the decision maker, from which he/she has to identify what that animal is and to decide what action to take against that. With incomplete observation of those evidences, the decision maker has to update his beliefs on the confirmation of the plausible hypothesis (i.e., animal classes of the encountering animals), while the role of the interface agent is how to display the information so that it can afford the decision maker's easy recognition as well as his/her naturalistic responses properly.

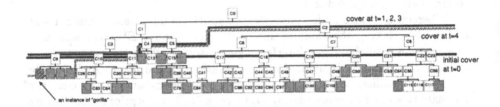

Figure 8: Dynamic shift of conceptual cover according to sequential observations of evidences.

The task of uncertainty management is performed based on the Bayes reasoning of the influence diagram, while the tasks of risk management and the resource management are controllable by a variety of definition related to the value node of the diagram. To explicate this value structure, we subcategorized an agent's comprehensive value into the one shown in Fig.7. In this modified model, the value node are split into three value component of "action effect utility (ACU)", "time criticality utility (TCC)", and "execution cost utility (ECU)", all of which are assumed to be measurable on the same dimension and the total comprehensive utility is assumed to be evaluated as a sum of those.

At first we construct an animal taxonomy using a conceptual formation technique of Cobweb as shown in Fig.8. Then, based on the information learned and stored within the individual classes of this taxonomy, we build an actor's decision model according to Fig.7. Finally, we present how the conceptual cover that is to be displayed to the user can shift according to the variety of the definition of the above-mentioned utility components. In Fig.8, a dynamic shift of conceptual covers according to sequential observations of evidences are shown.

6. Conclusions

In this report, we described about a series of works concerning with the architectures of agents that have to collaborate with other agents including a human user. We introduce an idea of *interface agent* as a sophisticated *associate* for a human user (i.e., an operator for a complex plant). The ideas mentioned in those are now extended to a general design principle for human-artifact symbiotic systems. Such an interface agent's reasoning tasks and its decision making styles are quite different from the ones of conventional expert systems and from conventional decision support systems in that they are severely bounded to the *realtime contexts and situations.*

References

[1] Fisher, D.: Knowledge Acquisition via Incremental Conceptual Clustering, *Machine Learning*, 2, pp.139-172, 1987.

[2] Howard, R.A. and Matheson, J.E.: Influence Diagrams, in Howard, R.A. and Matheson, J.E. (Eds.), *The Principles and Applications of Decision Analysis*, Strategic Decision Group, Menlo Park, CA., 1983.

[3] Klein, G.A. et al.: *Decision Making in Action: Models and Methods*, Ablex Pub. Corp., Norwood, NJ., 1993.

[4] Maes, P. and Kozierok, R.: Learning interface agents, *Proceedings of the Eleventh National Conference on Artificial Intelligence*, pp.459-465, 1993.

[5] Maes, P.: Agents that Reduce Work and Information Overload, *Communications of the ACM*, 37-7, pp.30-40, 1994.

[6] Sawaragi, T., Iwatsu, S., Katai, O.: Biologically-Inspired Dynamic Reconstruction of Perceptual States Using Concept Formation Technique, *Proc. of International Conference of IEEE System, Man and Cybernetics*, 1, pp.1394 - 1399, 1998.

[7] Sawaragi, T., Takada, Y., Katai, O. and Iwai, S.: Realtime Decision Support System for Plant Operators Using Concept Formation Method, *Preprints of International Federation of Automatic Control (IFAC) 13th World Congress*, Vol.L, pp.373-378, San Francisco, 1996.

[8] Sawaragi, T., Tani, N. and Katai, O.: Evolutional Concept Learning from Observations through Adaptive Feature Selection and GA-Based Feature Discovery, to be published in *Journal of Intelligent and Fuzzy Systems*, 7-3, pp.239-256, 1999.

[9] Shiose, T., Okada, M., Sawaragi, T. and Katai, O.: Emergent Mechanism of Sociality by Bi-Referential Model , *Cognitive Studies*, 6(1), pp.66-76, 1999 (in Japanese).

[10] Sawaragi , T. and Katai, O.: Modelling Recognition-Primed Decision Making under Time Pressure and Multiple Situational Contexts, *Preprints of 7th IFAC/IFIP/IFORS/IEA Symposium on Analysis, Design and Evaluation of Man-Machine Systems*, pp.269 - 274, 1998.

An Emergent Approach for System Designs

Hisashi Tamaki [1] and Shinzo Kitamura [2]

[1] Department of Electrical and Electronics Engineering, Faculty of Engineering
Kobe University, Rokko-dai, Nada-ku, Kobe 657-8501, Japan
Tel/Fax: +81-78-803-6102, E-mail: tamaki@eedept.kobe-u.ac.jp

[2] Department of Computer and Systems Engineering, Faculty of Engineering
Kobe University, Rokko-dai, Nada-ku, Kobe 657-8501, Japan
Tel/Fax: +81-78-803-6244, E-mail: kitamura@cs.kobe-u.ac.jp

Abstract: An emergent approach for system designs is introduced. First, the process of system design is reformulated in the framework of the set and mapping theory, and the problem of system design is defined as an inverse mapping from the set of specifications depending on a certain environment to the set of components and their connections. Next, the problems are classified from the viewpoint of the information on specifications and an environment. The concepts of evolution, adaptation, learning and coordination can be related to the classified problem. An emergent design procedure is defined in the framework of these concepts, especially, by taking the evolutionary computing techniques into account.

1. Introduction

For the design of existing artifacts (artificial systems), human designers have played a main role. We have encountered many difficulties recently, however, in designing process for complex and large scaled systems. In order to cope with such a situation, what kinds of technologies can contribute?

In this paper, an approach by employing the concept of emergence is introduced [1]. Design procedure of artifacts is first reformulated in the framework of the set and mapping theory [2]. In this formulation, the design problem is defined as an inverse problem. This inverse problem is normally solved by the iteration of forward mapping where in many cases human designers should play an important role. However, they fails sometimes in solving the problems adequately for the case of ambiguous environment or even ambiguous specifications. In such problems, the concepts of optimization, evolution, adaptation, learning and man-machine coordination will be effective. In the paper, an emergent design procedure is defined in this framework [3].

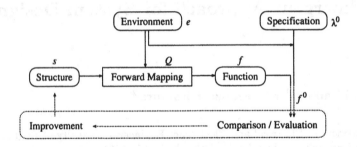

Figure 1 The framework of system design based on an iterative procedure.

Furthermore, the problem of system design is classified from the viewpoint of the information on specifications and an environment. Then, an emergent design procedure is defined, especially by taking the evolutionary computing techniques into account.

2. System Design

In this section, we reformulate the problem of designing artifacts (in the following, we simply say 'systems') such as electrical, mechanical, chemical systems and computer programs [1, 3].

We define first the concept of 'structure' and 'function' of a system to be designed.

(a) Structure: determined by specifying both components (or parts) and their connection.

(b) Function: determined by an interaction between the property associated with a structure and the environment around the system.

According to these definitions and assuming that a required function is given as a 'specification', we define the problem of system design as follows:

(c) System design: to synthesize the structure of a system so that it satisfies the prescribed specification under known or ambiguous environment.

2.1. Formal Definition

Let b, c, s, f and e represent a component, a connection, a structure, a function and an environment, respectively:

$$b \in B \quad (B : \text{the set of components}),$$
$$c \in C \quad (C : \text{the set of connection}),$$
$$s \in S \quad (S : \text{the set of structures}),$$

$f \in \mathcal{F}$ ($\mathcal{F}$: the set of function) ,

$e \in \mathcal{E}$ ($\mathcal{E}$: the set of environment) .

Let $\mathcal{B}_s$ be a subset of $\mathcal{B}$. Then, a structure s is generated by applying a connection c to $\mathcal{B}_s$, and this structure s (i.e., the system) exhibits a certain function f by the interaction with an environment e. This relation can be denoted as follows: Let Q represent a mapping from a structure $s = s\,(\mathcal{B}_s, c)$ to a mapping λ:

$$Q \,:\, (\mathcal{B}_s, c) \longmapsto \lambda. \tag{1}$$

where the mapping λ defines a specification. If an environment e is fixed, a function f with the structure s is determined by

$$Q(\mathcal{B}_s, c)(e) = \lambda(e) = f \in \mathcal{F}. \tag{2}$$

Based on this representation, the problem of system design is defined as follows.

(d) System design: to obtain an inverse mapping Q^{-1} (i.e., to solve an inverse problem).

In contrast to the system design, the system analysis (i.e., to trace the causality to clarify the function f of a system) is rather easy by applying related theories, by executing computer simulations, or by doing some experiments. Whereas, as described above, the system design is an inverse problem (i.e., to determine the inverse mapping Q^{-1}). In order to obtain an approximate solution of this problem, we need an iterative procedure where the forward mapping Q is repeatedly applied. Figure 1 shows this framework.

2.2. Difficulties in System Design

The difficulties in solving the problem of system design are considered in relation to the information on the environment and the specification. If the information on the environment e is complete (e.g., if its dynamics is completely described), we can separate a system to be designed from the environment. Moreover, if the specification given by human designers is complete, we can separate the system from the human designers. Then, the difficulties in the system design are categorized in the following three classes as shown in Fig. 2:

(1) Complete problems: Both the information on the environment and specification are complete. The problem in this case is reduced to the optimization procedure.

(2) Incomplete environment problems: The information on the environment is incomplete, while the specification is complete. Hence, we must consider a framework including the dynamical properties of unknown or ambiguous environment. The key point is adaptation and/or learning to cope with the changes of environment.

578

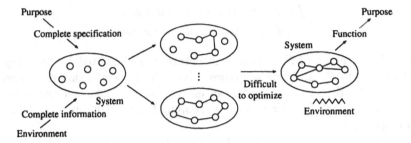

(a) Class 1: Complete problems. Emergent methodology adopted in this class includes evolutionary algorithms such as genetic algorithms, genetic programming and evolution strategies.

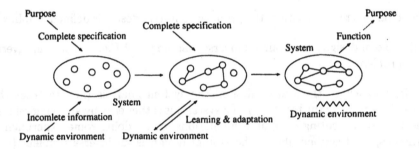

(b) Class 2: Incomplete environment problems. Emergent methodology adopted in this class includes reinforcement learning, adaptive strategy and self-organization.

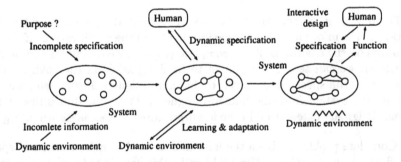

(c) Class 3: Incomplete purpose problems. Emergent methodology adopted in this class includes interactivity, multi-agent, co-evolution and cooperative behavior.

Figure 2 Three classes of difficulties in system designs.

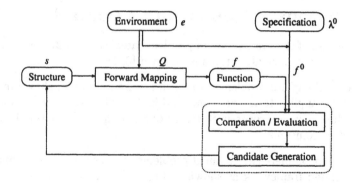

Figure 3 The basic framework of the emergent design.

(3) Incomplete specification problems: Both the specification as well as the information on the environment are incomplete. Hence, we must additionally consider the human-machine interaction as compared with the case of (2). The key point is the coordination between human designers and machines.

3. Emergent Design

The word 'emergence' is used, in this paper, in the following sense.

(a) Emergence: appearance of unpredictable functions or structures through the design process.

Here, the word 'emergence' has been favorably used by researchers in the field of artificial life [4], and the above implication is almost same to their notion.

The processes as evolution, adaptation, learning and coordination are considered to be essential for revealing the emergent property. According to the definition of the emergent property, we extend the framework in Fig. 1 as follows.

(b) Emergent design: to find a structure satisfying (or being closer to) the specification by iterative procedures (i.e., by repeating the evaluation of the function generated, the comparison of it with the specification, and the generation of new candidates).

In Fig. 3, a scheme of emergent design is shown. The 'comparison/evaluation' and the 'candidate generation' blocks are essential for composing the emergent design procedure. Evolutionary computing techniques (e.g., the genetic algorithm and the genetic programming) [5] can be used as a tool to implement these blocks. In this approach, generated entity corresponds to the 'phenotype' and the information to specify the entity to the 'genotype'. The concept of 'genotype' is rather new in the design process of artifacts.

4. Evolutionary Computing Approach

To each class of the system design problem categorized in **2.2**, a framework of applying evolutionary algorithms is outlined [6, 7].

(1) Class 1: Complete problems. An ordinary framework as shown in Fig. 4 (a) can be applied to the problems in this class. The key point in adopting an evolutionary approach is efficiency of a search for the optimal solution (or better solutions).

(2) Class 2: Incomplete environment problems. To the problems in this class, an extended framework as shown in Fig. 4 (b) can be considered. Here, not only to find better solutions in the environment currently observed but also to identify dynamic environment or to measure the robustness of solutions to dynamic environment are necessary. The methodologies in this class includes adaptation, learning and self-organization.

(3) Class 3: Incomplete purpose problems. In this class, besides the extension of the framework in the case of Class 2, an human-machine coordination is necessary, and a framework including an interaction with a 'human' (e.g., a decision maker) as shown in Fig. 4 (c) can be considered. Here, the requirements for a search by evolutionary algorithms are (a) to keep many candidates of solutions and (b) to catch up with a specification given dynamically. The methodologies in this class includes multi-agent, co-evolution and cooperative behavior.

5. Conclusion

A bottom up approach for designing artifacts has been proposed under the framework of emergent property and of evolutionary computing. Of course, the advantage of this method is not apparent at present and we do not think it can replace the conventional design procedure in various engineering fields. However, it may show a scope for extending such methods to the design of complex systems.

In the paper, moreover, difficulties in system designs are categorized into three classes, with respect to the completeness of the information to describe environment and that of the specification to meet the purpose of systems. Then, the frameworks of emergent system designs based on evolutionary computation techniques are shown to the problems in each class of the difficulties.

Acknowledgement

This research has been pursued partially under the financial support through "Methodology for Emergent Synthesis" project (project number: 96P00702) in Research for the Future Program of the Japan Society for the Promotion of Science.

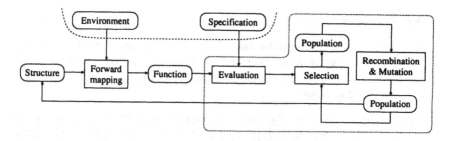

(a) Class 1: Complete problems.

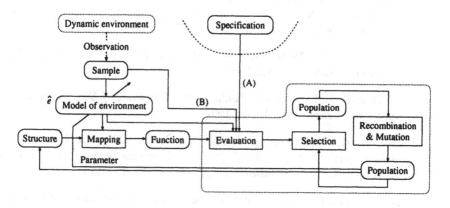

(b) Class 2: Incomplete environment problems.

A framework including a model of the dynamics of environment is shown.

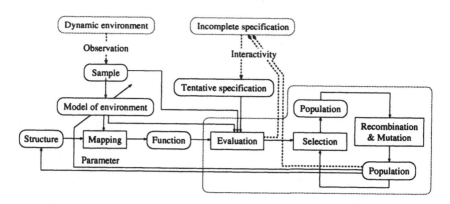

(c) Class 3: Incomplete purpose problems.

A framework including a model of the dynamics of environment is shown.

Figure 4 Frameworks of evolutionary computing approaches.

References

[1] S. Kitamura, "Interpretations of Emergent Properties in Engineering," *Newsletter for System Theory of Function Emergence (Scientific Research on the Priority Area from the Ministry of ESSC of Japan)*, No. 5, pp. 1–6, 1996 (in Japanese).

[2] H. Yoshikawa, "Introduction to General Design Theory," *J. of the Japan Society for Precision Engineering*, Vol. 45, No. 8, pp. 20–26, 1979 (in Japanese).

[3] S. Kitamura, Y. Kakuda and H. Tamaki, "An Approach to the Emergent Design Theory and Applications," *Proc. of 3rd Int. Symp. on Artificial Life and Robotics (AROB III '98)*, pp. 77–80, 1998.

[4] C. G. Langton, "Artificial Life," *Artificial Life* (C. G. Langton, Ed.), Addison-Wesley, pp. 1–48, 1989.

[5] T. Bäck, D. B. Fogel and Z. Michalewicz, *Handbook of Evolutionary Computation*, Oxford Univ. Press, 1997.

[6] H. Tamaki, "Difficulties in Engineering Optimization Problems and Genetic Algorithm Approaches," *Proc. of Australia-Japan Joint Workshop on Intelligent and Evolutionary Systems*, pp. 11–19, 1997.

[7] K. Ueda and H. Tamaki, "Aim of the Project 'Methodology of Emergent Synthesis'," *Proc. of 7th Design & Systems Conf.* (The Japan Society of Mechanical Engineers), pp. 121–124, 1997 (in Japanese).

Training of Fuzzy Rules in the Freehand Curve Identifier FSCI

Sato SAGA, Saori MORI and Toru YAMAGUCHI

Muroran Institute of Technology, 27–1 Mizumoto–cho Muroran 050–8585 JAPAN,
saga@csse.muroran–it.ac.jp

Abstract: This paper investigates the actual situation of training of Fuzzy Spline Curve Identifier (FSCI). FSCI is a primitive curve identification system which has been proposed to establish a general-purpose freehand interface for computer aided drawing (CAD) systems. It succeeded in distinguishing a freehand drawing into seven kinds of primitive curves which are indispensable for CAD. The key was the introduction of the fuzzy reasoning which embodied a strategy to try to find the simplest primitive curves based on user's drawing manner. Furthermore, a trainable version of FSCI has learning ability to adjust its fuzzy reasoning rules (which are materialized as a fuzzy neural network) and adapt itself to individual user's drawing manner. This paper sets up some experiment to train this FSCI, and demonstrates how the adjustment of the fuzzy neural network improves FSCI's curve class recognition rates. Then, through some detailed observations on a concrete training result from one user, it shows how the adjusted fuzzy neural network explicitly explains the effect of the training.

1. Introduction

Usual CAD entities are drawn as combinations of seven classes of primitive curves: line, circle, circular arc, ellipse, elliptic arc, closed free curve and open free curve. Accordingly, a general-purpose curve identifier should be required to have a capability to classify a freehand drawing into the seven kinds of primitive curves. However, the shape of a freehand drawing is not enough information to determine curve classes due to the inclusion relations among the primitive curve classes shown in Figure 1: line is a kind of circular arc, circular arc is a kind of elliptic arc, and so on.

The Fuzzy Spline Curve Identifier (FSCI)[1],[3] has overcome the difficulty by utilizing user's drawing manner as well as the curve shape. FSCI was designed to tend to classify roughly drawn curves as simple primitive curves, but carefully drawn curves as complex ones. This implies that a user can intend to draw a rather simple curve by drawing roughly but a rather complex curve by drawing carefully. Experimental results in [4] and [5] showed that the strategy was effective for expert users. However, since the strategy was realized as fuzzy reasoning with a fixed fuzzy rule set, new users needed quite a little drawing practice to master the characteristics of FSCI. A trainable FSCI was then proposed to adapt itself to each user's characteristics and reduce new user's burden in practice[6]. This was actualized by replacing the fixed fuzzy rule set in the original FSCI with a common feedforward 3-layer neural network. The learning of neural network carried plasticity into FSCI to improve the curve class recognition rates for the experienced but non-expert users. However, it lost FSCI the explicit representation of the original strategy. In [7], a new version of trainable FSCI was finally proposed by introducing a structured fuzzy neural network into the original FSCI in order that it would acquire learning ability while it would preserve the original strategy; and its fundamental function was confirmed.

This paper, first, gives the outline of the trainable version of FSCI proposed in [7]. Then, it sets up some experiment to demonstrate how the adjustment of the fuzzy neural network improves FSCI's curve class recognition rates. Furthermore, through some detailed observations on a concrete training result from one user, it shows how the adjusted fuzzy neural network explicitly explains the effect of the training.

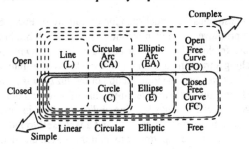

Fig. 1: Inclusion relations among primitive curve classes.

2. Outline of Trainable FSCI

Given a freehand curve, FSCI performs fuzzy reasoning to identify it as one of the seven kinds of primitive curves, and outputs a fuzzy primitive curve. The fuzzy primitive curve is concretely composed of seven membership grades (which are $\mu(L)$, $\mu(C)$, $\mu(CA)$, $\mu(E)$, $\mu(EA)$, $\mu(FC)$ and $\mu(FO)$) and seven sets of curve shape parameters which are associated with the curve classes. It is also regarded as seven different classes of primitive curve candidates ordered according to the grades.

The introduction of the fuzzy reasoning is essential for FSCI to tell the difference among the seven curve classes. The shape of a freehand curve is not enough information to determine the curve classes because of the inclusion relations shown in Figure 1: strictly speaking, all freehand curves should be categorized into open free curve as long as only the shape is taken into account. In order to overcome the problem, FSCI utilizes the drawing manner as well as the curve shape. So far as the membership grades are concerned, the schematic process of FSCI is illustrated as shown in Figure 2. First, FSCI performs *fuzzy spline interpolation* and models a freehand curve as a *fuzzy spline curve* which involves vagueness (associated with roughness in drawing) in their positional information. Secondly, it performs *possibility evaluation*,

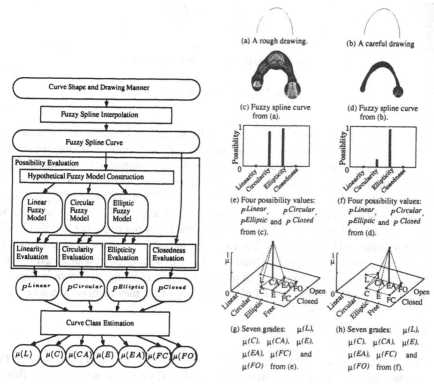

(a) A rough drawing.

(b) A careful drawing

(c) Fuzzy spline curve from (a).

(d) Fuzzy spline curve from (b).

(e) Four possibility values: p^{Linear}, $p^{Circular}$, $p^{Elliptic}$ and p^{Closed} from (c).

(f) Four possibility values: p^{Linear}, $p^{Circular}$, $p^{Elliptic}$ and p^{Closed} from (d).

(g) Seven grades: $\mu(L)$, $\mu(C)$, $\mu(CA)$, $\mu(E)$, $\mu(EA)$, $\mu(FC)$ and $\mu(FO)$ from (e).

(h) Seven grades: $\mu(L)$, $\mu(C)$, $\mu(CA)$, $\mu(E)$, $\mu(EA)$, $\mu(FC)$ and $\mu(FO)$ from (f).

Fig. 2: Schematic process flow by FSCI.

Fig. 3: Identification by FSCI.

where it estimates linearity, circularity, ellipticity and closedness[1] of the fuzzy spline curve taking account of the vagueness, and outputs four possibility values: p^{Linear}, $p^{Circular}$, $p^{Elliptic}$ and p^{Closed}. Thirdly, it performs *curve class estimation*, a sort of fuzzy reasoning, where it tries to find the simplest possible primitive curves based on the four possibility values, and outputs seven membership grades: $\mu(L)$, $\mu(C)$, $\mu(CA)$, $\mu(E)$, $\mu(EA)$, $\mu(FC)$ and $\mu(FO)$. Because even a simple primitive curve can be possibly found in the fuzzy spline curve when it is vague enough, a user is now given a way to let FSCI identify a simple primitive curve. This implies that a user can intend to draw a rather simple curve by drawing roughly but rather complex curve by drawing carefully (see Figure 3).

In the trainable version of FSCI, the curve class estimation process is realized as a fuzzy neural network so that FSCI may adapt itself to user's drawing manner.

2.1 Fuzzy Spline Interpolation

A drawn curve is given to the system as a sequence of a certain number of sampled points p_k and time stamps t_k. However, the sampled points are not always considered

---

[1]We use a term "closedness" to express the degree to which the fuzzy spline curve is closed.

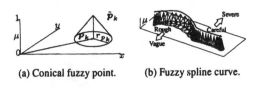

(a) Conical fuzzy point.   (b) Fuzzy spline curve.

Fig. 4: Fuzzy spline interpolation.

to have accurate positional information exactly reflecting the intention of the user. In general, the more roughly a curve is drawn, the vaguer its positional information will be. From this observation, each sampled point p_k is replaced by a conical fuzzy point model $\tilde{p}_k = <p_k, r_{p_k}>$ shown in Figure 4 (a), where the fuzziness r_{p_k} is generated according to the roughness in drawing. In FSCI, the value of r_{p_k} is simply set as $r_{p_k} = Q \times a_{p_k}$, where a_{p_k} is the acceleration at p_k and Q is a constant value. Then, a fuzzy spline curve that interpolates to the fuzzy points $\tilde{p}_k$ is generated by the method proposed in [1] and [2]. The fuzzy spline curve is defined as an extension of an ordinary spline curve and illustrated as a locus of a fuzzy point which travels while changing its vagueness according to the roughness in drawing, as shown in Figure 4 (b). It is utilized as a fuzzy model of the drawing which may involves vagueness.

2.2 Possibility Evaluation

First, FSCI constructs three hypothetical fuzzy models: the linear fuzzy model, the circular fuzzy model and the elliptic fuzzy model. They are obtained as fuzzy Bézier curves[2] whose parameters are adjusted so that they fit the given fuzzy spline curve as well as possible. Secondly, each hypothetical fuzzy model is compared with the original fuzzy spline curve and its validity is evaluated by a possibility value: P^{Linear}, $P^{Circular}$ or $P^{Elliptic}$ based on the possibility measure[9]. In other words, the degrees of linearity, circularity and ellipticity of the drawn curve are evaluated by P^{Linear}, $P^{Circular}$ and $P^{Elliptic}$ respectively. Thirdly, the accordance between the fuzzy end points of the fuzzy spline curve is checked and the closedness is evaluated by another possibility value P^{Closed}.

Now, it must be noted that three of the possibility values obtained in this process are always in a fixed order. Namely, P^{Linear} is always less than or equal to $P^{Circular}$, and $P^{Circular}$ is always less than or equal to $P^{Elliptic}$, as shown in Figure 3 (e) and (f). This is because of the inclusion relations among the primitive curve classes shown in Figure 1.

2.3 Curve Class Estimation

Due to the fixed order among the three possibility values, it is inconclusive to determine the curve class by simply comparing them. In addition, the closedness should be taken into account for FSCI to distinguish between closed primitive curves and open primitive curves (for example, between circle and circular arc). Therefore,

---

[2]A fuzzy Bézier curve is defined as a special case of the fuzzy spline curve.

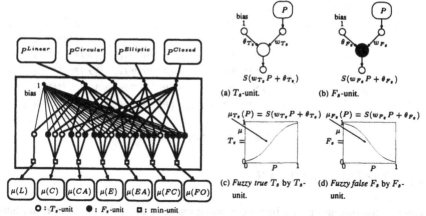

(a) T_s-unit.   (b) F_s-unit.

(c) *Fuzzy true T_s by T_s-*
unit.

(d) *Fuzzy false F_s by F_s-*
unit.

O : T_s-unit   ● : F_s-unit   □ : min-unit

Fig. 5: Curve class estimation by a fuzzy neural network.

Fig. 6: Two types of initial settings for a sigmoid unit and their corresponding fuzzy truth values.

FSCI performs the curve class estimation process that is embodied as a fuzzy neural network shown in Figure 5; and calculates the seven membership grades: $\mu(L)$, $\mu(C)$, $\mu(CA)$, $\mu(E)$, $\mu(EA)$, $\mu(FC)$ and $\mu(FO)$ from the four possibility values: P^{Linear}, $P^{Circular}$, $P^{Elliptic}$ and P^{Closed}. In the fuzzy neural network, each min-unit performs min operation that outputs the minimum value. On the other hand, both T_s-units and F_s-units are sigmoid units each of which has a function $S(x) = 1/(1 + e^{-x})$, and the ith sigmoid unit outputs $S(w_i P_i + \theta_i)$, where P_i ($\in \{$ P^{Linear}, $P^{Circular}$, $P^{Elliptic}$, P^{Closed} $\}$) is the input to the unit, w_i is the weight factor to the input, and θ_i is the bias term.

Let us see how this fuzzy neural network plays a role of fuzzy reasoning that tries to find the simplest possible curve class. Let us set w_{T_s} (= 6.6) and θ_{T_s} (= −3.3) to w_i and θ_i respectively for all T_s-units as shown in Figure 6 (a); and set w_{F_s} (= −6.6) and θ_{F_s} (= 3.3), for all F_s-units as shown in Figure 6 (b). Then, with this setting, each T_s-unit acts as a fuzzy proposition "P is T_s," where T_s is *fuzzy true* shown in Figure 6 (c); and each F_s-unit acts as a fuzzy proposition "P is F_s," where F_s is *fuzzy false* shown in Figure 6 (d). Considering that the min-unit can be regarded as a logical operator *and*, the fuzzy neural network can be translated into the fuzzy rule set which consists of the seven expressions shown in Figure 7, where $\wedge$ denotes the logical multiplication or the min-operator and the fuzzy truth values shown as membership functions are T_s or F_s respectively. Because the fuzzy rules regarding rather complex curve classes are severer than the ones regarding rather simple curve classes, it is now understood that the fuzzy neural network embodies the fuzzy reasoning that tries to find the simplest possible curve class.

The structure of the fuzzy neural network lets FSCI preserve the basic strategy: "Try to find the simplest possible primitive curves." On the other hand, the learning ability of the neural network makes FSCI trainable, as we discuss in the following section.

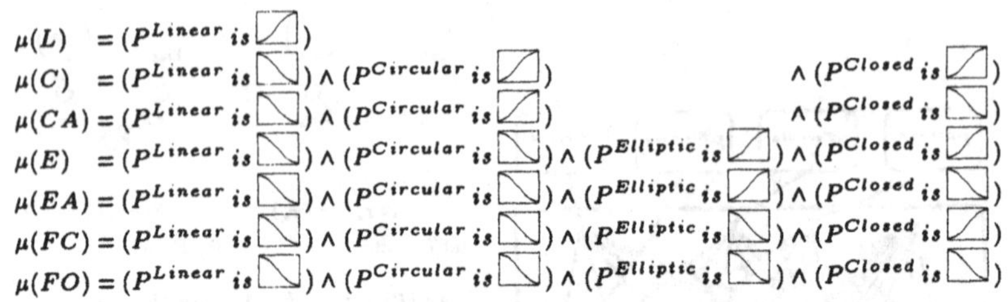

$\mu(L)\ =(P^{Linear}\ _{is}\ \boxed{\diagup})$

$\mu(C)\ =(P^{Linear}\ _{is}\ \boxed{\diagdown}) \wedge (P^{Circular}\ _{is}\ \boxed{\diagup}) \qquad\qquad \wedge (P^{Closed}\ _{is}\ \boxed{\diagup})$

$\mu(CA) =(P^{Linear}\ _{is}\ \boxed{\diagdown}) \wedge (P^{Circular}\ _{is}\ \boxed{\diagup}) \qquad\qquad \wedge (P^{Closed}\ _{is}\ \boxed{\diagdown})$

$\mu(E)\ =(P^{Linear}\ _{is}\ \boxed{\diagdown}) \wedge (P^{Circular}\ _{is}\ \boxed{\diagdown}) \wedge (P^{Elliptic}\ _{is}\ \boxed{\diagup}) \wedge (P^{Closed}\ _{is}\ \boxed{\diagup})$

$\mu(EA) =(P^{Linear}\ _{is}\ \boxed{\diagdown}) \wedge (P^{Circular}\ _{is}\ \boxed{\diagdown}) \wedge (P^{Elliptic}\ _{is}\ \boxed{\diagup}) \wedge (P^{Closed}\ _{is}\ \boxed{\diagdown})$

$\mu(FC) =(P^{Linear}\ _{is}\ \boxed{\diagdown}) \wedge (P^{Circular}\ _{is}\ \boxed{\diagdown}) \wedge (P^{Elliptic}\ _{is}\ \boxed{\diagdown}) \wedge (P^{Closed}\ _{is}\ \boxed{\diagup})$

$\mu(FO) =(P^{Linear}\ _{is}\ \boxed{\diagdown}) \wedge (P^{Circular}\ _{is}\ \boxed{\diagdown}) \wedge (P^{Elliptic}\ _{is}\ \boxed{\diagdown}) \wedge (P^{Closed}\ _{is}\ \boxed{\diagdown})$

Fig. 7: Initial fuzzy rule set.

3. Training of FSCI

Given drawings and user's intentions about curve classes, the parameters w_i and θ_i of the fuzzy neural network presented in Figure 5 are adjusted so as to adapt FSCI's identification results to the user's intentions as much as possible. The inputs to the neural network: P^{Linear}, $P^{Circular}$, $P^{Elliptic}$ and P^{Closed} are calculated from each of the given drawings by the possibility evaluation process following the fuzzy spline interpolation process shown in Figure 2. On the other hand, the desired outputs from the neural network: $\mu(L)$, $\mu(C)$, $\mu(CA)$, $\mu(E)$, $\mu(EA)$, $\mu(FC)$ and $\mu(FO)$ are directly set based on the user's intention. (For example, when the user's intention is CA, we set 1 to $\mu(CA)$ and 0 to all other grades.) Therefore, the commonly used back-propagation learning algorithm can be simply applied to train the network.

4. Experimental Results of Training

This section demonstrates how the adjustment of the fuzzy neural network improves FSCI's curve class recognition rates, and then shows how the adjusted fuzzy neural network explicitly explains the effect of the training.

4.1 Experimental Conditions

For the experiment, we gathered 840 drawing samples from each of six different users (named A, B, C, D, E and F). Each user was presented with the six kinds of patterns (each of which have seven curve shapes and their corresponding curve classes) shown in Figure 8 in turn; and requested to draw primitive curves similar to the ones in the patterns intending to let FSCI recognize the indicated curve classes. A set of presentation consisted of the six patterns of small size and the ones of large size (that is 12 patterns in total) and ten sets were presented to each user. Out of the ten sets of presentation (that is 840 drawing samples) to each user, seven sets (that is 588 drawing samples) were used for training and the other three sets (that is 252 drawing samples) were used for testing. Because the fuzzy neural network has the explicit representation as a fuzzy rule set, all the training could start with the meaningful initial setting shown in Figure 7.

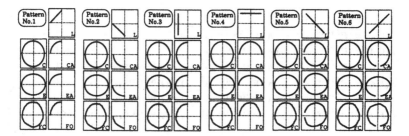

Fig. 8: Presented patterns.

Table 1: Curve class recognition rates.

User	Fuzzy Rule Set	Recognition Rates(%)			User	Fuzzy Rule Set	Recognition Rates(%)		
		1st Candidate	1st-2nd Candidates	1st-3rd Candidates			1st Candidate	1st-2nd Candidates	1st-3rd Candidates
A	Initial	78.869	91.369	94.345	D	Initial	76.786	94.048	97.917
	Trained	80.952	92.560	95.238		Trained	83.613	95.536	99.405
B	Initial	78.869	92.262	96.429	E	Initial	91.071	98.512	100.000
	Trained	79.762	92.857	96.726		Trained	93.451	99.405	99.702
C	Initial	62.500	80.060	88.095	F	Initial	70.833	87.500	94.048
	Trained	68.155	86.012	92.857		Trained	72.024	91.071	96.429

4.2 Improvement of Curve Class Recognition Rates

Table 1 shows the curve class recognition rates by FSCI with the initial fuzzy rule set in Figure 7 and ones by FSCI with fuzzy rule sets obtained after the training. In the table, the column labeled "1st Candidate" shows the recognition rates regarding the curve classes given the highest grades; "1st–2nd Candidates," the first and second highest grades; and "1st–3rd Candidates," the first through third highest grades. Although we evaluated the curve class recognition rates using the testing samples (without using the samples used for training), the results, from all of the six users, demonstrates the improvement of the curve class recognition rates after the training.

4.3 Discussions on a Concrete Example

Let us look at the case of the user D in detail. Figure 9 shows the fuzzy rule set (that is the explicit representation of the fuzzy neural network) obtained after the training for the user D. Table 2 shows the curve class recognition map by the trained fuzzy rule set, comparing it with the one by the initial fuzzy rule set. Now, Figure 9 tells us how the fuzzy rule set was adjusted so that it would adapt FSCI to the drawing manner of the user D.

For example, when we pay attention to the propositions with P^{Closed} for free curves (that is FC and FO), the *fuzzy true* got milder than the initial *fuzzy true* T_s while the *fuzzy false* got severer than the initial *fuzzy false* F_s. (Figure 10 emphasizes such propositions.) This implies that FSCI was trained so that the user D would easily close free curves. Indeed, it was difficult for the user D to get FC when FC was his intention as shown in Table 2 (a). However, it was improved after the training as

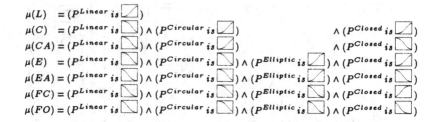

$$\mu(L) = (P^{Linear}\ is\ \square)$$
$$\mu(C) = (P^{Linear}\ is\ \square) \wedge (P^{Circular}\ is\ \square) \qquad\qquad \wedge (P^{Closed}\ is\ \square)$$
$$\mu(CA) = (P^{Linear}\ is\ \square) \wedge (P^{Circular}\ is\ \square) \qquad\qquad \wedge (P^{Closed}\ is\ \square)$$
$$\mu(E) = (P^{Linear}\ is\ \square) \wedge (P^{Circular}\ is\ \square) \wedge (P^{Elliptic}\ is\ \square) \wedge (P^{Closed}\ is\ \square)$$
$$\mu(EA) = (P^{Linear}\ is\ \square) \wedge (P^{Circular}\ is\ \square) \wedge (P^{Elliptic}\ is\ \square) \wedge (P^{Closed}\ is\ \square)$$
$$\mu(FC) = (P^{Linear}\ is\ \square) \wedge (P^{Circular}\ is\ \square) \wedge (P^{Elliptic}\ is\ \square) \wedge (P^{Closed}\ is\ \square)$$
$$\mu(FO) = (P^{Linear}\ is\ \square) \wedge (P^{Circular}\ is\ \square) \wedge (P^{Elliptic}\ is\ \square) \wedge (P^{Closed}\ is\ \square)$$

Fig. 9: Trained fuzzy rule set for the user D.

Table 2: Curve class recognition maps for user D.

(a) By initial fuzzy rule set.

Intentional Curve Class — Recognized Results (Number of Drawings)

Intentional Curve Class	L	CA	EA	FO	C	E	FC
L	48	0	0	0	0	0	0
C	0	0	0	0	46	2	0
CA	5	34	1	1	6	1	0
E	0	0	0	0	15	29	4
EA	0	13	24	7	1	3	0
FC	0	0	0	8	4	3	33
FO	0	3	1	44	0	0	0

(b) By trained fuzzy rule set.

Intentional Curve Class — Recognized Results (Number of Drawings)

Intentional Curve Class	L	CA	EA	FO	C	E	FC
L	48	0	0	0	0	0	0
C	0	0	0	0	42	4	0
CA	2	39	1	1	4	1	0
E	0	0	0	0	9	35	4
EA	0	6	30	9	0	3	0
FC	0	0	0	1	0	4	43
FO	0	2	1	44	0	0	1

shown in Table 2 (b). Figure 11 shows a drawing sample in the case.

On the other hand, when we regard the propositions with $P^{Circular}$ in Figure 9, we find that the *fuzzy true*'s got severer but the *fuzzy false*'s got milder after the training. (Figure 12 emphasizes two of such propositions.) This means that the fuzzy rule set was adjusted to get severer to C and CA; and, as a result, it came to tend to recognize E or EA rather than C or CA. This tendency obtained after the training considerably improved the curve class recognition rates for E and EA as shown in Table 2. Figure 13 shows drawing samples that are misidentified as C by the initial fuzzy rule set but identified as E by the trained one.

It will be noticed from these examples that the curve class estimation process realized as the fuzzy neural network tells us what was difficult for a specific user to deal with and how the fuzzy rules were adjusted to relieve the difficulties.

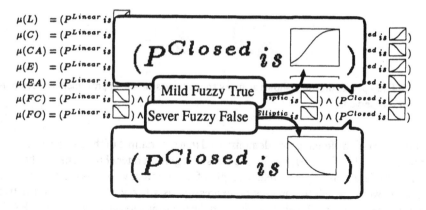

Fig. 10: Examples of fuzzy propositions that brought improvement about closedness.

Fig. 11: An example of drawing (shown as a fuzzy spline curve) where improvement about closedness was observed.

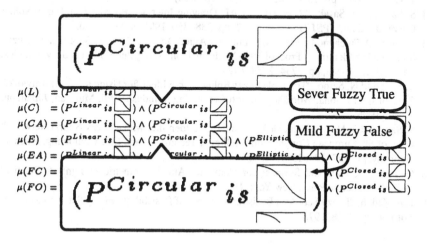

Fig. 12: Examples of fuzzy propositions that brought improvement about circularity.

5. Conclusions

This paper gave an outline of a trainable version of FSCI, in which a fuzzy neural network was embedded as the curve class estimation process. Then, experimental

Fig. 13: Examples of drawing (shown as fuzzy spline curves) where improvement about circularity was observed.

results, from six different users, demonstrated that the training of the fuzzy neural network improved FSCI in terms of curve class recognition rates. Furthermore, through some considerations on a concrete example of the training, we showed that the fuzzy neural network (which has an explicit expression as a fuzzy rule set) is informative for us to analyze users' drawing manner and the identification characteristics of FSCI. This is expected to give us hints for further improvement in the algorithms of FSCI.

References

[1] S. Saga and H. Makino, "Fuzzy Spline Interpolation and Its Application to On-Line Freehand Curve Identification," Proc. of 2nd IEEE Int. Conf. on Fuzzy Systems, San Francisco, USA, pp. 1183–1190, 1993.

[2] S. Saga, H. Makino and J. Sasaki, "A Method for Modeling Freehand Curves — the Fuzzy Spline Interpolation —," IEICE Trans., Vol.J77–D–II, 8, pp. 1610–1619, 1994 (in Japanese).

[3] S. Saga, H. Makino and J. Sasaki, "The Fuzzy Spline Curve Identifier," IEICE Trans., Vol.J77–D–II, 8, pp. 1620–1629, 1994 (in Japanese).

[4] S. Saga and J. Sasaki, "A Freehand CAD Drawing Interface Based on the Fuzzy Spline Curve Identifier," IPSJ Trans., Vol.36 2, pp. 338–350, 1995 (in Japanese).

[5] S. Saga, "A Freehand Interface for Computer Aided Drawing Systems Based on the Fuzzy Spline Curve Identifier," Proc. of 1995 IEEE Int. Conf. on Systems, Man and Cybernetics, Vancouver, Canada, pp. 2754–2759, 1995.

[6] S. Saga and N. Seino, "Trainable Fuzzy Spline Curve Identifier Using a Neural Network," Proc. of International Workshop on Soft Computing in Industry '96, Muroran, Japan, pp. 41–46, 1996.

[7] S. Saga and S. Mori, "Trainable Freehand Curve Identifier with a Fuzzy Neural Network," Proc. of 5th European Congress on Intelligent Techniques and Soft Computing, Aachen, Germany, pp. 127–131, 1997.

[8] G. Farin, "Curves and Surfaces for Computer Aided Geometric Design: A Practical Guide," Academic Press, New York, 1998.

[9] L.A. Zadeh, "Fuzzy Sets As a Basis for a Theory of Possibility," Fuzzy Sets and Systems, Vol.1, 1, pp. 3–28, 1978.

Intelligent real-time control of moulding mixtures composition in foundries

Jan Voracek

Department of Information Technology, Lappeenranta University of Technology
P.O. Box 20, Lappeenranta, FIN-53851 Finland
Phone: +358-5-621 3458, Fax: +358-5-621 3456, E-mail: Jan.Voracek@lut.fi

Abstract: The implementation of the adaptive knowledge-based system to the foundry industry is presented in this paper. Primarily the problem of moulding mixture composition was solved because majority of casting defects are due to inappropriate mould quality. The original learning method and system architecture was designed to ensure simple interaction and maintenance of knowledge base as well as unambiguous interpretation of its content. Elimination of the human factor was achieved with fully automated real-time data acquisition. The method proposed was successfully tested in standard industrial conditions during a period of two years.

1. Introduction

Foundry technology is a suitable branch for employment of artificial intelligence techniques, because the latest scientific results are still heavily combined with personal experience there. Possible application areas are summarised, e.g., in [1-5]. The role of expert systems is emphasised in [6-8]; computerised control of single stages of melting process is presented in [9-11]. Unfortunately, the number of hitherto realised and especially regularly employed applications is still insufficient. The aim of this paper is to clarify some common reasons of the above mentioned situation and to describe design, realisation and practical operational experience with original learning system controlling the quality of permanently circulating foundry moulding mixture. Alternative solutions of this task could be found also in [12] and [13].

Besides the general problem overview, also the structure of developed deterministic knowledge-based system is described. Its learning phase is entirely inductive and the developed knowledge base includes multiyear operational data from jolt squeeze moulding production line. Raw measurements are completed with personal knowledge and verified by a board of experts. System adaptation is fast in comparison with time constants of the controlled process. Therefore, data so far unseen can be regularly introduced to the knowledge base and the sequentially updated system can handle new observations correctly.

An essential part of the system presented is the real-time technological computer network collecting all required values and realising resulting control actions. Straight connection between intelligent system and technology enables to minimise human errors and ensures accuracy and repeatability of output behaviour. Outstanding long-term industrial results document suitability and practical applicability of the method presented.

2. Sand mixture quality control in foundries

2.1 Overview

Sand moulds based technology represents the widespread method for production of commercial castings. The typical inorganic green sand includes:

- an opening material (natural sand, silica sand etc.),
- a binder (bentonite, organic binder, water-glass etc.),
- an auxiliary compounds and additives (coal dust, hardeners, graphite etc.),
- the water.

The basic raw materials are cheap, ordinary and the final surface of manufactured castings is acceptable. Using this technology, a wood pattern is inserted into separated mould boxes, embedded with moulding mixture and closely rammed. After its removal single parts of hollow mould are assembled and the whole package is transported to the casting field. There a molten metal of a temperature around 1500°C is poured into and let to cool. After the proper solidification, boxes are disassembled, raw castings are knocked out and fettled. Degraded and typically hot used moulding mixture can be optionally reclaimed and finally stored into sand silos for another application. Situation is illustrated in Fig. 1(a). If an intelligent control system is used, real-time data acquisition is the essential condition for credible development of conjoint knowledge base. Typical architecture of industrial computer network affecting directly technological procedures is in Fig. 1(b).

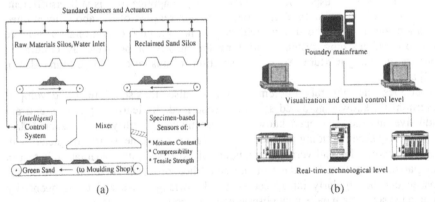

(a)   (b)

Fig. 1. (a) Moulding mixture circulation and revitalisation. (b) Real-time data acquisition in foundry computers' hierarchy.

Physical-chemical processes accompanying single technological steps significantly change properties of a mixture. The proper stuff recovery represents a complex control problem, because even small improper changes of mixture parameters can result in unpredictable accumulation of casting defects. Though the global ratio of rejects considerably varies according to the concrete technology, typically more then one-half of overall rejection rate is caused by unbalanced composition of moulding material. To summarise previous information, an applicable mixture must show the following characteristics:

- shape preserving,
- heat stability,
- reclamation possibility,
- reusability

and the properly designed control process should guarantee them in accordance with a set of observable variables. Foundry sand laboratories are supporting this process and providing following sets of mixture parameters:

- physical (grain size, shape or surface, moisture content, adhesion and cohesion, porosity etc.),
- chemical (chemical composition, pH-value, lustrous carbon, ignition loss, acidity, alkalinity etc.),
- thermal (refractoriness, temperature, gas content, combustibility, thermal expansion etc.),
- technological (adhesion or binding power, compressibility, mouldability, plasticity etc.) or
- mechanical (compression strength, bending strength, mould hardness, green bond, etc.).

2.2 Traditional Methods

Nevertheless the theoretical number of applicable mixture descriptors is high, each foundry collects regularly only several variables depending on the production technology and, of course, personal experience and traditions. The following methods can be, for example, used for determination of bentonite amount in a moulding mixture:

(a) amount of so called active clay $Clay_M$, based on the speed of the methylene blue absorption,

(b) content of so called applicable clay $Clay_A$ which is defined [14] as follows:

$$Clay_A = (0.435 P_t W \log S + 362)/109 \qquad (1)$$

where P_t means the tensile strength, W means the moisture content and S means the compressibility.

(c) effect of so called partial clays $Clay_1$, $Clay_2$, $Clay_3$ calculated according to [15] represents a refinement of (1):

$$\log Clay_1 = \log W - 0.3807 \log S + 0.9321, \qquad (2)$$

$$\log Clay_2 = (\log(0.145 P_t W \log S) - 0.35281)/2.0875, \qquad (3)$$

$$\log Clay_3 = (\log(0.145 P_t) + \log\log S + 0.3807 \log S - 1.285)/1.0875, \qquad (4)$$

with the same meaning of single variables as in (1).

General control strategy for moulding mixtures can include, e.g., :
- raw-materials and used sand dosing,
- mixing time and intensity,
- sand reclaiming parameters
- sand circulation speed and technological delays

and presented $Clay_x$ parameter is a comprehensive quantity enabling to formulate appropriate conclusions. In our co-operating foundry was, for example, primarily used the internal criterion U , derived from (2)-(4) as follows:

$$U = 89.1103 - 11.4939(Clay_1 - Clay_3).$$ (5)

Then the mixture was considered as proper if $U \in \langle 85,95 \rangle$, improper otherwise.

Suggested standard way of data processing has, however, at least two main disadvantages:
- explicitly defined formulas cannot incorporate all mutual interactions among single parameters and represent only a general estimation of the system behaviour,
- feedback via sand laboratory is slow and does not reflect the actual situation; methylene blue based estimation of bentonite content requires, e.g., units of hours to provide the result,

hence the applicability and expedience of advanced computational methods is evident.

2.3 Intelligent Approach

Majority of practical problems related to the production of castings can be formulated as multivariate, non-deterministic tasks, very suitable for application of advanced soft computing principles. Intelligent algorithms supported with automated data collection can successfully replace traditional intuitive procedures. Because there is no general guarantee of the final fruitfulness, detailed analysis including parallel comparison of the both approaches must be performed. The possibility of instant regress to the original methodology is also mandatory. Sophisticated algorithms can be designed using the local expert knowledge and archived past values enable to refine them with inductive learning. Beyond the presented example, there have great practical importance e.g., general prediction of casting defects, optimal computation of furnace charge or exact enumeration of production costs per single casting.

3. Built-in Computing Method

The model of controlled process should primarily follow existing control mechanisms. Though the formal approach expressed with (5) is available, plant operators typically formulate own conclusions considering even other measurable variables and appending their frequently subjective feelings and experience. Hereunder it is difficult to formulate even very general heuristics or formalise control mechanisms in detail. Therefore we decided to apply inductive learning approach capable to unify particular methods and concentrate available information inside a knowledge base. For its implementation the hierarchical classification structure was selected. Working responses are realised by means of binary decision tree with simple neurones in inner nodes. Conditional pre-clustering of training data makes it absolutely convergent, tree topology guarantees fast processing time. Mathematically clear structure allows backtracking of wrong decisions and goal-directed structural or parametrical modifications. All mentioned changes can be done either automatically or interactively, depending on the working mode of the system. Foundry plants represent typical systems with delayed response, in which the detailed interaction with human expert is necessary only during initial realisation stages. Hence a subsequent knowledge base updating can be performed automatically, because the computation itself is faster than the lowest time constant of the process.

Employed classifier generates a piecewise linear discrimination function to separate n real output classes ($n=2$ in our application). Such approach is applied to the original feature space and preserves consequently real problem background. The learning process includes two main phases.

During the first one, an unsupervised learning principle is applied to divide the normalised feature space into the corresponding number w of working classes ($w \geq n$). Relation between w and n is given by the parameter Similarity Level (SL), which defines the initial estimation of uniformity of a working class region. Using, e.g., the Euclidean metric, $SL \in \langle 0,1 \rangle$ and it represents the radius of circle A surrounding the currently processed training sample x_A. For each x_A located inside A and belonging to the same output class n the etalon x_E is calculated and conjoint samples are removed. Such procedure is repeated until all samples are replaced with etalons. Note that the value of quantity w can vary from n for $SL=1$ to the cardinality of the training set C for $SL=0$. Then all etalons are recursively separated and the final structure is stored as a binary tree. Euclidean metric can be used again as a distance measure.

In the second phase, the topology of resultant tree structure (dendrogram) is verified and the exact form of discrimination functions is searched. Each internal tree node represents one etalon and is realised with single standard neurone [16]. The appropriate subset of real training samples x_I determined in the previous step is dichotomised with the anti-gradient algorithm, the validity of each division line is limited with its predecessors from upper nodes towards to the root.

Computational realisation of learning procedure ensures standard shape of error function and allows to break the adaptation if the linear separation fails in any node. In such a case, the SL value must be decreased and the learning process has to start with a new pre-clustering. Otherwise leaf nodes are recursively pruned (if $w_{LEAF_1} = w_{LEAF_2}$ for the same parent), renumbered ($w \rightarrow n$) and the algorithm is successfully finished.

Verified tree topology proper weights in single nodes is used for the routine classification. Weights' updating for a real data is fast, because new instances are typically located in the neighbourhood of existing ones. Detailed presentation of the learning principle was published in [17] and the introduction to the system structure in [18]. The structure of the learning phase is in Fig. 2.

4. Real-time technological computer network

An adequate quantity of reliable information is the essential premise for correct control decisions and fully automated data acquisition represents the efficient way to guarantee it in practice. The main positives of such conception include:

- significant restriction of a human error,
- real-time context of data acquired,
- de-centralised local control,
- archiving and advanced decision making possibilities.

Though the data in some applications can be extracted easily and cheaply, foundry technology is a different case. An approximate list of reasons contains the following items:

- unsuitable external conditions for any computerisation (conductive and dusty environment, heavy electrical disturbance due to electric furnaces feeding, unqualified operators),
- difficult direct measurement of important technological variables (the evaluation of moulding mixture is mainly based on defined destruction of a test specimen),
- general personal conservativeness inhibiting from advanced non-foundry technicalities.

The industrial computer network was realised to prove the functionality and demonstrate overall merit of real-time interaction between computer and foundry technology. Tens of analogous and digital signals were recorded and visualised in order to specify the most suitable set of future application areas. For the presented moulding mixture problem the integrated sensor capable to measure all required quantities was constructed. Analysed mixture was taken from the mixer and transported to the storage bin. There the moisture content was measured. In the next step, a cylindrical specimen was pneumatically created and its compressibility was determined. During the final destruction of a sample we obtained the tensile strength value.

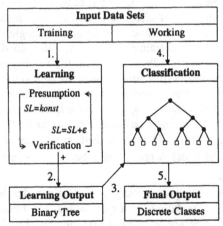

Fig.2. Block structure of presented algorithm.

The whole measurement is completed within three minutes, which represents a significant improvement in comparison with a traditional laboratory method. During our future experiments we are planing to process data concerning following activities:

- preparation and on-line testing of moulding mixture characteristics,
- control of metallurgical processes and charge composition,
- evaluation of laboratory results (advanced properties of moulding materials, final mechanical parameters of castings or metallographical results),
- monitoring and control of overall power consumption in the foundry.

5. Results

The realisation presented followed two main goals. At first, we wanted to demonstrate applicability and convenience of direct low-level employment of computers in foundry technology. This idea had mainly edifying character and was oriented at the top foundry management. The observed part of technology was visualised and related data were graphically interpreted. Complete information including statistical summaries and predictions was available in local company network. Here especially the real-time data background was highly appreciated

The second goal of our research included the following particular tasks:
- to prepare an integral knowledge base summarising individual control approaches to moulding mixture preparation,
- to specify a sufficient set of process variables among of all data observed,
- to realise an interactive user interface of the knowledge base to support operator control decisions,
- to formulate and apply general control heuristics.

During the implementation process we had to accept all local conventions and use only an exactly specified set of inputs, regardless their possible drawbacks. The testing sand plant was put into service in January 1994 and five-year history of mixture quality is shown in Fig. 3(a). Excluding trial operational period in years 1994 and 1995, the relative accuracy A_l calculated from U criterion for production year 1996 was only $A_l = 0.28\%$. Low output quality and frequent problems with conjoint casting defects such as expansion scabs, gas holes or hard spots (see [19] for details) compelled foundry management to search for an alternative method for keeping the mixture composition within the acceptable tolerance area.

We started with knowledge base preparation and testing since September 1996 and presented computer-aided decision system was fully used since January 1997. During the whole industrial testing period only the tensile strength, the moisture content and the compressibility values were processed regularly. The positive effect of proposed method is clear from Fig. 3(a) again; concrete values are summarised in Table 1. Beyond the aforementioned set of variables, also other parameters are measured in foundry sand laboratories. Their concrete usage is optional, frequently depending on the current plant operator. Auxiliary variables are typically taken into account only in special situations, e.g., if new types of raw materials were purchased, the current production includes different types of castings or if it is necessary to refresh the plant service after a breakdown.

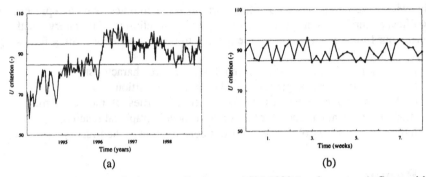

Fig. 3. (a) Development of mixture quality in years 1994-1998 (week averages). Composition is acceptable within the marked interval. (b) Advanced mixture control from January and February 1998

Control actions were entered interactively using the dialog window. Simple automatic mode affecting in sequence the water dosing and the mixing parameters was also successfully tested. During the standard control session made our system one measurement per 5 minutes and the applicable mixture was released to the moulding shop immediately. Otherwise an operator was warned of this situation and asked for a correction. To compare characteristics of standard and extended input data sets, an advanced control task was formulated and tested in the course of January and February 1998. Corresponding model and results are shown in Table 2.

Table 1: Basic set of input parameters

Variables				Cardinality of applied data sets		Classification accuracy A_{II}
		Range				
Name	Unit	Min.	Max.	Training	Working	
Moisture content	%	3.17	4.32	1340	680	84.23 %
Tensile strength	kPa	137.5	190.3			
Compressibility	%	36.2	51.3			

Table 2: Extended set of input parameters

Variables				Cardinality of applied data sets		Classification accuracy A_{III}
		Range				
Name	Unit	Min.	Max.	Training	Working	
Moisture content	%	3.17	4.32	720	40	92.78 %
Tensile strength	kPa	137.5	190.3			
Compressibility	%	36.2	51.3			
Compression strength	kPa	16.8	24.1			
Clay wash analysis	%	9.93	14.76			
Moisture absorption	mm	11.2	17.3			
Lustrous carbon	%	1.71	3.63			
Permeability	u.p.	131.9	218.7			

From their graphical interpretation in Fig. 3(b) it can be seen that complicated manual gathering was compensated with excellent accuracy. Though we used the same criterion U for the output classification, there was no explicit formula available in our testing foundry. In contrast of the previous example, human operators were not able to find all possible relations among wider set of input parameters and their decisions were based entirely on computer's recommendations.

During the standard control session made our system one measurement per 5 minutes and the applicable mixture was released to the moulding shop immediately. Otherwise an operator was warned of this situation and asked for a correction.

The both described experiments represent more or less decision support utilities than real control applications. This fact results partly from already mentioned specifics of foundry production. Moreover, there exist hidden correlations among single technological steps and the traditional specification of input, state and output variables is difficult. Because of inter-operational delays, it is almost impossible to assign the exact composition of the moulding mixture or the molten metal to the defective product. Castings, however, represent real output of each foundry and the location of possible technological bottlenecks is the logical complement of our research effort. It also means that the problem of mixture parameters cannot be solved separately, but as a part of significantly complicated structure. The complexity of such task is suggested in Fig.4.

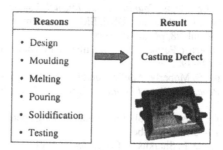

Fig. 4. Possible resources of low-class production of castings.

5. Conclusions

Theoretical principles, industrial realisation and concrete results of intelligent control method for foundry sand plants were described. Proposed system was developed and realised in a medium-size cast iron commercial foundry. After two years of the routine industrial service it can be evaluated as useful and reliable. The number and the character of mould originated casting defects were reduced evidently during this period. The system is permanently developed and especially the prediction of concrete casting defects derived from advanced set of input parameters is of a great interest. Our future plans also expect to exclude the operator from the direct control process completely and assign him only a supervisory role.

For successful completion of this task, a set of high-level heuristics must be formulated to specify exactly the correspondence between observed technological variables and resulting control actions. Such improved intelligent control system has a real chance to become a powerful tool for the total quality management process in modern foundries.

References

[1] D. L. Trevor, "The application of computers to foundry operations,", *Foundry International*, pp. 38-44, March, 1996.

[2] P. Bartelt, S. Moberly, "Applying Artificial Intelligence to the Modern Foundry, *Modern Casting*, vol. 86. pp. 52-55, February, 1996.

[3] G. P. Moynihan, Data Resource Analysis and Design of Knowledge Based System, *AFS Transactions*, vol. 103, pp. 239-242, 1995.

[4] R. V. Sillen, Using AI in the Foundry, *Modern Casting*, vol. 81, pp. 34-37, December, 1991.

[5] R. V. Sillen, Optimizing Iron Quality through Artificial Intelligence, *Modern Casting*, vol. 86, pp. 43-46, November 1996.

[6] G. P. Moynihan, Expert systems and their applications to the foundry, in *Industrial Engineering in the Foundry*, S. P. Thomas Ed., pp. 341-349, 1994.

[7] B. Yoberd, P. M. Stephenson, J. W. D. Todd,. An expert system for diagnosis of aluminium alloy casting defects, *Foundry Trade Journal*, 164, pp. 837-838, 1990.

[8] P. Hairy, Y. Hemon, J. Y. Marmier, DIADEM - un systéme expert en fonderie sous pression, *Fonderie*, vol. 98, pp. 38-44, 1990.

[9] D. L. Schroeder, Electric Furnace Control Utilizing Personal Computers, *Transactions AFS*, vol. 95, pp. 493-498, 1987.

[10] P. Bartelt, N. Bliss, S. Moberly, Application of AI to Power Input Control in the Modern Foundry, in *Proc. 61st. World Foundry Congress CIATF*, Beijing, China, pp. 215-221, 1995.

[11] R. J. Snook, Computerised control of a cupola charging operation, *Foundryman*, vol. 86, 1993, pp. 403-408, October 1993.

[12] J. R. Luckenbaugh, D. P. Sharkus, A Computer Controlled Green Sand System, *Transactions AFS*, vol. 95, pp. 117-122, 1987.

[13] H. L. Roes, F. Satmer, Essential controls for greensand quality, *Foundry Trade Journal*, vol. 163, pp. 422-424, 1989.

[14] R. W. Heine, J. S. Schumacher, Compressibility and Binding Clay in Green sand, *Giesserei-Praxis*, pp. 38-48, March 1979.

[15] G. R. Strong, Computerized Sand Control, *Modern Casting*, vol. 73, pp. 18-21, March 1983.

[16] Y. H. Pao, *Adaptive Patter Recognition and Neural Networks*. Addison-Wesley, 1989.

[17] J. Voracek, J. Voracek, Note on Adaptation of Nets by Backpropagation, in *Proc. IASTED Int. Conference on Robotics and Manufacturing*, Oxford, England, pp. 147-149, 1993.

[18] J. Voracek, Tree neural classifier for character recognition, in *Proc. 6th Int. Conf. on Computer Analysis of Images and Patterns - CAIP*, Prague, Czech Republic, pp. 631-636, 1995.

[19] coll., *International Atlas of Casting Defects*, American Foundrymen Society, USA, 1993.

Chapter 9: New Frontiers of Soft Computing

Papers:

Test Feature Classifiers and a 100% Recognition Rate
V. Lashkia, and S. Aleshin

Evaluation of the Modified Parzen Classifier in Small Training Sample Size Situations
Y. Muto, H. Nagase, and Y. Hamamoto

Soft Limiting in Adaptive Notch Filtering
O. Vainio

An Algorithm for Induction of Possibilistic Set-Valued Rules by Finding Prime Disjunctions
A. A. Savinov

Dataflow Realizes a Diagrammatic Programming Method
S. Kawaguchi and H. Shirasu

DIFFOBJ – A Game for Exercising Teams of Agents
T. Hirst and T. Kalus

Test Feature Classifiers and a 100% Recognition Rate

V. Lashkia[1] and S. Aleshin[2]

[1]Department of Information and Computer Eng.,Okayama University of Science, 1-1 Ridai-cho, Okayama, 700-0005 Japan
[2]Department of Mathematical Theory of Intellectual Systems, Moscow State University

Abstract: In this paper, we present a class of test feature classifiers. Introducing kernels and a rejection option, we discuss the properties and performance of the proposed classifiers. We describe cases when a 100% recognition rate (on a test data) can be achieved. To test the performance of the classifiers, we apply them to a well-known phoneme database. Conventional classifiers and even their combinations can not deal with the phoneme classification problem and have low recognition rates. Our experimental results show that the proposed classifiers have a high ability of recognition and suggest that they can be used in a variety of pattern recognition applications.

1 Introduction

There is much interest in the field of pattern recognition on trainable pattern classifiers, as seen, for example, in the growth of research in neural networks and fuzzy neurons [1, 2]. Many classifiers in the learning phase require optimization methods, and have problems with convergence, stability and time efficiency. Statistical and structural methods learn badly when exact statistical or structural knowledge is not available. Any method based on metrics uses the hypothesis that the proximity in the data space generally expresses the membership of the same class, and therefore a data set which does not satisfy this condition cannot be treated by such an approach.

In this paper, we present class of combinatorical-logical classifiers called test feature classifiers. These classifiers allow us to avoid the above drawbacks. Classifiers are generated directly from training samples using so-called *tests*, sets of features that are sufficient to distinguish patterns from different classes of training samples. The concept of the test was first introduced in [3] for the purpose of digital logic circuit analysis. It was then realized that tests could be very useful in pattern recognition. First, as the pattern recognition tool tests were used in [4]. The concept of test feature classifiers was presented in [5].

Test feature classifiers are m-degree polynomials, and can be used for partitioning the n-dimensional feature space, $m \leq n$, whose features are assumed to be binary-valued. Optimization methods, statistical, structural or metrical characteristics of patterns are not required. The method is desirable when statistical or structural information is not available. In this work, we have attempted to address some issues relevant to test feature classifiers which have not previously received attention. The performance of the proposed classifiers depend on the selection of tests and training samples. Introducing kernels and a rejection option we discuss the properties and performance of the proposed classifiers. We show that for almost all classes with a polynomial number of elements a 100% recognition rate on any test data can be achieved.

We applied the proposed classifiers to a well-known phoneme data. The phoneme database is one of the largest available in the Internet, and it is known to be quite difficult classification problem [6]. We compare the performance of the proposed classifiers to the performance of the conventional classifiers and standard CMC (Combination of Multiple Classifiers) algorithms. Recently, viewing that a single classification algorithm generally could not yield a very low error rate, many researches have turned their attention to the use of multiple classifiers [7, 8, 9]. However in many cases the multiple classifiers do not improve the performance significantly enough to make a convincing argument for using multiple classifiers (which are more time consuming) rather then a single one. Our experimental results show that much research still should be done with single classifiers. Our simulations show that the proposed test feature classifiers have much better performance than conventional methods and standard CMC.

The phoneme database consists from real-valued features. Analyzing set of tests of the training samples, we obtained the following interesting result related to the structure of the phoneme data. The phoneme data can be sufficiently recognized even when we use only the decimal parts of features.

The objective of this work is to present a high generalization ability of test feature classifiers. This paper is organized as follows: in Section 2, we introduce test feature classifiers and in Section 3 we discuss their properties and performance. In Section 4, we present experiments with real data.

2 Test Feature Classifiers

Assume that P is an n-dimensional feature space, $P = \{\bar{t} = (t_1, ..., t_n)\}$, and each pattern is represented as a binary-valued feature vector in this space $t_i \in \{0, 1\}$. Let us also assume that there are 2 possible classes I_1 and I_2. The problem of designing a classifier for pattern recognition can be stated as follows: a function V must be found such that a pattern $\bar{x}$ is in the class I_1 (in the class I_2) if and only if $V(\bar{x}) \geq 0$ $(V(\bar{x}) < 0)$.

Let us denote $B_1 = \{\bar{x}^1, ..., \bar{x}^{m_1}\}$ as a set of training samples from the class I_1 and $B_2 = \{\bar{y}^1, ..., \bar{y}^{m_2}\}$ as a set of training samples from the class I_2, where $\bar{x}^j = (x_1^j, ..., x_n^j), j = 1, ..., m_1$, $\bar{y}^j = (y_1^j, ..., y_n^j), j = 1, ..., m_2$, and $I_1 \cap I_2 = \emptyset$. A collection of features, $\tau = \{i_1, ..., i_k\}, (1 \leq k \leq n)$ is called a *test feature* (or test) of B_1 and B_2 if for any p $(1 \leq p \leq m_1)$ and any q $(1 \leq q \leq m_2)$ there exist some $i_s \in \tau$ $(1 \leq s \leq k)$ such that $x_{i_s}^p \neq y_{i_s}^q$. In other words, a test is a collection of features which is sufficient to distinguish vectors from different classes of training samples. If for a test τ, the set $\tau - \{i_s\}$ is not a test for any s $(1 \leq s \leq k)$, then τ

is called a *prime test feature* (or prime test).

A test $\tau = \{i_1, ..., i_k\}$, can be considered as an n-tuple vector, $\bar{\tau} = (\tau_1, ..., \tau_n)$, where τ_i is 1 if $i \in \{i_1, ..., i_k\}$, and 0 otherwise. A test is a collection of features for discriminating training samples of different classes and it can be used for the classification of unknown patterns. Introducing tests allows us to construct different types of classifiers for the purpose of pattern recognition.

For a given test $\bar{\tau}$ we can measure the degree of similarity of an unknown pattern $\bar{t}$ to the training pattern $\bar{x}$ by

$$\prod_{i=1}^{n}(1 - \tau_i|t_i - x_i|) \tag{1}$$

This expression takes the value 1 if and only if $\bar{t}$ and $\bar{x}$ coincide in the features defined by test $\bar{\tau}$, and takes the value 0 otherwise. In this case, no metric is used and and only an equivalence relation is required. The degree of similarity can also be measured in the classical way using distance

$$d(\bar{\tau} \circ \bar{t}, \bar{\tau} \circ \bar{x}), \tag{2}$$

where d is a distance metric, and the operation $\circ$ has the following meaning, $\bar{a} \circ \bar{b} = (a_1 \cdot b_1, ..., a_n \cdot b_n)$.

Denote T to be a set of tests. Taking (1) as a measure of similarity we calculate votes $V_1(\bar{t})$ and $V_2(\bar{t})$ for the classes I_1 and I_2 in the following way

$$V_1(\bar{t}) = \frac{1}{m_1} \sum_{\bar{\tau} \in T} \sum_{\bar{x} \in B_1} \prod_{i=1}^{n}(1 - \tau_i|t_i - x_i|)$$

$$V_2(\bar{t}) = \frac{1}{m_2} \sum_{\bar{\tau} \in T} \sum_{\bar{y} \in B_2} \prod_{i=1}^{n}(1 - \tau_i|t_i - y_i|).$$

We call a classifier based on the discriminant function $V(\bar{t}) = V_1(\bar{t}) - V_2(\bar{t})$ as *test feature classifier* [5] and denote it by TF_T. We extend TF to reject patterns $\bar{t}$ for which $V(\bar{t}) = 0$, and denote TF_T classifier as TFR_T for the rejection option [10].

Similarly, taking (2) as a measure of similarity and using the nearest neighbor concept we introduce a new classifier. For each test $\bar{\tau} \in T$ we calculate votes $W_1^{\bar{\tau}}(\bar{t})$ and $W_2^{\bar{\tau}}(\bar{t})$ for the classes I_1 and I_2 by the nearest neighbor rule, using (2) as a measure of similarity. Let

$$W_1(\bar{t}) = \frac{1}{m_1} \sum_{\bar{\tau} \in T} W_1^{\bar{\tau}}(\bar{t})$$

$$W_2(\bar{t}) = \frac{1}{m_2} \sum_{\bar{\tau} \in T} W_2^{\bar{\tau}}(\bar{t}).$$

We denote the classifier based on the discriminant function $W(\bar{t}) = W_1(\bar{t}) - W_2(\bar{t})$ by $TFNN_T$.

Since for $t_i, x_i \in \{0, 1\}$, $|t_i - x_i| = (1 - 2x_i)t_i + x_i$, V is a polynomial of n variables with a degree less than or equal to n. For big values of n, the evolution of TF becomes time consuming. We can improve the time requirement by extracting important (for classification) features and reducing the dimension of a feature space. Denote $\hat{T}$ to be the set of all prime tests. Denote $\hat{T}_i, i = 1, ..., n$, to be the set of all prime tests containing the i-th feature. Let $p_i = |\hat{T}_i|/|\hat{T}|$. We call vector $\bar{p} = (p_1, ..., p_n)$ an info vector and p_i an info weight [5]. We assume that the more the prime tests contain the ith feature, the more the ith feature is important for the classification purpose. The info weight can be considered as a measure of the feature importance. Therefore, features can be sorted by their importance and we can reduce their number by removing features with small info weights.

Denote the number of features (1s) in a test $\bar{\tau}$ as $|\bar{\tau}|$. We say $|\bar{\tau}|$ is the length of $\bar{\tau}$.

3 Performance

It is easy to prove that TF, TFR and $TFNN$ have no error on the training samples. As seen from the definition of the test feature classifier the classification performance of TF on the test samples depends on the set of tests T, and on the set of training samples B_1 and B_2. Let us call a test (a prime test) of I_1 and I_2 a *kernel* (a prime kernel). To provide the maximum classification rate for TF we need to find a set of kernels. If $\bar{\kappa}$ is a kernel then obviously $\bar{\kappa}$ is a test for B_1 and B_2 and the following relation holds

$$\bar{\kappa} \in \bigcap_{B_1 \in I_1, B_2 \in I_2} T \qquad (3)$$

where T is the set of all tests for B_1 and B_2. In general, it is impossible to find a set of kernels for unknown I_1 and I_2, but we can estimate it from the training sets using relation (3). Suppose that a set of kernels K for unknown I_1 and I_2 is found. It is easy to see that TFR_K has a 100% recognition rate on any test samples for any training set. Even if we know a kernel, we need an appropriate training set to obtain a recognition rate of 100% for TF classifiers.

We say that pair (B, K), $B = B_1 \cup B_2$, covers a pattern $\bar{z}$ if there exist $\bar{x} \in B$ and $\bar{\kappa} \in K$ such that $\bar{z} \cdot \bar{\kappa} = \bar{x} \cdot \bar{\kappa}$. Denote by $C(B, K)$ the set of all $\bar{z}$ that are covered by (B, K). We call a set B as a *prototype* set for K if $C(B, K) \supseteq I_1 \cup I_2$. It is easy to see that if B is a prototype set for K then TF_K will have a 100% recognition rate on any test samples.

Let us describe two important properties of coverings. If $K_1 \subseteq K_2$ then $|C(B, K_1)| \leq |C(B, K_2)|$. If $|\bar{\kappa}_1| \leq |\bar{\kappa}_2|$ and $\bar{\kappa}_1 \cdot \bar{\kappa}_2 = \bar{\kappa}_1$, then $|C(B, \bar{\kappa}_2)| \leq |C(B, \bar{\kappa}_1)|$. From these properties we can conclude that in general to achieve a high recognition rate for test feature classifiers we should select a set of short tests using (3), and the shorter tests we have the better the recognition rate is. After selecting a set of tests T the training set B can be reduced by determining a set $B' \subset B$ such that $C(B, T) = C(B', T)$.

The approach with kernels is suitable when the set of all kernels is different from the trivial set $\{(1, 1..., 1)\}$. Below, we estimate the number of classes which have non-trivial sets of kernels.

The classes I_1 and I_2 can be represented as a Boolean partial function $f(x_1, ..., x_n)$ such that $f(\bar{x}) = 0$ when $\bar{x} \in I_1$, and $f(\bar{x}) = 1$ when $\bar{x} \in I_2$. The ith variable of f is called an essential variable if there exists $(a_1, ..., a_n)$ and $(b_1, ..., b_n)$

such that $a_1 = b_1,..., a_{i-1} = b_{i-1}$, $a_i \neq b_i$, $a_{i+1} = b_{i+1},..., a_n = b_n$ and $f(a_1, ..., a_n) \neq f(b_1, ..., b_n)$. It can be easily proved [5] that the ith variable of f is an essential variable if and only if the info weight $p_i = 1$.

Denote $S(n)$ to be the number of all partial functions with at least one essential variable defined on m number of n-tuples. Denote $S_{m_1 m_2}(n)$ to be the number of all partial functions with at least one essential variable that takes the value 0 m_1 times, and the value 1 m_1 times. Let $m = m_1 + m_2$. First, we estimate $S_{m_1 m_2}(n)$.

A set of $m_1 - 1$ number of n-tuples $\bar{\alpha}_1, ..., \bar{\alpha}_{m_1-1}$ with the property $f(\bar{\alpha}_i) = 0$ can be selected in $\binom{2^n}{m_1-1}$ different ways. From the remaining $2^n - (m_1 - 1)$ tuples, a set of $m_2 - 1$ number of n-tuples $\bar{\beta}_1, ..., \bar{\beta}_{m_2-1}$ with the property $f(\bar{\beta}_i) = 1$ can be selected in $\binom{2^n - (m_1-1)}{m_2-1}$ different ways. Suppose that the remaining two tuples (where f is defined) guarantee the essential dependence on one of the variables $x_1, ..., x_n$. The essential variable x_i can be selected in n different ways. Let $\bar{\gamma}$ be an n-tuple from the remaining $2^n - (m_1 + m_2 - 2)$ tuples. $f(\bar{\gamma})$ can be equal to 0 or 1. Since x_i is an essential variable then the last remaining $(m_1 + m_2)$th n-tuple $\bar{\delta}$ will be determined uniquely. In other words, if $\bar{\gamma} = (\gamma_1, ..., \gamma_{i-1}, \gamma_i, \gamma_{i+1}, ..., \gamma_n)$ then $\bar{\delta} = (\gamma_1, ..., \gamma_{i-1}, \delta_i, \gamma_{i+1}, ..., \gamma_n)$ where $\gamma_i \neq \delta_i$. Thus

$$S_{m_1 m_2}(n) \leq \binom{2^n}{m_1-1}\binom{2^n-(m_1-1)}{m_2-1} \times$$
$$2n(2^n - (m_1 + m_2 - 2))$$

Note that the above expression is an upper bound, since we take into account even impossible cases when for example $\bar{\delta}$ is already selected between $(m_1 + m_2 - 2)$ n-tuples.

Denote $R(n)$ to be the number of all partial functions defined on m tuples. It is easy to see that $R(n) = \binom{2^n}{m}2^m$. The following lemma gives asymptotic growth ratios of $S(n)$ and $R(n)$ in the cases, when the domain of partial functions consists of n^k tuples, $k > 1$, i. e., $m_1 + m_2 = n^k$.

Lemma.

$$\lim_{n \to \infty} \frac{S(n)}{R(n)} \to 0.$$

Proof. Without loss of generality we consider $k = 2$.

$$\frac{S(n)}{R(n)} \leq$$

$$\frac{\sum_{m_1+m_2=n^2} 2n(2^n - (m_1 + m_2 - 2))\binom{2^n}{m_1-1}\binom{2^n-m_1+1}{m_2-1}}{\binom{2^n}{n^2}2^{n^2}} =$$

$$\frac{2n(2^n - n^2 + 2)\sum_{m_1+m_2=n^2}\binom{2^n}{m_1-1}\binom{2^n-m_1+1}{m_2-1}}{\binom{2^n}{n^2}2^{n^2}}.$$

Since

$$\sum_{m_1+m_2=n^2}\binom{2^n}{m_1-1}\binom{2^n-m_1+1}{m_2-1} =$$

$$\sum_{m_1=0}^{n^2} \binom{2^n}{m_1-1}\binom{2^n-m_1+1}{n^2-m_1-1} \le$$

$$n^2 \max_{0\le m_1\le n^2}\left(\binom{2^n}{m_1-1}\binom{2^n-m_1+1}{n^2-m_1-1}\right) \le$$

$$n^2\binom{2^n}{n^2/2-1}\binom{2^n-n^2/2+1}{n^2/2-1},$$

it follows that

$$\frac{S(n)}{R(n)} \le \frac{2n(2^n-n^2+2)n^2\binom{2^n}{n^2/2-1}\binom{2^n-n^2/2+1}{n^2/2-1}}{\binom{2^n}{n^2}2^{n^2}} =$$

$$\frac{2n^3(2^n-n^2+2)(n^2)!}{2^{n^2}(2^n-n^2+2)(2^n-n^2+1)(n^2/2-1)!(n^2/2-1)!} =$$

$$\frac{2n^3(n^2)!}{2^{n^2}(2^n-n^2+1)((n^2/2-1)!)^2}.$$

From the well known Stirling formula we have

$$\frac{(n^2)!}{2^{n^2}} \sim \frac{\sqrt{2\pi n^2}(n^2)^{n^2}}{e^{n^2}2^{n^2}} = \frac{\sqrt{2\pi n^2}(n^2/2)^{n^2}}{e^{n^2}} =$$

$$\frac{\sqrt{2\pi n^2}(n^2/2)^{n^2/2}(n^2/2)^{n^2/2}}{e^{n^2/2}e^{n^2/2}} \sim \frac{\sqrt{2}(n^2/2)!(n^2/2)!}{\sqrt{\pi n}}.$$

Thus,

$$\frac{S(n)}{R(n)} \preceq C\frac{n^3((n^2/2)!)^2}{n(2^n-n^2+1)((n^2/2-1)!)^2} =$$

$$C\frac{n^3(n^2/2)(n^2/2)}{n(2^n-n^2+1)} = C'\frac{n^6}{2^n-n^2+1}.$$

Since $2^{(1-\epsilon)n} \preceq 2^n - n^2 + 1$ then

$$\frac{S(n)}{R(n)} \preceq C'\frac{n^6}{2^{(1-\epsilon)n}} \to 0. \qquad \square$$

From the lemma we can conclude that almost all functions (with a domain of n^k tuples, $k > 1$) do not have essential variables. Therefore, for each ith feature the info weight $p_i \ne 1$ and the set of kernels is non-trivial. Thus, in the cases when $|I_1 \cup I_2| \sim n^k, k > 1$, (these are cases which we encounter in reality when n becomes large) almost all classes have a non-trivial set of kernels and if we are able to select a set of kernels (from a set of tests) we will have a 100% recognition rate.

Num	Features					Class
1	1.239670	0.874530	-0.20510	-0.078137	0.066867	0
2	0.268281	1.351780	1.035080	-0.331522	0.216897	0
......	......					
5404	0.136604	0.714084	1.349810	0.972467	-0.630074	1

a) Phoneme database.

Class	0	1
0	0.0	0.62
1	0.62	0.0

b) Dispertion matrix computed on the phoneme database.

K-Nearest Neighbor	87.76%
Neural Network	79.21%
C4.5 Decision Tree	83.92%
Quadratic Bayes	75.41%
Linear Bayes	73.00%

c) The best results for each classifier.

DCS-LA: Local Class Acc.	88.49%
DCS-LA: Overall Accuracy	87.64%
Classifier Rank	87.31%
Modified Classifier Rank	88.75%
Behavior Knowledge Space	85.68%

d) Results for CMC algorithms.

Table 1: Phoneme database and recognition results of conventional methods.

4 Experiments with Real Data

We applied the proposed classifiers to the phoneme data (Table 1.a) which is available via ftp at: ftp.dice. ucl.ac.be. The classification problem of the phoneme data is to distinguish between nasal and oral vowels. There are 5404 samples in this database. The phoneme database is known to be a difficult classification problem [6]. The dispersion matrix computed on the phoneme database is given at Table 1.b, and Fischer's coefficient is equal to 0.0756. As seen from dispersion matrix and from value of Fischer's coefficient classes are very overlapped.

There many papers related to the phoneme data recognition problem (see [6]). In Table 1.c we list the results of the different classifiers, which were obtained in [7]. Each classifier was optimized with respect to selecting "good" values for the parameters which govern its performance. The database was divided into two equal halves. One half was used as training set and the classification accuracy was then evaluated using the other half. The Nearest Neighbor is on the top, almost 88% recognition rate. Other classifiers have quite low recognition rates. In Table 1.d we list results for standard CMC algorithms [7]. The Modified Classifier Rank

Num.	Features	Class
1	1.24 0.87 -0.21 -0.08 0.07	0
2	0.27 1.35 1.04 -0.33 0.22	0
....	....	
5404	0.14 0.71 1.35 0.97 -0.63	1

a) Phoneme database with rounded features.

Num.	Features	Class
1	2397 8746 2051 781	0
2	2683 3518 351 3315	0
....	....	
5404	1366 7141 3498 9725	1

c) The reduced phoneme database.

TF_{T_1} $|T_1|=400$	89.32%
TFR_{T_1}	89.52%, 176 rejected
$TFNN_{T_1}$	95.65%
TF_{T_2} $|T_2|=800$	90.27%
TFR_{T_2}	90.34%, 125 rejected
$TFNN_{T_2}$	95.58%
$TF_{T_1^*}$ $|T_1^*|=400$	96.41%
$TFR_{T_1^*}$	100%, 728 rejected
$TF_{T_2^*}$ $|T_2^*|=800$	99.09%
$TFR_{T_2^*}$	100%, 309 rejected

TF_T $|T|=500$	86.82%
TFR_T	86.72%, 335 rejected
$TFNN_T$	95.93%

d) Results for TF classifiers on the reduced phoneme database.

b) Results for TF classifiers on the phoneme database with rounded features.

Table 2: The reduced phoneme database and recognition results of TFs.

Algorithm and DCS-LA show slightly better performance than the Nearest Neighbor and the other CMC algorithms failed to improve upon the performance of the Nearest Neighbor classifier. The recognition results in Table 1 show that conventional methods encounter difficulties on large complex data set like phoneme.

To apply TF to the phoneme data features were rounded to two decimal places (Table 2.a), and we used a binary representation of them. Two decimal places is a minimal representation of the features by which patterns from different classes can be distinguished. We divided the phoneme database into equal halves for training and testing, keeping same conditions as in [7]. In Table 2.b we show TFs recognition rates, where T_i is a set of randomly chosen short prime tests (with no kernels) of training samples, and T_i^* is a set of kernels. Even when a lot of information is lost due to the rounding feature space, recognition rate of TFs are much higher than conventional ones. TFs can achieve high recognition rate even with no kernels. The second half of Table 2.b shows what would happen if we were able to select even a small part of kernels. The more short kernels we have the better is the performance of TF and the less rejections we obtain for TFR.

Calculating the info vector on the training samples we obtained that the decimal part of the first, second third and fourth features are most important. This information suggests a nonstandard reduction of feature space. We reduce the phoneme data by taking four decimal places of first, second, third and fourth features, see Table 2.c (the fifth feature was completely removed). Even this reduction of feature space looks quite unnatural the recognition rates of TFs are still high (see Table 2.d, where T is a set of randomly chosen short prime tests with no kernels). It is interesting that phoneme data can be sufficiently recognized even if we use only decimal parts of the first, second third and fourth features.

5 Conclusions

The classifiers that have been presented are simple and robust. Optimization methods, statistical, structural or metrical characteristics of patterns are not required. We discuss the properties and performance of the proposed classifiers. We show that almost all classes with a polynomial number of elements have nontrivial set of kernels and in the cases when we are able to select a set of kernels we will achieve a 100% recognition rate on any test data. To test the performance of the classifiers, we apply them to a well-known phoneme database. Conventional classifiers and even their combinations can not deal with the phoneme classification problem and have low recognition rates. Experiments on real data show that TFs can achieve a high recognition rate even with no kernels. Especially high and stable performance is shown by $TFNN$ classifier which give almost a 96% of recognition rate on the phoneme data. Our experimental results show that the proposed classifiers have a high ability of recognition and suggest that they can be used in a variety of pattern recognition applications.

References

[1] R. Lippmann: Pattern Classification Using Neural Networks, IEEE Commun. Mag., Nov. 47-64, 1989.

[2] S. Horikawa, T. Furuhashi, Y. Uchikawa, "On Fuzzy Modeling Using Fuzzy Neural Networks with the Back-Propagation Algorithm," IEEE Trans. on Neural Network, Vol. 3, No. 5, 801-806, 1992.

[3] I. Chegis and S. Yablonsky, "Logical Methods for Controlling Electric Circuit Function", Proceedings of V. A. Steklov Inst. of Maths., Vol 51 (in Russian), 1958.

[4] Yu. Zhuravlov, A. Dmitriev and F. Krendelev, "Mathematical Principles of the classification of Objects and Scene", Discrete Analyse, Vol 7 (in Russian), 1966.

[5] S. V. Aleshin, Recognition of Dynamical Objects, Moscow University Press, (in Russian) 1996.

[6] Enhanced Learning for Evolutive Neural Architecture, ESPRIT Basic Research Project, No. 6891, 1995.

[7] K. Woods, W. Kegelmeyer, and K. Bowyer: "Combination of Multiple Classifiers Using Local Accuracy Estimates", IEEE Trans. PAMI-19, 405-410, 1997.

[8] J. Cao, M. Ahmadi and M. Shridhar, Recognition of Handwritten Numerals with Multiple Features and Multistage Classifier, Pattern Recognition, 28, 153-160, 1995.

[9] A. Hojjatoleslami and J. Kittler, Strategies for Weighted Combinations of Classifiers Employing Shared and Distinct Representations, IAPR ICPR'98, 339-341, 1998.

[10] V. Lashkia and S. Aleshin: "Test Feature Classifiers: Performance and Application", IAPR ICPR'98, 341-343, 1998.

Evaluation of the Modified Parzen Classifier in Small Training Sample Size Situations

Yoshihiko Muto[1], Hirokazu Nagase[2] and Yoshihiko Hamamoto[2]

[1] Ube National College of Technology, Ube, Yamaguchi 755-8555, Japan,
muto@ube-k.ac.jp
[2] Faculty of Engineering, Yamaguchi University, Ube, Yamaguchi 755-8611, Japan,
hamamoto@csse.yamaguchi-u.ac.jp

Abstract: In this paper, we discuss the effects of the sample size on the generalization ability of Parzen classifiers. When the sizes of samples per class are much unequal, the performance of the Parzen classifier is further degraded. In order to overcome this problem, we propose to use the Toeplitz estimator and bootstrap samples in designing Parzen classifiers. Experimental results show that these techniques are very effective means for designing Parzen classifiers, particularly when the sizes of samples per class are much unequal.

1 Introduction

Gaussian kernel functions have been used for the Parzen classifier in practice. The Gaussian kernel function requires the estimation of a covariance matrix. If possible, one should estimate the covariance matrix by using a large number of training samples. However, in practice, the covariance matrix must be estimated from a finite number of training samples. Its estimation error due to the finite training samples leads to the degradation of the performance of the resulting Parzen classifier. It is particularly important to note that when the ratio of the training sample size to the dimensionality (i.e. the number of features) is significantly small, the covariance matrix becomes singular. If the covariance matrix is singular, then it is impossible to design the Parzen classifier with the Gaussian kernel functions. This problem, which is called the small sample size problem [1], becomes extremely severe, particularly when the sizes of samples per class are much unequal.

In order to overcome this problem, we propose to use the Toeplitz estimator [2] and bootstrap samples [3] in designing Parzen classifiers. The Toeplitz estimator is used to reduce directly the estimation error of the covariance matrix [4]. To reduce the influence of outliers which distort the distribution, on the other hand, the bootstrap samples are used [3]. The use of bootstrap samples may be considered as the smoothing of a density function. Experimental results show that these techniques lead to the improvement of the performance of Parzen classifiers.

2 Parzen Classifier

The Parzen classifier allows us to obtain complex nonlinear decision boundaries. First, we briefly describe the Parzen classifier. In the following, for class ω_k, we assume that we are given N_k training samples, $x_1^k, x_2^k, \cdots, x_{N_k}^k$. The Parzen classifier depends on the kernel function and on the value of the window-width. In this paper we will use the most commonly used n-dimensional Gaussian kernel function which leads to the following density estimate [5] :

$$p(x|\omega_k) = \frac{1}{N_k} \sum_{j=1}^{N_k} \left[\frac{1}{(2\pi)^{n/2} h_k^n |\hat{\Sigma}_k|^{1/2}} \exp\left\{ -\frac{1}{2h_k^2} (x - x_j^k)^T \hat{\Sigma}_k^{-1} (x - x_j^k) \right\} \right],$$

$$(1)$$

where $\hat{\Sigma}_k$ is the sample covariance matrix of class ω_k and h_k is the window-width of class ω_k. In this paper, for simplicity, we assume that $h = h_1 = h_2 = \cdots = h_m$, where m is the number of classes. As previously mentioned, $\hat{\Sigma}_k$ must be estimated only from the training set $X_{N_k} = \{x_1^k, x_2^k, \cdots, x_{N_k}^k\}$. In the conventional Parzen classifier, the sample covariance matrix is usually used, which is estimated by

$$\hat{\Sigma}_k = \frac{1}{N_k - 1} \sum_{j=1}^{N_k} (x_j^k - \hat{\mu}^k)(x_j^k - \hat{\mu}^k)^T,$$

$$(2)$$

where

$$\hat{\mu}^k = \frac{1}{N_k} \sum_{j=1}^{N_k} x_j^k.$$

$$(3)$$

In the Parzen approach, a pattern x is classified into ω_k if

$$P(\omega_k)p(x|\omega_k) > P(\omega_j)p(x|\omega_j), \quad k \neq j$$

$$(4)$$

where $P(\omega_k)$ denotes the a priori probability of class ω_k.

It is difficult to decide the optimal value of window-width h in terms of the error rate theoretically[5]. In this paper, parameter h is optimized as follows:

1. For each parameter value, estimate the error rate by the leave-one-out method.

2. Find the optimal value of the parameter which minimizes the error rate.

3 Proposed Method

3.1 Toeplitz estimator

In this section, for simplicity, the subscript k representing the class is omitted. We follow Fukunaga's[2] notations. The sample covariance matrix is described by

$$\hat{\Sigma} = \begin{bmatrix} \hat{\sigma}_1^2 & \hat{c}_{12} & \cdots & \hat{c}_{1n} \\ \hat{c}_{12} & \hat{\sigma}_2^2 & & \\ \vdots & & \ddots & \\ \hat{c}_{1n} & & & \hat{\sigma}_n^2 \end{bmatrix}, \tag{5}$$

where $\hat{\sigma}_i^2$ is the sample variance of x_i and $\hat{c}_{ij}$ is the sample covariance between x_i and x_j. It is convenient to express c_{ij} by

$$\hat{c}_{ij} = \hat{\rho}_{ij}\hat{\sigma}_i\hat{\sigma}_j, \tag{6}$$

where $\hat{\rho}_{ij}$ is the sample correlation coefficient between x_i and x_j. Then:

$$\hat{\Sigma} = \hat{\Gamma}\hat{R}\hat{\Gamma}, \tag{7}$$

where

$$\hat{\Gamma} = \begin{bmatrix} \hat{\sigma}_1 & & & 0 \\ & \hat{\sigma}_2 & & \\ & & \ddots & \\ 0 & & & \hat{\sigma}_n \end{bmatrix} \quad \text{and} \quad \hat{R} = \begin{bmatrix} 1 & \hat{\rho}_{12} & \cdots & \hat{\rho}_{1n} \\ \hat{\rho}_{21} & 1 & & \\ \vdots & & \ddots & \\ \hat{\rho}_{n1} & & & 1 \end{bmatrix}. \tag{8}$$

We will call $\hat{R}$ a sample correlation matrix.

By assuming the Toeplitz form for the correlation matrix, the correlation coefficient between x_i and x_j depends only on $|i-j|$. Expressing the covariance matrix as $\hat{\Sigma} = \hat{\Gamma}\hat{R}\hat{\Gamma}$, the inverse matrix and determinant are:

$$\hat{\Sigma}^{-1} = \hat{\Gamma}^{-1}\hat{R}^{-1}\hat{\Gamma}^{-1} \tag{9}$$

$$|\hat{\Sigma}| = |\hat{\Gamma}||\hat{R}||\hat{\Gamma}| = \left(\prod_{i=1}^{n}\hat{\sigma}_i\right)^2 |\hat{R}|. \tag{10}$$

Thus, we can focus our attention on $\hat{R}^{-1}$ and $|\hat{R}|$. A particular form of the Toeplitz matrix has the following closed forms for the inverse and determinant as

$$\hat{R} = \begin{bmatrix} 1 & \hat{\rho} & \cdots & \hat{\rho}^{n-1} \\ \hat{\rho} & 1 & & \vdots \\ \vdots & & \ddots & \hat{\rho} \\ \hat{\rho}^{n-1} & \cdots & \hat{\rho} & 1 \end{bmatrix}, \tag{11}$$

$$\hat{R}^{-1} = \frac{1}{1-\hat{\rho}^2} \begin{bmatrix} 1 & -\hat{\rho} & 0 & \cdots & & 0 \\ -\hat{\rho} & 1+\hat{\rho}^2 & \ddots & \ddots & & \vdots \\ 0 & & & & & 0 \\ \vdots & & & \ddots & 1+\hat{\rho}^2 & -\hat{\rho} \\ 0 & & \cdots & 0 & -\hat{\rho} & 1 \end{bmatrix}, \tag{12}$$

$$|\hat{R}| = (1-\hat{\rho}^2)^{n-1}. \tag{13}$$

Thus, the estimation process of a covariance matrix is given as follows:

1. Estimate the sample variance $\hat{\sigma}_i^2$.

2. Estimate the sample covariance $\hat{c}_{i,i+1}$ and divide $\hat{c}_{i,i+1}$ by $\hat{\sigma}_i \hat{\sigma}_{i+1}$ to obtain the estimate of $\hat{\rho}_{i,i+1}$.

3. Average $\hat{\rho}_{i,i+1}$ over $i = 1, \cdots, n-1$, to obtain $\hat{\rho}$.

4. Insert $\hat{\rho}$ into equation (11) to form $\hat{R}$.

Note that in the Toeplitz approach, only $(n+1)$ parameters, $\hat{\sigma}_i (i = 1, \cdots, n)$ and $\hat{\rho}$, are used to estimate a covariance matrix. This means that the computational cost is not so severe even if the dimensionality is large and that the estimation error of a covariance matrix is considerably reduced. Substituting equation (12) into (9), we finally obtain $\hat{\Sigma}^{-1}$. On the other hand, the determinant $|\hat{\Sigma}|$ is obtained by substituting equation (13) into (10).

3.2 Bootstrap Technique

Bootstrapping is similar to other resampling schemes such as cross-validation and jackknifing. We proposed a bootstrap approach for nearest neighbor classifier design [3]. We will try to apply the bootstrap method to the Parzen classifier design. The procedure to generate a bootstrap sample set $Y_{N_k} = \{y_1^k, y_2^k, \cdots, y_{N_k}^k\}$ of size N_k is as follows.

1. Select a training sample $x_{s_0}^k$ from X_{N_k} randomly.

2. Find the r nearest neighbor samples $x_{s_1}^k, x_{s_2}^k, \cdots, x_{s_r}^k$ of $x_{s_0}^k$, using the Euclidean distance.

3. Compute a bootstrap sample $y_1^k = \sum_{j=0}^r w_j x_{s_j}^k$ where w_j is a weight. The weight w_j is given by

$$w_j = \frac{\Delta_j}{\sum_{c=0}^r \Delta_c}, \quad 0 \le j \le r, \tag{14}$$

where Δ_j is chosen from a uniform distribution on $[0, 1]$. Note that $\sum_{j=0}^r w_j = 1$.

4. Repeat steps 1., 2. and 3. N_k times.

When bootstrap samples are used, the resulting density function is given by

$$p^B(x|\omega_k) = \frac{1}{N_k} \sum_{j=1}^{N_k} \left[\frac{1}{(2\pi)^{n/2} h^n |\hat{\Sigma}_k|^{1/2}} \exp\left\{ -\frac{1}{2h^2}(x - y_j^k)^T \hat{\Sigma}_k^{-1}(x - y_j^k) \right\} \right], \tag{15}$$

where $p^B(x|\omega_k)$ denotes the density function estimated by using bootstrap samples. Note that in the bootstrap-based Parzen classifier, parameters r and h are simultaneously optimized by using the leave-one-out method . The comparison of the proposed Parzen classifier to the conventional Parzen classifier is shown in Figure 1.

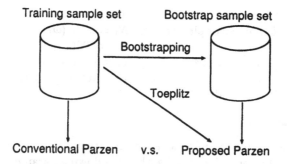

Figure 1: The comparison of the proposed Parzen classifier to the conventional Parzen classifier

Table 1: Parameter values of the $I - \Lambda$ data set

i	1	2	3	4	5	6	7	8
μ_i	3.86	3.10	0.84	0.84	1.64	1.08	0.26	0.01
λ_i	8.41	12.06	0.12	0.22	1.49	1.77	0.35	2.73

4 Experimental Results

The purpose of this experiment is to show that the proposed Parzen classifier is superior to the conventional Parzen classifier in terms of the error rate. The error rate is estimated by using a large number of test samples. Test samples must be statistically different from training samples.

First, we describe three data sets (two artificial data sets and one real data set), which are used in our experiment.

(a) $I - \Lambda$ data set [2]

The available samples were independently generated from 8 - dimensional Gaussian distributions $N(\mu_k, \Sigma_k)$ with the following parameters:

$$\mu_1 = [0, \cdots, 0]^T, \ \mu_2 = [\mu_1, \mu_2, \cdots, \mu_8]^T$$

$$\Sigma_1 = \begin{pmatrix} 1 & & 0 \\ & \ddots & \\ 0 & & 1 \end{pmatrix}, \ \Sigma_2 = \begin{pmatrix} \lambda_1 & & 0 \\ & \ddots & \\ 0 & & \lambda_8 \end{pmatrix}.$$

Parameter values are shown in Table 1. In the $I - \Lambda$ data set, both the mean vectors and the covariance matrices differ. Experimental conditions are as follows:

Dimensionality	:	$n = 8$
Training sample size	:	$N_1 = 200P(\omega_1)$,
		$N_2 = 200(1 - P(\omega_1))$
Test sample size	:	$T_1 = 10000P(\omega_1)$,
		$T_2 = 10000(1 - P(\omega_1))$
No. of trials	:	100

(b) Ness data set [6]

The available samples were independently generated from n - dimensional Gaussian distributions $N(\mu_k, \Sigma_k)$ with the following parameters:

$$\mu_1 = [0, \cdots, 0]^T, \quad \mu_2 = [\Delta/2, 0, \cdots, 0, \Delta/2]^T$$

$$\Sigma_1 = \begin{pmatrix} 1 & & 0 \\ & \ddots & \\ 0 & & 1 \end{pmatrix}, \quad \Sigma_2 = \begin{pmatrix} I_{n/2} & 0 \\ 0 & \frac{1}{2}I_{n/2} \end{pmatrix},$$

where I_n is the $n \times n$ identity matrix and Δ is the Mahalanobis distance between class ω_1 and class ω_2. Experimental conditions are as follows:

Dimensionality	:	$n = 10$
Mahalanobis	:	$\Delta = 6.0$
Training sample size	:	$N_1 = 200P(\omega_1)$,
		$N_2 = 200(1 - P(\omega_1))$
Test sample size	:	$T_1 = 10000P(\omega_1)$,
		$T_2 = 10000(1 - P(\omega_1))$
No. of trials	:	100

(c) Zernike data set

Zernike moment features [7] have been used to recognize handwritten numeral characters. Zernike moment features are only rotation invariant. To achieve scale and translation invariancy, the image is first normalized. Then, the Zernike features are extracted. We used the normalization technique presented by Khotanzad and Lu. Translation invariancy was achieved by moving the origin to the centroid of the shape. Scale invariancy was achieved by enlarging or reducing each shape such that its *zeroth* moment was set equal to a predetermined value, β. From preliminary experiments, the value of β was 300. Then, 12th order Zernike moments were used. This means that 47 Zernike features were utilized. The character images come from the ETL1 database [8]. From the ETL1 database, two pairs of characters, "2 and 3" and "2 and 5" were used. Experimental conditions are as follows:

Dimensionality	:	$n = 47$
Training sample size	:	$N_1 = 400P(\omega_1)$,
		$N_2 = 400(1 - P(\omega_1))$
Test sample size	:	$T_1 = 1000P(\omega_1)$,
		$T_2 = 1000(1 - P(\omega_1))$
No. of trials	:	10

In this experiment, 2 Parzen classifiers were compared: the first one is the conventional Parzen classifier, the second one is the Parzen classifier with the Toeplitz estimator and bootstrap samples. In order to highlight the difference in the performance of two classifiers, the comparative experiment was conducted for each of 7 different values of $P(\omega_i)$ ranging from 0.2 to 0.8 in steps of 0.1. For comparison, the performance of the Parzen classifier only with the Toeplitz estimator is also presented.

Results are shown in Figures 2-5. Regardless of $P(\omega_k)$ and the used data sets, the Parzen classifier with the Toeplitz estimator and bootstrap samples provides almost the smallest error rate. Especially when the sizes of samples per class are much unequal, the effect of combining the Toeplitz estimator with bootstrap samples becomes very clear. Therefore, these techniques should be considered in the design of Parzen classifiers, particularly in situations where the a priori probabilities are much different.

Acknowledgment

We would like to thank the Electrotechnical Laboratory in Japan for providing the ETL-1 database.

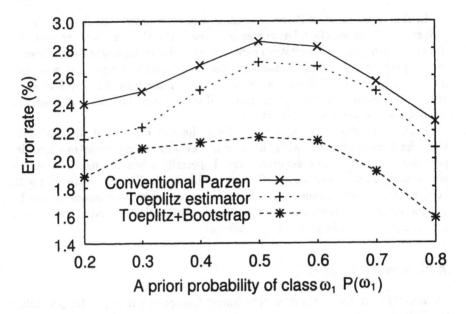

Figure 2: Comparison of the proposed Parzen classifiers and the conventional Parzen classifier on the $I - \Lambda$ data set.

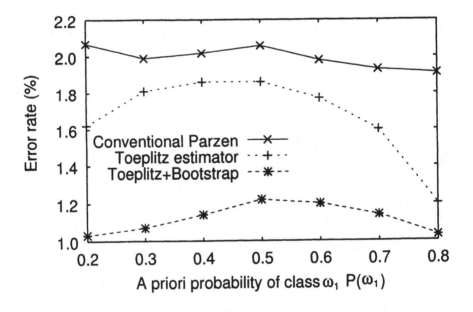

Figure 3: Comparison of the proposed Parzen classifiers and the conventional Parzen classifier on the Ness data set.

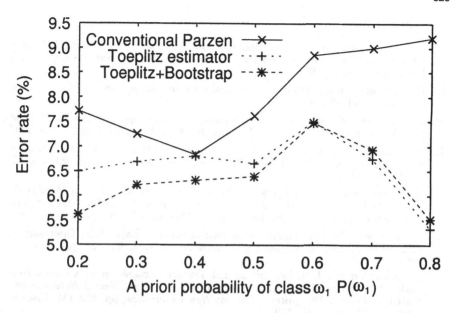

Figure 4: Comparison of the proposed Parzen classifiers and the conventional Parzen classifier for 2 and 3 characters.

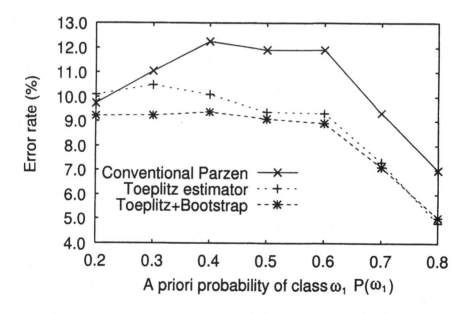

Figure 5: Comparison of the proposed Parzen classifiers and the conventional Parzen classifier for 2 and 5 characters.

References

[1] S. J. Raudys and A. K. Jain, "Small Sample Size Effects in Statistical Pattern Recognition: Recommendations for Practitioners", *IEEE Trans. PAMI*, Vol. 13, No. 3, pp. 252–264, 1991.

[2] K. Fukunaga, *Introduction to Statistical Pattern Recognition*, Academic Press, New York, 2nd edition, 1990.

[3] Y. Hamamoto, S. Uchimura and S. Tomita, "A Bootstrap Technique for Nearest Neighbor Classifier Design", *IEEE Trans. PAMI*, Vol. 19, No. 1, pp. 73–79, 1997

[4] Y. Hamamoto, Y. Fujimoto and S. Tomita, "On the Estimation of a Covariance Matrix in Designing Parzen Classifiers", *Pattern Recognition*, Vol. 29, No. 10, pp. 1751–1759, 1996.

[5] A. K. Jain and M. D. Ramaswami, "Classifier Design with Parzen Windows", In E. S. Gelsema and L. N. Kanal, editors, in *Pattern Recognition and Artificial Intelligence*, pp. 211–228, Elsevier Science Publishers B.V., North-Holland, 1988.

[6] J. Van Ness, "On the Dominance of Non-parametric Bayes Rule Discriminant Algorithms in High Dimensions", *Pattern Recognition*, Vol. 12, pp. 355–368, 1980.

[7] A. Khotanzad and J. H. Lu, "Shape and Texture Recognition by a Neural Network", In I. K. Sethi and A. K. Jain, editors, *Artificial Neural Networks and Statistical Pattern Recognition: Old and New Connection*, pp. 109–131, Elsevier Science Publishers B.V., 1991.

[8] T. Saito, H. Yamada and S. Mori, "An Analysis of Hand-Printed Character Data Base III", *Bul. Electrotech. Lab*, Vol. 42, No. 5, pp. 385–434, 1978.

Soft Limiting in Adaptive Notch Filtering

Olli Vainio

Tampere University of Technology, Signal Processing Laboratory, P.O. Box 553, FIN-33101 Tampere, Finland

Abstract: An adaptive digital notch filter is considered for 50/60 Hz line frequency signal processing. An adaptive algorithm allows tracking of the main frequency component such that the notch frequency is the only changing parameter. Since the power line frequency is known to within a narrow tolerance band, it is possible to apply soft or hard limiting on the notch frequency. Soft and hard limiting are compared for Laplacian noise, and the effect of filtering is investigated for a signal with strong harmonic contents.

1. Introduction

Filtering of narrow-band signals is often needed, for example, in the synchronization of power converters [9-11,15] and active filters for power conditioning [12].

Digital notch filtering is a computationally efficient method of separating a single-frequency or a very narrow-band signal from superimposed noise and disturbances. Since the digital implementation does not suffer from component tolerance, aging, or temperature effects, highly selective notches can be constructed. This naturally requires that the signal frequency is predefined, or can be accurately estimated in real time with an adaptive algorithm. In the case of filtering 50 Hz or 60 Hz line frequency signals, the expected frequency deviation is relatively small, typically less than ±2% in Western European countries. Therefore, quite robust filtering algorithms can be used. In stand-alone systems, on the other hand, larger frequency variations are encountered, therefore adaptive algorithms can be useful [13].

The noise encountered in industrial and power electronics is typically impulsive by nature. Deep commutation notches and other disturbances in the waveform cause problems for level-crossing detection and other synchronization tasks. Impulsive noise is often modeled with a Laplacian or double exponential distribution [14] with the probability density function

$$f(x) = \frac{1}{2}e^{-|x|}, \quad -\infty < x < \infty \tag{1}$$

2. The Notch Filter

For the purpose of implementing a single selective notch, a recursive digital filter is computationally more efficient than a transversal filter [10]. A second-order notch filter has the z-domain transfer function

$$H(z) = \frac{1 - 2r_z \cos(\omega_1)z^{-1} + r_z^2 z^{-2}}{1 - 2r_p \cos(\omega_1)z^{-1} + r_p^2 z^{-2}}, \tag{2}$$

which has zeros at $r_z e^{\pm j\omega_1}$ and poles at $r_p e^{\pm j\omega_1}$. When $r_z = 1$ and $r_p < 1$, the zero is placed on the unit circle, and the filter passes all frequencies except the narrow notch at ω_1. This was the configuration used in [7]. In this work, we focus on the case $r_z < r_p < 1$, which produces a bandpass notch type filter. The selectivity (Q-value) depends on r_z and r_p.

The phase response [8] of the notch filter section is given by

$$\arg H(e^{j\omega}) = \arctan\left[\frac{r_z \sin(\omega - \omega_1)}{1 - r_z \cos(\omega - \omega_1)}\right] + \arctan\left[\frac{r_z \sin(\omega + \omega_1)}{1 - r_z \cos(\omega + \omega_1)}\right]$$

$$- \arctan\left[\frac{r_p \sin(\omega - \omega_1)}{1 - r_p \cos(\omega - \omega_1)}\right] - \arctan\left[\frac{r_p \sin(\omega + \omega_1)}{1 - r_p \cos(\omega + \omega_1)}\right]. \tag{3}$$

We can see that $\arg H(e^{j\omega_1}) = 0$ if we choose $\omega_1 = \pi/2$. Synchronous filtering is therefore possible, but requires strict control of the computational latency since a sampling period corresponds to a quarter of the sine wave cycle.

Since $\arg H(e^{j\omega}) > 0$ for $\omega < \omega_1$, short-step prediction is possible by tuning ω_1 to be slightly higher than the nominal signal frequency ω_n. In this way, the implementation latency can be compensated to a certain extent. The prediction length in units of phase is given by $\arg H(e^{j\omega_n})$. The feasible prediction length depends on the Q-value and the allowed gain peak for out-of-band frequencies.

The frequency response of a notch filter is shown in Fig. 1 with $\omega_1 = 0.5\pi$, $r_z = 0.5$ and $r_p = 0.9375$.

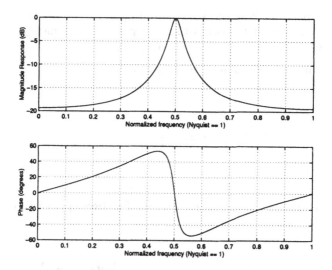

Fig. 1. Frequency response of a notch filter.

However, a fixed notch frequency is not acceptable in 50/60 Hz line frequency signal processing with a high-Q filter because of possible frequency deviations in power distribution networks. A number of researchers have investigated adaptive algorithms for notch filters [1-7]. A true least squares single section adaptive notch filter was described by Strobach [7]. The algorithm for locating the notch is summarized in (4)-(9) for $r_z = 1$, $r_p < 1$.

$$\epsilon(n) = x(n) + x(n-2) - r_p^2 e(n-2) \tag{4}$$

$$\alpha(n) = 2\cos(\omega_1(n)) = A(n)/B(n) \tag{5}$$

$$A(n) = \lambda A(n-1) + (1-\lambda)\xi(n-1)\epsilon(n) \tag{6}$$

$$B(n) = \lambda B(n-1) + (1-\lambda)\xi(n-1)\xi(n-1) \tag{7}$$

$$e(n) = \epsilon(n) - \alpha(n)\xi(n-1) \tag{8}$$

$$\xi(n) = x(n) - r_p e(n). \tag{9}$$

λ is a positive exponential forgetting factor close to 1.

3. Soft Limiting of the Notch Frequency

In the above algorithm, the parameter $\alpha(n) = 2\cos(\omega_1(n))$ is a measure of the instantaneous notch frequency $\omega_1(n)$. This is an advantage compared to a transversal adaptive filter [16], since the notch frequency can be monitored, and soft or hard limiting on $\alpha(n)$ can be applied. Consequently the notch location is guaranteed to stay within the permitted frequency band even if there are disturbances in the filter input signal.

Soft limiting is based on a sigmoidal function:

$$\sigma(x) = \frac{1}{1 + e^{-\beta x}}. \tag{10}$$

Denoting $\alpha_n = 2\cos(\omega_n)$, the soft limiting function applied to $\alpha(n)$ is given by

$$\alpha_s(n) = \alpha_n - (\alpha_n - \alpha_{min}) + \frac{2(\alpha_n - \alpha_{min})}{1 + e^{-\beta(\alpha(n)-\alpha_n)}} \tag{11}$$

which is shown in Fig. 2 for $\beta = 45$, $\alpha_n = 0$ and $\alpha_{min} = -0.0628$. Consequently $\omega_1(n) = \arccos(\alpha_s(n)/2)$.

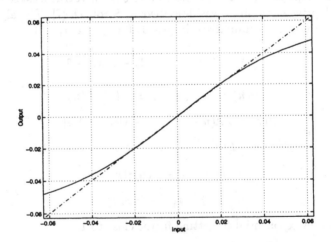

Fig. 2. The soft limiter function (solid line).

Behavior of the parameter $\alpha_s(n)$ in the adaptive notch filter algorithm is shown in Fig. 3 (dash-dotted line) when the input is sinusoidal with stepwise changes in the frequency. The solid line is a plot of $2\cos(\omega(n))$, where $\omega(n)$ is the input frequency. Because a soft limiter is used, the notch angular frequency remains within the $\pm 2\%$ range, which for $\omega_n = \pi/2$ corresponds to $-0.0628 < \alpha(n) < 0.0628$.

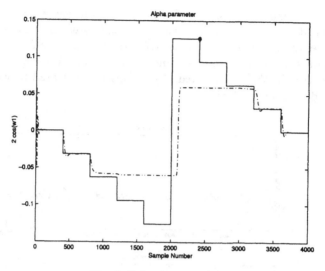

Fig. 3. Behavior of $\alpha_s(n)$.

Next we compare the error in $\alpha(n)$ for soft and hard limiting. The input is a sinusoid which is contaminated by zero-mean Laplacian noise. The same input signal is given to both soft and hard limited notch filters with otherwise the same parameters, and the accumulated squared errors are computed in steady-state operation. The input frequency and the standard deviation of the Laplacian noise are the varying parameters against which the results are plotted in Fig. 4. Negative values of the function indicate smaller squared error in the soft limited case.

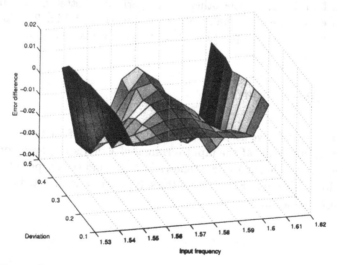

Fig. 4. Comparison of **hard and soft** limiting applied to $\alpha(n)$.

As can be expected, soft limiting has the advantage when the input frequency deviates from the nominal central value, because in such a case, strong disturbances are more likely to occasionally try to move the frequency parameter $\alpha(n)$ outside of the permitted range, and the amplitude of a disturbance will be smaller in the soft limited case. This is illustrated in Fig. 5 where frequency deviation is $+1.8\%$.

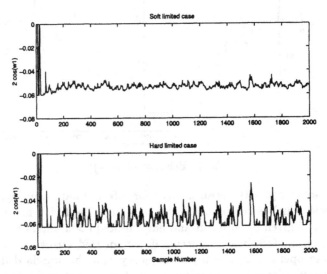

Fig. 5. Comparison of hard and soft limiting applied to $\alpha(n)$.

Hard limiting, on the other hand, performs better at the band edges, which are not easily reachable with the function of Fig. 2. No limiting is needed for low noise amplitudes where the surface is nearly flat.

4. A Test Case

As a test case, we have used an approximative replica of the test signal observed by Weidenbrüg et al. at the input terminal of a thyristor power converter [15]. This signal, sampled at 1 kHz, is shown in Fig. 6 and the output of the adaptive notch filter is shown in Fig. 7. Here the soft limiter was adjusted to allow $\pm10\%$ frequency variation.

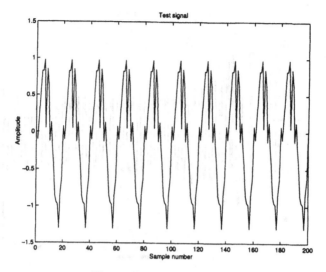

Fig. 6. Test input signal.

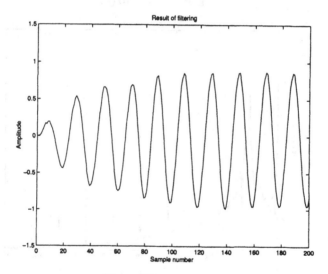

Fig. 7. Filter output.

In this case, the input frequency is nominally equal to 0.1π, which corresponds to $\alpha = 1.90$. A trace of $\alpha_s(n)$ in this experiment is shown in Fig. 8. Finally, the frequency spectrum of the output signal is shown in Fig. 9, calculated using the Fast Fourier Transform (FFT).

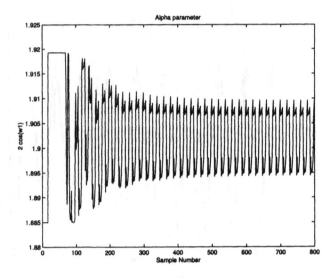

Fig. 8. Trace of $\alpha_s(n)$.

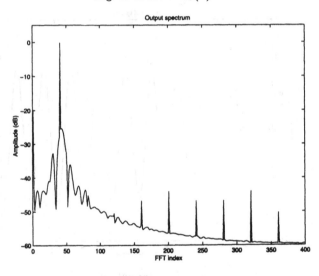

Fig. 9. Spectrum of the filter output.

5. Conclusions

The adaptive notch filter has certain advantages in line frequency signal processing compared to transversal adaptive filters. Since the actual notch frequency can be easily extracted in the algorithm, soft or hard limiting on the parameter can be applied. In this work, the notch filter was considered for smoothing of 50/60 Hz line frequency signals in industrial and power electronics, where the noise is typically impulsive by character.

References

[1] D. V. B. Rao and S. Y. Kung, "Adaptive notch filtering for the retrieval of sinusoids in noise," *IEEE Trans. Acoust., Speech, Signal Processing*, vol. ASSP-32, pp. 791-802, Aug. 1984.

[2] A. Nehorai, "A minimal parameter adaptive notch filter with constrained poles and zeros," *IEEE Trans. Acoust., Speech, Signal Processing*, vol. ASSP-33, pp. 983-996, Aug. 1985.

[3] P. Stoica and A. Nehorai, "Performance analysis of an adaptive notch filter with constrained poles and zeros," *IEEE Trans. Acoust., Speech, Signal Processing*, vol. 36, pp. 911-919, June 1988.

[4] B. Rao and R. Peng, "Tracking characteristics of the constrained IIR adaptive notch filter," *IEEE Trans. Acoust., Speech, Signal Processing*, vol. 36, pp. 1466-1479, Sept. 1988.

[5] J. M. Travassos-Romano and M. Bellanger, "Fast least squares adaptive notch filtering," *IEEE Trans. Acoust., Speech, Signal Processing*, vol. 36, pp. 1536-1540, Sept. 1988.

[6] P. Tichavsky and P. Händel, "Two algorithms for adaptive retrieval of slowly time-varying multiple cisoids in noise," *IEEE Trans. Signal Processing*, vol. 43, pp. 1116-1127, May 1995.

[7] P. Strobach, "Single section least squares adaptive notch filter," *IEEE Trans. Signal Processing*, vol. 43, pp. 2007-2010, Aug. 1995.

[8] A. V. Oppenheim and R. W. Schafer, *Discrete-Time Signal Processing*. Englewood Cliffs, NJ: Prentice-Hall, 1989.

[9] O. Vainio and S.J. Ovaska, "Digital filtering for robust 50/60 Hz zero-crossing detectors," *IEEE Trans. Instrum. Meas.*, vol. 45, no. 2, pp. 426-430, Apr. 1996.

[10] O. Vainio, "A synchronous digital filter for 50 Hz line frequency signal processing," in *Proc. the 23rd Int. Conf. on Industrial Electronics, Control, and Instrumentation*, New Orleans, LA, Nov. 1997, pp. 228-233.

[11] F. P. Dawson and L. Klaffke, "Frequency adaptive digital filter for synchronization signal aquisition and synchronized event riggering," in *Proc. IEEE Power Electron. Spec. Conf.*, Fukuoka, Japan, May 1998, pp. 342-347.

[12] H. Akagi, "New trends in active filters for power conditioning," *IEEE Trans. Ind. Appl.*, vol. 32, pp. 1312-1322, Nov./Dec. 1996.

[13] I. Kamwa and R. Grondin, "Fast adaptive schemes for tracking voltage phasor and local frequency in power transmission and distribution systems," *IEEE Trans. Power Delivery*, vol. 7, pp. 789-795, April 1992.

[14] B. C. Arnold, N. Balakrishnan, and H. N. Nagaraja, *A First Course in Order Statistics*. New York, NY: John Wiley & Sons, 1992.

[15] R. Weidenbrüg, F. P. Dawson, and R. Bonert, "New synchronization method for thyristor power converters to weak AC-systems," *IEEE Trans. Ind. Electron.*, vol. 40, pp. 505-511, Oct. 1993.

[16] O. Vainio and S.J. Ovaska, "Multistage adaptive filters for in-phase processing of line-frequency signals," *IEEE Trans. Ind. Electron.*, vol. 44, no. 2, pp. 258-264, April 1997.

An Algorithm for Induction of Possibilistic Set-Valued Rules by Finding Prime Disjunctions

Alexandr A. Savinov

GMD — German National Research Center for Information Technology
Schloss Birlinghoven, Sankt-Augustin, D-53754 Germany
E-mail: savinov@gmd.de, http://ais.gmd.de/~savinov/

Abstract. We present a new algorithm, called Optimist, which generates possibilistic set-valued rules from tables containing categorical attributes taking a finite number of values. An example of such a rule might be "IF HOUSEHOLDSIZE={Two OR Tree} AND OCCUPATION={Professional OR Clerical} THEN PAYMENT_METHOD={CashCheck (Max=249) OR DebitCard (Max=175)}. The algorithm is based on an original formal framework generalising the conventional boolean approach in two directions: (i) finite-valued variables and (ii) continuos-valued semantics. Using this formalism we approximate the multidimensional distribution induced from data by a number of possibilistic prime disjunctions (patterns) representing the widest intervals of impossible combinations of values. The Optimist algorithm described in the paper generates the most interesting prime disjunctions for one pass through the data set by means of transformation from the DNF representing data into the possibilistic CNF representing knowledge. It consists of generation, absorption and filtration parts. The set-valued rules built from the possibilistic patterns are optimal in the sense that they have the most general condition and the most specific conclusion. For the case of finite-valued attributes and two-valued semantics the algorithm is implemented in the Chelovek rule induction system for Windows 95.

1. Introduction

The field of knowledge discovery in databases or data mining has been paid a lot of attention during recent years as large organisations has accumulated huge databases and begun to realise the potential value of the information that is stored their. One specific data mining task is the mining of dependencies in the form of rules. The task is to determine hidden patterns that characterise the problem domain behaviour from a large database of previous records and then to represent them in the form of rules. The rules can then be used either for description or for prediction purposes.

The problem of rule induction can be stated as follows: given a database consisting of a number of records, where each record is a sequence of attribute values; find rules which by their conditions select wide intervals in the multidimensional space where the distribution of values in conclusion is highly inhomogeneous, i.e., contains large quantity of information. In the case where variables in condition and conclusion of such rules may take only one value we obtain well known association rules [1-3], e.g.:

IF $x_1 = a_{13}$ AND $x_2 = a_{25}$ THEN $x_3 = a_{32}$ (Support $= s$, Confidence $= c$)

If the variables are allowed to take as a value any subset of the domain then we obtain so called set-valued rules having the form:

IF $x_1 = \{a_{13}, a_{14}\}$ AND $x_2 = \{a_{21}, a_{27}\}$ THEN $x_3 = \{a_{33}, a_{36}\}$

Here a_{ij} are values of the i-th variable. Each variable in such a rule may take any value from the corresponding subset, e.g., x_1 may be equal to either a_{13} or a_{14}.

In the paper we consider the problem of mining set-valued rules where all variables have a finite number of values, the universe of discourse is equal to the Cartesian product of all sets of the values, and the semantics is represented by a frequency distribution over the universe of discourse (the number of observations that belong to each point). The condition of such rules selects a subspace in the form of conjunctive interval within the universe of discourse. Then the conclusion consolidates all information about the problem domain from this subspace by projecting the semantics restricted within this interval onto the goal variable. The problem is that the number of all possible conjunctive intervals of the multidimensional space is extremely large and it is necessary to have some criterion of interestingness for the rules. One obvious criterion is that the more general the rule condition is, i.e., the larger interval it selects is, the more interesting the rule is. However, if we take too general interval and use it for the rule condition, it may well happen that the rule will not be interesting since the conclusion is not surprising, i.e., it does not contain much information. For example, the rule

IF $x_1 = \{a_{13}, a_{14}\}$

THEN $x_3 = \{a_{31}$ (Max $= 151$), a_{32} (Max $= 152$), a_{33} (Max $= 153$)$\}$

is not interesting since it says that under these conditions the variable x_3 takes any of its 3 values with approximately the same possibility (almost constant possibility distribution in conclusion does not bear much information and hence the conclusion does not impose real constraints on the values of the goal variable). On the other hand, the rule

IF $x_1 = \{a_{13}, a_{14}\}$

THEN $x_3 = \{a_{31}$ (Max $= 151$), a_{32} (Max $= 0$), a_{33} (Max $= 49$)$\}$

is much more interesting (informative) since it says that, contrary to our expectations, the value a_{32} is absolutely impossible within this interval while the value a_{33} has much less possibility than a_{31}. Thus informally, the more general the rule condition is (the wider the interval selected by the condition) and the more specific the conclusion is (the closer the conclusion is to the singular form) the more interesting the rule is. Thus for assessing rules we proceed from the criterion of informative interestingness rather than from their classification power in relation to some target attribute. In particular, we do not impose constraints onto the choice of the target attribute or the form of the conclusion — the main criterion is the quantity of information (or generally the interestingness) rather than the form of representation.

To find such maximally general in condition and specific in conclusion rules we use an approach according to which any multidimensional possibility distribution can be formally represented by a set of special logical constructions called possibilistic disjunctions. Each such disjunction is made up of several one-dimensional possibility distributions (propositions in logical terms) over the values of individual variables combined with the operation OR (interpreted as maximum) and represents some distribution over the universe of discourse (the disjunction semantics). One disjunction cannot represent any multidimensional possibility distribution but it can represent it within some interval of this space where it sets maximal possible values the whole distribution may take. With the help of a set of disjunctions combined with the operation AND (interpreted as minimum) we can represent any multidimensional distribution by approximating it separately within various intervals (generalisation of CNF). Thus any one disjunction sets an upper bound for the distribution in some subspace while the overall semantics is equal to the sum of these constraints, i.e., the overall distribution is pressed down by individual disjunctions (Fig. 1).

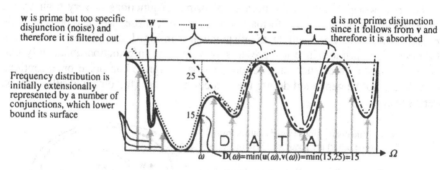

Fig. 1. The data set semantics is approximated by a number of disjunctions which upper bound the multidimensional distribution induced from data.

The distribution itself is initially represented by singular conjunctions combined with the connective OR and lower bounding the distribution surface (generalisation of DNF). Each such conjunction corresponds to one record (one combination of attribute values) along with the number of its occurrences. This extensional representation can be intensionalised by merging singular conjunctions into more general ones, which lower bound the distribution within wide intervals instead of only one point. Conceptually this approach is applied in many existing rule induction algorithms. Our algorithm is based on the dual approach where the initial representation is transformed into its dual intensional form, i.e., the set of singular conjunctions is transformed into the set of general disjunctions (Fig. 1).

Obviously, only the disjunctions, which follow from the initial distribution, can be used to represent it. For example, disjunctions **d, u, v, w** follow from the multidimensional distribution shown in Fig. 1 as bold line. Although all these disjunctions can be used to represent the semantics and to form rules, some of them are not interesting since they impose too weak constraints, which are weaker than those imposed by some other disjunctions. For example, it makes no sense to use

638

the disjunction **d** for representation since it follows from **v** (Fig. 1,2). Moreover, if we have **v** then we can always build from it all weaker disjunctions including **d**. Thus we come to a very important notion of prime disjunction: the disjunction is said to be prime if it does not follow from any other disjunction (except for itself). Note that prime disjunctions are always defined in relation to some distribution for which they are the strongest.

The key point of the proposed rule induction algorithm is that we generate possibilistic prime disjunctions for the semantics represented by a set of records and then transform these disjunctions into the form of rules. Since prime disjunctions cannot be strengthened (they by definition impose the strongest constraints) both our criteria — maximal generality of condition and maximal specificity of conclusion — are reached when building rules. The prime disjunctions themselves are also interesting since they can be interpreted as negative associations, i.e., they intensionally represent the intervals in the universe of discourse where combinations of values have low degree of possibility (intervals of incompatible attribute values).

For real world problems the number of prime disjunctions is very high so we have to find only the most informative of them, i.e., those, which represent wide negative intervals. For example, the disjunction **w** in Fig. 1 and 2 is not informative (although it is prime) since it represents very specific information about only one point. In many cases such disjunctions can be interpreted as noise or exceptions and should not be generated during the induction process (the obtained rules will be very specific).

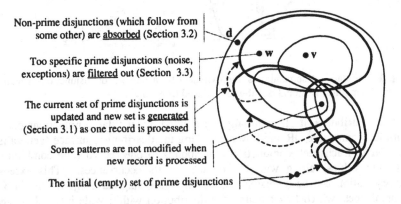

Non-prime disjunctions (which follow from some other) are <u>absorbed</u> (Section 3.2)

Too specific prime disjunctions (noise, exceptions) are <u>filtered</u> out (Section 3.3)

The current set of prime disjunctions is updated and new set is <u>generated</u> (Section 3.1) as one record is processed

Some patterns are not modified when new record is processed

The initial (empty) set of prime disjunctions

Fig. 2. Search in the space of all disjunctions.

In [4] an approach to mining set-valued rules based on a generalised covering technique was proposed. However it has one big disadvantage that the whole data set has to be in memory to reflect the state of the cover. In this paper we propose another algorithm, called Optimist, which builds all prime disjunctions simultaneously while processing all records in succession. Formally, it uses an explicit transformation of possibilistic DNF representing data into CNF consisting of prime disjunctions (knowledge). An advantage is that the set of prime disjunctions is built for one pass through the data set. The algorithm assumes that at each moment the semantics is equal to the number of processed records and the current set of prime disjunctions is updated each time new record is processed (Fig. 2). At the very beginning there are no prime disjunctions since no conjunctions are processed. After the first step the set of prime disjunctions is semantically equivalent to the first conjunction which has been processed. Then the set of prime disjunctions is equivalent to two data elements and so on. Thus at each step all current prime disjunctions are updated so that they include the semantics of one more data element. To keep the set of prime disjunctions tractable the most specific of them are removed since they are not interesting and cannot be generalised (the process goes from general to specific disjunctions).

Prime disjunctions can be regarded as a sort of patterns representing significant multidimensional regularities characteristic of the possibility distribution. Once the set of interesting patterns has been found they can be used to form rules by inverting the propositions about variables which should be in condition part. The only problem here is that we obtain possibilistic conditions, which are not very comprehensible and should be transformed into the conventional crisp form. As a result we might obtain the following rule:

$$\text{IF } x_1 = \{a_{12}, a_{15}\} \text{ THEN } x_3 = \{a_{31} \, (\text{Max} = 151), a_{33} \, (\text{Max} = 49)\}$$

where (as well as in the whole paper) absent values are supposed to have the degree of possibility 0 and the weight Max=151 means that within this condition interval the corresponding value occurs maximally 151 times. Note that if we built rules in the form of prime conjunctions then the weight would be interpreted dually as the necessity degree, e.g., Min=37 (and the condition intervals would be different). Since this rule is built from prime disjunction it is guaranteed that the condition cannot be generalised (strengthened) without generalising the conclusion and vice versa, the conclusion cannot be made more specific since it is an exact possibilistic projection of the distribution from the condition interval onto the goal variable. Once such rules have been found it is easy to fill them with other semantics, e.g., the sum of observations. Then the rule might look like

$$\text{IF } x_1 = \{a_{12}, a_{15}\}$$

$$\text{THEN } x_3 = \{a_{31} \, (\text{Max} = 151, \text{Sum} = 753), a_{33} \, (\text{Max} = 49, \text{Sum} = 238)\}$$

The notion of prime disjunction (conjunction, implicant etc.) and algorithms for finding them has received a lot of attention in various fields, especially in classical cybernetics (generation of prime implicants [5]) and combinatorics. In particular, the notion of prime implicant has been successfully used for efficient mining association rules [6,7] by finding maximum frequent itemsets. In this paper, however, we use this notion in an original possibilistic form generalising the conventional boolean analogue onto the case of (i) finite-valued variables (instead of only two values 0 and 1), and (ii) continuos-valued semantics (instead of only true and false).

For representing data and knowledge we use a so called method of sectioned vectors and matrices originating from the paper [8] and later generalised onto fuzzy case [9]. The idea of transformation from fuzzy DNF into fuzzy CNF and finding rules was proposed in [10]. A fuzzy version of this rule induction algorithm, which is based on the covering method, was described in [4].

2. Data and Knowledge Representation

Let some problem domain at the syntactic level be described by a finite number of *variables* or *attributes* $x_1, x_2, ..., x_n$ each of which takes a finite number of *values* and corresponds to one column of data table:

$$x_i \in A_i = \{a_{i1}, a_{i2}, .., a_{in_i}\}, \quad i = 1, 2, .., n$$

where n_i is the number of values of the i-th variable and A_i is its set of values. The *state space* or the *universe of discourse* is defined as the Cartesian product of all sets of the values: $\Omega = A_1 \times A_2 \times .. \times A_n$. The universe of discourse is a finite set with the structure of a multidimensional space. Each syntactic object (state) from the universe of discourse is represented by a combination of values of all variables: $\omega = \langle x_1, x_2, ..., x_n \rangle \in \Omega$. The number of such objects is equal to the power of the universe of discourse: $|\Omega| = n_1 \times n_2 \times .. \times n_n$.

Formally the problem domain *semantics* is represented by a frequency distribution over the state space, i.e., a natural number or 0 (the number of observations) is assigned to each combination of attribute values (syntactic object). The frequency 0 is interpreted as the absolute impossibility of the corresponding object while all positive numbers are interpreted as various degrees of possibility. If we map this distribution into the continuous interval [0,1] then 1 is interpreted as an uncertainty, complete possibility. The absence of information means that the distribution is equal to 1 in any point of the universe of discourse, while the more information we have the lower the distribution is, i.e., some points are impossible (disabled, prohibited). Thus the possibilistic interpretation has negative meaning. Yet for the problem of rule induction it is simpler to work directly with frequencies interpreted as possibilities so we will not use the mapping into [0,1]. Note that there is also the dual (positive) interpretation where 0 is absolute uncertainty while positive frequencies represent a degree of necessity.

We will use a special technique of sectioned vectors and matrixes to represent the semantics. Each construction of this mechanism along with interpretation rules imposes constraints of certain form on possible combinations of attribute values. Depending on the logical connective used in combining these constructions they bound either minimal or maximal values. The sectioned constructions are written in bold font with the two lower indexes corresponding to the number of variable and to the number of value, respectively.

The *component* $\mathbf{u}_{ij}$ of the sectioned vector $\mathbf{u}$ is a natural number assigned to j-th value of i-th variable. The *section* $\mathbf{u}_i$ of the sectioned vector $\mathbf{u}$ is an ordered sequence of n_i components assigned to i-th variable and representing some distribution over all values of one variable. For example, $\mathbf{u}_i = \{7, 0, 83\}$ means that three values of the i-th variable have frequencies 7, 0, and 83. The sectioned *vector* $\mathbf{u}$ is an ordered sequence of n sections for all variables. Thus the total number of components in sectioned vector is equal to $n_1 + n_2 + .. + n_n$. For example, the constructions 01.567.0090 and $\{0,1\}.\{5,6,7\}.\{0,0,9,0\}$ represent the same sectioned vector written in different ways (with sections separated by dots). The sectioned *matrix* consists of a number of sectioned vectors written as its lines.

Each section represents an elementary proposition about the corresponding variable by assigning degrees of possibility to its values while the whole vector can be interpreted either as conjunction or as disjunction. If the sectioned vector **k** is interpreted as conjunction then it defines the distribution, which is equal to the minimum of the vector components corresponding to the point coordinates:

$$\mathbf{k}(\omega) = \mathbf{k}(\langle x_1, x_2, .., x_n \rangle) = \mathbf{k}_1(x_1) \wedge \mathbf{k}_2(x_2) \wedge .. \wedge \mathbf{k}_n(x_n) = \min_{i=1,..,n} \mathbf{k}_i(x_i)$$

(The minimum is taken among n components — one from each section.) If the sectioned vector **d** is interpreted as disjunction then it defines the distribution, which is equal to the maximum of the vector components corresponding to the point coordinates:

$$\mathbf{d}(\omega) = \mathbf{d}(\langle x_1, x_2, .., x_n \rangle) = \mathbf{d}_1(x_1) \vee \mathbf{d}_2(x_2) \vee .. \vee \mathbf{d}_n(x_n) = \max_{i=1,..,n} \mathbf{d}_i(x_i)$$

Sectioned matrixes are analogues of the conventional DNF and CNF depending on their interpretation. If the matrix **K** is interpreted as DNF then its sectioned vector-lines are combined with the connective $\vee$ and interpreted as conjunctions (disjunction of conjunctions). In the dual way, if the matrix **D** is interpreted as CNF then its sectioned vector-lines are combined with the connective $\wedge$ and interpreted as disjunctions (conjunction of disjunctions). Thus DNF/CNF defines the distribution which is equal to the maximum/minimum of the distributions represented by its lines.

The data can be easily represented in the form of DNF so that each conjunction represents one record along with the number of its occurrences in the data set. The conjunction corresponding to one record consists of all 0's except for one component in each section, which is equal to the number of record occurrences. For example, conjunction $\{5,0\} \wedge \{5,0,0\} \wedge \{5,0,0,0\}$ or shortly $50.500.5000$ represents the distribution value in the point $\omega = \langle a_{11}, a_{21}, a_{31} \rangle$ where it equals 5.

One distribution is said to be a *consequence* of another if it *covers* the second distribution, i.e., its values in all points of the universe of discourse are greater than or equal to the values of the second distribution. The consequence relation for conjunctions, disjunctions, DNF, CNF and other representation constructions is defined as the consequence relation for the corresponding distributions.

The operation of *elementary induction* consists in increasing one component of a disjunction so that it becomes weaker. Since one component corresponds to one dimension in the space of all vectors this operation can be used to search through this space by moving along separate dimensions, e.g., to find some interesting patterns.

By dependency we mean any information restricted by a *simple* form of representation. In our approach we consider dependencies in the form of possibilistic disjunctions which are thought of as *patterns* representing the distribution regularities. The disjunction is referred to as *prime* if it is a consequence of the distribution but is not a consequence of any other disjunction except for itself. Thus prime disjunctions are the strongest among those which follow from the distribution. In particular, any other disjunction can be obtained from a prime one by weakening (increasing) its components. On the other hand, if any component of prime disjunction is decreased then it already is not a consequence of the distribution. Thus formally the problem of finding dependencies in our case is reduced to the problem of generating possibilistic prime disjunctions given a distribution represented extensionally by a number of conjunctions.

3. Search for Possibilistic Patterns

3.1. Generation of Disjunctions

To add the conjunction $\mathbf{k}$ to the matrix of CNF $\mathbf{D}$ it is necessary to add it to all m disjunctions of the matrix:

$$\mathbf{k} \vee \mathbf{D} = \mathbf{k} \vee (\mathbf{d}^1 \wedge \mathbf{d}^2 \wedge .. \wedge \mathbf{d}^m) = (\mathbf{k} \vee \mathbf{d}^1) \wedge (\mathbf{k} \vee \mathbf{d}^2) \wedge .. \wedge (\mathbf{k} \vee \mathbf{d}^m)$$

Addition of conjunction to disjunction is carried out by the formula:

$$\mathbf{k} \vee \mathbf{d} = (\mathbf{k}_1 \vee \mathbf{d}) \wedge (\mathbf{k}_2 \vee \mathbf{d}) \wedge ... \wedge (\mathbf{k}_n \vee \mathbf{d}) =$$

$$(\mathbf{k}_1 \vee (\mathbf{d}_1 \vee \mathbf{d}_2 \vee .. \vee \mathbf{d}_n)) \wedge$$

$$(\mathbf{k}_2 \vee (\mathbf{d}_1 \vee \mathbf{d}_2 \vee .. \vee \mathbf{d}_n)) \wedge$$

$$(\mathbf{k}_n \vee (\mathbf{d}_1 \vee \mathbf{d}_2 \vee .. \vee \mathbf{d}_n)) =$$

$$(\mathbf{k}_1 \vee \mathbf{d}_1 \quad \vee \quad \mathbf{d}_2 \qquad \vee .. \vee \quad \mathbf{d}_n) \wedge$$

$$(\mathbf{d}_1 \qquad \vee \quad \mathbf{k}_2 \vee \mathbf{d}_2 \quad \vee .. \vee \quad \mathbf{d}_n) \wedge$$

$$(\mathbf{d}_1 \qquad \vee \quad \mathbf{d}_2 \qquad \vee .. \vee \quad \mathbf{k}_n \vee \mathbf{d}_n) \wedge$$

and in general case n new disjunctions are generated from one source disjunction. Each new disjunction is generated from the source one by applying the elementary induction, i.e., by increasing one component.

If $\mathbf{k}$ is covered by $\mathbf{d}$ then its addition to $\mathbf{d}$ does not change the semantics: $\mathbf{k} \vee \mathbf{d} = \mathbf{d}$. In this case the disjunction can be simply copied to the new matrix with no modifications. Thus the whole set of new disjunctions can be divided into two subsets: modified and non-modified.

For example, let us suppose that we have two conjunctions $\mathbf{k}^1 = 05.005.0005$ and $\mathbf{k}^2 = 03.003.0030$ which have to be transformed into disjunctions. Each new matrix is obtained from the previous one as follows: $\mathbf{D}^i = \mathbf{D}^{i-1} \vee \mathbf{k}^i$, where $i = 1,2$, and $\mathbf{D}^0 = \mathbf{d}^0 = 00.000.0000$. Thus after processing the first conjunction we obtain:

$$\mathbf{D}^1 = \mathbf{D}^0 \vee \mathbf{k}^1 = 00.000.0000 \vee 05.005.0005 = \begin{array}{|l|l}
\underline{05}.000.0000 & \mathbf{1} \\
00.\underline{005}.0000 & \mathbf{2} \\
00.000.\underline{0005} & \mathbf{3}
\end{array}$$

where increased components (to which elementary induction has been applied) are underlined and three resulted disjunctions are denoted with bold numbers. After processing the second conjunction we obtain:

$$
\mathbf{D}^2 = \mathbf{D}^1 \vee \mathbf{k}^2 = \begin{vmatrix} 05.000.0000 \\ 00.005.0000 \\ 00.000.0005 \end{vmatrix} \vee 03.003.0030 = \begin{array}{l|l} 05.000.0000 & 1 \\ 00.005.0000 & 2 \\ 0\underline{3}.000.0005 & 3.1 \\ 00.00\underline{3}.0005 & 3.2 \\ 00.000.00\underline{35} & 3.3 \end{array}
$$

where for convenience we separate the lines produced by different parents. Note that the disjunctions **1** and **2** are not modified since they cover $\mathbf{k}^2$ and only three new disjunctions **3.1, 3.2,** and **3.3** have been generated from their parent **3**.

3.2. Absorption of Disjunctions

As new disjunctions are generated and added to the new matrix the absorption procedure should be carried out to remove weak lines, i.e., the lines which are not prime and follow from others. In general, each new disjunction can either be absorbed itself or absorb other lines. Thus the comparison of lines has to be fulfilled in both directions. To check for the consequence relation between two disjunctions we have to reduce them (see [9] for more information about reduced forms) and then compare all their components.

For example, if we add new conjunction $\mathbf{k}^3 = 05.005.0500$ to the matrix $\mathbf{D}^2$ (section 3.1) then we obtain

$$
\mathbf{D}^3 = \mathbf{D}^2 \vee \mathbf{k}^3 = \begin{vmatrix} 05.000.0000 \\ 00.005.0000 \\ 03.000.0005 \\ 00.003.0005 \\ 00.000.0035 \end{vmatrix} \vee 05.005.0500 = \begin{array}{l|l} 05.000.0000 & 1 \\ 00.005.0000 & 2 \\ 0\underline{5}.000.0005 & 3.1.1 \supseteq 1 \\ 03.00\underline{5}.0005 & 3.1.2 \supseteq 2 \\ 03.000.0\underline{5}05 & 3.1.3 \\ 0\underline{5}.003.0005 & 3.2.1 \supseteq 1 \\ 00.00\underline{5}.0005 & 3.2.2 \supseteq 2 \\ 00.003.0\underline{5}05 & 3.2.3 \\ 0\underline{5}.000.0035 & 3.3.1 \supseteq 1 \\ 00.00\underline{5}.0035 & 3.3.2 \supseteq 2 \\ 00.000.0\underline{5}35 & 3.3.3 \end{array}
$$

where 6 lines are absorbed and therefore the final matrix is:

$$
\mathbf{D}^3 = \begin{array}{l|l} 05.000.0000 & 1 \\ 00.005.0000 & 2 \\ 03.000.0505 & 3.1.3 \\ 00.003.0505 & 3.2.3 \\ 00.000.0535 & 3.3.3 \end{array}
$$

Several properties, which are formulated below, significantly simplify the absorption process.

Property 1. The disjunctions, which cover the current conjunction and hence are not modified, cannot be absorbed by any other disjunction.

This property follows from the fact that the matrix of disjunctions is always maintained in the state where it contains only prime disjunctions, which do not absorb each other. Hence if some disjunction has not been absorbed earlier then without modifications it will not be absorbed in new matrix as well since this new matrix contains only the same or weaker disjunctions.

Let us suppose that $\mathbf{u}$ is non-modified disjunction while $\mathbf{v}$ was modified on the component $\mathbf{v}_{rs}$, and $\mathbf{v}'_{rs}$ is old value of the modified component ($\mathbf{u}_{ij} = \mathbf{u}'_{ij}$ since $\mathbf{u}$ was not modified). Then the following property takes place.

Property 2. If $\mathbf{u}_{rs} \leq \mathbf{v}'_{rs}$ then $\mathbf{v}$ does not follow from $\mathbf{u}$. (This property is valid only if the constant of $\mathbf{v}$ has not been changed. More about constants and reduced forms read in [9].)

To use this property each line has to store information on the old value $\mathbf{v}'_{rs}$ of modified component and its number (r and s). There are analogous properties for comparing two modified disjunctions but they are not formulated here since they are a little more complicated and have several different cases. These properties are valuable since frequently they allow us to say that one line is not consequence of another by comparing only one pair of components.

Property 3. If the sum of components in $\mathbf{v}$ or in any of its sections $\mathbf{v}_i$ is less then the corresponding sum in the disjunction $\mathbf{u}$ then $\mathbf{v}$ does not follow from $\mathbf{u}$.

To use this property we have to maintain the sums of the vector and section components in the corresponding headers. If all these necessary conditions are satisfied then we have to carry out a component-wise comparison of two vectors in the loop consisting of $n_1 + n_2 + .. + n_n$ steps.

3.3. Filtration of Disjunctions

In spite of using various methods to increase the performance of the transformation from DNF into CNF, it remains too computationally difficult for real world problems. However for the task of rule induction it is not necessary to carry out this transformation in complete form since usually it is required to find only the strongest dependencies among the attributes. Therefore the algorithm has been modified so that the number of lines in the matrix of prime disjunctions is limited by a special parameter while the lines themselves are ordered by a criterion of interestingness. Thus only a limited number of the most interesting disjunctions are stored and processed whereas those disjunctions, which according to their criterion do not go into it, are removed.

The procedure is organised as follows. Before a new disjunction is to be generated we calculate its criterion value (the degree of interestingness), which is compared with that of the last line of the matrix. If the new disjunction does not go into the matrix, it is simply not generated. Otherwise, if it is interesting enough, it is first generated, then checked for absorption, and finally inserted into the corresponding position in the matrix (the last line is removed).

There may be different criteria of pattern ordering determining the induction process direction. The Optimist algorithm uses the criterion of interestingness in the form of the impossibility interval size. Informally, the more points of the distribution have smaller values, the more general and strong the corresponding disjunction is.. Formally the following formula is used to calculate this parameter:

$$H = \frac{1}{n_1} \sum_{j=1}^{n_1.} \mathbf{d}_{1j} + \frac{1}{n_2} \sum_{j=1}^{n_2.} \mathbf{d}_{2j} + .. + \frac{1}{n_n} \sum_{j=1}^{n_n.} \mathbf{d}_{nj}$$

according to which H is equal to the weighted sum of components, and the less this value the stronger the disjunction. In particular, in boolean case changing one component from 0 to 1 in two-valued section is equivalent to changing three components from 0 to 1 in six-valued section. For example, in the matrix $\mathbf{D}^3$ (section 3.2) two disjunctions **3.1.3** and **3.2.3** can be transformed into rules (three other disjunctions are degenerated and represent the distribution projection on individual variables). However, the second of them is more interesting (informative) since it has larger interval of impossibility:

$$H(\mathbf{3.1.3}) = \tfrac{1}{2}3 + \tfrac{1}{3}0 + \tfrac{1}{4}10 = 4 > 3.5 = \tfrac{1}{2}0 + \tfrac{1}{3}3 + \tfrac{1}{4}10 = H(\mathbf{3.2.3})$$

Generally, each attribute or even each attribute value may have their own user-defined weights, which reflect their informative importance or subjective interestingness for the user. This mechanism provides the capability of more flexible control over the process of rule induction. There may be also other mechanisms of filtration. For example, restricting the number of non-0 components for the target attribute allows us to find only the rules with one value in conclusion and set-valued conditions.

4. Transformation of Disjunctions into Rules

The transformation from DNF into CNF is the most difficult part of the algorithm, and once the prime disjunctions have been generated they can be used to build the possibilistic set-valued rules. Formally, rules are obtained in the conventional way by negating the propositions (sections) which should be in the condition and thus obtaining an implication, for example:

$$\mathbf{d}_1 \vee \mathbf{d}_3 \vee \mathbf{d}_5 \quad \Leftrightarrow \quad \overline{\mathbf{d}}_1 \wedge \overline{\mathbf{d}}_3 \rightarrow \mathbf{d}_5$$

The sections, which consist of all 0's, should not be considered since they are equivalent to the absence of proposition about the corresponding attribute. The only problem here is that it is more desirable to have crisp conditions instead of possibilistic ones resulted from the negation. Thus we need a mechanism for a meaningful removing uncertainty from the negated sections.

Let us suppose that $\mathbf{d}_{min} = \max_i \min_j \mathbf{d}_{ij}$, and $\mathbf{d}_{max} = \max_{i,j} \mathbf{d}_{ij}$ are minimal and maximal components of the disjunction, respectively. The maximal component is the same for all disjunctions (1 when mapped into the [0,1] interval). The most straightforward way is to negate the section $\mathbf{d}_i$ as follows:

$$d_{ij} = \begin{cases} d_{max}, & \text{if } d_{ij} \le d_{min} \\ d_{min}, & \text{otherwise} \end{cases}$$

i.e., all components, which are greater than the disjunction constant d_{min}, are mapped into d_{min} while only the components, which are less than or equal to d_{min}, are included into the condition. For example, the disjunction $d = \{0,1\} \vee \{0,6,0\} \vee \{0,2,9,5\}$ with $d_{min} = 0$ and $d_{max} = 9$ can be transformed into the implication $\{9,0\} \wedge \{9,0,9\} \to \{0,2,9,5\}$, which is interpreted as the possibilistic rule

IF $x_1 = \{a_{11}\}$ AND $x_2 = \{a_{21}, a_{23}\}$

THEN $x_3 = \{a_{31} : 0, a_{32} : 2, a_{33} : 9, a_{34} : 5\}$

where weights in the conclusion are interpreted in possibilistic sense, i.e., they define maximal possible values of the distribution within this interval (since we used prime disjunction it is guaranteed that these values cannot be decreased).

Generally, instead of d_{min} we can use any user-defined or pattern-dependent (automatically calculated) threshold. In the above example, we see that the pattern d involves a weak section d_1, i.e., the section with small maximal component (very close to trivial). In such cases we can obtain more simple rules by adding to the disjunction the maximal component of the weak section so that the weak section becomes constant. Then we apply the negation to desired sections of this weaker disjunction and in our case obtain the implication $\{9,1,9\} \to \{1,2,9,5\}$ which is interpreted as the rule

IF $x_2 = \{a_{21}, a_{23}\}$ THEN $x_3 = \{a_{31} : 1, a_{32} : 2, a_{33} : 9, a_{34} : 5\}$

Note that this rule is more general on the condition but less informative (weaker) on the conclusion. It is important that when generating rules in such a way we always loose some information and it is why, in particular, different patterns may produce the same rules. However, the optimality of the rules in the sense that the conclusion cannot be strengthened without weakening the condition remains (i.e., maximal frequencies in conclusion are exact).

The rules can be easily filled in with statistical information in the form of the sum of occurrences within the rule condition interval (for one additional pass through the data set). Then we might obtain the rule like the following

IF $x_2 = \{a_{21}, a_{23}\}$

THEN $x_3 = \{a_{31} (\text{Sum} = 2, \text{Max} = 1), a_{32} (\text{Sum} = 4, \text{Max} = 2), ...\}$

The Optimist algorithm for the case of multi-valued attributes and two-valued semantics has been implemented in the Chelovek for Windows 95 rule induction system[*] (Fig. 3). Since the number of patterns is limited (by a special parameter) the procession time is linear to the table length. For several hundreds of patterns the procession time per record is rather low (less than the overheads such as loading the record and parsing the attribute values). When the number of patterns exceeds 1000 the combinatorial part becomes significant. We also noticed that the algorithm works much better with ordered data.

---

[*] Developed at the Institute of Mathematics & Informatics, Moldavian Academy of Sciences.

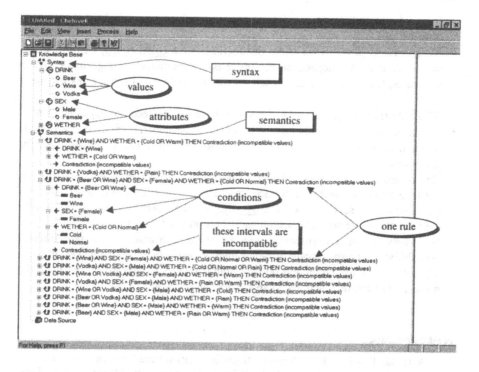

Fig. 3. . Screenshot of the Chelovek rule induction system

5. Conclusion

Below we summarise our rule induction algorithm characteristic features, advantages and disadvantages.

- The algorithm is based on the <u>original formal framework</u> generalising the conventional boolean approach on the case of (i) finite-valued attributes and (ii) continuous-valued semantics.
- To build rules the patterns in the form of possibilistic <u>prime disjunctions</u> are used, which represent the widest intervals of impossibility in the multidimensional space of all value combinations. In particular, it allows us to reach <u>optimality</u> of the rules in the sense of maximal generality of condition and specificity of conclusion.
- The filtration mechanism guarantees finding all the <u>most interesting</u> patterns according to some criterion while too specific ones are not generated.
- The algorithm is <u>iterative</u> in the sense that it processes all records for one pass through the database.
- The patterns in the form of possibilistic prime disjunctions as well as rules generated by the algorithm have <u>clear and easily interpretable semantics</u> in the

form of possibilistic constraints (upper bound) on the maximal number of observations.

- All generated possibilistic prime disjunctions have <u>equal rights</u>, i.e., the whole semantics does not depend on the order of disjunctions (as, e.g., in [11]). In particular, the interpretation of each rule is independent of other rules and their order.
- All attributes have <u>equal rights</u>, particularly, we do not need the target attribute.
- The knowledge base in the form a set of the most general prime disjunctions is approximately <u>equivalent</u> to the database and therefore can be easily used for prediction purposes.
- One minus of the algorithm is a large number of generated rules especially for dense distributions with fine surface structure when it tries to reflect all details of the too complex surface (a kind of <u>overfitting</u>). This problem can be solved with the help of a more powerful search and filtration mechanism.
- For many problem domains it may be more desirable to generate directly probabilistic set-valued patterns rather than possibilistic ones. However, this is a separate highly important and rather difficult problem since we do not have the notions of prime disjunctions, DNF etc. for the probabilistic case.

References

1. R. Agrawal, T. Imielinski, A. Swami. Mining association rules between sets of items in large databases. Proc. of the ACM SIGMOD Conference on Management of Data, Washington, D.C., May 1993, 207–216.
2. R. Agrawal, R. Srikant. Fast Algorithms for Mining Association Rules. Proc. of the 20th Int'l Conference on Very Large Databases, Santiago, Chile, Sept. 1994.
3. H. Mannila, H. Toivonen, A.I. Verkamo. Efficient algorithms for discovering association rules. In KDD-94: AAAI Workshop on Knowledge Discovery in Databases, Seattle, Washington, July 1994, 181–192.
4. A. Savinov. Application of multi-dimensional fuzzy analysis to decision making. In: Advances in Soft Computing — Engineering Design and Manufacturing. R. Roy, T. Furuhashi and P.K. Chawdhry (eds.), Springer-Verlag London, 1999.
5. J. R. Slagel, C.-L. Chang, and R. C. T. Lee. A New Algorithm for Generating Prime Implicants. IEEE Trans. on Computers, C-19(4):304–310, 1970.
6. D. Lin, and Z. M. Kedem. Pincer-Search: A New Algorithm for Discovering the Maximum Frequent Set. In Proc. of the Sixth European Conf. on Extending Database Technology, 1998.
7. R. J. Bayardo Jr. Efficiently Mining Long Patterns from Databases. In Proc. of the 1998 SIGMOD Conf. on the Management of Data, 1998.
8. A.D. Zakrevsky, Yu.N. Pechersky and F.V. Frolov. DIES — Expert System for Diagnosis of Technical Objects. Preprint of the Institute of Mathematics and Computer Center, Academy of Sciences of Moldova, Kishinev, 1988 (in Russian).
9. A.A. Savinov. Fuzzy Multi-dimensional Analysis and Resolution Operation. Computer Sci. J. of Moldova 6(3), 252-285, 1998.
10. A. Savinov. Forming Knowledge by Examples in Fuzzy Finite Predicates. Proc. conf. "Hybrid Intellectual Systems", Part 1, Rostov-na-Donu—Terskol, 177–179, 1991 (in Russian).
11. P. Clark and T. Niblett. The CN2 Induction Algorithm. Machine Learning 3(4):261–283, 1989.

Dataflow Realizes
a Diagrammatic Programming Method

Susumu Kawaguchi, Telecommunications System Group, Hitachi Co. Ltd.,
216 Totsukamachi, Totsukaku, Yokohama, 244-8567, Japan
Prof. Hirotoshi Shirasu, Kogakuin University,
Nishishinjuku Tokyo, 163-8677, Japan, E-mail: jbh03235@nifty.ne.jp

Abstract
A large database of connections between software parts replaces the procedural programs to realize a virtually-wired (visual) logic and to eliminate the regular work of programming. Having a knowledge database within a drawing environment allows programmers to easily call out program parts using a mouse on the screen. This paper presents the system concept from the aspects of soft computing, as well as giving an overview of functional principles, discussing its actual use over the past ten years and raising issues for further study about it.

1. Introduction

Project managers and design engineers involved in a system development often complain that, "software is harder than hardware." Telephone switching systems, one of the largest type of control systems, were originated in the pre-computer era; these were revolutionized in the 70s with the change of control technology from wired logic (relay) control to program (computer) control. The revolution, however, resulted in a large increase in the number of development engineers needed for the control software, even though the same services were provided. Even today, the situation seems to be still very evident and unchanged as it is in other fields. Indeed most people recognize this significant gap between hardware and software development, but its origin has been explored rather little.

Referring to the fundamental difficulty in programming, Buckus claimed that the problem stems from the von Neumann computer architecture [1]. From a similar viewpoint, DeMarco recommended that a dataflow diagram is the best means to directly express functions of an application [2]. Then Stevens claimed that abandoning the von Neumann style and adopting the dataflow architecture was the only means to achieve high software productivity comparable with that of ordinary hardware design [3]. Shirasu and his colleagues in Hitachi, stimulated by these arguments, introduced a visual design concept and developed a software design method named DDL (Data Driven Logic) [4-6]. DDL has been successfully used for a decade in the call processing program of Hitachi's digital PABX system. In this method, program logic is expressed with a dataflow diagram which is fundamentally easier to understand, compared with usual language description, due to its visuality and coherency of its two-dimensional logic. Diagrams are converted to object modules and executed by an ordinary von Neumann processor with an interpreter emulating the dataflow mechanism. The improvement in understandability is achieved at the expense of an increase in memory required for databases and in processing time for the emulation.

This paper introduces the DDL method emphasizing aspects with regard to the soft computing method. First, a significant amount of program codes in the regular von Neumann style can be eliminated by introducing a large dataflow connection database. The principle and design practice of a DDL (dataflow) diagram, its source description, is quite similar to that of a wired logic circuit diagram. The new style of software becomes, accordingly, "soft" enough to diminish the above-mentioned hardware/software gap. Secondly, the DDL source description is to depict a DDL diagram on a screen, where FEs (Function Elements) are searched out from a knowledge database. Programmers can complete their work simply by connecting FEs to designate dataflows.

Thirdly, the DDL diagram contains, with regard to the processing order of FEs for the serial processing machine, an inherent uncertainty or fuzziness, though this is different from that of the fuzzy theory. This is due to the functional parallelism of dataflow, which is essential to exactly represent the system requirements. The fuzziness is, therefore, important and should be preserved in the description but has to be eliminated in conventional programming through manual and arbitrary decisions to line up codes in one-dimensional processing order. The processing order in DDL, on the contrary, is decided ultimately and automatically during the emulation.

Fourthly, the DDL diagram can integrate a state transition diagram, a useful description to understand FSM (Finite State Model) such as of communication software, taking advantage of their common nature [7-8]. A switching program and its state transition diagram of FSM specifications are normally produced as separate documents. The state transition, however, can be described as the first layer of the DDL diagram by implementing a state as a special kind of DDL program parts, thereby being unified with the program as its highest layer. The first layer DDL program is the state transition diagram but includes the second layer DDL program macros, whose details are described in separate DDL diagrams. The integration further makes the source program description easier to understand and excludes estrangement problem between both kinds of descriptions.

Recently more attention is paid to OOP (Object Oriented Programming) in all fields of software. Finally, the DDL method can use OOP technology to refine its system structure, for the DDL method inherently contains OOP principles. The OOP method solves various software problems, except for making programs more understandable; this problem can be solved by combining both methods.

2. Databases "Soften" Software

Fig. 1(a) shows an example of the DDL or dataflow diagram. The boxes represent FEs (Function Elements), which are program parts driven by multiple data inputs. In dataflow, the system consists of many program parts, which send/receive data between themselves, while each program part fires only when all its input data has arrived. The principle is similar to asynchronous wired logic, which is naturally visual and understandable. An

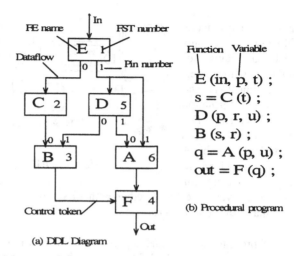

(a) DDL Diagram

Function Variable

E (in, p, t) ;
$s = C$ (t) ;
D (p, r, u) ;
B (s, r) ;
$q = A$ (p, u) ;
out $= F$ (q) ;

(b) Procedural program

Fig. 1 Which is easier?

important point is that the logic retains two-dimensionality in the original application. Fig. 1(b) shows the typical procedural program that corresponds to Fig. 1(a) in the C like style of programming. The difference in understandability is apparent. The latter accords with the serial machine architecture, which necessitates additional conversion of the two-dimensional logic into a one-dimensional procedure. For this purpose, the conversion extensively uses the variables, which are versatile in use but foreign to the original logic and rather troublesome. The difference may originate first in the von Neumann style and secondly in a difference between human and computers in their ways of cognition.

The FE has a mostly elemental function less than 50 steps, or sometimes a larger dedicated function, written in assembly language or C. FEs are prepared as pooled subroutines and called out to the boxes in Fig. 1(a), where an actual entity of the box is a data template named FST (Fire Setting Table). An FST stores its FE name to call the subroutine and all the necessary data to emulate the dataflow mechanism. FSTNs (FST Numbers), indicated in the boxes, do not imply any processing order but only identify each FST. Pin numbers are used to produce a connection list and to specify input/output ports of the FST.

Fig. 2 shows the process for producing FSTs. The DDL programmer draws a DDL diagram on a screen with a mouse, using a two-dimensional graphical editor named DFDraw. DFDraw includes an FE library, which is called out on request. The library covers FEs from very fundamental or standard ones to highly knowledgeable or application-oriented ones so that the FE library itself can be regarded as a knowledge database. The practice is similar to a logical CAD system and does not require conducting any programming work but only making connections between FEs (FSTs), as if designing a wired logic. This method also does not need a large effort to prepare FEs, because most of FE programs are small and ready-made. DFDraw outputs DDL diagrams together with their connection lists and FE usage information. FIDAS (Firing Design Automation), a compiler, combines them to build an integrated connection

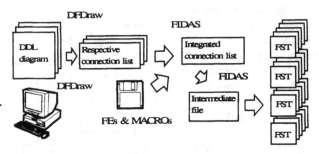

Fig. 2 From DDL diagram to FST

list and converts the list together with FEs and macros into an intermediate file and finally into many FSTs.

A programmers manual is prepared to instruct how to use each FE, specifying functional attributes of the FE together with data entering and going out through its input/output pins. This system can exclude arbitrary or uncontrolled description of programs, because the project manager can take strict control over the FE library, without which there is no way to write programs.

Fig. 3 shows the FST format. The format comprises necessary data to define the dataflow such as the token counter, the number of token arrivals to fire, the FE name, connection data to the upper and lower reaches, immediate input data, and the output buffer area. Every FST corresponds to every place where the FE is used.

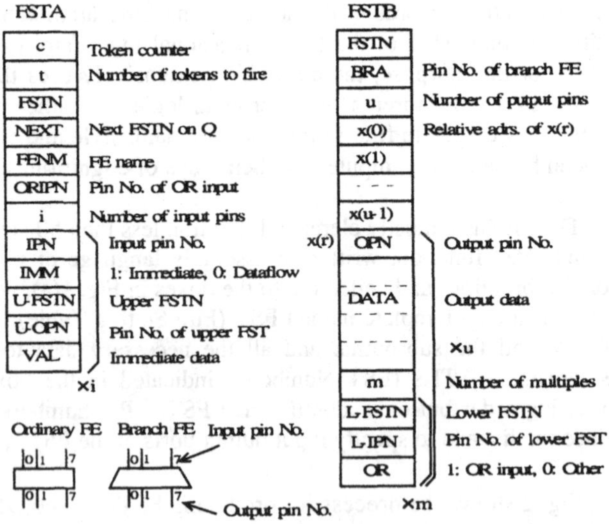

Fig. 3 FST format

It is interesting to note that the memory required for all FSTs within the system occupies, for example, some 95% of the object program module, while FEs occupy only the other portion of the module. This unique composition demonstrates the exclusion of the procedural program (except FEs) by replacing it with FST data. The FST data which has such an extensive and intelligent content as to replace the role of a conventional main program might be regarded as another knowledge database which contains the system framework. The two knowledge databases form the whole system as a result.

The execution of DDL program is an emulation of the dataflow mechanism with a conventional processor. To minimize the emulation overhead, an FE sends a token instead of actual data to the following FEs and stores it in the FST output buffer which is read by the following FEs when needed. Fig. 4 shows details of the emulation control program FIRER, which is rather simple

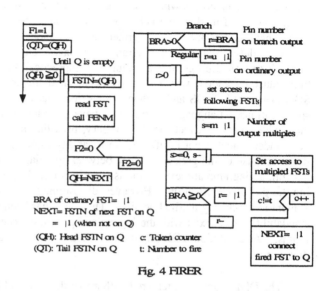

Fig. 4 FIRER

(only 250 steps). FIRER gets the FST from the queue head and calls the FE subroutine whose name is designated in the FST. The FE processes its own logic using the input data in the buffers of the preceding FSTs accessible through the FSTA data. Then the FE puts output data in its own buffer in the FSTB and the FIRER increments the token counters in the following FSTs which are to receive the data. When any token counter reaches the specified number of token arrivals, FIRER determines the FST to fire and places it into the queue. FIRER, then, takes out the next FST from the queue, repeating the process until the queue becomes empty. The purpose of flags F1 and F2 is described later.

DDL supports a method for introducing a layered structure in the DDL design method to improve the understanding and handling of DDL diagrams. Automatic compiling of DDL macros by FIDAS (Firing Design Automation System) is devised to integrate them into a unified DDL object module. The execution of the unified DDL object module must be separated corresponding to DDL macro diagrams, which are independently defined. Therefore, the execution of FEs inside a macro must be done to the end prior to those outside the macro without designating their execution order. Accordingly, special purpose FEs are employed at both start and end of each macro to inform the FIRER when the macro's execution is entered and exited. The FIRER is also devised to establish an independent execution queue for each macro, which is not shown in the Fig. 4 for simplicity. The number of possible layers is unlimited, while the last layer is the FE layer.

3. Integrating State Transition Diagram

In telecommunication software, the state transition diagram or SDR-GR, the international standard, is widely used as a specification description of FSM to ease understanding of its dynamic behavior of functions. SDR-PR, the related method with enhanced describability of further detail, is another kind of language but it is not as widely used as the diagrammatic one. The automatic conversion of SDL-PR to a final program is available but impractical, because too many descriptions must be added, which destroy the original understandability. Other than SDL methods, LOTOS and Esttele offer stricter description of telecommunication specifications. Languages of this kind have poor understandability, which limits their wider use. Petrinet, on the contrary, uses the diagrammatic method and is very understandable but insufficient in its ability to describe detailed system functions, because of the greater distance between the diagram and final program. In general, a diagrammatic expression is superior to the language style in understanding the global specification. However, diagrammatic expression is usually inconvenient in describing its final details. Thus there has been no way in which understandability can coexist with the logical completeness necessary for complete programs. As a solution DDL offers the method to unify the state design with a DDL diagram.

The DDL diagram is featured with a two-dimensional diagram, which directly represents functional algorithm. The state transition diagram is also featured with the same two-dimensional diagram which depicts transition of call states having an endless structure. The proposed method offers the way to embody this endless structure on the dataflow diagram and realize its execution method using a conventional von Neumann processor. As a result, the proposed method offers the following advantages.

1. Excellent understandability equivalent to SDL-GR
2. Complete functional description as the final program
3. Consistency between program and FSM specification
4. Simple software management of a single document

Fig. 5 explains the structure to integrate the state transition diagram with DDL. To describe the diagram in DDL a special kind of FE named STA is defined and used for the state function. The diagram includes many STAs (circles) and DDL tasks which are bridges between STAs, describing many endless processes. Here, an important design condition is given, i.e., the state number (STN) of an STA directly corresponds to its FSTN. Thereby the state number brought by a new event can decide the first FST that is the entry of the DDL task. As shown in Fig. 5, STA includes two functional modes, i.e., the starting process and the ending process of a DDL task. F1 flag is set by FIRER to let the first FST (STA) act in the starting mode and reset by itself. F2 flag is set by the STA in the ending mode to let the FIRER cancel dataflows and reset by the FIRER.

The line finder routine, shown in Fig. 5, is the base level program which is continuously interrogating the new event hopper at the switching peripheral interface. Upon detecting a new event, the routine saves its terminal number with the received message information in its terminal table that stores the call processing data, and gets its current state number from the table. The rou-

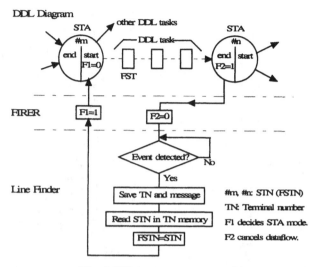

Fig. 5 STA integrates state transition

tine, then, puts the FST (STA) whose FSTN equals to the STN into the head of DDL queue and goes to FIRER. The routine is restarted when FIRER is finished the DDL task. The mechanism, thus, integrates the state transition diagram into the first layer of a DDL call processing diagram, and the line finder routine is concealed behind STA circles in the DDL diagram.

The STA, after started by FIRER, first distinguishes the processing mode with the F1 flag. In the starting mode, STA fetches new event data, which has already been saved by the line finder routine. The data contains the terminal number together with received information from the terminal such as on/off hook and received digits. Then, STA selects its output pin number out of one or more possibilities according to the event data. The state number, or the FST number, with the selected pin number, determines one of the DDL tasks to be executed. This is done by the output pin connection of the STA in the first layer diagram. Around the end of the task execution, FIRER fires an FST that represents the next state, the STA being started in the ending mode according to the F1 flag (0). STA is to be fired on an OR basis so that every FST corresponding to every state is fired with the first token arrival among its multiple inputs (multiple task ends). In the ending mode, the STA stores the FST number, or the new STN, in the terminal table. Then STA sets the F2 flag, which requests FIRER to halt driving of the following FSTs, because the driving must be started by the next event.

Fig. 6 describes the first layer of DDL diagram, integrating the state transition, for an ISDN intra-office connection for a trial. In the figure those boxes which have a double-sided edge are DDL macros, while trapezoidal ones are selective macros. They are described in the second layer DDL diagrams. There can be three or more layers of DDL macro, where the last layer is the FE layer.

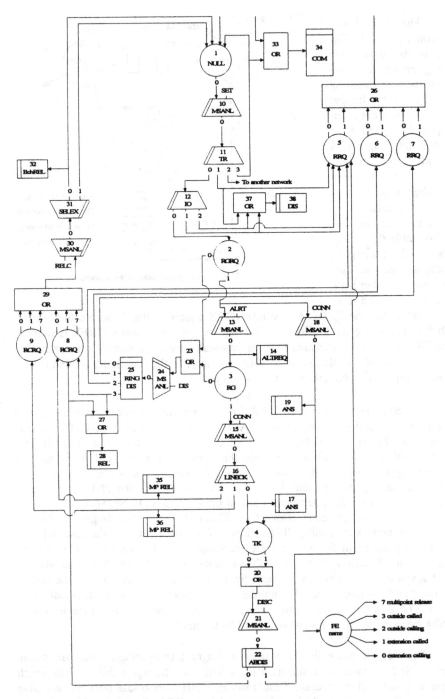

Fig. 6 First layer DDL diagram for ISDN intra-office connection

4. Practical Application

Hitachi Ltd. adopted the DDL method in the call processing program of its CX5000 series PABX for business and hotel-motel customers, which was first developed in 1986. With thousands of products being shipped to a variety of customers in the Japanese and U.S. markets, the DDL method has been ensuring the company a steady and well-managed production of switching software, which was in continual trouble before. To show facts in figures, the number of types of FEs, which is indicative of the real work of programming, is presented as follows.

For the early shipment of the products, 110 types of FEs were used. However, the types of FEs had been increased to support sophisticated service features for a variety of PBX market segments and also to reduce the execution overhead. In 1990, 92 types of general-purpose FEs (average size: 40 steps in assembler language) and 138 types of dedicated FEs (average size: 320 steps), for a total of 230 types were being used. For a recent instance of the largest customer ever delivered, 94 types of general-purpose FEs (average size: 43 steps) and 222 types of dedicated FEs (average size: 320 steps), for a total of 316 types are used, where a total of 150,393 FSTs are produced. The number of FE types and its increment are fairly small, considering such a large size of FST module and the elapse of eight years. The integration with the state transition diagram has not yet been actually used.

5. Issues for Further Study

Applying the DDL method to a firmware design has been studied, because the scale of firmware has become larger and larger, resulting in a problem for the company. The study group has been organized in cooperation with Kogakuin University and they adopted the firmware of Hitachi's ATM switching equipment as its trial model. The original firmware program was written in 7000 steps in the usual way using the C language. That firmware program was then rewritten in 1000 FSTs, where thirty-eight kinds of FEs and fourteen sheets of DDL diagrams were developed. The resulted documents demonstrate very high understandability compared with the original program. This trial has proved the potential of applying the DDL method to a wider field of applications.

The DDL method originally involves something of the OOP (Object Oriented Programming) principle, because every FST, which corresponds to every dataflow node, can be regarded as an encapsulated object module. The module includes not only its FE method but also connection data, dataflow-buffer, and other attributes of the dataflow node. Also arbitrary data accessing is inhibited, because programmers can access data only through prepared FEs. Taking advantage of the support environment of the recent OOP language, however, will achieve a significant refinement in the DDL architecture. According to the ordinary concept of OOP, some system elements holding system data must be encapsulated with their accessing methods as system objects. In addition, FEs/FSTs can be defined as objects of a special category, namely DFN objects (DataFlow Node), which communicate with the system objects. Here, the control function of dataflow emulation can be regarded as their super-class. The structural advantages in both types of technology can be compatible, this allowing application designers to work on DDL diagram basis without losing its high understandability, which OOP itself does not bring about. Thus the introduction of OOP technology to the DDL will bring enhancements especially in its reliability, maintainability and versatility to fit a wide variety of applications.

Acknowledgement

Authors wish to express their gratitude to engineers in Hitachi Ltd. for their helpful suggestions. Also thanks are due to graduates of Kogakuin University. This research is supported by a grant from Support Center for Advanced Telecom Technology Research, Hitachi Central Research Laboratory and Kogakuin University.

References

[1] J. Buckus, "Can Programming Be Liberated from the von Neumann Style? A Functional Style and its Algebra of Programs," Commun. ACM, 21, 8, pp. 613-641, 1978

[2] Tom DeMarco, "Structured Analysis and System Specification," 1979, Prentice-Hall, Inc.,

[3] W. P. Stevens, "How dataflow can improve application development productivity," IBM System Journal, vol. 21, 2, Aug. 1982

[4] H. Shirasu, "Innovative Approach to Switching Software Design Using Dataflow Concept," ISS '87, 2, B-4-3

[5] H. Shirasu, "Dataflow Brings Innovation in Switching Software," 5th World Telecom. Forum '87, 2-3, 3, pp. 365-369

[6] Y. Maejima, "A Graphical Programming Environment for Switching Software," IEICE Trans. Comm., Vol. E75-B, No.10, '92

[7] H. Shirasu, "Integrating the State Transition Diagram with Call Processing Program Using Dataflow Technique," ICIN '94

[8] H. Shirasu, "Integrating the State Transition diagram with Call Processing Program Using Dataflow Technique," Transaction of IEICE, vol. J80-D-1 No. 4, '97, pp. 389-394

DIFFOBJ - A Game for Exercising Teams of Agents

Tony Hirst[1] and Tony Kalus[2]

1,2 Intelligent Agent Group, Department of Computer Science, University of Portsmouth, Mercantile House, Hampshire Terrace, Portsmouth, PO1 2EG. {Tony.Hirst, Tony.Kalus}@port.ac.uk

Abstract- Despite the increasing amount of research in the use of theories of teamwork as an organising principle for multi-agent systems, there are few, if any, simple standard problems for testing such systems. Afetr describing the essential characteristics of a team of agents, we present one such test problem - DIFFOBJ - in which the members of a team must each collect a different object. At least initially, the particular object each individual should collect is left open. After describing DIFFOBJ, we demonstrate how it revealed certain weaknesses in a pre-existing model of teamwork for the Soar architecture. Finally, we show how DIFFOBJ tests those characteristics identified as being essential properties of individual agents who are also team members.

1. Introduction

In recent years, there has been an increasing amount of interest in multi-agent systems (MAS). One form of organisation that is of particular interest is a team-based organisation [1][2]. In this paper, we shall present a simple game, DIFFOBJ, that may be used as a testbed for team behaviour. The game may be played at various levels - for example, testing coordination of activity (such as initiating team behaviour) or belief (such as the maintenance of 'team awareness').

To date, work in the artificial agent community that has explicitly used teamwork as organising principle has tended to test team behaviour within complex domains (for example, for playing robot football, or commanding aircraft within large scale military simulations [3]). When it comes to identifying simple scenarios appropriate for testing the core priciple underlying team activity, there is little to draw on from within the literature. We address this need by describing a simple game - DIFFOBJ - that requires team activity. The game is offered as a pedagogical tool which may be easily implemented but which nevertheless provides a basis for quite complex testing of team behaviour.

In the next section, we briefly describe the notion of an agent team, and comment upon some of their more significant properties. We then go on to describe DIFFOBJ, and show how it tests the team properties already identified. Finally, we report on how DIFFOBJ was used to test a legacy model of teamwork, and in doing so revealed several difficulties with it as well as providing a testbed for the subsequently revised model.

2. An Introduction to Teams of Agents (ToA)

A team of agents (ToA) is a group of two or more interdependent agents that act cooperatively, and in a coordinated way, in order to attain a mutually held (team) goal [2]. In addition to the 'standard' properties of agent (autonomy, reactivity, goal-directed behaviour etc.), team member agents should also be capable of:

1. maintaining mutual beliefs;
2. maintaining joint goals;
3. maintaining joint intentions (intention in the sense of actively pursuing a goal);
   These in turn suggest that the agents are capable of:
4. communicating with each other (i.e. are social within the team);
5. co-ordinating belief and behaviour across the team;
6. adhering to some form of control structure.

These properties are discussed further in [2], although with the various phases of activity that are likely to occur in a team exercise.

Teams of agents are typically organised in terms of a single, distinguished leader presiding over an unspecified number of peer members. In fact, team members may either be individual agents or subteams organised in a similar way to the parent team. At the outset, team members will identify themselves to the leader and the leader will familiarise the members of the team with the team plan. Plans typically comprise of a sequence of goals. The goal currently (i.e. actively) being pursued is termed an intention. Particular roles (and the individual plans associated with them) are then assigned to the team members. Just how member agents should be selected for inclusion within the team and how the role assignment should be achieved is a live research area. The team leader then coordinates team activity through the various steps in the plan, although it is the responsibility of each team member that the rest of the team is kept aware of anything that member believes may jeopardise the team plan. In order to simplify matters, formal team protocols are initiated by the team leader and often proceed according to some predetermined speaking order; elsewhere, we have shown how a modification to the speaking order protocol will handle the loss of a team member without causing communications to halt at that point in the speaking order [4].

It is the role of the leader to act as a focus for the team. Although the leader is responsible for initiating team action, it may do so at the requent of one of the team members. Under certain implementations (such as Tambe's STEAM model of teamwork for the Soar architecture [3]), all team members may be capable of 'terminating' team goals under according to predefined success, failure or irrelevancy conditions.

In the next section we shall describe a simple game - DIFFOBJ - that tests the coordination of activity and belief within an agent team. For a full test of team behaviour, assignment of the team plan and its adoption by the team should be achieved in a preceding phase.

3. DIFFOBJ - A Game for Testing Teamwork

Prequel: in the following description of the game, we assume that all team members know that they are about to play the game, and the order in which they are to take turns. No mechanism is provided for coping with the loss of a team member. Each and every team member is assumed to know a) the team goal, b) their own (ill-specified) individual goal. An intial 'adopt goal' turn may be used to test the ability of the team to adopt a common goal.

Suppose there is a world containing a team, T, made of several individual agents, m_i; suppose further that the world contains many instances (ANOBJ) of several different object types, ANYOBJ (3 oranges, 2 apples etc.). ANOBJ is taken to be a variable that may be set to a member, ANYOBJ, of the set of objects. The team goal is for each agent to collect a single, different object, hence the name of the game - DIFFOBJ. Although the game is described using a standard logic of belief, we have tried to present it in terms that will be generally accessible. For the moment, let us assume that each individual, $m_i \in T$, is charged with collecting an object by the adoption of a simple achievement goal:

$$GOAL(m_i \; CARRIES(m_i \; ANOBJ)) \tag{1}$$

That is to say, individual m_i has the goal of collecting the object ANOBJ.

In addition, the individual must ensure that this object is different to any objects that the individual believes are already in the possession of other team members:

$$m_i, \; \forall m_j \in T, \; {\sim}BEL(m_i \; CARRIES(m_j \; ANOBJ)): m_j \neq m_i \tag{2a}$$

where BEL(x p) reads 'x believes that p' and ~ reads 'NOT'. That is, a member does not believe that another member has already collected ANOBJ. In other words (via the serial accessibility relation): an individual does believe that no other individual has already collected ANOBJ. That is:

$$m_i, \; \forall m_j \in T, \; BEL(m_i \; {\sim}CARRIES(m_j \; ANOBJ)): m_j \neq m_i \tag{2b}$$

The team goal is satisfied when:

$$\forall m_i, m_j \in T, \; MB(T \; (CARRIES(m_i \; ANOBJ) \; \& \; {\sim}CARRIES(m_j \; ANOBJ)): m_j \neq m_i \tag{3}$$

MB(T p) is the mutual belief across the team, T, that p. Mutual beliefs are typically defined according to a fixed point relation:

$$MB(T, p) \equiv \forall m_i \in T, \; BEL(m_i \; (p \; \& \; MB(T \; p))) \tag{4}$$

The notion of mutual belief in bounded (i.e. real agents) is discussed in the appendix.

Team goals are essentially mutual goals - goals that are mutually believed to be held:

$$MG(T \; p) \equiv \forall \; m_i \in T, \; MB(T \; GOAL(m_i \; p)) \tag{5}$$

The systemic state of mutual belief is created by forcing an individual to inform the rest of the team about any object that it has collected. This may be achieved by using a weak achievement goal rather than the simple individual achievement goal suggested above:

$$WG(m_i \; T \; CARRIES(m_i \; ANOBJ)) \tag{6}$$

A weak goal held by an individual in a team context requires that when the goal is achieved, becomes unachievable or is rendered irrelevant, this fact is communicated to the rest of the team.

Note that if an individual is the last member of the team to pick up an object, then the team goal will have been achieved. Communication thus provides the basis for deciding whether or not the team has achieved the goal, as well as constraining the behaviour of individuals not yet in possession of an object.

Under this view, the game is minimally defined as follows:

$$DIFFOBJ(m_i, m_j) \equiv (CARRIES(m_i \; ANOBJ) \; \& \sim CARRIES(m_j \; ANOBJ))) \tag{7a}$$

$$\textit{team goal}: \forall m_i, m_j \in T, MG(T \; MB(T \; DIFFOBJ(m_i, m_j))): m_j \neq m_i \tag{7b}$$

$$BELDIFF(m_i, m_j) \equiv CARRIES(m_i \; ANOBJ) \; \& \sim BEL(m_i \; CARRIES(m_j \; ANOBJ)) \tag{7c}$$

$$\textit{individual goals}: \forall m_i, m_j \in T, WG(m_i \; T \; BELDIFF(m_i, m_j)): m_j \neq m_i \tag{7d}$$

We note that the way the individual weak goal is defined encompasses a degree of uncertainty with respect to which individual should collect which object. That is, an individual may collect any object, as long as it believes that no other agent has already done so. When any one individual collects a valid object, it announces this fact to the rest of the team via a communicative goal. A consequence of this is a reduction in the uncertainty in the way in which other individuals may satisfy their individual goal. The communicative goal itself is raised by the satisfaction of an achievement functional model.

The game itself may be played by mobile agents spatially located in a 2d-grid world, with objects scattered randomly over it. DIFFOBJ may also be played 'out of time' by assigning to each object a probability that it will 'occur' during any given turn. By generating a random number at the start of each turn, an individual may be presented with a particular (or no) object accordingly. In order to simplify the game, we prevent communicative race conditions by supposing that the game is played in strict turn order. Turns pass through the team according to this (predefined) order. When the 'last' team member has finished its turn, the 'first' team member has the next turn. Turns proceed as follows: if the individual is not already carrying an object, it 'looks' to see if there is a valid object available; if there is, it picks up the object and informs the rest of the team that it has done so (for example with a broadcast message "To all team agents, from agent x, I have collected object y"); otherwise, it does nothing; at this point, its turn ends. When all team members are in possession of a unique object, and they are all aware of this fact, the team goal has been satisified.

The complexity of the game may be increased by splitting each 'team turn' into three phases. In the first phase, each individual identifies in turn order whether or not there is an object available and notifies the rest of the team accordingly. In the second phase, individuals reason about an appropriate course of action for phase 3. For example: in a 2 player game with similarly defined agents, suppose agent 1 is carrying object x, and has object y available; suppose further that agent 2 is empty handed and has object x available. The team goal may be satisfied immediately if agent 1 puts down object x in favour of object y; and agent 2 picks up object x. Given the same information is available to both agents, and they both have the same reasoning capabilities, there is no reason to suppose they won't both adopt the optimal course of action independently of the other. There is no need to 'negotiate' in a strict sense of the word, since although several courses of action are possible, there is only one that satisfies the team goal (and hence only one worth pursuing). However, to guarantee this course of action, confirmatory dialogue should occur.

4. Using DIFFOBJ to Test Team Behaviour

DIFFOBJ was originally developed as a spatially located game for testing the STEAM ruleset [3] for the Soar production system architecture [5]. Soar itself was originally developed as a cognitive modelling architecture, although it is increasingly being used as an agent architecture [6], and to develop MAS [3][7]. The STEAM ruleset implements a model of teamwork based on the joint intention theory of Cohen & Levesque [1][8]. Although developed for simulated helicopter teams, a significant portion of the STEAM ruleset was successfully reused in the robot football domain, demonstrating its generality.

In order to familiarise ourselves with the STEAM ruleset we had obtained (version 1.1), a spatial variant of DIFFOBJ was developed. The game world was a 2 dimensional grid. Various objects were distributed randomly over the grid, along with the agent team. The exercise required the team leader to a) confirm each individual's commitment to the team goal (i.e. to achieve the DIFFOBJ goal); b) achieve that goal.

The initial act of confirming the team's adoption of the team goal is automatically handled by STEAM, assuming that a fixed speaking order is prearranged. However, if there is a breakdown in communication (such as the loss of a team member) then the team leader will repeatedly call for the team to adopt the goal but its adoption will be thwarted by the lack of response of the team member. By silencing one of the agents within the game, this situation was easily simulated and consequently revealed the need for a modification to the initial team commitment protocol.

As far as the second requirement goes, and according to the analysis above, each individual must hold a weak individual goal with respect to collection of a unique object in order to satisfy the team goal. That is, communication is required in the event of an individual achieving an individual goal. As far as we could determine, the version of STEAM we were using supported individually held weak team goals but not weak individual goals. In addition, the weak team goals had side effects associated with them that terminated similar goals in other team members. That is, if one individual believed the weak team goal to be achieved, unachievable or irrelevent, then it would inform the rest of the team and they in turn would drop the goal.

Despite its simplicity, and easy implementation, DIFFOBJ was thus able to provide a quite a complex testing domain for the STEAM model, and in doing so identified several weaknesses in the model (these, along with possible solutions, are discussed more completely in [4]).

5. DIFFOBJ as a Formal Test of Team Member Characteristics

To what extent does DIFFOBJ test the characteristics (identifed above and repeated here for clarity) that we require of team members? Namely:

1. maintaining mutual beliefs: MB(T p);
2. maintaining mutual/joint goals: MG(T q);
3. maintaining joint intentions (where an intention is taken to be an actively pursued goal): JI(T q)=MG*(T q);
4. communicating with each other (i.e. are social within the team);
5. co-ordinating belief (i.e. mental state information) and behaviour across the team;
6. adhering to some form of control structure.

In terms of the formal statement of DIFFOBJ, eq. (7), there is obviously a need for mutual beliefs. It is not sufficient for each member to believe that the goal has been achieved - it is also necessary that each team member believes that the rest of the team believes that the goal has been satisified. The extent to which mutual beliefs may be approximated in real agents is considered in the appendix. Whilst mutual goals are also required within DIFFOBJ, the statement of the game does not *require* that a joint intention is held. In order to enforce this condition, we therefore require an additional statement along the lines of:

$$\forall m_i, m_j \in T, INT(m_i\ p)\ \&\ INT(m_j\ p): m_j \neq m_i \tag{7e}$$

where INT(m p) reads 'm intends p' and requires GOAL(m p). This would guarantee that each team member is actively pursuing the current team goal.

Given that each agent maintains its own private mental state independently of other agents, some mechanism is required for coordinating mental states. Communication provides a way of making private mental state public. Coordination (of mental state) requires that an individual somehow has an internal model of others' mental state (i.e. it should be capable of holding beliefs about the beliefs of all the other team members (see appendix for a discussion of these 'social beliefs')); in addition, it should be able to update its social beliefs on the basis of communications received. Coordination of activiy (i.e. behaviour) requires not only that agents are in an appropriate intentional state, but also that some cue is provided as to when to commence that intended behaviour.

6. Summary

In this paper, we have described a simple game - DIFFOBJ - for testing the essential properties of a team agents. The game is scalable in that it will cope with teams of any size and is amenable to the introduction of various 'failures', such as the loss of a team member. DIFFOBJ has been shown to be useful for identifying several weaknesses in a legacy model of teamwork for the Soar architecture. In addition, through its statement of several test conditions, DIFFOBJ implicitly presents a simply stated set of minimal requirements for future teams of agents.

References

[1??] Cohen, P. R., Levesque, H. R., & Smith, I., 1999. On team formation. In Hintikka, J. and Tuomela, R. (Eds.) Contemporary Action Theory. Synthese.

[2] Hirst, A.J. & Kalus, A., *in prep.*, "An Introduction to Teams of Agents".

[3] Tambe, M., 1997, Agent architectures for flexible, practical teamwork. In *Proceedings of the National Conference on Artificial Intelligence*, pages 198—202. AAAI Press

[4] Hirst, A.J. & Kalus, A., 1999, 'Flexible Communication within Agent Teams', to be presented at IJCAI99 Workshop on Team Behaviour, Stockholm, August 1999.

[5] Newell, A., 1990, *Unified Theories of Cognition.* Harvard University Press, Cambridge Massachusetts.

[6] Rosenbloom, P.S., Laird, J.E., Newell, A. and McCarl, R., 1991, *A Preliminary Analysis of the Soar Architecture as a Basis for General Intelligence.* Artificial Intelligence (47).

[7] Kalus, A. & Hirst, A.J., 1998, "Soar Agents for OOTW Mission Simulation. " Presented at *Command & Control Decision Making in Emerging Conflicts - 4th International Command & Control Research & Technology Symposium*, Näsby Park, Sweden.

[8] Smith, I.A. & Cohen, P.R., 1996. Towards Semantics for an agent communication language based on speech acts. In *Proceedings of the National Conference on Artificial Intelligence (AAAI'96)*, pages 24—31, Menlo Park, California. AAAI Press.

Appendix - A Note on Mutual Belief

Typically, mutual beliefs (beliefs that hold over all members, m_i, in a team, T) are represented according to fixed point definition:

$$MB(T, p) = \forall m_i \in T, BEL(m_i\ p\ \&\ MB(T\ p)) \tag{A1}$$

However, in a bounded agent, that holds local (private) beliefs, we never reach the fixed point. At first glance, it may appear that minimally we may approximate a mutual belief with a belief common to all team members. This latter case may be phrased as a 'common belief' across the team:

$$\forall m_i \in T, BEL(m_i\ p) = BEL(T(m_i)\ p) = CB(T\ p) \tag{A2}$$

Note that we may now write (A1) as:

$$MB(T, p) = BEL(T(m_i) \, p \& MB(T \, p)) = BEL(T(m_i) \, p) \ \& \ BEL(T(m_i) \, MB(T \, p))$$

$$= CB(T \, p) \ \& \ CB(T \, MB(T \, p))$$

$$= CB(T \, p) \ \& \ CB(T \, CB(T \, p)) \ \& \ CB(T \, CB(T \, MB(T \, p))) = \dots \tag{A3}$$

Mutual and common belief states describe the state of the system as a whole and are not directly held by individuals. Under mutual belief, an agent not only believes a thing, it also believes that the other agents believe the thing, and that they believe the one believes the thing, and so on. In a common belief, all individuals share a belief, but do not necessarily believe that they share the belief. We suggest the minimum requirement of an approximation to a mutual belief is that a belief about a common belief is also held.

Now, an agent may hold certain beliefs about what other agents believe; these beliefs about (other agents') beliefs we term *social beliefs*:

$$SB(m_i \, m_j \, p) = BEL(m_i \, BEL(m_j \, p))) \tag{A4}$$

Note that for true sociality, we require $m_j \neq m_i$, but we shall find it more convenient to allow a belief about one's own belief to be classified as a social belief also.

To what extent may we regard a mutual belief state as holding if we are dealing with bounded agents that can at best support a social belief state? What happens if each individual holds a social belief relative to the rest of the team? That is:

$$\forall m_i, m_j \in T: SB(m_i \, m_j \, p) = SB(T \, T \, p) \tag{A5}$$

We also have:

$$\forall m_i, m_j \in T: SB(m_i \, m_j \, p) = BEL(T(m_i) \, BEL(T(m_j) \, p))) \ = CB(T \, CB(T \, p)) \tag{A6}$$

We suggest a first order approximation to a mutual belief in the form of a social common belief, in which all the agents in a group believe a thing and they all believe that every other agent in the group believes the same thing:

$$SCB(T \, p) = CB(T \, p) \ \& \ [\forall m_i, m_j \in T: SB(m_i \, m_j \, p)] \tag{A7}$$

Hence (from (A5)):

$$SCB(T \, p) = CB(T \, p) \ \& \ SB(T \, T \, p) \tag{A8}$$

and also (from (A6)):

$$SCB(T \, p) = CB(T \, p) \ \& \ CB(T \, CB(T \, p)) \tag{A9}$$

Note the equivalence with the low order terms in the expansion of a mutual belief (A3). To summarise:

- A common belief is a zero'th order approximation to a mutual belief *from the point of view of others* (zero because this may be true before the first order agent's view of mutual belief (i.e. belief of a common belief) holds);

- Belief of a common belief ($BEL(m_i \, CB(T \, p))$) (which 'includes' $BEL(m_i \, BEL(m_i \, p))$) is a first order approximation to mutual belief *from the point of view of an individual itself.*

- The SCB is a first order approximation to a mutual *belief from the point of view of others.*

Author Index